Handbook of Modern Sensors: Emerging Technologies

Handbook of Modern Sensors: Emerging Technologies

Edited by
Isabella Edwards

MURPHY & MOORE
www.murphy-moorepublishing.com

Published by Murphy & Moore Publishing,
1 Rockefeller Plaza,
New York City, NY 10020, USA

ISBN: 978-1-63987-288-6

Cataloging-in-Publication Data

Handbook of modern sensors : emerging technologies / edited by Isabella Edwards.
p. cm.
Includes bibliographical references and index.
ISBN 978-1-63987-288-6
1. Detectors. 2. Sensor networks. 3. Detectors--Technological innovations. I. Edwards, Isabella.
TK7872.D48 H36 2022
681.2--dc23

For information on all Murphy & Moore Publications
visit our website at www.murphy-moorepublishing.com

MURPHY & MOORE

Contents

Preface

A sensor is a device that detects changes in the environment and sends the information to a processor or other electronics. A sensor is said to be ideal if it is sensitive to the measured quantity and insensitive to any other property, and does not influence the former. It is considered to be sensitive if the output changes relative to each small deviation in the input quantity. The use of sensors is seen in items of everyday usage, such as touch-sensitive elevator buttons or lamps that brighten or dim upon touching the base. Due to the advancement in micromachinery and user-friendly microcontroller platforms, sensors are no longer restricted to the measurements of temperature, pressure or flow alone. They are increasingly being used in airplanes, cars, robotics, manufacturing, machinery and aerospace. Sensors are also used in biotechnology and biomedicine to detect analytes in in-vitro and in-vivo applications. This book outlines the working principles and applications of sensors in detail. It elucidates emerging sensor technologies and their applications in a multidisciplinary manner. It is appropriate for students seeking detailed information in this area as well as for experts.

All of the data presented henceforth, was collaborated in the wake of recent advancements in the field. The aim of this book is to present the diversified developments from across the globe in a comprehensible manner. The opinions expressed in each chapter belong solely to the contributing authors. Their interpretations of the topics are the integral part of this book, which I have carefully compiled for a better understanding of the readers.

At the end, I would like to thank all those who dedicated their time and efforts for the successful completion of this book. I also wish to convey my gratitude towards my friends and family who supported me at every step.

Editor

Preface

Editor

1

Influence of Frequency on the Resolution of Magnetostrictive Bio-Inspired Whisker Sensors

Ran Zhao, [1,2] **Bo-wen Wang,** [2] **Quan-guo Lu,** [1] **Jian-wu Yan,** [1] **and Xiao-cui Yuan** [1]

[1]*Jiangxi Province Key Laboratory of Precision Drive & Control, Nanchang Institute of Technology, Nanchang, China*
[2]*State Key Laboratory of Reliability and Intelligence of Electrical Equipment, Hebei University of Technology, Tianjin, China*

Correspondence should be addressed to Ran Zhao; zhaoran@nit.edu.cn

Academic Editor: Maria Luz Rodriguez-Mendez

Magnetostrictive biomimetic whiskers have been used as tactile and flow sensors. Compared to other types of whiskers, such whiskers have the advantage of being able to perform static and dynamic measurements. For dynamic measurement, the whisker's resolution changes with varying vibration frequency; however, the mechanism for this influence has not been studied yet. Thus, the aim of this study is to investigate the resolution–frequency correlation. First, the structure and operation principle of the whisker were analyzed. Then, the Euler–Bernoulli beam theory was employed to establish the sensing model of the magnetostrictive whisker. Finally, the mapping relationship between sensor resolution and frequency was obtained. The eigenfrequency analysis was implemented by FEM to obtain the frequency response of the whisker. A vibration experimental system was built for dynamic testing. The experimental results were in good agreement with the theoretical calculations. Furthermore, it was noted that the resolution was positively correlated with frequency, and the maximum resolution was attained at the natural frequency (two peak values appeared at the first-order and second-order eigenfrequencies). Our research reveals the manner in which a whisker sensor's resolution is affected by the vibration frequency. The theoretical model can be used to predict the resolution of magnetostrictive whisker sensors.

1. Introduction

In nature, several animals use their whiskers to collect ambient information, including obstacles, distance, and force and vibration, to overcome any limitations of vision and hearing. For example, seals use their whiskers to perceive hydrodynamics, and rats rely on whiskers to recognize obstacles. Many researchers have studied the operation mechanism of animal whiskers and developed artificial whisker systems. As a near-range sensor, the biomimetic whisker can be a useful supplement for traditional sensing technologies.

In the last two decades, research on bio-inspired whiskers has progressed greatly. The principle, materials, measurement methods, and applications of biomimetic whiskers have been studied extensively. Among them, Lucianna et al and Kan et al. investigated the mechanism and elastic modulus of a rat whisker sensory system [1, 2]. Bovet et al. established the simulation models of rigid and flexible whiskers [3]. Kaneko and Tsuji first used a flexible beam coupling with a torque sensor to detect the hole shape in machining processes; the whisker had a high sensitivity of 5 μm [4, 5]. Ju and Ling developed a piezoelectric whisker and applied it to tactile and geometry measurements [6, 7]. Another kind of tactile sensor, a PVDF beam-based whisker was reported, and its frequency response was studied by Tiwana et al. [8]. Pearson et al. and N'Guyen et al. applied the whisker sensory system in a robotic rat for obstacle identification [9, 10]. However, all the abovementioned whiskers are based on elastic or piezoelectric materials, and their application range is limited.

It is known that magnetostrictive devices exhibit excellent static and dynamic performances [11–13]. This characteristic makes them potentially applicable in the sensing field. To this end, Raghunath and coworkers developed a magnetostrictive whisker [14]. On the basis of magnetoelastic effect mechanism, a magnetostrictive whisker can measure both stress and vibration. As a novel whisker-type tactile sensor, it can sense static and dynamic information, and it demonstrates satisfactory environmental adaptability. In these studies, the whisker's magnetoelastic bending model and design optimization problems (the location of the bias magnetic field and the relation between sensor sensitivity and beam thickness [15, 16]) were investigated; however, the dynamic performance (such as dynamic sensitivity and resolution) has not been studied yet. In [17], we discussed the influence of frequency on the sensitivity of the magnetostrictive whisker, but the mapping relation between the vibration frequency and whisker response is still unclear.

To realize the quantitative description of a whisker's resolution, we fabricated, analyzed, and modeled a galfenol-(FeGa alloy-) based magnetostrictive whisker sensor. The frequency response of the whisker was obtained by using a finite element software. A dynamic experimental system was established to test the dynamic performance of the whisker. The details are presented in the following sections.

2. Design

2.1. Structure and Fabrication. Figure 1 shows a typical magnetostrictive whisker consisting of four parts: a magnetostrictive cantilever beam, a bias magnet, and a magnetic sensor. Several magnetostrictive materials, such as galfenol and alfenol (Fe–Al alloy) [15, 16], can be used as raw materials of the whisker beam. The beam was clamped at the base and magnetized by the bias magnet. The change in magnetic flux was measured by a magnetic sensor. Either a Hall sensor or a GMR sensor can be used as the magnetic sensors. Here, we selected a Hall sensor, the WSH138 (manufactured by Winson Lt. Co.) to measure the change in magnetic flux [18].

In the design process, a high magnetostrictive coefficient and flexibility are the two mainly considered factors. Thus, we selected a galfenol thin plate to fabricate the beam. The magnetostrictive coefficient of galfenol is higher than that of alfenol (their saturation magnetostrictions are 250 and 120 ppm, respectively, reported in [19]). The rolling process ensures that this material demonstrates excellent elasticity. The geometric and physical parameters of the magnetostrictive whisker are listed in Table 1.

2.2. Operation Principle. Figure 2 illustrates the operation principle of the proposed whisker. The bias magnet provides a magnetic field to orient the magnetic domains along the y-axis direction. The linkage flux is detected by a Hall sensor fixed next to it (Figure 2(a)). Raghunath et al. observed the change in magnetic domain in a bending magnetostrictive beam using a Kerr microscope [20]. It was found that the domain rotation depends on the inner stress distribution. As shown in Figure 2(b), the magnetic domains on the right side of the natural centerline have the maximum deflection angle. Thus, the linkage flux increases. In contrast, when stress is applied to the beam rightwards, this value decreases (Figure 2(c)). In summary, when stress is applied from different directions, the Hall sensor outputs an alternating voltage.

The measurement method of the proposed whisker allows the sensor to measure both static and dynamic stresses. Furthermore, the microsecond response speed of

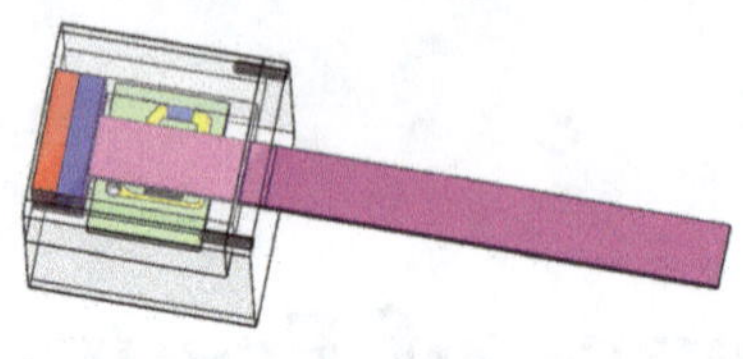

FIGURE 1: 3D model of the magnetostrictive whisker.

galfenol and Hall sensor can help the whisker realize real-time monitoring tasks.

2.3. Definition of Resolution. In our design, the whisker transforms the displacement signal to a voltage signal. In addition, its displacement resolution is dependent on the minimum conversion voltage of the measurement circuit. This means that the displacement resolution of the whisker system is limited by the resolution of the Hall sensor. Herein, we introduce the beam's displacement-linkage flux transmission coefficient α and the Hall's magnetic resolution r_H. The system resolution can be expressed as

$$r = \alpha \cdot r_H, \tag{1}$$

where α is defined as dy/dB, that is, the differential of the displacement to the magnetic flux, and the magnetic resolution of WSH138 is 0.1 mT [18]. From (1), it is known that the system resolution depends on the system sensitivity and the resolution of the Hall sensor.

3. Analytical Modeling

3.1. Classic Beam Theory. In this section, a classical beam theory, in the form of the Euler–Bernoulli beam equation, is introduced to model the proposed magnetostrictive whisker. Considering that the designed whisker has a low working frequency, we use a pseudostatic analysis method. For this method, the base vibration excitation (along the direction of the z-axis) is equivalent to a dynamic load applied to the free end of the beam [21]. In this way, the static beam equation can be used to model the whisker [22].

According to the Euler–Bernoulli theory, the stress in the x-direction of the beam is defined as

$$\sigma_x = -z\frac{M}{I}, \tag{2}$$

where M is the bending moment, and $M = F(L-x)$; I represents the second moment of area. Substituting $M = F(L-x)$ into (2), the stress can be rewritten as

$$\sigma_x(t) = -z\frac{(L-x)}{I}F(t), \tag{3}$$

where z is the position along the thickness direction, and $F(t)$ represents the equivalent dynamic load, reflecting the effect of the ambient vibration excitation.

Table 1: Parameters of the magnetostrictive whisker.

Material	Name	Symbol	Value	Unit
Bias magnet	Magnetic field strength	H_b	300	mT
	Dimension	$L \times W \times H$	$8.5 \times 3.5 \times 2$	mm^3
Galfenol	Saturation magnetostriction	λ_s	170	ppm
	Young's modulus	E	60	GPa
	Poisson's ratio	ν	0.3	—
	Relative permeability	μ_r	200	—
	Dimension	$L \times W \times H$	$80 \times 4 \times 0.15$	mm^3

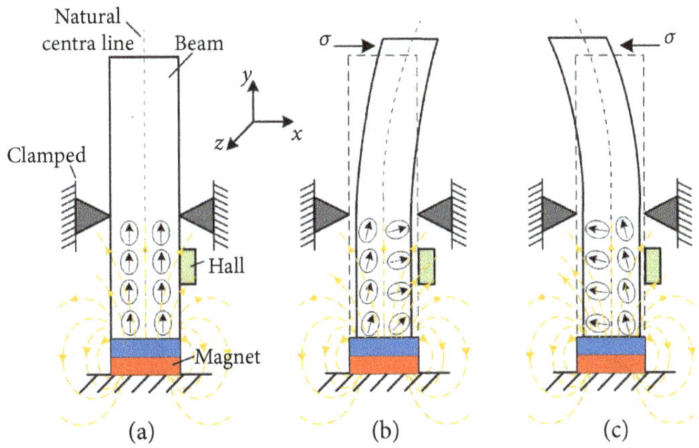

Figure 2: 3D model of the magnetostrictive whisker.

As mentioned in [21], the vibration excitation is described as an input harmonic base motion of

$$y_b(t) = Y_b \sin \omega_b t, \ddot{y}_b(t) = -Y_b \omega_b^2 \sin \omega_b t. \quad (4)$$

The tip deflection, $y(t)$, can be obtained by the motion equation of the beam due to a base excitation for acceleration $-Z_b \omega_b{}^2$ at a frequency of ω_b as

$$y(t) = Y_b a \, \mathbf{sin} \, (\omega_b t - \varphi), \quad (5)$$

where ω_1 is the beam's first-order natural frequency; $\varphi = \tan^{-1}(2\varsigma\omega_1\omega_b/(\omega_n{}^2 - \omega_b{}^2)) + \tan^{-1}(\omega_1/2\varsigma\omega_b)$, ς represents the beam's damping ratio, and $a = \omega_1[\omega_1{}^2 + 4\varsigma^2\omega_b{}^2/((\omega_1{}^2 - \omega_b{}^2)^2 + 4\varsigma^2\omega_1{}^2\omega_b{}^2)]^{1/2}$.

The equivalent force can be expressed as

$$F(t) = \frac{3EI}{L^3}(y(t) - y_b(t)). \quad (6)$$

By substituting (4), (5), and (6) into (3), we obtain

$$\sigma_x(t) = Y_b \frac{3(L-x)zE}{L^3}\{\sin \omega_b t - a \sin (\omega_b t - \varphi)\}. \quad (7)$$

Most resonant-based energy harvesters demonstrate their highest performance when excited near their resonant frequency. The ith resonance frequency of a uniform beam can be estimated using

$$\omega_n = \beta_n^2 \sqrt{\frac{EI}{mL^4}}, \quad (8)$$

where β_n is the solution of the equation $\cosh(\beta_n L)\cos(\beta_n L) + 1 = 0$.

3.2. Magnetoelastic Sensing Model. The magnetoelastic effect is the basis for the magnetostrictive whisker to realize force sensing. In general, the behavior of magnetostrictive materials is described by the piezomagnetic equation. Using this linear equation, Downey and Flatau derived the expression for the magnetic flux of a bending beam under a bias magnetic field [23]; the formula for which is

$$B_i(t) = d_{33}^* \sigma_x(t) - d_{33}^* d_{33} E H_i + \mu H_i, \quad (9)$$

where B_i is the internal magnetic flux of the beam. d_{33} and $d^*{}_{33}$ are the piezomagnetic and inverse piezomagnetic coefficients, respectively; they are defined as $d_{33} = \partial\varepsilon/\partial H$ and $d^*{}_{33} = \partial B/\partial H$. H_i is the internal magnetic field of the beam, and μ is the permeability of the magnetostrictive material.

Assuming that the bias magnetic and flux are H_0 and B_0, B_l is the linkage flux in the space; B_t is the linkage flux at the

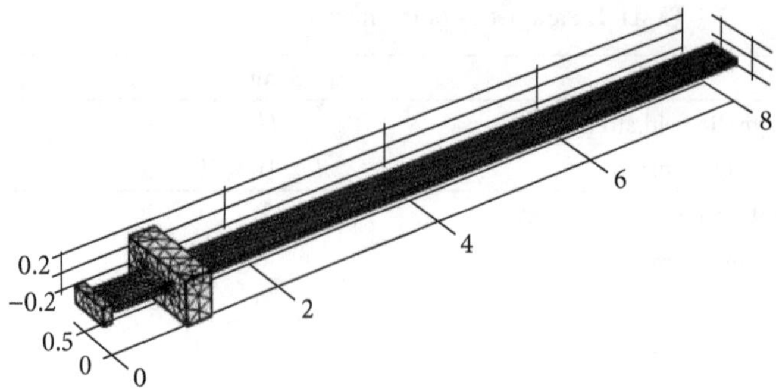

FIGURE 3: 3D model and mesh of the magnetostrictive whisker.

measurement point; and $B_0 = B_i + B_l$; the linkage flux B_t can be expressed as

$$B_t(t) = \frac{1}{2}k_m(B_0 - B_i(t)), \tag{10}$$

where k_m is the linkage magnetic coefficient related to the location of the Hall sensor, relative permeability of galfenol, and the magnetic field distribution. According to (4), (10), and (11), the transform coefficient α can be given by

$$\alpha(t) = \frac{dy_b(t)}{dB_t(t)} = \frac{1}{6}k_m\frac{L^3}{(x-L)zE} \cdot \frac{1}{\sqrt{(1-a\cos\varphi)^2 + a^2\sin^2\varphi}} \cdot \frac{\cos\omega_b t}{\cos(\omega_b t + \phi)}, \tag{11}$$

where $\phi = \tan^{-1}(a\sin\varphi/(1 - a\cos\varphi))$, and $x = 0$.

Considering that $\alpha(t)$ is a time-varying function, in order to obtain the frequency characteristics of the whisker resolution, we perform a Laplace transform of $\alpha(t)$ and note that $s = j\omega$. Then, we achieve the modular $|\alpha(j\omega)|$; substitute it into (2), and the whisker's displacement resolution can be obtained as follows:

$$r = |\alpha(j\omega)| \cdot r_H = \frac{1}{6}k_m\frac{L^2}{zE} \cdot \frac{1}{\sqrt{(1-a\cos\varphi)^2 + a^2\sin^2\varphi}} \cdot r_H. \tag{12}$$

4. Finite Element Analysis

For the finite element analysis, a more sophisticated model of the whisker was established in the COMSOL. The eigenfrequency analysis was implemented using structural mechanics, dynamics, and vibration modules. Figure 3 shows the 3D model and mesh creation of the whisker model. A fine mesh was created with 5501 tetrahedral elements, 3976 boundary triangular elements, 572 mesh points, and 34,302 degrees of freedom. The minimum element quality was set to 0.066. The model was created using the Lagrange quadratic element type. The computations employed second-order polynomials, which was a good tradeoff between the memory and result accuracy. The use of finer meshes did not have a significant effect on the results. The simulation was performed to calculate the eigenfrequencies for the first six modes.

The COMSOL model also enables the analysis of the shape of the whisker at different modes, as shown in Figure 4. Using the model, the natural frequency of the whisker can be set to any value between 0.5 and 500 Hz by choosing different materials and physical dimensions.

The theoretical value of the eigenfrequency can be obtained using an analytical model based on (8). The simulated and theoretical values for the six modes are listed and compared in Table 2. The relative error shows that at the first-order natural frequency, theoretical calculation results do not agree well with the simulation result. With the increase in frequency, the results of theoretical models and finite element analysis approach equivalence.

5. Experiment

An experimental system was established to test the resolution performance of the proposed magnetostrictive whisker. Figure 5 shows the vibration test system. In this system, the sinusoidal signal was generated by the dSPACE (ds1103) and then amplified by an amplifier (AE7224) to drive the vibration exciter in order to simulate the ambient vibration. The vibration amplitude of the vibration excitation was measured by a laser displacement sensor (Ltc-025). The data from the Hall sensor and laser sensor were acquired using the dSPACE platform. The vibration frequency was set in the range of 0.5–60 Hz. To measure the minimum displacement, for each frequency, the voltage amplitude of the Hall sensor was required to be maintained at 8.3 mV.

In the experiment, some parameters required for the theoretical model were also determined. The damping coefficient ς of the whisker system was determined as 0.014 using the Rayleigh model. The linkage magnetic coefficient is $k_m = 0.6$. Further, according to the data reported in [24], the inverse piezomagnetic coefficient is $d^*_{33} = 20$ T/N.

6. Results and Discussion

Figure 6 shows the dynamic performance of the magnetostrictive whisker sensor. The voltage amplitude of the Hall sensor was limited to 8.3 mV. Then, we obtained the amplitude of the ambient vibration at different frequencies. At 0.5 Hz, which represents the quasi-static condition, the amplitude of the ambient vibration was 1.22 mm. When the frequency became 60 Hz, this value changed to 0.18 mm. It

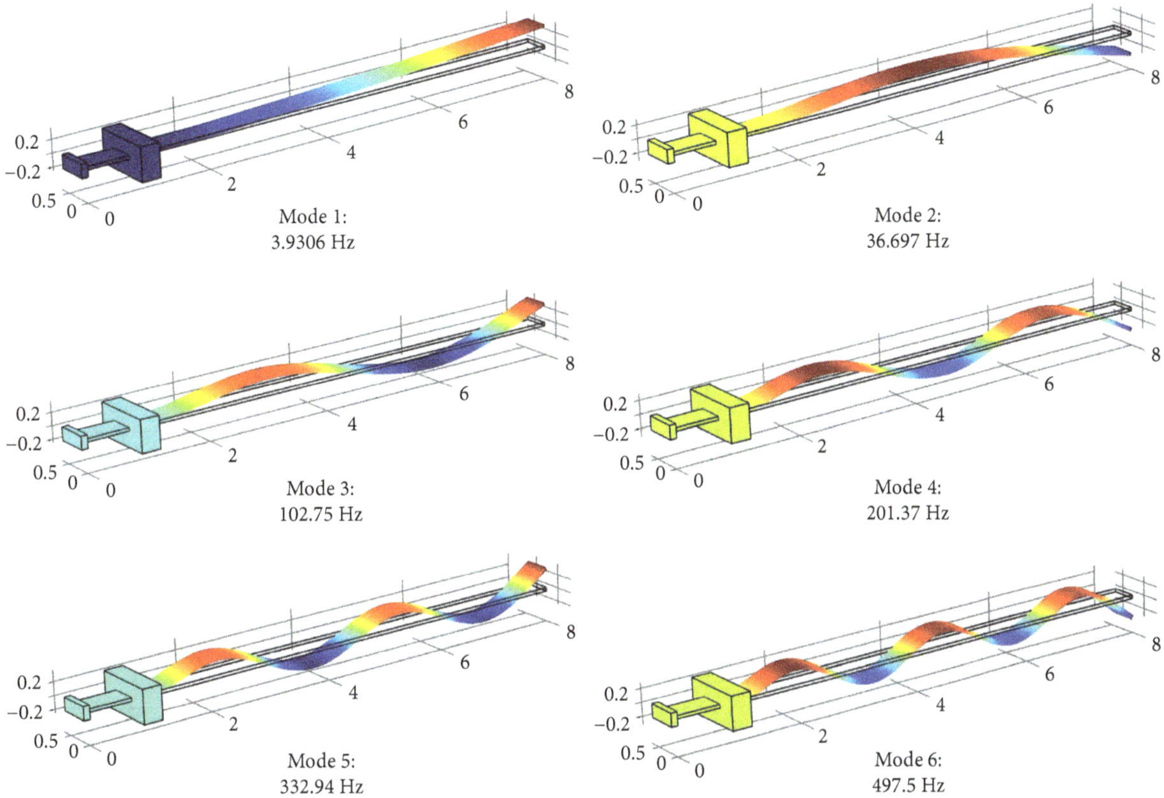

FIGURE 4: Shape of the whisker beam at the first six modes.

TABLE 2: Eigenfrequency calculated using analytical model and finite element software.

Mode	Simulated (Hz)	Theoretical (Hz)	Error (%)
1st	3.9306	4.297	9.32
2nd	36.697	35.374	3.60
3rd	102.75	99.049	3.60
4th	201.37	194.09	3.62
5th	332.94	322.29	3.19
6th	497.5	483.52	2.81

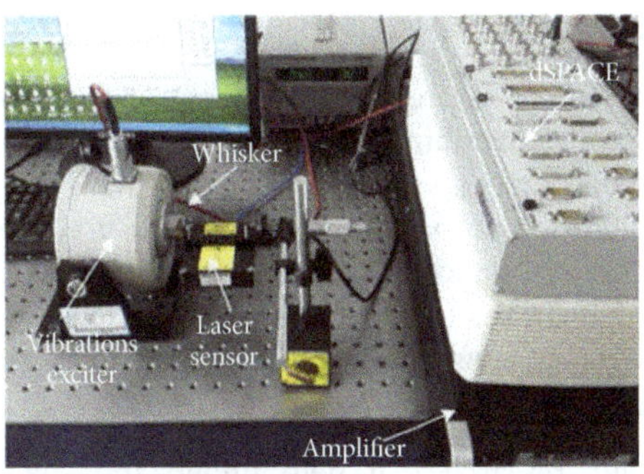

FIGURE 5: Vibration test system for magnetostrictive whisker.

was noted that with increasing frequency, the displacement amplitude sensed by the magnetostrictive whisker decreased gradually. It is thus concluded that the frequency and resolution are positively correlated.

The theoretical value and the experimental results are shown in Figure 7. The figure displays a clear change in the resolution, depending on the vibration frequency. The results of the experiment show the existence of two peaks: P_1 (0.18 mm) at 3.5 Hz and P_2 (0.06 mm) near 40 Hz. These two peaks, located near the 1st and 2nd natural frequencies of the whisker, are the maximum values of the displacement resolution. This means that at natural frequencies, the whisker demonstrates a larger response amplitude than that at other frequencies. The theoretical calculation is in good agreement with the actual value, but only P_1 occurs in the theoretical curve. This is because our analytical model does not consider additional vibration modes (besides the 1st order natural frequency) for the cantilever beam.

7. Conclusion

This study aimed to determine the influence of the vibration frequency on the displacement resolution of the whisker sensor. We established an analytical model based on the Euler–Bernoulli beam theory to obtain the mapping relation of the sensor resolution and frequency. To analyze the frequency response, we used the finite element method. Further, according to the vibration mode analysis, the first-order to sixth-order natural frequencies were identified. A vibration test system was constructed to determine the dynamic performance of the whisker. Finally, the following conclusions were derived:

(i) The displacement resolution of the whisker sensor is positively correlated. Moreover, the theoretical

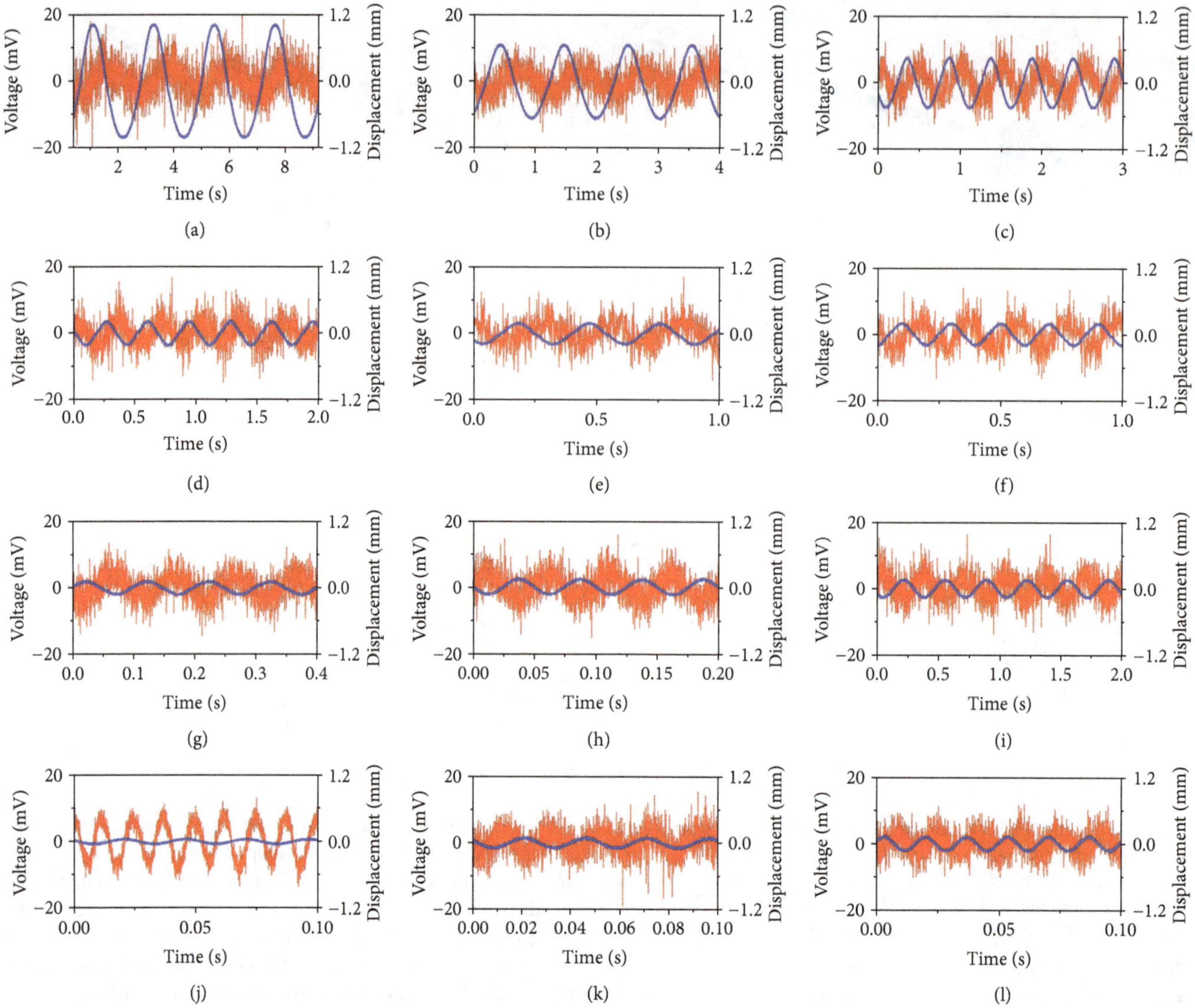

FIGURE 6: Sensing voltage and vibration amplitude at various frequencies: (a) 0.5, (b) 1, (c) 2, (d) 3, (e) 3.5, (f) 5, (g) 10, (h) 20, (i) 30, (j) 40, (k) 50, and (l) 60 Hz.

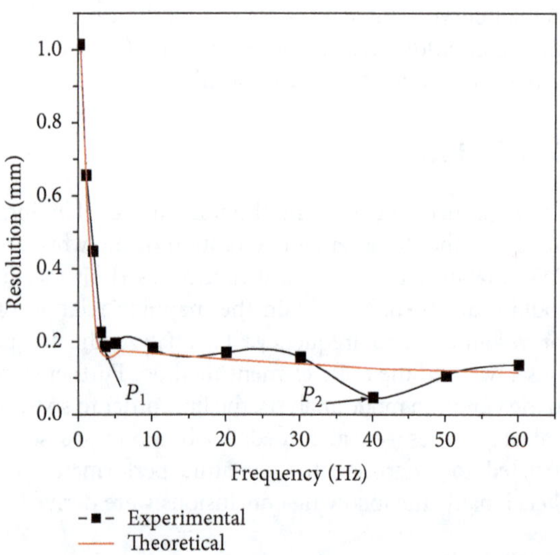

FIGURE 7: Vibration amplitude resolution dependent on frequency.

model presented in this work appropriately conformed to this trend.

(ii) When operating at the resonance frequencies, the resolution attains a local maximum value. For example, in the experiment, two peaks were noted in the vibration amplitude-frequency curve at the 1st and 2nd resonance frequencies of the whisker sensor. That means that the resolution performance of the whisker sensor can be improved by a reasonable frequency design.

(iii) Our research can help understand the dynamic behavior of magnetostrictive whisker sensors and predict its dynamic resolution.

Conflicts of Interest

The authors declare that they have no conflicts of interest.

Acknowledgments

This work was supported by the Science and Technology Project of Jiangxi Province Education Department of China (Grant no. GJJ161105), Natural Science Foundation of Hebei Province (Grant no. E2017202035), the University Science and Technology Projects of Jiangxi Province (Grant nos. KJLD14094 and KJLD14096), and the Open Fund of Jiangxi Province Key Laboratory of Precision Drive & Control (Grant no. KFKT-201601).

References

[1] F. A. Lucianna, A. L. Albarracín, S. M. Vrech, F. D. Farfán, and C. J. Felice, "The mathematical whisker: a review of numerical models of the rat's vibrissa biomechanics," *Journal of Biomechanics*, vol. 49, no. 10, pp. 2007–2014, 2016.

[2] Q. Kan, R. Rajan, J. Fu, G. Kang, and W. Yan, "Elastic modulus of rat whiskers—a key biomaterial in the rat whisker sensory system," *Materials Research Bulletin*, vol. 48, no. 12, pp. 5026–5032, 2013.

[3] S. Bovet, M. Fend, and R. Pfeifer, "Simulating whisker sensors - on the role of material properties for morphology, behavior and evolution," *Journal of the American Veterinary Medical Association*, vol. 166, no. 5, pp. 481–486, 2004.

[4] M. Kaneko and T. Tsuji, "A whisker tracing sensor for manufacturing application," *IFAC Proceedings Volumes*, vol. 33, no. 26, pp. 383–388, 2000.

[5] M. Kaneko and T. Tsuji, "A whisker tracing sensor with 5 μm sensitivity," in *Proceedings 2000 ICRA. Millennium Conference. IEEE International Conference on Robotics and Automation. Symposia Proceedings (Cat. No.00CH37065)*, pp. 3907-3912, San Francisco, CA, USA, April 2000.

[6] F. Ju and S. F. Ling, "Bioinspired active whisker sensor for geometry detection of high aspect ratio microholes with simultaneous actuation and sensing capability," *Smart Materials and Structures*, vol. 24, no. 3, article 035010, 2015.

[7] F. Ju and S. F. Ling, "Bioinspired active whisker sensor for robotic vibrissal tactile sensing," *Smart Materials and Structures*, vol. 23, no. 12, article 125003, 2014.

[8] M. I. Tiwana, M. I. Tiwana, S. J. Redmond, N. H. Lovell, and J. Iqbal, "Bio-inspired PVDF-based, mouse whisker mimicking, tactile sensor," *Applied Sciences*, vol. 6, no. 10, pp. 297–312, 2016.

[9] M. J. Pearson, A. G. Pipe, C. Melhuish, B. Mitchinson, and T. J. Prescott, "Whiskerbot: a robotic active touch system modeled on the rat whisker sensory system," *Adaptive Behavior*, vol. 15, no. 3, pp. 223–240, 2007.

[10] S. N'Guyen, P. Pirim, and J. Meyer, "Texture discrimination with artificial whiskers in the robot-rat Psikharpax," in *Biomedical Engineering Systems and Technologies. BIOSTEC 2010. Communications in Computer and Information Science, vol 127*, A. Fred, J. Filipe, and H. Gamboa, Eds., pp. 74–81, Springer, Berlin, Heidelberg, 2010.

[11] H. Zhang, T. Zhang, and C. Jiang, "Magnetostrictive actuators with large displacement and fast response," *Smart Materials and Structures*, vol. 21, no. 5, article 055014, 2012.

[12] T. Zhang, C. Jiang, X. Liu, and H. Xu, "Dynamic magnetostrain properties of giant magnetostrictive alloy actuators for damping," *Smart Materials and Structures*, vol. 14, no. 4, pp. N38–N41, 2005.

[13] H. Zhang, T. Zhang, and C. Jiang, "Design of a uniform bias magnetic field for giant magnetostrictive actuators applying triple-ring magnets," *Smart Materials and Structures*, vol. 22, no. 11, article 115009, 2013.

[14] G. Raghunath, A. B. Flatau, S. M. Na, and B. Barkley, "Development of a bio-inspired tactile magnetostrictive whisker sensor using alfenol," in *ASME Conference on Smart Materials, Adaptive Structures and Intelligent Systems*, Newport, RI, USA, September 2014.

[15] P. R. Downey and A. B. Flatau, "Bending of iron-gallium (galfenol) alloys for sensor applications," in *Proceeding SPIE 6174, Smart Structures and Materials 2006: Sensors and Smart Structures Technologies for Civil, Mechanical, and Aerospace Systems*, San Diego, CA, USA, April 2006.

[16] S. M. Na, M. Rice, G. Raghunath, V. Klimchenko, and A. B. Flatau, "Magnetostrictive alfenol whisker sensor performance and sensitivity to whisker thickness," *IEEE Transactions on Magnetics*, vol. 50, no. 11, pp. 1–4, 2014.

[17] R. Zhao, Q. Yuan, J. Yan, and Q. Lu, "The static and dynamic sensitivity of magnetostrictive bioinspired whisker sensor," *Journal of Nanotechnology*, vol. 2018, Article ID 2591080, 6 pages, 2018.

[18] http://www.winson.com.tw/Data%20Sheet/WSH138.pdf.

[19] N. Srisukhumbowornchai and S. Guruswamy, "Large magnetostriction in directionally solidified FeGa and FeGaAl alloys," *Journal of Applied Physics*, vol. 90, no. 11, pp. 5680–5688, 2001.

[20] G. Raghunath, M. Marana, S.-M. Na, and A. Flatau, "Kerr microscopy studies of the effects of bending stress on galfenol," *Journal of Applied Physics*, vol. 115, no. 17, article 17E310, 2014.

[21] J.-H. Yoo and A. B. Flatau, "A bending-mode galfenol electric power harvester," *Journal of Intelligent Material Systems and Structures*, vol. 23, no. 6, pp. 647–654, 2012.

[22] C. Qinghua, C. Dingfang, L. Quanguo et al., "Sensor performance of cantilevered magnetostrictive beam," *International Journal on Smart Sensing and Intelligent Systems*, vol. 7, no. 3, pp. 1221–1238, 2014.

[23] P. R. Downey and A. B. Flatau, "Bending behavior of iron-gallium (galfenol) alloys for sensor applications," in *Proceeding SPIE 5764, Smart Structures and Materials 2005: Smart Structures and Integrated Systems*, pp. 120–129, San Diego, CA, USA, May 2005.

[24] S. Datta and A. B. Flatau, "Application of iron-gallium alloy as magnetostrictive sensors," in *Proceedings SPIE 5764, Smart Structures and Materials 2005: Smart Structures and Integrated Systems*, pp. 110–110, San Diego, CA, USA, May 2005.

Active Thermal Sensor for Improved Distributed Temperature Sensing in Haptic Arrays

D. Cheneler[1] and M. C. L. Ward[2]

[1]*Department of Engineering, Lancaster University, Lancaster, UK*
[2]*School of Engineering and the Built Environment, Birmingham City University, Birmingham, UK*

Correspondence should be addressed to D. Cheneler; d.cheneler@lancaster.ac.uk

Academic Editor: Stefano Stassi

The efficacy of integrating temperature sensors into compliant pressure sensing technologies, such as haptic sensing arrays, is limited by thermal losses into the substrate. A solution is proposed here whereby an active heat sink is incorporated into the sensor to mitigate these losses, while still permitting the use of common VLSI manufacturing methods and materials to be used in sensor fabrication. This active sink is capable of responding to unknown fluctuations in external temperature, that is, the temperature that is to be measured, and directly compensates in real time for the thermal power loss into the substrate by supplying an equivalent amount of power back into the thermal sensor. In this paper, the thermoelectric effects of the active heat sink/thermal sensor system are described and used to reduce the complexity of the system to a simple one-dimensional numerical model. This model is incorporated into a feedback system used to control the active heat sink and monitor the sensor output. A fabrication strategy is also described to show how such a technology can be incorporated into a common bonded silicon-on-insulator- (BSOI-) based capacitive pressure sensor array such as that used in some haptic sensing systems.

1. Introduction

There are many variants of silicon-based sensors available today [1–3]. Many of the compliant sensors, in particular strain and pressure sensors, require temperature compensation in order to mitigate thermal expansion effects [2]. As such, many designs incorporate some manner of structure to prevent/correct for thermal expansion [4] or some manner of temperature sensor, usually a temperature-sensitive resistor, to facilitate correction of the sensor output [5, 6]. In general, these integrated temperature sensors do not provide a measure of the actual ambient temperature, but rather the temperature of the compliant sensor, which is a function not only of the ambient temperature but also of thermal losses through the substrate. However, there are several instances where one is interested in the temperature not to simply correct other sensors, but to provide an additional sensing modality. A typical scenario where this applies is with haptic arrays where one may wish to incorporate an array, sometimes hundreds, of compliant pressure sensors into some manner of robotic end effector to monitor contact and facilitate the handling of objects [6–9]. Temperature sensors are often incorporated into these arrays to provide a more biomimetic sensing modality [10, 11].

As with the pressure sensors, there are many variants of thermal sensor available [12]. One of the most common devices for temperature measurement is the temperature-sensitive resistor or resistance temperature detector (RTD) [13, 14]. Generally consisting of a thin wire, a film of highly conductive metal such as platinum, or an area of highly doped silicon, RTDs have been found to offer better accuracy and repeatability over alternative sensors such as thermocouples [15]. The resistance ideally varies linearly with temperature, although the self-heating effects need to be considered [16]. However, if every pressure sensor in a large haptic array requires individual conventional temperature sensors or a commercial RTD located next to it, then the spatial resolution of the tactile sensor will be severely undermined. By integrating a micro thermal sensor into the same substrate as the pressure sensor, the temperature-dependent error of the

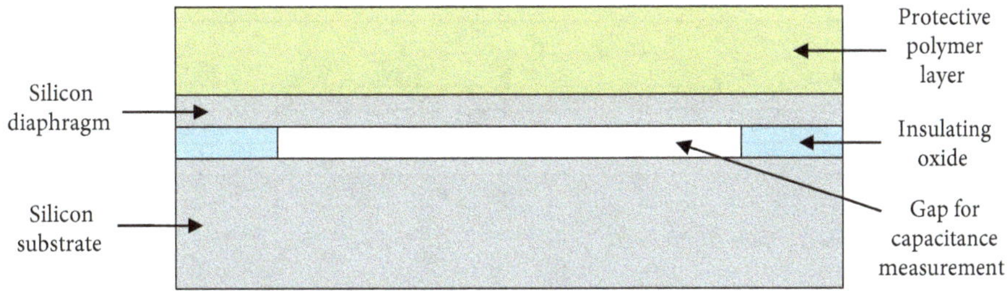

FIGURE 1: A schematic of a typical capacitive pressure sensor (not to scale).

pressure sensor can be minimised without dramatically affecting the resolution of the array and still allow an accurate measurement of the ambient temperature. Using existing designs, such a sensor, however, will be very sensitive to its environment, in particular the substrate, due to the previously mentioned thermal losses.

This paper is aimed at addressing this issue through the incorporation of a microactive heat sink that acts as an intermediate structure between the thermal sensor and the substrate. The system is designed to be suspended away from the substrate in a similar manner to many silicon-based pressure sensors, especially capacitive diaphragm-based sensors [4, 17, 18], so that it can be fabricated in much the same way. The fabrication method is detailed below. Both the thermal sensor and active sink are thermoelectric elements that are sensitive to the thermal energy that flows in and out of the respective elements; however, the active sink actively responds to losses from the sensor into the substrate and compensates accordingly by providing the equivalent power back to the sensor in real time. This allows the thermal sensor to behave as though it is thermally isolated from the substrate below, ensuring its output is only a function of the external ambient temperature/heat flux. The system is analytically and numerically modelled below to elucidate its thermodynamic behaviour and to justify simplified one-dimensional expressions that are then used in the feedback system. This feedback system is used to control the output of the active heat sink automatically and monitor the thermal sensor output.

2. Basic Design

The pressure sensors used in a tactile array are often diaphragm-based capacitive sensors [17–20] whereby the external contact force deflects the diaphragm changing the gap between it and the substrate, resulting in a change in capacitance between the two. A typical silicon-based design is shown in Figure 1. While many equivalent designs are possible, for simplicity it is assumed here that the pressure sensor was fabricated on a bonded silicon-on-insulator (BSOI) wafer coated with a polymer layer for protection with the diaphragm being released through selective etching of the insulating oxide layer. The theory that follows is applicable for many, not necessarily silicon-based, equivalent designs, provided that the sensor consists of two conductive elements (the diaphragm and substrate) separated by an electrically, but not thermally, insulting layer (the oxide layer in this example).

The thermal sensor would need to be as close as possible to the pressure sensor in order to reduce the size of the array and to make accurate measurements without introducing too many additional fabrication steps. Therefore, the thermal sensor will ideally be fabricated on the same layer as the silicon diaphragm. However, a simple temperature sensor comprising of a thin wire of silicon etched out of the device layer of the BSOI will not suffice. This is due to the high thermal conductivity of both silicon and its oxide. If the silicon substrate is maintained at room temperature, any heat energy supplied through the polymer layer will be lost immediately to the substrate and the temperature sensor will not work as intended. A heating element at the same temperature as the ambient temperature located between the sensor and the substrate would mean that the temperature gradient across the sensor would be zero and no heat will be lost. This can be achieved if the sensor is fabricated on top of an "active sink." Both the sensor and the active sink will be maintained at a constant, slightly elevated, temperature by monitoring their resistance and controlling the voltage passing through them so that their resistance is maintained at a predefined value. A schematic of a typical temperature sensor is shown in Figure 2.

This schematic is the simplest configuration of the active sink design. In practice, one would wish the thermal sensor and its sink to be much stiffer than the pressure sensor to avoid contact with the substrate while under load. This can be achieved by making the sensor short. This will reduce the resistance and so one may wish to have several of these elements acting together like a long thermal sensor with higher resistance supported by several intermediate islands of oxide. These designs are all equivalent, and the theory below applies to them all.

3. Theory

It has been known for a long time that the passage of an electric current through a conductor releases heat. The phenomena was first studied by James Joule in 1841 [21], and the equation for the temperature in a thin wire through which a constant current is flowing was found by Verdet in 1872 [22]. It was discovered that the mechanisms behind both electrical and thermal conduction were inextricably linked. It is now known that Joule heating, or resistive heating, is caused by interactions between the moving current and the charged atoms that make up the body of the conductor. Charged particles in the electric circuit are accelerated by an electric field but give up some of their kinetic energy each

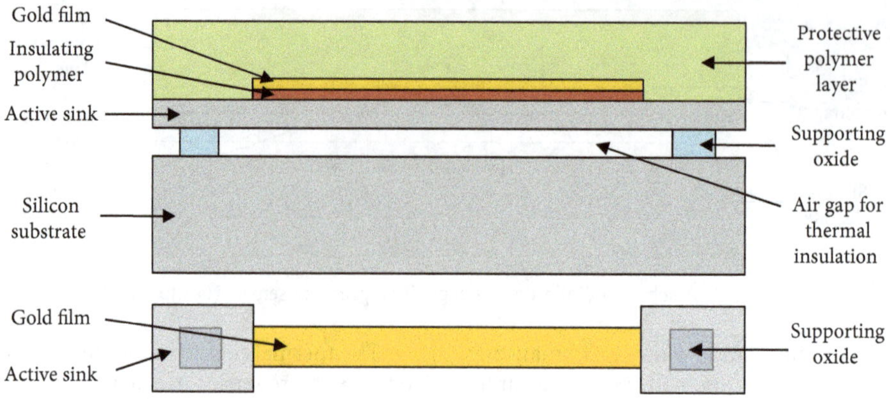

Figure 2: A schematic of a temperature sensor with active sink. The gold film acts as the sensing element. Top: a cross-sectional view showing the different layers. Bottom: a top view without protective polymer layer and substrate.

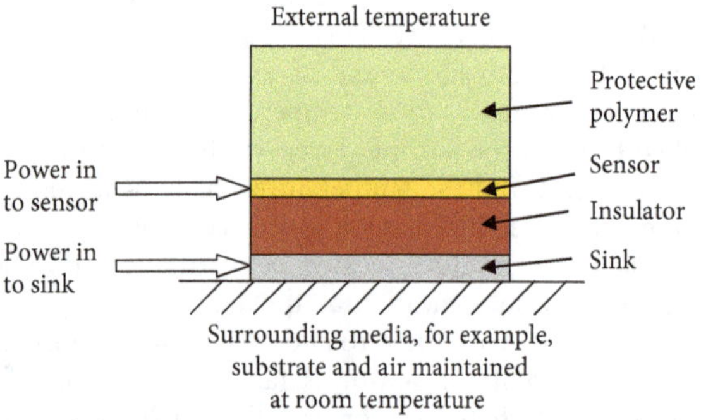

Figure 3: A schematic of the one-dimensional representation of the temperature sensor with active sink. The temperature variation will be in the vertical direction.

time they collide with an ion. The increase in the kinetic or vibrational energy of the ions manifests itself as heat and a rise in the temperature of the conductor. Hence, energy is transferred from electrical power to the conductor and any materials with which it is in thermal contact [23]. This transfer of energy through molecular interaction is very similar to that accomplished through heat conduction, except in this case the motion is not necessarily caused by an electric field but through momentum exchange from a molecule with a higher energy level [24].

In general, this is a three-dimensional thermoelectric problem. However, as will be shown, the problem reduces to a simpler one-dimensional problem. The schematic for the simplified model is shown in Figure 3.

As can be seen in Figure 3, the device can be considered to be a simple layered structure. The top layer, the protective polymer, is an insulating layer between a surface set at the external temperature and a surface at the temperature of the sensor. The sensor is a highly conductive thin layer which has a current flowing through it. There is heat energy being generated within the sensor due to electrical resistance and energy exchange through the polymer and insulator layers. The insulating layer is similar in nature to the polymer layer except that the external temperatures are dictated by the temperature of the sensor and active sink, respectively. The active sink is another highly conductive thin layer with a current flowing through it. Again, heat energy is generated due to the electrical resistance and energy is exchanged through the insulation layer. However, an additional complication is that energy is also lost from the ends of the sink through the oxide as well as from the surrounding air.

The method for solving this problem was to treat the four layers as four coupled partial differential equations, which could then be solved numerically. The governing differential equation for a one-dimensional transient conduction problem with internal energy generation is given by [25]

$$k\frac{\partial^2 T}{\partial x^2} + \dot{q} = \rho C_p \frac{\partial T}{\partial t}, \tag{1}$$

where k is the thermal conductivity of the layer in $\mathrm{W/(m \cdot K)}$, T is the temperature distribution within the layer in K, $\dot{q}$ is the rate of heat energy generated within the layer per unit volume, ρ is the density in $\mathrm{kg/m^3}$, and C_p is the specific heat of the material in $\mathrm{J/(kg \cdot K)}$. This essentially means, considering just a small element of material within the layer,

$$\begin{aligned}&\left\{\begin{matrix}\text{Rate of energy}\\ \text{conduction into}\\ \text{element}\end{matrix}\right\} - \left\{\begin{matrix}\text{Rate of energy}\\ \text{conduction out}\\ \text{of element}\end{matrix}\right\}\\ &+\left\{\begin{matrix}\text{Rate of energy}\\ \text{generation}\\ \text{within element}\end{matrix}\right\} = \left\{\begin{matrix}\text{Rate of accumu-}\\ \text{lation of energy}\\ \text{within element}\end{matrix}\right\}.\end{aligned} \tag{2}$$

In the protective polymer layer, there is no internal heat generation and the temperature gradient within is completely determined by its initial temperature distribution and boundary conditions. The initial temperature distribution is assumed to be constant and at room temperature as will be the case before the device is turned on, and any changes to the external temperatures have been made. The boundary conditions are the temperatures of the outer

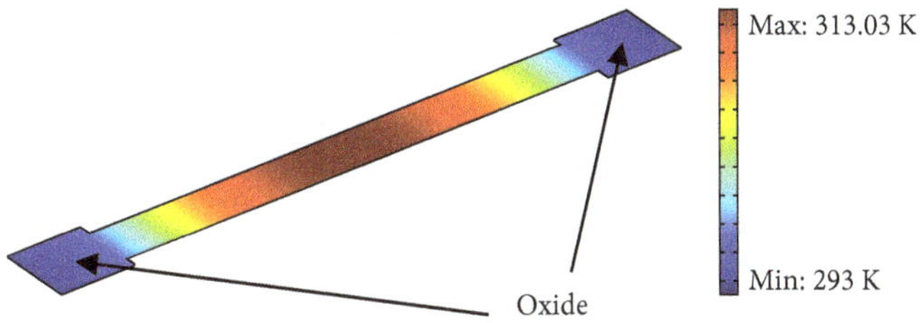

FIGURE 4: The steady-state temperature distribution of the active sink when 1 V is applied across it, as found using FEA in Comsol.

surface and of the sensor. Given this, the differential equation governing this layer is

$$\frac{\partial T}{\partial t} = \alpha \frac{\partial^2 T}{\partial x^2}, \quad (3)$$

where $\alpha = k/\rho C_p$ and is known as the thermal diffusivity and is expressed in m^2/s. This equation is known as Fick's second law and is essentially the diffusion equation [24]. Analytical solutions exist for this equation as shown in Appendix A; however, numerical integration of the highly oscillatory integrals found in the solution means that the solution required extensive computation time and so is not very convenient for further numerical analysis. It is far more convenient in this case to solve (3) fully numerically using the Crank-Nicolson method as described in Appendix B [26]. The solution for the transient temperature distribution for the insulation layer is found in exactly the same way, with the end temperatures set by the temperature of the active sink and sensor, respectively.

The sensor has a high thermal conductivity and is very thin compared to the surrounding materials. This means that it can be assumed that the temperature at any instant is effectively constant throughout the layer. In this case, the layer is known as being "thermally thin" [27]. From (2), it can be seen that the governing equation becomes

$$\rho C_p v \frac{\partial T}{\partial t} = \dot{q}_{\text{in}} - \dot{q}_{\text{p}} + \dot{q} v, \quad (4)$$

where v is the volume of the layer, $\dot{q}_{\text{in}}$ is the energy entering via the insulator, and $\dot{q}_{\text{p}}$ is the energy being dissipated through the polymer layer. Naturally, the signs of $\dot{q}_{\text{in}}$ and $\dot{q}_{\text{p}}$ depend on the instantaneous temperature distributions within the respective layers and so may change. The energy exchange, $\dot{q}_i$, through the relevant layers is due to conduction which is described by the Fourier rate equation:

$$\dot{q}_i = kA \frac{\partial T}{\partial x}, \quad (5)$$

where A is the area of the layer in thermal contact with the other layer. The rate of generation of heat due to electrical resistance in a conductor can be calculated by Ohm's law:

$$P = \dot{q} v = VI = \frac{V^2}{R}, \quad (6)$$

where P is power in watts, V is the voltage across the sensor, and I and R are the current through the sensor and the resistance of the sensor, respectively. As no other work is being done, all the energy generated as electrical power is dissipated as heat. It is assumed that the resistance will be linearly dependent on temperature as shown in (7). This assumption is reasonable over small ranges of temperature.

$$R = \frac{\rho_e L}{A_c} (1 + a\Delta T), \quad (7)$$

where ρ_e is the initial resistivity of the sensor at room temperature measured in $\Omega \cdot$ m, L is the length of the conductor in m, A_c is its cross-sectional area in m^2, a is the temperature coefficient of resistance for the layer measured in 1/K, and ΔT is the change in temperature of the layer from room temperature. Therefore, the equation to solve for the response of the sensor can be shown to be

$$\frac{\partial T}{\partial t} = \frac{1}{\rho C_p v} \left[\left(kA \frac{\partial T}{\partial x} \right)_{\text{in}} - \left(kA \frac{\partial T}{\partial x} \right)_{\text{p}} + \frac{V^2}{(\rho_e L/A_c)(1 + a\Delta T)} v \right]. \quad (8)$$

As was mentioned above, the active sink behaves in a manner similar to the sensor. This means the governing equation for the sink is the same as (4). It is also assumed that the heat exchange to the insulation layer and to the air is of the same form as (5) and that the resistance of the sink varies in the same manner as (7). The difference lies in the exchange of heat to the substrate. Whereas the sensor is in complete thermal contact along its length with the polymer and insulator layer, the active sink is only in contact to the substrate at the ends. This was deliberate in order to reduce the amount of heat loss from the sink. The consequence of this decision is that there is a temperature distribution along the length of the sink. For example, consider an active sink comprised of silicon, with a potential of 1 V applied across it, of dimensions $1000 \times 100 \times 2\,\mu$m connected to a substrate maintained at room temperature via silicon blocks of $150 \times 150 \times 2\,\mu$m on top of silicon dioxide blocks of $50 \times 50 \times 2\,\mu$m as shown in Figure 4. The silicon was assumed to be doped to saturation with arsenic to make it electrically conductive. The properties of the silicon and its oxide are given in Table 1.

As can be seen in Figure 4, the thermoelectric problem for the active sink has been solved using COMSOL 4.3, a commercial FEA package [28]. The results show that there

Table 1: The properties of doped silicon and oxide used in the active sink.

	Silicon	Oxide
Thermal conductivity (W/(m · K))	130	1.38
Electrical resistivity (Ω·m)	0.00006	10e12
Temperature coefficient of resistance (1/K)	0.0005	0
Specific heat (J/(kg · K))	712	703
Density (kg/m³)	2330	2200

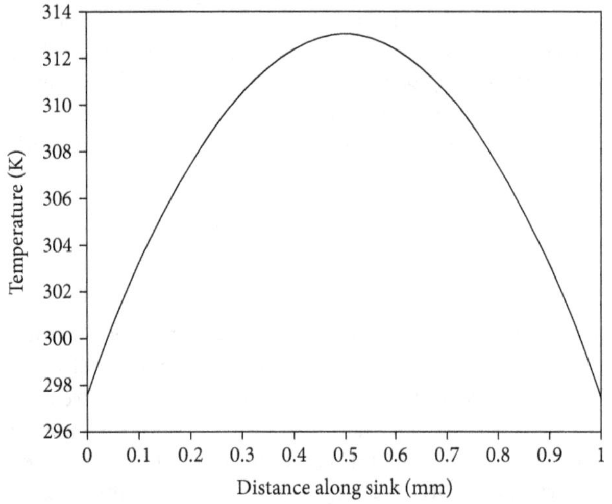

Figure 5: The one-dimensional steady-state temperature distribution of the active sink when 1 V is applied across it, as found using Comsol. Note room temperature was set to 293 K.

is a temperature distribution along the length of the sink in a direction perpendicular to the assumed direction of heat flow as dictated in Figure 3. This is because the dominant heat loss mechanism in the active sink is conduction through the ends of the sink to the substrate below, which is maintained at room temperature. In order to reduce this problem to the required equivalent one-dimensional problem, it is necessary to ensure that the relevant heat exchange mechanisms are the same for both cases. Plotting the steady-state temperature distribution for the active sink in a more convenient form, see Figure 5, shows that the temperature distribution is essentially parabolic in nature. This means the temperature distribution can be assumed to have the form

$$dT_d(y) = dT_0\left[1 + 4\left(\frac{y}{L} - \frac{1}{2}\right)^2\right] + dT_{end}, \tag{9}$$

where $dT_d(y)$ is the horizontal temperature distribution in the horizontal direction, y, along the length, L, of the active sink. dT_{end} is the difference between the temperature of the ends of the sink and room temperature, and dT_0 is the difference between the peak temperature in the sink and T_{end}.

In order to be consistent, the energy converted from electrical to heat needs to be the same in the two-dimensional and one-dimensional cases. The electrical power passing through the sink is given by combining (6) and (7). In the one-dimensional case, it is assumed that the temperature difference is constant throughout the sink, that is, $\Delta T = dT_c$, and so the power can be given as

$$P = I^2 \frac{\rho_e L}{A_c}(1 + a\,dT_c). \tag{10}$$

In the two-dimensional case, the temperature distribution has to be taken into account. In this situation, the power can be expressed as

$$P = \int_0^L I^2 \frac{\rho_e L}{A_c}\left(1 + a\left(dT_0\left[1 + 4\left(\frac{y}{L} - \frac{1}{2}\right)^2\right] + dT_{end}\right)\right)dy. \tag{11}$$

Equation (11) can be solved to give

$$P = I^2 \frac{\rho_e L}{A_c}\left(1 + a\left[\frac{2}{3}dT_0 + dT_{end}\right]\right). \tag{12}$$

In order that the energy generated due to electrical resistance is the same in both cases, (10) and (12) must be equal. Therefore, the relationship between the assumed constant temperature of the active sink and the actual parameters of the temperature distribution is

$$dT_c = \frac{2}{3}dT_0 + dT_{end}. \tag{13}$$

However, in assuming that the temperature difference of the sink is dT_c, the temperature difference at the ends of the sink is higher than they should be, that is, dT_{end}. This would result in apparently more heat energy being lost to the substrate than would actually be the case. In order to ensure that the heat loss is correct, it is necessary to find the temperature at the end of the sink as a function of the constant temperature alone. The rate of heat loss due to conduction was given in (5). As in this case, the heat has to flow through two layers in series to the substrate and so (5) needs to be rephrased slightly to give the heat loss through one end of the sink as [24]

$$\dot{q}_{end} = \frac{dT_{end}}{(u_o/k_{Si}A_{Si}) + (t_o/k_o A_o)}, \tag{14}$$

where k_{Si} is the thermal conductivity of silicon, A_{Si} is the cross-sectional area of the silicon block, u_o is the distance from the end of the sink to the oxide, k_o is the thermal conductivity of the oxide layer, t_o is its thickness, and A_o is the area of the oxide in contact with the substrate. For the example given, $\dot{q}_{end}$ is approximately $1.9 \times 10^6 dT_{end}$. Due to conservation of energy, the heat flow out of one end of the active sink must equal the heat flow into the silicon block as given by (14). Applying (5) to the end of the active sink gives

$$\dot{q}_{\text{end}} = k_{\text{Si}} A_{\text{c}} \frac{\partial \left[T_0 \left[1 + 4((y/L) - (1/2))^2 \right] + T_{\text{end}} \right]}{\partial y}. \tag{15}$$

By solving (15) for the two ends of the active sink, we obtain

$$\dot{q}_{\text{end}} = \begin{cases} 4k_{\text{Si}} A_{\text{c}} \dfrac{dT_0}{L}, & y = 0, \\[2ex] -4k_{\text{Si}} A_{\text{c}} \dfrac{dT_0}{L}, & y = L. \end{cases} \tag{16}$$

Equating (16) for the case of $y = 0$ with (14) gives the relationship

$$dT_0 = \left(\frac{dT_{\text{end}}}{(u_{\text{o}}/k_{\text{Si}} A_{\text{Si}}) + (t_{\text{o}}/k_{\text{o}} A_{\text{o}})} \right) \frac{L}{4k_{\text{Si}} A_{\text{c}}}. \tag{17}$$

This can be substituted into (13) to give the temperature of the end of the active sink as a function of the presumed constant temperature

$$dT_{\text{end}} = dT_{\text{c}} \left[\left(\frac{1}{(u_{\text{o}}/k_{\text{Si}} A_{\text{Si}}) + (t_{\text{o}}/k_{\text{o}} A_{\text{o}})} \right) \frac{L}{6k_{\text{Si}} A_{\text{c}}} + 1 \right]^{-1}. \tag{18}$$

Therefore, the heat loss due to conduction to the substrate can be given as

$$\dot{q}_{\text{end}} = 2 \frac{dT_{\text{c}}}{(u_{\text{o}}/k_{\text{Si}} A_{\text{Si}}) + (t_{\text{o}}/k_{\text{o}} A_{\text{o}})} \cdot \left[\left(\frac{1}{(u_{\text{o}}/k_{\text{Si}} A_{\text{Si}}) + (t_{\text{o}}/k_{\text{o}} A_{\text{o}})} \right) \frac{L}{6k_{\text{Si}} A_{\text{c}}} + 1 \right]^{-1}. \tag{19}$$

Here, a factor of two is included because, of course, the heat is being lost out of both ends of the active sink. In a similar fashion to (8), the equation governing the response of the active sink can be shown to be

$$\frac{\partial T}{\partial t} = \frac{1}{\rho C_p v} \left[\left(kA \frac{\partial T}{\partial x} \right)_{\text{in}} - 2 \frac{dT_{\text{c}}}{(u_{\text{o}}/k_{\text{Si}} A_{\text{Si}}) + (t_{\text{o}}/k_{\text{o}} A_{\text{o}})} \cdot \left[\left(\frac{1}{(u_{\text{o}}/k_{\text{Si}} A_{\text{Si}}) + (t_{\text{o}}/k_{\text{o}} A_{\text{o}})} \right) \frac{L}{6k_{\text{Si}} A_{\text{c}}} + 1 \right]^{-1} + \frac{V^2}{(\rho_e L/A_{\text{c}})(1 + a_{\text{Si}} dT_{\text{c}})} v \right]. \tag{20}$$

For validation of this last result, (20) was solved in isolation using a fourth order Runge-Kutta algorithm in Simulink. For clarity, the term related to the heat exchange through the insulator was omitted. This was then compared to the transient response of the isolated active sink calculated using FEA in Comsol shown in Figure 4. In

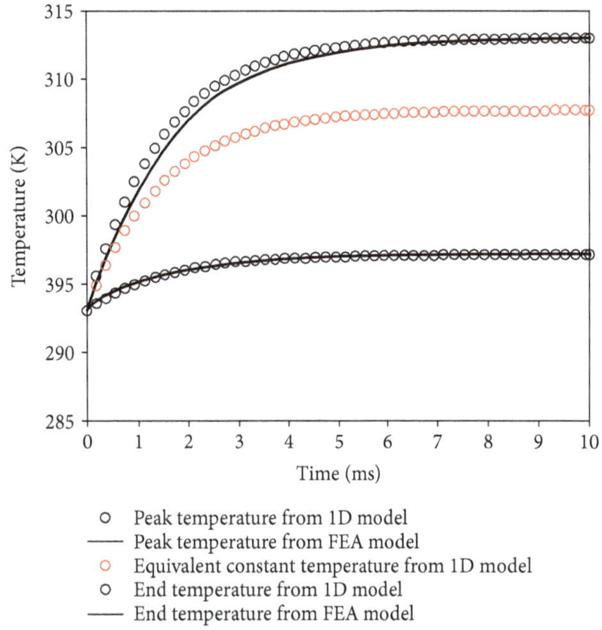

Figure 6: The transient response of the end and peak temperature of the active sink when 1 V is passed it on its own. Note that the 1D model corresponds well to the full 3D model solved using FEA.

Figure 6, the peak temperature and the end temperature of the active sink as calculated using FEA are compared to the temperatures as inferred from the results of the Simulink simulation and (13) and (18). The equivalent constant temperature as calculated from the Simulink model is also plotted for comparison. It is important to note that it took over twenty minutes to solve the full 3D model of the active sink with FEA and less than ten seconds to solve the equivalent 1D case in Simulink.

In Figure 7, the transient response of the current flow through the active sink as calculated by the FEA model assuming that the full three-dimensional temperature distribution and the equivalent one-dimensional distribution assumed in the Simulink model are compared. As the transient response of the current and the calculated end temperatures of the sink are the same in both models, it shows that the two are equivalent with the same amount of energy being generated and lost in both cases. The advantage of using the one-dimensional case instead of using FEA directly is the reduction of computational time and the ease of including feedback systems in the model without reducing accuracy significantly. Discrepancies are likely to be due to the assumption that the temperature distribution is parabolic which is only approximately true.

To summarise, it has been shown that the heat exchange through the insulator and protective skin layer can be expressed as (3). It has also been shown that the thermoelectric response of the sensor layer and the active sink can be given as (8) and (20), respectively. As the response of one layer is directly affected by the response of neighbouring layers, the governing equations are coupled and hence an analytical solution is unlikely. Fortunately, now the problem is in this reduced form, a numerical solution is possible.

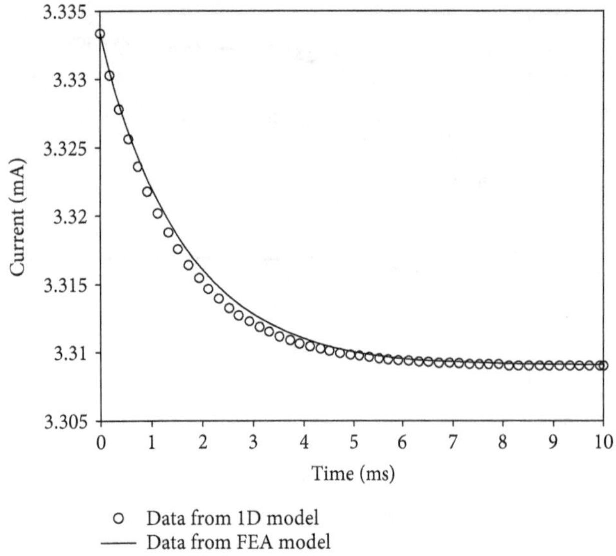

FIGURE 7: The transient response of the current when 1 V is passed through the active sink on its own. Note that the 1D model corresponds well to the full 3D model solved using FEA.

4. Feedback System

The purpose of using an active sink as part of the device is to prevent heat loss from the temperature sensor to the substrate. This allows the heat flux through the polymer layer due to changes in external temperature to be directly measured by the temperature sensor. For this to work, the temperature of the active sink and the temperature sensor has to set to the same constant value. This can be seen if (2) is applied to the temperature sensor. If the temperature is kept constant, the rate of accumulation of energy is zero. Similarly, if the active sink is at the same temperature as the sensor, there is no loss to the substrate. Therefore, if the accumulation and loss of energy is zero, the rate of energy coming into the sensor must equal the rate of energy being generated. Put more simply, the change in heat flux through the polymer will equal the change in electrical power needed by the sensor to maintain its temperature; that is, if the external temperature increases, the power through the sensor will decrease.

The feedback control to the active sink and temperature sensor consists of a simple proportional controller. By measuring each element's resistance at room temperature and temperature coefficient of resistance, (7) can be used to determine the resistance of each element at the set temperature. In the case of this device, the initial specified temperature is to be set at 37.5°C to be comparable to the body temperature. During the experiment, the instantaneous resistance of each element is compared to the required resistance and normalised and the voltage is changed by the normalised resistance multiplied by a gain factor, that is,

$$V(t) = V(t - dt) + dV = V(t - dt) + G\frac{R_{\text{req}} - R(t)}{R_{\text{req}} - R_o}, \quad (21)$$

where dV is the change in applied voltage, G is the gain, R_{req} is the required resistance, that is, the resistance calculated using (7), R_o is the resistance at room temperature, and $R(t)$ is the instantaneous resistance. In this way, the resistance is kept constant and hence each element's temperature. The flux coming through the polymer layer is then monitored by measuring the electrical power through the temperature sensor. The power through the sink will be essentially constant except at times of rapid external temperature change where there will be a small change before it returns to its usual level. Clearly implementation of more sophisticated feedback systems is quite simple, but not necessary.

4.1. Solution of the Model. As was mentioned previously, the system model for the entire device can be solved using Simulink; see Figure 8 for the model. The algorithm is as follows:

(1) Initially, it is assumed that all layers are in thermal equilibrium and are at substrate temperature. There is initially no power going to the active sink and temperature.

(2) In the first time step, the resistance of the temperature sensor and active sink is measured and compared to the required resistance equivalent to the initial specified elevated temperature (37.5°C) and the voltage is increased accordingly.

(3) This results in a change in temperature of the sink and sensor. Given these temperatures as boundary conditions, the temperature profile throughout the insulation and polymer layer is found by solving (3) using the Crank-Nicolson method as discussed in Appendix B. This is achieved using an embedded Matlab code in Simulink. The heat flux being transferred to/from the sink and sensor can be found by differentiating the temperature profile in the insulation/polymer layer, respectively, at the interface as per (5). This then determines the heat flow into and out of the electrically heated layers.

(4) The instantaneous temperature of the temperature sensor is then found by integrating (8) using the fourth-order Runge-Kutta solver with discreet timestepping built into Simulink.

(5) Similarly, the instantaneous temperature of the active sink is found by integrating (20) in the same manner as the sensor.

(6) The temperatures are then used to find the resistances of the active sink/temperature sensor using (7).

(7) Given the resistance and the applied voltage, the electrical power passing through the sensor is measured using (6).

(8) In the next time step, the resistances are compared again to the desired resistances and the voltages being applied to the active sink/temperature sensor are again changed accordingly. The sequence from step 3 to 8 is repeated for the necessary number of time steps.

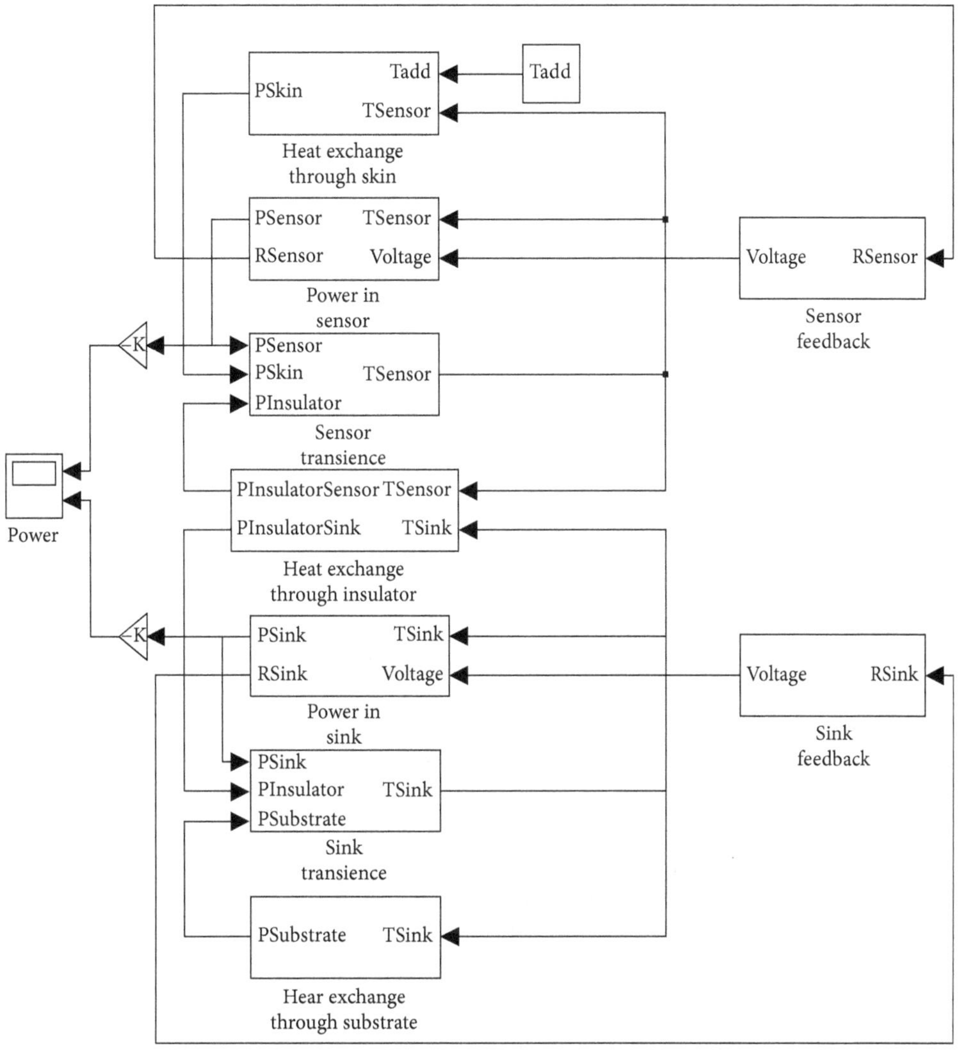

FIGURE 8: The Simulink model depicting how the different layers are linked together.

(9) At some juncture, it is possible that the response to a change in external temperature is needed. The algorithm just described is the same except that the boundary conditions stated in step 3 for the polymer layer will change as the external temperature changes.

5. Results

For the purposes of simulation, it will be assumed that the temperature sensor will be made out of a gold layer 200 nm thick and with the same width and length as the active sink. The insulation layer and polymer layer will be assumed to have the same properties as PDMS for simplicity. The insulation layer will be 10 μm thick, and the protective polymer layer will be 100 μm. The properties of the layers are shown in Table 2.

It should be noted that gold was considered as it is compatible with a number of MEMS fabrication facilities. Other materials, perhaps with a higher temperature coefficient of resistance or electrical resistivity, could be used if convenient. The time, dt, for each integration step is 1 μs. The gain for the active sink proportional controller was set to 1 and for the temperature sensor was set to 0.05. It was assumed that the controllers would measure the resistances of the pertinent layers and change their voltages accordingly every 0.1 ms. The initial sensor temperature was set to 310.5 K, as before, and the external temperature was set to room temperature (293 K) until 0.25 s when the device was instantaneously brought into contact with an object at 273 K. The response for the active sink and temperature sensor can be that shown in Figure 9.

TABLE 2: The properties of gold and PDMS as used in the temperature sensor, insulation, and polymer layers.

	Gold	PDMS
Thermal conductivity (W/(m · K))	318	0.15
Electrical resistivity (Ω·m)	$2.23E-8$	$10E-12$
Temperature coefficient of resistance (1/K)	$3.768E-3$	—
Specific heat (J/(kg · K))	129	1460
Density (kg/m^3)	19,320	970

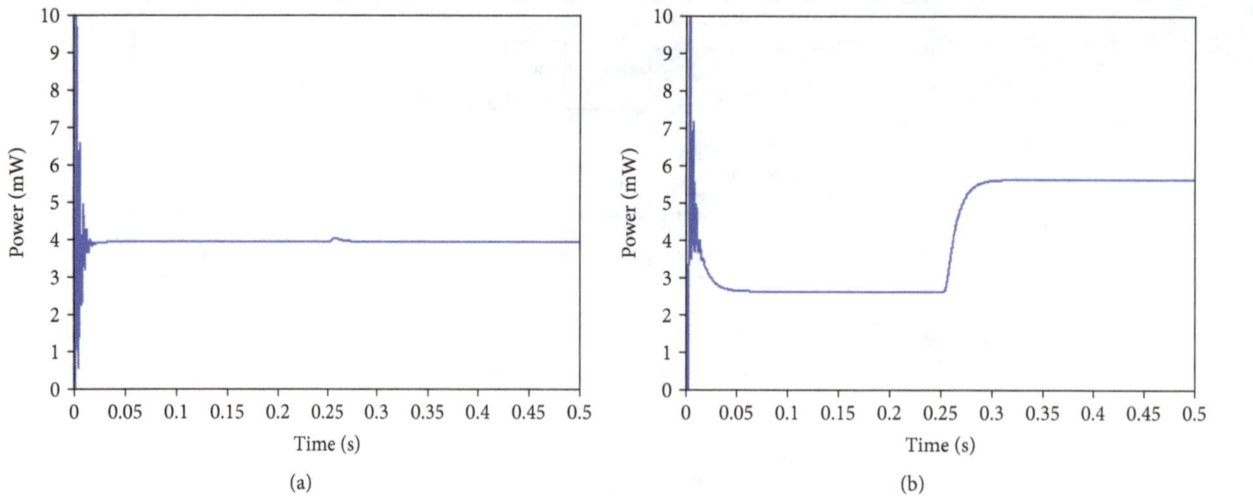

Figure 9: (a) The response of the active sink to a change in external temperature of −20 K and (b) the reponse of the temperature sensor.

There are several features that are worth noting in the response of the active sink and temperature sensor. The first is the initial transience. As the sink and sensor are starting from room temperature and need to be heated to 310.5 K, there is tendency for the contoller to overshoot before settling down to the correct temperature, especially if the sampling time of the controller is large. It is also important to note that while the temperature was changed instantly, it took the temperature sensor about 0.1 s to reach steady state. This is important because the external temperature can only be calculated analytically if the temperature gradient throughout the polymer layer is known. Also, while it can be predicted using solutions such as that given in Appendix A, in reality it is likely only to be known exactly when the gradient is constant, that is, when the sensor is at steady state. It is interesting to note the small increase in power in the active sink before returning to normal as it compensates for the sudden heat loss in the temperature sensor.

It should not be that such changes in power and even the small changes in the resistances of the gold layer are easily achieved using modern digital multimeters, such as NI 4070 [29], which is capable of measuring 1 V at 1 μV resolution and 20 mA at 10 nA resolution and therefore can measure resistances to ca. 60 μ·Ω and powers to 23 nW at the appropriate level to seven digits at 5 S/s, which is more than sufficient. Cheaper options can be created using multimeter ICs, such as MAX1367 [30], to provide a sufficient resolution.

Figure 9(b) shows how the electrical power in the temperature sensor changes with external power. Comparing Figure 9(b) with Figure 10 which shows how the heat energy flowing into the polymer layer changes, it is clearly seen that the electrical energy is completely converted into heat which is lost through the polymer layer. This verifies the theory that the active sink, if maintained at the same temperature as the sensor, does prevent any heat being lost to the substrate.

The effect of the control mechanism can be quite profound. If the sampling rate is too low or the gain is too high, abhorrent behaviour can occur masking the effects of changing the external temperature. For example, in Figure 11 the response of the active sink and temperature sensor is given again for the system described above but with the sampling time increased to 5 ms. Even with the gains for the active sink and temperature sensor controller reduced to 0.5 and 0.05, respectively, the behaviour is not ideal. This will dictate what technology is used for the controllers.

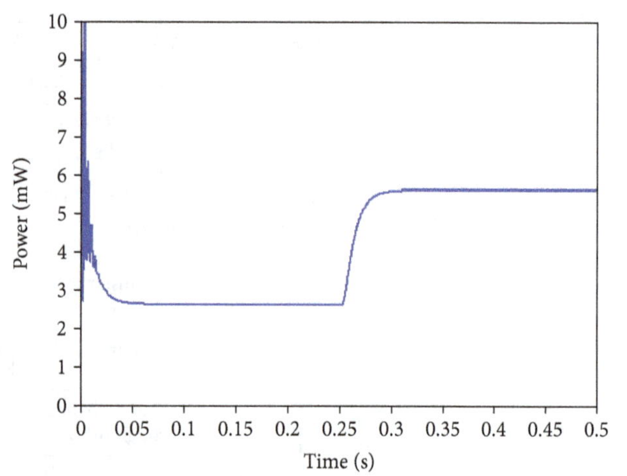

Figure 10: The heat flux passing from the sensor into the polymer layer. Note how it increases when the temperature gradient decreases.

The device temperature was set to 310.5 K which is comparable to the body temperature and is therefore able to be handled safely. The reason for this high temperature is that the steady-state temperature difference between this temperature and room temperature is 17.5 K which sets the limit for the highest external temperature measureable. It can be seen in Figure 12 that if the external temperature is decreased relative to room temperature, the amount of electrical power converted to heat is increased. Conversely, if the external temperature is increased, the electrical power needed by the sensor to maintain its temperature is decreased. Eventually, the external temperature will be so high that even if no electrical power was passing through the sensor, it would still be

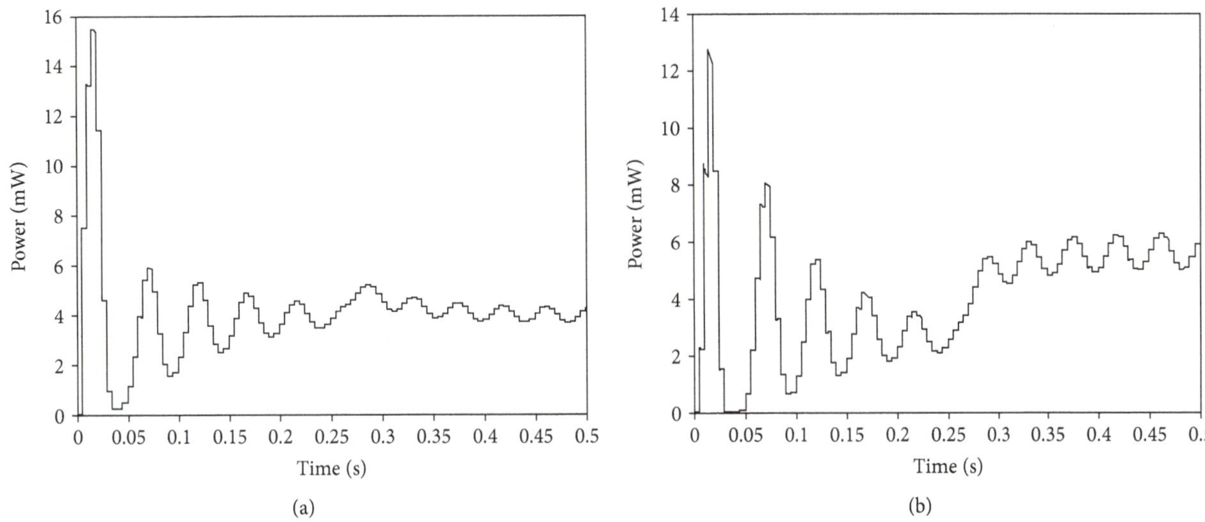

FIGURE 11: (a) The response of the active sink to a change in external temperature of −20 K given a reduction in the sampling rate of the feedback controller to 5 ms and (b) the reponse of the temperature sensor.

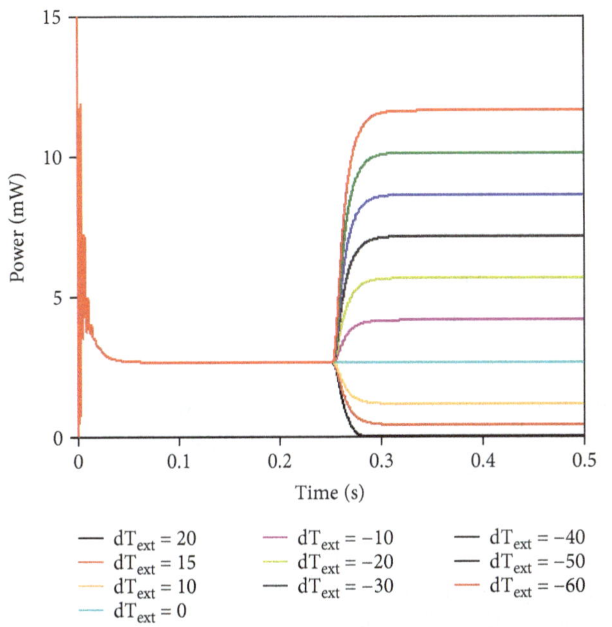

FIGURE 12: The responses of the temperature sensor due to changes in the external temperature. dT_{ext} indicates the difference from room tempertaure (293 K). Note how the time to reach steady state is the same in all cases.

at a higher temperature than required and no more increases in temperature can be measured as the power would be zero for all higher temperatures. The minimum external temperature measureable by the sensor will be determined by how much power the control circuit can provide.

The responses given in Figure 12 show a trend which is more clearly seen in Figure 13. Figure 13 shows how the steady-state response changes when the external temperature is changed from room temperature. It can be seen that the difference in the steady-state power levels in the sensor as compared to the steady-state power level at room temperature

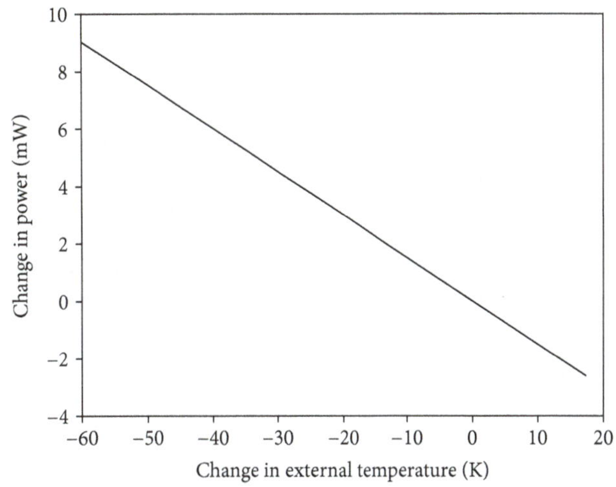

FIGURE 13: The relationship between change in external temperature from room temperature and the change in electrical power used by the temperature sensor. Note how the relationship is linear.

is a linear function of the change in external temperature. This is likely to be due to the assumption that the resistance of the sensor has a linear dependance on temperature as given in (7).

6. Fabrication

As mentioned above, while there are several designs that are possible, they are all equivalent to each other provided that the sensor consists of two conductive elements (the diaphragm and substrate) separated by an electrically, but not thermally, insulting layer (the oxide layer in this example). The design configuration below provides an additional common ground between the detecting layer and the active sink. If the pressure sensor was as described above, that is,

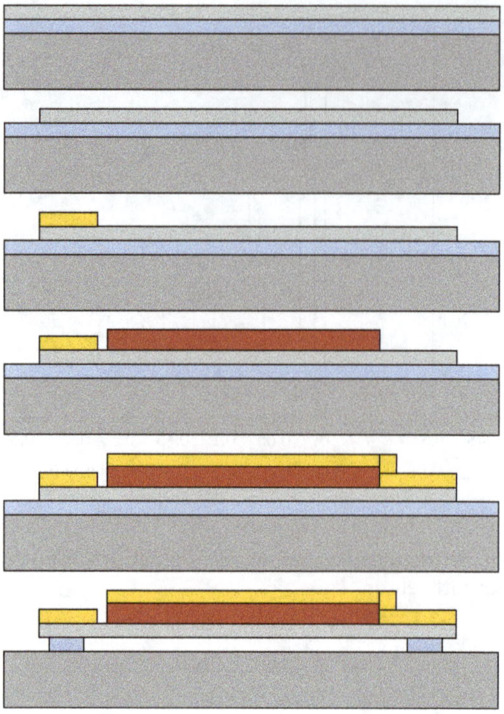

FIGURE 14

fabricated on a bonded silicon-on-insulator (BSOI) wafer coated with a polymer layer for protection with the diaphragm being released through selective etching of the insulating oxide layer, as is quite common, then a simple process plan to fabricate the temperature and capacitive sensors together is possible. A possible plan is given in Figure 14.

7. Conclusions

A temperature sensor capable of measuring temperatures from 40°C to less than −60°C has been described. The sensor has been designed with the expressed intent of being able to be integrated with existing compliant capacitive pressure sensors. It was noted that if these sensors are silicon-based, heat loss to the substrate is significant. Therefore, the sensor incorporates an active heat sink. The active sink and temperature sensor have individual proportional controllers allowing their temperatures to be maintained at a constant equal value, regardless of the external temperature. By keeping the active sink and temperature sensor at an equal temperature, it has been shown that heat loss from the temperature sensor to the substrate is prevented, so that the change in heat flux due to change in the external temperature can be directly and accurately measured.

Even though it is inherently a three-dimensional structure, the thermoelectrodynamics has been simplified into a one-dimensional Simulink model which allows for the response of the sensor to be known. The model incorporates the effect of the temperature profile in the insulation layers as well as the transient nature of heat generation. In this manner, the heat exchange mechanisms between the layers have been fully modelled. The fabrication plan for these sensors has also been described.

Appendix

A. Analytical Solution for Temperature Distribution in Polymer Layer

As was previously mentioned, (3) does have an analytical solution. In the context of the numerical integration scheme discussed above, only the change in the temperature distribution, $T(x)$, during each fixed time step, dt, is required as this gives the heat flux flowing in/out of the layer's surface. In general, the temperature distribution and the associated end temperatures, $T_{1,2}$, of the insulation/polymer layer are continuous functions of time. However, in the numerical scheme, it is assumed that the end temperatures are constant for the duration of the fixed time step. Therefore, the boundary conditions to be used in the solution of (3) are:

$$\begin{aligned} T(x,t) &= T_1\left(t'\right), \quad \text{when } x = 0, \\ T(x,t) &= T_2\left(t'\right), \quad \text{when } x = L, \end{aligned} \tag{A.1}$$

where t' is the time during each time step, that is, $0 \leq t' \leq dt$. The initial condition is the temperature distribution, $g(x)$, inherited from the solution of the previous time step. When $t = 0$, it is assumed that the whole structure is in thermal equilibrium with the environment and therefore the initial condition can be stated as

$$\begin{aligned} T(x,0) &= T_R, \quad t = 0, \\ T(x,0) &= g(x), \quad t > 0, \end{aligned} \tag{A.2}$$

where T_R is the room temperature in K. As surface conditions are assumed to be independent of time during each time step, the general problem can be simplified to that of two problems, one of steady temperature and one of variable temperature with prescribed initial temperature and zero surface temperature [25]. Put

$$T = u + w, \tag{A.3}$$

where u and w satisfy the following equations:

$$\frac{d^2u}{dx^2} = 0\ (0 < x < L), \tag{A.4}$$

$$\begin{aligned} u &= T_1, \quad \text{when}\ x = 0, \\ u &= T_2, \quad \text{when}\ x = L, \end{aligned} \tag{A.5}$$

$$\frac{\partial w}{\partial t} = \alpha \frac{\partial^2 w}{\partial t^2} \quad (0 < x < L), \tag{A.6}$$

$$w = 0, \quad \text{when}\ x = 0\ \text{and}\ x = L, \tag{A.7}$$

$$w(x, 0) = g(x) - u. \tag{A.8}$$

Straightaway, the solution to (A.4) can be given as

$$u = \frac{T_1 + (T_2 - T_1)x}{L}. \tag{A.9}$$

If the initial temperature distribution, $g(x)$, can be expanded in the sine series,

$$g(x) \cong \sum_{n=1}^{\infty} a_n \sin \frac{n\pi x}{L}, \tag{A.10}$$

where

$$a_n = \frac{2}{L} \int_0^L g(x') \sin \frac{n\pi x'}{L} dx'. \tag{A.11}$$

Then,

$$w(x, 0) = \sum_{n=1}^{\infty} b_n \sin \frac{n\pi x}{L}, \tag{A.12}$$

where

$$b_n = \frac{2}{L} \int_0^L \left[g(x') - T_1 + (T_2 - T_1)\frac{x'}{L} \right] \sin \frac{n\pi x'}{L} dx'. \tag{A.13}$$

It is clear that the solution to (A.6) must be

$$w(x, t') = \sum_{n=1}^{\infty} b_n \sin \frac{n\pi x}{L} e^{-\alpha n^2 \pi^2 (t'/L^2)}. \tag{A.14}$$

Therefore,

$$\begin{aligned} T(x, t') = {} & T_1 + (T_2 - T_1)\frac{x}{L} + \frac{2}{\pi} \sum_{n=1}^{\infty} \frac{T_2 \cos(n\pi) - T_1}{2n + 1} \sin \frac{n\pi x}{L} e^{-\alpha n^2 \pi^2 (t'/L^2)} \\ & + \frac{2}{L} \sum_{n=1}^{\infty} \sin \frac{n\pi x}{L} e^{-\alpha n^2 \pi^2 (t'/L^2)} \int_0^L g(x') \sin \frac{n\pi x'}{L} dx'. \end{aligned} \tag{A.15}$$

From (A.15), the temperature gradient at the surfaces of the layer can be found and hence the heat flux exchange into the layer in accord with (5). When $g(x)$ is a simple, known function, an exact answer for (A.15) can be found. However, in general, $g(x)$ will be the solution of (A.15) for the previous time step and will only be known numerically. This means that in general (A.15) will need to be solved using numerical integration methods. The problem is that (A.15) tends to converge slowly and so the integral becomes highly oscillatory. While there are numerical integration methods that can be used effectively, such as that given in [31], an accurate numerical solution to (A.15) is prohibitively computationally expensive and so the method detailed in Appendix B is used instead.

B. Numerical Solution for Temperature Distribution in Polymer Layer

Equation (3) is a parabolic partial differential equation frequently encountered in heat transfer problems. It is well known that boundary and initial conditions of the form given in (B.1) are sufficient to ensure that the solution is unique [26].

$$\begin{aligned} T(0, t) &= f_0(t), \\ T(L, t) &= f_L(t), \\ T(x, 0) &= g(x). \end{aligned} \tag{B.1}$$

In this case, a solution is required for the region $0 \le x \le L, 0 \le t \le dt$, where dt is the time step in the system numerical integration scheme. To solve this problem, a finite difference approach is used. Here, a rectangular grid is superimposed on this region, with equal increments, Δx, in the space coordinate, and Δt in the time coordinate. The grid points are then

$$\begin{aligned} x_i &= i\Delta x, \quad i = 0, \ldots, N_x, \Delta x = \frac{L}{N_x}, \\ t_j &= j\Delta t, \quad j = 0, \ldots, N_t, \Delta t = \frac{dt}{N_t}. \end{aligned} \tag{B.2}$$

The Crank-Nicolson method can be used to solve this problem and has the advantage of being unconditionally stable and being second-order accurate in both Δx and Δt [26].

The finite difference expressions can be shown to be

$$\frac{1}{2}\left(\frac{T_{i+1,j+1}-2T_{i,j+1}+T_{i-1,j+1}}{(\Delta x)^2}+\frac{T_{i+1,j}-2T_{i,j}+T_{i-1,j}}{(\Delta x)^2}\right) = \alpha\frac{T_{i,j+1}-T_{i,j}}{\Delta t}. \quad \text{(B.3)}$$

By defining

$$\gamma = \frac{\Delta t}{\alpha(\Delta x)^2}. \quad \text{(B.4)}$$

Equation (B.3) can be rearranged by collecting all the $j+1$ terms on one side to give

$$\frac{\gamma}{2}T_{i+1,j+1}-(1+\gamma)T_{i,j+1}+\frac{\gamma}{2}T_{i-1,j+1} = -\frac{\gamma}{2}T_{i+1,j}+(\gamma-1)T_{i,j}-\frac{\gamma}{2}T_{i-1,j}. \quad \text{(B.5)}$$

These equations can be collected together for a specific value of j into a single matrix equation:

$$\boldsymbol{A}(\gamma)\overrightarrow{T}_{j+1}=\boldsymbol{A}(\gamma)\overrightarrow{T}_{j+1}+\overrightarrow{b}, \quad \text{(B.6)}$$

where

$$\boldsymbol{A}(\gamma)=\begin{vmatrix} 1+\gamma & -\frac{\gamma}{2} & 0 & \cdots & 0 \\ -\frac{\gamma}{2} & 1+\gamma & -\frac{\gamma}{2} & \cdots & 0 \\ 0 & -\frac{\gamma}{2} & 1+\gamma & \cdots & 0 \\ \vdots & \vdots & \vdots & \ddots & \vdots \\ 0 & 0 & 0 & \cdots & 1+\gamma \end{vmatrix}, \quad \text{(B.7)}$$

$$\overrightarrow{\boldsymbol{b}}=\begin{vmatrix} \frac{1}{2}[f_0(t_0)+f_0(t_1)] \\ 0 \\ \vdots \\ 0 \\ \frac{1}{2}[f_L(t_0)+L(t_1)] \end{vmatrix}. \quad \text{(B.8)}$$

As the complete thermoelectric problem is being solved using a 4th-order Runge-Kutta algorithm with a small but fixed time step dt, N_t can be quite small and still produce accurate results. In this case, $N_t = 10$, and $N_x = 25$. Even though the subsequent matrices are quite small and could be solved using matrix division quite easily, due to the size of the complete problem and the resulting demands on computer memory, (B.6) is solved using sparse matrix methods to reduce computation times [26].

Conflicts of Interest

The authors declare that they have no conflicts of interest.

Acknowledgments

This research was undertaken within the FP7-NMP NANOBIOTOUCH project (contract no. 228844). The authors are grateful for the financial support provided by the European Commission.

References

[1] R. Bogue, "MEMS sensors: past, present and future," *Sensor Review*, vol. 27, no. 1, pp. 7–13, 2007.

[2] W. P. Eaton and J. H. Smith, "Micromachined pressure sensors: review and recent developments," *Smart Materials and Structures*, vol. 6, no. 5, pp. 530–539, 1997.

[3] S. S. Kumar and B. D. Pant, "Design principles and considerations for the 'ideal' silicon piezoresistive pressure sensor: a focused review," *Microsystem Technologies*, vol. 20, no. 7, pp. 1213–1247, 2014.

[4] Y. S. Lee and K. D. Wise, "A batch-fabricated silicon capacitive pressure transducer with low temperature sensitivity," *IEEE Transactions on Electron Devices*, vol. 29, no. 1, pp. 42–48, 1982.

[5] C. Pramanik, T. Islam, and H. Saha, "Temperature compensation of piezoresistive micro-machined porous silicon pressure sensor by ANN," *Microelectronics Reliability*, vol. 46, no. 2–4, pp. 343–351, 2006.

[6] V. Maheshwari and R. Saraf, "Tactile devices to sense touch on a par with a human finger," *Angewandte Chemie International Edition*, vol. 47, no. 41, pp. 7808–7826, 2008.

[7] D. J. Beebe, A. S. Hsieh, D. D. Denton, and R. G. Radwin, "A silicon force sensor for robotics and medicine," *Sensors and Actuators A: Physical*, vol. 50, no. 1-2, pp. 55–65, 1995.

[8] A. M. Almassri, W. Z. Wan Hasan, S. A. Ahmad et al., "Pressure sensor: state of the art, design, and application for robotic hand," *Journal of Sensors*, vol. 2015, 12 pages, 2015.

[9] N. Besse, S. Rosset, J. J. Zarate, and H. Shea, "Flexible active skin: large reconfigurable arrays of individually addressed shape memory polymer actuators," *Advanced Materials Technologies*, vol. 2, no. 10, 2017.

[10] M. F. P. Cruz, E. Fiedler, O. F. C. Monjarás, and T. Stieglitz, "Integration of temperature sensors in polyimide-based thin-film electrode arrays," *Current Directions in Biomedical Engineering*, vol. 1, no. 1, pp. 529–533, 2015.

[11] J. Park, M. Kim, Y. Lee, H. S. Lee, and H. Ko, "Fingertip skin-inspired microstructured ferroelectric skins discriminate static/dynamic pressure and temperature stimuli," *Science Advances*, vol. 1, no. 9, e1500661, 2015.

[12] A. Tong, "Improving the accuracy of temperature measurements," *Sensor Review*, vol. 21, no. 3, pp. 193–198, 2001.

[13] J. Kim, J. Kim, Y. Shin, and Y. Yoon, "A study on the fabrication of an RTD (resistance temperature detector) by using Pt thin film," *Korean Journal of Chemical Engineering*, vol. 18, no. 1, pp. 61–66, 2001.

[14] T. K. Maiti, "A novel lead-wire-resistance compensation technique using two-wire resistance temperature detector," *IEEE Sensors Journal*, vol. 6, no. 6, pp. 1454–1458, 2006.

[15] D. W. Osborne, H. E. Flotow, and F. Schreiner, "Calibration and use of germanium resistance thermometers for precise heat capacity measurements from 1 to 25°K high purity copper for interlaboratory heat capacity comparisons," *Review of Scientific Instruments*, vol. 38, no. 2, pp. 159–168, 1967.

[16] D. Cheneler, J. Teng, M. Adams, C. J. Anthony, E. L. Carter, and M. Ward, "Printed circuit board as a MEMS platform for focused ion beam technology," *Microelectronic Engineering*, vol. 88, no. 1, pp. 121–126, 2011.

[17] V. Rochus, B. Wang, H. A. C. Tilmans et al., "Fast analytical design of MEMS capacitive pressure sensors with sealed cavities," *Mechatronics*, vol. 40, pp. 244–250, 2016.

[18] G. Blasquez, Y. Naciri, P. Blondel, N. Ben Moussa, and P. Pons, "Static response of miniature capacitive pressure sensors with square or rectangular silicon diaphragm," *Revue de Physique Appliquée*, vol. 22, no. 7, pp. 505–510, 1987.

[19] Y. M. Chen, S. M. He, C. H. Huang et al., "Ultra-large suspended graphene as a highly elastic membrane for capacitive pressure sensors," *Nanoscale*, vol. 8, no. 6, pp. 3555–3564, 2016.

[20] M. Y. Cheng, X. H. Huang, C. W. Ma, and Y. J. Yang, "A flexible capacitive tactile sensing array with floating electrodes," *Journal of Micromechanics and Microengineering*, vol. 19, no. 11, p. 115001, 2009.

[21] J. T. Bottomley, "Scientific worthies," *Nature*, vol. 26, no. 678, pp. 617–620, 1882.

[22] E. Verdet, *Théorie Méchanique de la Chaleur, Vol. 2*, Lacroix éditeur, 1872.

[23] E. J. Davies, *Conduction and Induction Heating*, IEE Power Engineering Series, IET, 1990.

[24] J. R. Welty, C. E. Wicks, R. E. Wilson, and G. Rorrer, *Fundamentals of Momentum, Heat and Mass Transfer*, Wiley & Sons, 4th edition, 2001.

[25] H. S. Carslaw and J. C. Jaeger, *Conduction of Heat in Solids*, Clarendon Press, 2nd edition, 1959.

[26] G. J. Borse, *Numerical Methods with MatLab: a Resource for Scientists and Engineers*, PWS Publishing Co., 1997.

[27] N. J. Siakavellas and D. P. Georgiou, "1D heat transfer through a flat plate submitted to step changes in heat transfer coefficient," *International Journal of Thermal Sciences*, vol. 44, no. 5, pp. 452–464, 2005.

[28] Multiphysics COMSOL, *Version 4.2, COMSOL. Inc.*, 2011, http://www.comsol.com.

[29] Specifications PXI-4070, "Report 371304J-01," 2017, http://ni.com. June 2018, http://www.ni.com/pdf/manuals/371304j.pdf.

[30] MAX1365/MAX1367, *Maxim*, 2006, June 2018, https://datasheets.maximintegrated.com/en/ds/MAX1365-MAX1367.pdf.

[31] A. I. van de Vooren and H. J. van Linde, "Numerical calculation of integrals with strongly oscillating integrand," *Mathematics of Computation*, vol. 20, no. 94, pp. 232–245, 1966.

3

Improving Accuracy in the Readout of Resistive Sensor Arrays

José A. Hidalgo-López, Raquel Fernández-Ramos, Jorge Romero-Sánchez, José F. Martín-Canales, and Francisco J. Ríos-Gómez

Departamento de Electrónica, Andalucía Tech, Universidad de Málaga, Campus de Teatinos, 29071 Málaga, Spain

Correspondence should be addressed to José A. Hidalgo-López; jahidalgo@uma.es

Academic Editor: Oleg Lupan

The most efficient way to carry out the reading of a set of resistive sensors is to organize them in an array form. This reduces the number of wires to the sum of the number of rows, M, and columns, N, and reading can be carried out using just $M+N$ multiplexers and a single operational amplifier. The drawback in this procedure is the appearance of crosstalk (the influence of the resistance of some sensors on the measurements of others). Although different proposals have been presented in the literature to reduce this phenomenon, errors in determining the resistance values of each sensor still exist in all the proposals. This article presents a new method to determine these values, which considerably reduces errors without the need for any hardware other than the simplest reading circuits. The method consists of a very fast recursive convergence procedure. The results show that the new proposal outperforms other solutions in the first or second step.

1. Introduction

A wide range of applications uses resistive sensor arrays, such as gas detectors [1–3], nuclear electronics [4], temperature sensors [5, 6], thermal anemometry [7], foot plantar application [8], and tactile sensors or artificial electronic skins [9–14]. These arrays are organized in M rows and N columns, so each sensor is connected to the other sensors of the same row through one of its terminals and to the sensors of the same column through the other one. The sensor structure therefore only uses a total of $M+N$ wires, which also allows the multiplexing circuits necessary to read each sensor's information individually to be reduced. This reduction in reading electronics comes at the cost of the appearance of crosstalk from parasitic parallel paths. Errors caused by this phenomenon in each sensor reading have been widely studied in the literature [15–20] and can reach values as high as 30% for a 4×4 array [21].

The main cause of the appearance of crosstalk in these arrays is the resistance of the different multiplexers and/or buffers used to select rows and columns in the reading electronics, which cause parasitic parallel paths to appear. A second group of causes is related to the nonidealities of the OAs (operational amplifiers) that are usually part of the reading circuits [19, 20, 22]. Naturally, the magnitude of the errors also depends on the type of reading circuit designed. Two excellent reviews of scanning approaches of two-dimensional resistive sensor arrays can be found in [23, 24]. There are two main ways to proceed when reading an array of resistive sensors: a reading circuit can be used for each array column, or a single circuit for the whole array. The first procedure allows faster reading of the array, but requires N OAs and M multiplexers. The second procedure uses a single OA and $M+N$ multiplexers. This second procedure is slower but reduces both power consumption and total hardware. In [25], crosstalk is reduced using the first procedure one additional OA. However, crosstalk can be eliminated using the first reading procedure by adding calibration resistors and an extra OA to the hardware and carrying out additional array measurements, as shown in [19, 20].

The second reading procedure with a single OA leads to two fundamental types of circuit. In the first one, the OA is used in its inverting configuration, only one column wire is connected to the inverter input of the OA, and only one-row wire is connected to V_{DD} (the supply voltage of the array). On the other hand, the rest of the wires are connected to ground. This is the zero potential method (ZPM). In the second type of circuit, the OA output (in noninverting configuration) is used as a feedback for the array; this is known as the voltage feedback method (VFM). However, the appearance of crosstalk is inevitable in both types of circuit. References [17, 18] evaluate errors in the array reading using the VFM structure, and Liu et al. [16] compare errors made by circuits that use different variations of the ZPM and VFM, reaching the conclusion that, in general and under the same operating conditions, errors in circuits based on the ZPM are lower than those based on the VFM.

In order to reduce crosstalk in the circuits based on the ZPM and VFM, Wu et al. present two designs: one based on the ZPM [26] using two OAs and an additional resistor (errors in this method are evaluated in [27]) and another based on the VFM [28, 29], which requires two additional resistors ([30] uses this approach for a piezoresistive composite sensor array). Although the results obtained improve measurement accuracy, they still show errors due to crosstalk. The increase in hardware in these proposals is offset by the fact that the value of the $M \cdot N$ resistances of the sensors can be found using uncomplicated algebraic equations, maintaining limited error.

This paper will present a method to obtain the $M \cdot N$ values of the resistances of the sensors which, using the simplest versions of the ZPM and VFM, allows errors due to crosstalk to be almost completely eliminated without having to add further elements.

The paper is structured in sections as follows: Section 2 presents the resistive sensor array reading circuits with a single OA that appears in the literature. Section 3 presents the new method proposed to analyse these arrays, which uses the simplest versions of the ZPM and VFM circuits. Section 4 sets out the simulation results and their discussion. The final section provides a summary of the conclusions.

2. Reading an Array of Resistive Sensors Using ZPM and VFM Circuits

The simplest circuit for reading an array of resistive sensors based on the ZPM is shown in Figure 1. The reading procedure for this circuit consists of selecting, using the row selection switches, a positive voltage for one of the wires of row $i\,(1 \le i \le M)$, while the rest are connected to ground. Likewise, all column wires except one, $j\,(1 \le j \le N)$, are connected to ground. Meanwhile, the column switch j connects the wire of the jth array column to the inverting input of the OA. Using this switch layout (if the switches were ideal), there is only current flowing through the sensor connected to row i and column j, since voltages Vr_i of the row wires and Vc_j of the column wires of the array of Figure 1 would be V_{DD} or ground, depending on the selection in the switches. This current could be calculated by the following expression:

$$R_{ij} = -\frac{V_{\mathrm{DD}}}{\mathrm{Vo}(i,j)}\mathrm{Rf}, \tag{1}$$

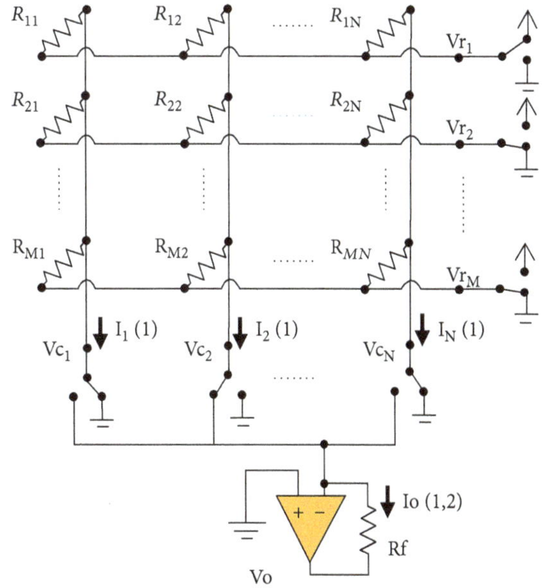

FIGURE 1: Circuit based on the ZPM for reading the array of resistive sensors. The position of the switches allows the reading of R_{12}.

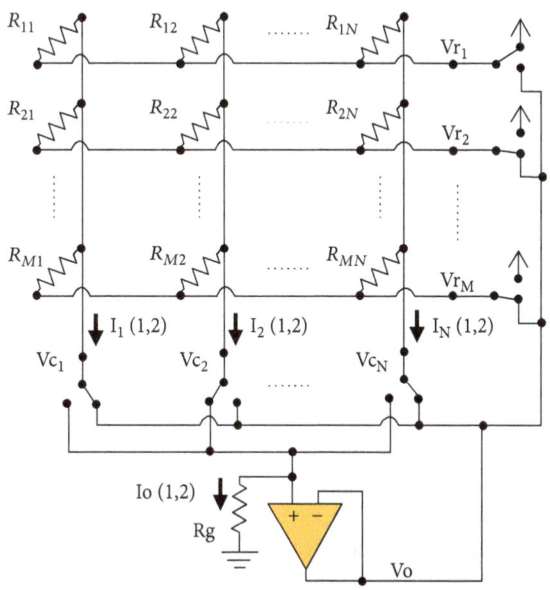

FIGURE 2: Circuit based on the VFM for reading the array of resistive sensors. The position of the switches allows the reading of R_{12}.

where $\mathrm{Vo}(i,j)$ is the output voltage of the OA when the switch of row i selects V_{DD} and the switch of column j connects this column's wire to the inverting input terminal of the OA. Unfortunately, (1) is merely an approximation to the real value of R_{ij}, since voltages Vr_i and Vc_j do not match V_{DD} or ground due to the resistances of the switches, Rs.

On the other hand, Figure 2 shows a circuit that uses the VFM for the array reading. This circuit is a slightly modified version of the classical VFM circuits that uses a single resistor, Rg (as shown in Figure 2), rather than the one with the same value for each of the columns of the array, such as in [16]. The readout procedure consists of using the row selection switches to select a positive voltage, V_{DD}, for one of the row wires, i, while the rest are connected to the output of the OA. Moreover, all column switches except one, j, are connected to the output voltage of the OA. The column switch, j, connects the wire of this column to the resistor, Rg, which in turn is connected to the noninverting input terminal of the OA. Once again, if the row and column switches were ideal, only the current would flow through the resistor R_{ij}, the value of which could be calculated by the following expression:

$$R_{ij} = \left(\frac{V_{DD}}{\mathrm{Vo}(i,j)} - 1\right) \cdot \mathrm{Rg}, \quad (2)$$

where $\mathrm{Vo}(i,j)$ is the output voltage of the OA when the switch of row i selects V_{DD} and the switch of column j connects this column's wire to the noninverting input terminal of the OA.

As in the case of the ZPM circuit, (2) would only be valid if the resistors of the column and row switches, Rs, are 0. The error that occurs in estimating R_{ij} using these equations will depend on Rs, on the size of the array (M and N), and the range of values of the resistances of the sensors [16, 26].

To reduce the errors due to Rs in (1), Wu et al. [31] propose modifying the circuit based on the ZPM of Figure 1 by adding a second OA and a resistor (Figure 3). For this circuit, Wu et al. propose calculating R_{ij}, with the value of Rs known, according to the following equation:

$$R_{ij} = -\left(V_{DD} + \mathrm{Vcg} \cdot \frac{\mathrm{Rs}}{\mathrm{Rcg}}\right) \frac{\mathrm{Rl}}{\mathrm{VL}_{ij}} - 2\mathrm{Rs}, \quad (3)$$

Vcg and VL_{ij} are the output voltages of the OAs in Figure 3 and Rl and Rcg are the feedback resistors of these OAs (as one example in Figure 3, multiplexers are selected to the readout of R_{11}).

Although the circuit in Figure 3 reduces the number of parasitic parallel paths due to Rs, these are not completely eliminated and, therefore, expression (3) still shows errors in estimating R_{ij}. These errors again depend on Rs, on the size of the array, and on the resistors to be measured.

Based on the VFM, in [26], Wu et al. propose a new design (Figure 4) in order to reduce crosstalk. Wu et al. propose evaluating R_{ij} according to the following equation:

$$R_{ij} = \left(\frac{V_{DD}}{\mathrm{Vo}(i,j)} - 1\right) \cdot \mathrm{Rg} + \frac{V_{DD}}{\mathrm{Vo}(i,j)} \cdot \mathrm{Rs}. \quad (4)$$

The second term of the member on the right is the only difference from (2). Although, for some ranges of R_{ij} and Rs, the errors using (4) for the circuit of Figure 4 are smaller than those obtained using (2) for the circuit (Figure 2), these errors do not disappear.

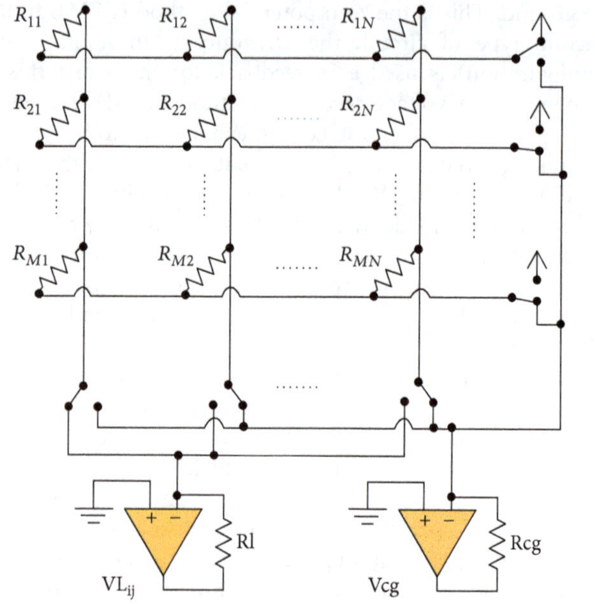

Figure 3: Modified circuit based on the ZPM proposed by Wu et al. [31] to reduce the influence of Rs in calculating R_{ij}.

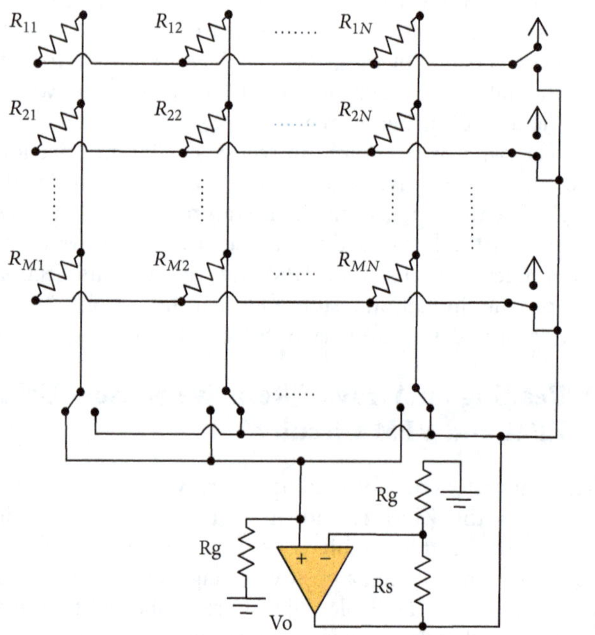

Figure 4: Modified circuit based on the VFM proposed by Wu et al. [26] to reduce the effects of crosstalk.

3. Method for Reducing Crosstalk Errors in Reading an Array of Resistive Sensors

To reduce errors, due to the resistances of the switches that appear in (1), (2), (3), and (4), a new method to calculate R_{ij} is presented in this section. The method is the same, with minor modifications, for the simplest version of the ZPM- and VFM-based circuits (Figures 1 and 2). In essence, the method uses the estimates values of R_{ij} by using (1) and (2) as the starting point to obtain the voltages in the row and column wires of the array (Vr_i and Vc_j) that will be used to, in successive steps, improve the estimation of R_{ij}. The method is described below in its two variants.

3.1. Method for Reducing Crosstalk Errors in Reading Circuits Based on ZPM. For the circuit of Figure 1, the proposed method begins by obtaining an initial estimation of all R_{ij}, $R_{ij}^{(0)}$, through (1):

$$R_{ij}^{(0)} = -\frac{V_{\mathrm{DD}}}{\mathrm{Vo}(i,j)}\mathrm{Rf}, \quad (5)$$

where the superscript will indicate the current step of the process. Next, if resistors of multiplexers, Rs, are considered for improving accuracy in the readout of R_{ij}, it is necessary to analyse the equivalent circuit in Figure 5. In this circuit, the current flowing towards the Rf resistor when connecting the wire of row i to V_{DD} and the wire of column j to the OA, $\mathrm{Io}(i,j)$, is

$$\mathrm{Io}(i,j) = -\frac{\mathrm{Vo}(i,j)}{\mathrm{Rf}} = I_j(i), \quad (6)$$

where $I_j(i)$ is the current flowing through the multiplexer of the jth column wire (see Figure 5). It should be noted that $I_j(i)$ does not depend on the column selected for reading since, in any selection state of the column switches, these connect the columns of the array at a voltage of 0 V. This current is used to calculate the voltage values of the column wires in Figure 5,

$$\mathrm{Vc}_j(i) = I_j(i)\cdot\mathrm{Rs} = -\mathrm{Vo}(i,j)\frac{\mathrm{Rs}}{\mathrm{Rf}}, \quad (7)$$

where $\mathrm{Vc}_j(i)$ is the voltage of the wire of column j when connecting row i to V_{DD}.

The recursive process begins by calculating the voltages of any row wire in the array, k, when V_{DD} is connected row wire i: $\mathrm{Vr}_k^{(1)}(i)$ (note that Vr_k do not depend on which column is selected). This voltage can be calculated using Millman's theorem [32] and (7) as

$$\mathrm{Vr}_k^{(1)}(i) = \frac{V_{\mathrm{DD}}\cdot\mathrm{Ys}\cdot\delta_{ik} + \sum_{p=1}^{N}\mathrm{Vc}_p(i)\cdot Y_{kp}^{(0)}}{\sum_{p=1}^{N} Y_{kp}^{(0)} + \mathrm{Ys}} \quad (8)$$

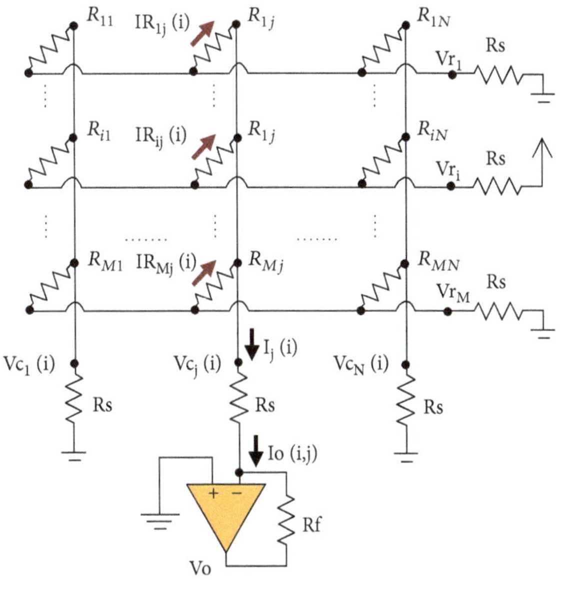

Figure 5: Equivalent circuit, based on the ZPM, including the resistor of multiplexers Rs for improving accuracy in the readout of R_{ij}.

and replacing $\mathrm{Vc}_p(i)$ from (7):

$$\mathrm{Vr}_k^{(1)}(i) = \frac{V_{\mathrm{DD}}\cdot\mathrm{Ys}\cdot\delta_{ik} - (\mathrm{Rs}/\mathrm{Rf})\cdot\sum_{p=1}^{N}\mathrm{Vo}(i,p)Y_{kp}^{(0)}}{\sum_{p=1}^{N} Y_{kp}^{(0)} + \mathrm{Ys}}, \quad (9)$$

$Y_{kp}^{(0)}$ and Ys are the inverse of $R_{kp}^{(0)}$ and Rs, respectively, while δ_{ik} is the Kronecker delta. On the other hand, the current passing through R_{ij} in this first step, $\mathrm{IR}_{ij}^{(1)}$ (see Figure 5), is

$$\begin{aligned}\mathrm{IR}_{ij}^{(1)} &= \mathrm{Io}(i,j) - \sum_{k=1,k\neq i}^{M}\mathrm{IR}_{kj}^{(1)} \\ &= \mathrm{Io}(i,j) - \sum_{k=1,k\neq i}^{M}\left(\mathrm{Vr}_k^{(1)}(i) - \mathrm{Vc}_j(i)\right)\cdot Y_{kj}^{(0)} \\ &= -\frac{\mathrm{Vo}(i,j)}{\mathrm{Rf}}\left(1 + \mathrm{Rs}\cdot\sum_{k=1,k\neq i}^{M} Y_{kj}^{(0)}\right) - \sum_{k=1,k\neq i}^{M}\mathrm{Vr}_k^{(1)}(i)\cdot Y_{kj}^{(0)}.\end{aligned} \quad (10)$$

Now, $R_{ij}^{(1)}$ can be found using the results of (5), (6), (7), (8), and (9):

$$R_{ij}^{(1)} = \frac{\mathrm{Vr}_i^{(1)}(i) - \mathrm{Vc}_j(i)}{\mathrm{IR}_{ij}^{1}} = \frac{\mathrm{Vr}_i^{(1)}(i) + \mathrm{Vo}(i,j)(\mathrm{Rs}/\mathrm{Rf})}{-(\mathrm{Vo}(i,j)/\mathrm{Rf})\left(1 + Rs\cdot\sum_{k=1,k\neq i}^{M} Y_{kj}^{(0)}\right) - \sum_{k=1,k\neq i}^{M}\mathrm{Vr}_k^{(1)}(i)\cdot Y_{kj}^{(0)}}. \quad (11)$$

Equations (9) and (11)are the recursive equations of the proposed method. Thus, in the qth step, new values of $\mathrm{Vr}_k^{(q)}(i)$ would be calculated using (9), with $R_{ij}^{(q-1)}$ from the $q-1$ step (instead of $R_{ij}^{(0)}$). The step finishes using (11) to calculate $R_{ij}^{(q)}$ with $\mathrm{Vr}_k^{(q)}(i)$ instead of $Vr_k^{(1)}(i)$.

3.2. Method for Reducing Crosstalk Errors in Reading Circuits Based on VFM. For the circuit in Figure 2, the method is very similar to the one presented in the previous subsection, beginning with the calculation of all the $R_{ij}^{(0)}$ from (2). Again, if Rs is considered, it is necessary to analyse the circuits of Figure 6 for improving accuracy. In this circuit, the current flowing towards the Rg resistor when switching row i to V_{DD} and connecting column j to the OA, $\mathrm{Io}(i,j)$, is calculated:

$$\mathrm{Io}(i,j) = \frac{\mathrm{Vo}(i,j)}{\mathrm{Rg}}. \tag{12}$$

For this circuit (if any row i is selected), it is possible to find the current flowing through the switch of a selected column t, $I_t(i,t)$ as a function of the current flowing through the switch of the same column, when another column j was selected:

$$\frac{I_t(i,j)}{I_t(i,t)} = \frac{V_{\mathrm{DD}} - \mathrm{Vo}(i,j)}{V_{\mathrm{DD}} - \mathrm{Vo}(i,t)}. \tag{13}$$

This equation shows the effect of the voltage scaling that has taken place in the array when column t or column j was selected (keeping i as the selected row). Having in mind that $I_t(i,t) = \mathrm{Io}(i,t)$, (12) and (13) can be used to calculate the voltage in the wire of the column t when selecting row i and column j, $\mathrm{Vc}_t(i,j)$:

$$\begin{aligned}\mathrm{Vc}_t(i,j) &= I_t(i,j)\cdot \mathrm{Rs} + \mathrm{Vo}(i,j)\\ &= \mathrm{Vo}(i,t)\cdot\frac{\mathrm{Rs}}{\mathrm{Rg}}\cdot\frac{V_{\mathrm{DD}} - \mathrm{Vo}(i,j)}{V_{\mathrm{DD}} - \mathrm{Vo}(i,t)} + Vo(i,j).\end{aligned} \tag{14}$$

The recursive process for this circuit starts by using Millman's theorem and (14) to recalculate the voltages of kth row of the array when selecting for reading row i and column j, $Vr_k^{(1)}(i,j)$:

$$\begin{aligned}\mathrm{Vr}_k^{(1)}(i,j) &= \frac{\sum_{t=1}^{N}\mathrm{Vc}_t(i,j)\cdot Y_{kt}^{(0)}}{\sum_{t=1}^{N}Y_{kt}^{(0)} + \mathrm{Ys}}\\ &\quad + \frac{(V_{\mathrm{DD}}\cdot\delta_{ik} + \mathrm{Vo}(i,j)\cdot(1-\delta_{ik}))\cdot\mathrm{Ys}}{\sum_{t=1}^{N}Y_{kt}^{(0)} + \mathrm{Ys}}\end{aligned} \tag{15}$$

or

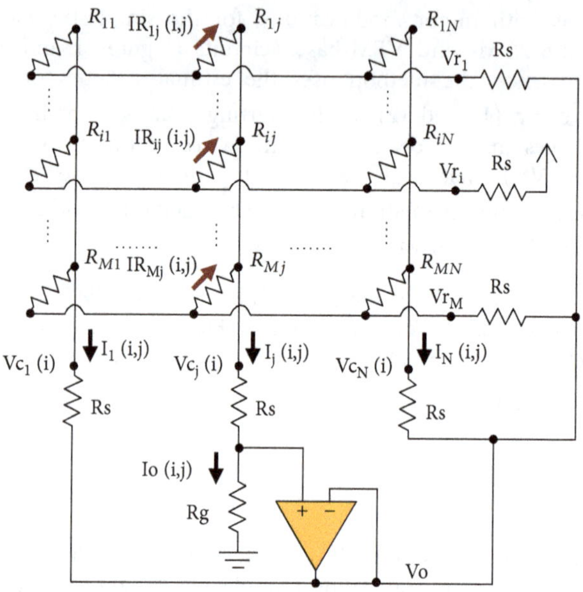

FIGURE 6: Equivalent circuit, based on the VFM, including the resistor of multiplexers Rs for improving accuracy in the readout of R_{ij}.

$$\begin{aligned}Vr_k^{(1)}(i,j) &= Vo(i,j)\\ &\quad\cdot\left(1 + \frac{Rs/Rg\cdot\sum_{t=1}^{N}(V_{DD} - Vo(i,j)/V_{DD} - Vo(i,t))\cdot Vo(i,t)/Vo(i,j)\cdot Y_{kt}^{(0)}}{\sum_{t=1}^{N}Y_{kt}^{(0)} + Ys}\right)\\ &\quad + \frac{(V_{DD} - Vo(i,j))\cdot\delta_{ik}\cdot Ys}{\sum_{t=1}^{N}Y_{kt}^{(0)} + Ys}.\end{aligned} \tag{16}$$

The current passing through R_{ij} when selected row i and column j (in the first step of the recursive method) $\mathrm{IR}_{ij}^{(1)}(i,j)$ is (see Figure 6):

$$\begin{aligned}\mathrm{IR}_{ij}^{(1)}(i,j) &= \mathrm{Io}(i,j) - \sum_{k=1,k\neq i}^{M}\mathrm{IR}_{kj}^{(1)}(i,j)\\ &= \mathrm{Io}(i,j) - \sum_{k=1,k\neq i}^{M}\left(\mathrm{Vc}_j(i,j) - \mathrm{Vr}_k^{(1)}(i,j)\right)\cdot Y_{kj}^{(0)}\\ &= \frac{\mathrm{Vo}(i,j)}{\mathrm{Rg}} - \mathrm{Vo}(i,j)\cdot\left(1 + \frac{\mathrm{Rs}}{\mathrm{Rg}}\right)\cdot\sum_{k=1,k\neq i}^{M}Y_{kj}^{(0)}\\ &\quad + \sum_{k=1,k\neq i}^{M}\mathrm{Vr}_k^{(1)}(i,j)\cdot Y_{kj}^{(0)}.\end{aligned} \tag{17}$$

Thus, $R_{ij}^{(1)}$ can be calculated as

$$R_{ij}^{(1)} = \frac{\mathrm{Vr}_i^{(1)}(i,j) - \mathrm{Vc}_j(i,j)}{\mathrm{IR}_{ij}^{(1)}(i,j)} = \frac{\mathrm{Vr}_i^{(1)}(i,j) - \mathrm{Vo}(i,j)\cdot(1 + (\mathrm{Rs}/\mathrm{Rg}))}{(\mathrm{Vo}(i,j)/\mathrm{Rg}) - \mathrm{Vo}(i,j)\cdot(1 + (\mathrm{Rs}/\mathrm{Rg}))\cdot\sum_{k=1,k\neq i}^{M}Y_{kj}^{(0)} + \sum_{k=1,k\neq i}^{M}\mathrm{Vr}_k^{(1)}(i,j)\cdot Y_{kj}^{(1)}}. \tag{18}$$

TABLE 1: Comparison of the results obtained in an 8 × 8 array with different values of R_{ij} if Rs = 1 Ω and the rest of the resistors of the array, Rns, take different values. (a) Classical approach. (b) ZPM circuit proposed by Wu et al. (c) Results for the proposed method in the first step. (d) Results for the proposed method in the second step.

(a) ZPM classical approach

Rs = 1 Ω	Error (%), R_{ij} (Ω)					
Rns (Ω)	100	500	2000	5000	7500	10000
100	16.00	12.39	5.48	5.68	13.30	19.78
200	9.10	7.19	5.95	4.07	2.59	1.16
300	6.73	5.02	4.42	3.77	3.27	2.79
400	5.55	3.88	3.44	3.11	2.88	2.65
600	4.36	2.73	2.38	2.22	2.12	2.03
800	3.77	2.15	1.82	1.70	1.64	1.58
1200	3.18	1.57	1.25	1.15	1.11	1.08
1800	2.79	1.18	0.86	0.78	0.74	0.72
3000	2.47	0.87	0.55	0.47	0.44	0.42
6000	2.24	0.63	0.32	0.24	0.21	0.19
10000	2.14	0.54	0.23	0.15	0.12	0.10

(b) ZPM-Wu

Rs = 1 Ω	Error (%), R_{ij} (Ω)					
Rns (Ω)	100	500	2000	5000	7500	10000
100	97.34	68.28	92.43	97.00	98.00	98.50
200	15.26	84.06	95.65	98.21	98.80	99.10
300	10.83	72.51	91.38	96.37	97.55	98.15
400	18.78	54.82	83.04	92.46	94.85	96.08
600	1.16	1.15	1.09	0.99	0.90	0.81
800	0.87	0.87	0.84	0.78	0.73	0.68
1200	0.58	0.58	0.56	0.53	0.50	0.48
1800	0.39	0.39	0.37	0.35	0.33	0.32
3000	0.23	0.23	0.22	0.20	0.19	0.17
6000	0.12	0.11	0.11	0.09	0.07	0.06
10000	0.07	0.07	0.06	0.04	0.03	0.01

(c) ZPM proposed method, 1st step

Rs = 1 Ω	Error (%), R_{ij} (Ω)					
Rns (Ω)	100	500	2000	5000	7500	10000
100	1.83	1.22	0.35	3.30	5.62	7.84
200	0.60	0.45	0.28	0.01	0.26	0.50
300	0.32	0.22	0.16	0.06	0.02	0.10
400	0.21	0.13	0.09	0.03	0.02	0.06
600	0.12	0.06	0.04	0.00	0.03	0.06
800	0.08	0.04	0.01	0.02	0.04	0.07
1200	0.06	0.02	0.00	0.03	0.05	0.07
1800	0.04	0.01	0.01	0.03	0.05	0.07
3000	0.03	0.00	0.01	0.03	0.05	0.06
6000	0.02	0.00	0.01	0.03	0.05	0.06
10000	0.02	0.00	0.01	0.03	0.04	0.06

(d) ZPM proposed method, 2nd step

Rs = 1 Ω	Error (%), R_{ij} (Ω)					
Rns (Ω)	100	500	2000	5000	7500	10000
100	0.24	0.14	0.13	0.64	1.06	1.48
200	0.04	0.02	0.02	0.10	0.17	0.23
300	0.02	0.00	0.02	0.07	0.12	0.16
400	0.01	0.00	0.02	0.06	0.10	0.13
600	0.00	0.00	0.02	0.05	0.08	0.11
800	0.00	0.00	0.02	0.05	0.07	0.09
1200	0.00	0.00	0.02	0.04	0.06	0.08
1800	0.00	0.00	0.01	0.04	0.06	0.07
3000	0.00	0.00	0.01	0.03	0.05	0.07
6000	0.00	0.00	0.01	0.03	0.05	0.06
10000	0.00	0.00	0.01	0.03	0.04	0.06

For VFM-based circuits, (16) and (18) are the recursive equations of the proposed method. Thus, in the qth step, new values of $\mathrm{Vr}_k^{(q)}(i,j)$ would be calculated using (16), with $R_{ij}^{(q-1)}$ founded in the $q-1$ step (instead of $R_{ij}^{(0)}$). The step finishes using (18) to calculate $R_{ij}^{(q)}$ with $\mathrm{Vr}_k^{(q)}(i,j)$ instead of $\mathrm{Vr}_k^{(1)}(i,j)$.

4. Results and Discussion

To compare the results provided by the proposed methods to those provided by the classical equations of the ZPM and VFM circuits, (1) and (2), and also the results of the circuits proposed by Wu et al., (3) and (4), a batch of simulations has been carried out using Cadence Orcad-Pspice 16.6. A general purpose OA, the Texas Instrument OPA4188, was used for the readout circuit in every test with the power supplies adjusted to +5 V and −5 V. The OA is designed with autozeroing techniques to provide low offset voltage (25 μV, maximum) and near zero-drift over time and temperature. The DC gain is 136 dB, and the input bias current is 16 pA. Errors due to the nonidealities of the OA are minimized thanks to these characteristics. If, for example, the offset voltage of the OA was higher, the technique reported in [22] could be used to reduce its influence. The sensing resistor R_{ij} range used for simulation was [100 Ω, 10 kΩ]. This range is typically for touch sensors [33]. Another reason to use this range of resistors is that it shows the difference in operation and limitations of circuits based on the ZPM and VFM, as will be shown in the results. For this range of sensor resistors, Rf = 75 Ω, Rg = 350 Ω, Rl = 75 Ω, and Rcg = 75 Ω have been selected in order not to saturate the output voltages of the OA and to ensure the difference between the maximum and the minimum Vo which is approximately equal in all circuits.

In general, simulations will be presented to compare the operation of the circuits of Figures 1 and 2 (evaluated, resp., by (1) and (2) and hereinafter classical approach), the two circuits of Figures 3 and 4 proposed by Wu et al. (evaluated, resp., by (3) and (4) and hereinafter ZPM-Wu and VFM-Wu), and the proposed method presented in the previous section.

TABLE 2: Comparison of the results obtained in an 8×8 array with different values of R_{ij} if Rns = 5 kΩ and the resistances of the switches take different values. (a) Classical approach. (b) ZPM circuit proposed by Wu e al. (c) Results for the proposed method in the first step. (d) Results for the proposed method in the second step.

(a) ZPM classical approach

Rns = 5 kΩ	Error (%), R_{ij} (Ω)					
Rs (Ω)	100	500	2000	5000	7500	10000
1	2.28	0.68	0.37	0.29	0.26	0.24
2	4.57	1.36	0.75	0.61	0.57	0.54
4	9.17	2.73	1.51	1.25	1.18	1.13
6	13.79	4.10	2.27	1.89	1.79	1.73
9	20.76	6.17	3.42	2.85	2.70	2.62
15	34.86	10.35	5.73	4.77	4.52	4.38
20	46.77	13.87	7.67	6.36	6.03	5.84
25	58.82	17.42	9.61	7.96	7.54	7.29

(b) ZPM-Wu

Rns = 5 kΩ	Error (%), R_{ij} (Ω)					
Rs (Ω)	100	500	2000	5000	7500	10000
1	0.14	0.14	0.13	0.11	0.10	0.08
2	0.28	0.28	0.27	0.25	0.24	0.22
4	0.56	0.56	0.55	0.53	0.51	0.49
6	0.84	0.84	0.82	0.80	0.78	0.77
9	1.26	1.25	1.24	1.21	1.19	1.17
15	2.09	2.09	2.06	2.02	1.98	1.95
20	2.79	2.78	2.75	2.68	2.63	2.57
25	3.48	3.47	3.42	3.33	3.26	3.18

(c) ZPM proposed method, 1st step

Rns = 5 kΩ	Error (%), R_{ij} (Ω)					
Rs (Ω)	100	500	2000	5000	7500	10000
1	0.03	0.00	0.01	0.03	0.05	0.06
2	0.10	0.01	0.01	0.03	0.04	0.06
4	0.38	0.04	0.01	0.02	0.03	0.05
6	0.80	0.09	0.03	0.00	0.02	0.03
9	1.67	0.20	0.08	0.04	0.02	0.00
15	4.07	0.55	0.23	0.16	0.14	0.11
20	6.56	0.94	0.41	0.31	0.27	0.24
25	9.39	1.43	0.63	0.48	0.44	0.40

(d) ZPM proposed method, 2nd step

Rns = 5 kΩ	Error (%), R_{ij} (Ω)					
Rs (Ω)	100	500	2000	5000	7500	10000
1	0.00	0.00	0.01	0.03	0.05	0.06
2	0.00	0.00	0.01	0.03	0.05	0.06
4	0.02	0.00	0.01	0.03	0.05	0.06
6	0.06	0.00	0.01	0.03	0.05	0.06
9	0.17	0.00	0.01	0.03	0.05	0.06

TABLE 2: Continued.

Rns = 5 kΩ	Error (%), R_{ij} (Ω)					
Rs (Ω)	100	500	2000	5000	7500	10000
15	0.63	0.03	0.00	0.03	0.04	0.06
20	1.26	0.07	0.01	0.02	0.03	0.05
25	2.12	0.14	0.03	0.00	0.02	0.04

Table 1 shows simulation results for the different circuits and methods presented based on the ZPM. The aim of this series of simulations is to analyse the influence of non-scanned resistors, Rns, on the measurement of the resistor, R_{ij}. In all cases, it is an 8×8 array in which the resistance of the switches is set to 1 Ω (a resistance that is relatively easy to obtain for switches). Different values of Rns are used for each value of the resistor to be measured, R_{ij}. Both Rns and R_{ij} vary in the selected range [100 Ω, 10 kΩ].

In the first place, it should be noted that the errors for Rns < 600 Ω are very large in the case of the ZPM-Wu circuit, Table 1(b). This is due to the additional OA decreasing its output voltage as the current flowing through the parasitic parallel paths increases (which occurs for low values of Rns), reaching a point at which its output becomes saturated. On the other hand, the four tables of Table 1 show how the errors are greater as Rns decreases. This is justified again by the increase in the current in the parasitic parallel paths with lower values of Rns. With the exception of cases in which Rns is low, the errors in the ZPM-Wu circuit are always lower than those that appear in the ZPM evaluated using the classical approach (Table 1(a)). For its part, the proposed method outperforms the previous methods. It can be seen how the convergence of this method is very fast, since better results are obtained in the first step (Table 1(c)) than with the other two methods (except for errors below 0.03% in ZPM-Wu). In the second step, the maximum error obtained is only 1.48%, more than an order of magnitude lower than that obtained by the classical approach, and with better results in all situations. It also improves the results of the ZPM-Wu circuit in all situations for errors over 0.06%. The results provided in Table 1(a) quite closely match those reported in [10] for a 10×10 array, in which the aim is to measure the value of a sensor when the rest of the sensors has resistances between 3 kΩ and 19 kΩ.

The aim of the simulations presented in Table 2 is to analyse the influence of the resistances of the switches, Rs, on the reading of R_{ij}. In the first place, it is observed how the errors increase for the classical approach and for ZPM-Wu when Rs is increased. The same happens when the value of R_{ij} decreases. The proposed method also shows these trends, except when the errors are very low (below 0.06%). For any combination of Rns and Rs, the results obtained by the proposed method in the second step outperform the results of the other methods. Even in the first step, except for R_{ij} = 100 Ω and Rs > 9 Ω, the results are better than in the other methods. It should be noted that the maximum error in the second step is only 2.12% when the resistances of the switches are a quarter of the value to be measured.

TABLE 3: Results obtained for methods based on the VFM in an 8 × 8 array with different values of R_{ij} if Rs = 1 Ω and the rest of the resistors in the array, Rns, take different values. (a) Classical approach. (b) VFM circuit proposed by Wu et al. (c) Results for the proposed method in the first step. (d) Results for the proposed method in the second step.

(a) VFM classical approach

Rs = 1 Ω	Error (%), R_{ij} (Ω)					
Rns (Ω)	100	500	2000	5000	7500	10000
100	13.85	80.64	95.03	98.00	98.67	99.00
200	9.17	54.87	88.01	95.14	96.75	97.56
300	6.79	14.46	76.08	90.19	93.42	95.05
400	5.60	3.92	51.04	79.41	86.12	89.53
600	4.40	2.76	2.43	2.30	2.24	2.18
800	3.81	2.18	1.86	1.78	1.75	1.72
1200	3.21	1.59	1.28	1.22	1.21	1.20
1800	2.81	1.19	0.89	0.84	0.83	0.82
3000	2.50	0.88	0.58	0.52	0.52	0.51
6000	2.26	0.64	0.34	0.29	0.28	0.28
10000	2.16	0.55	0.25	0.19	0.18	0.18

(b) VFM-Wu

Rs = 1 Ω	Error (%), R_{ij} (Ω)					
Rns (Ω)	100	500	2000	5000	7500	10000
100	0.19	73.03	93.09	97.22	98.14	98.61
200	0.54	37.21	83.35	93.25	95.49	96.61
300	0.66	0.19	66.77	86.38	90.87	93.13
400	0.69	0.12	31.97	71.40	80.71	85.45
600	0.72	0.09	0.27	0.36	0.41	0.47
800	0.72	0.08	0.24	0.29	0.31	0.33
1200	0.73	0.07	0.22	0.26	0.26	0.27
1800	0.73	0.07	0.22	0.25	0.25	0.25
3000	0.74	0.07	0.22	0.25	0.25	0.25
6000	0.74	0.07	0.22	0.25	0.25	0.25
10000	0.74	0.07	0.22	0.25	0.25	0.25

(c) VFM proposed method, 1st step

Rs = 1 Ω	Error (%), R_{ij} (Ω)					
Rns (Ω)	100	500	2000	5000	7500	10000
100	27.48	83.53	95.76	98.30	98.86	99.15
200	0.31	58.05	88.85	95.48	96.98	97.73
300	0.04	18.45	77.19	90.64	93.72	95.28
400	0.03	0.17	52.71	80.11	86.59	89.88
600	0.12	0.09	0.05	0.02	0.07	0.08
800	0.12	0.06	0.05	0.06	0.08	0.02
1200	0.08	0.04	0.04	0.04	0.05	0.05
1800	0.07	0.02	0.03	0.03	0.03	0.04
3000	0.05	0.02	0.02	0.02	0.03	0.03
6000	0.05	0.01	0.01	0.02	0.02	0.02
10000	0.04	0.01	0.01	0.01	0.02	0.02

(d) VFM proposed method, 2nd step

Rs = 1 Ω	Error (%), R_{ij} (Ω)					
Rns (Ω)	100	500	2000	5000	7500	10000
100	29.45	83.93	95.87	98.34	98.89	99.17
200	1.00	58.27	88.91	95.50	96.99	97.74
300	0.38	18.63	77.24	90.66	93.74	95.29
400	0.19	0.04	52.77	80.14	86.60	89.89
600	0.00	0.03	0.03	0.03	0.04	0.04
800	0.04	0.02	0.01	0.03	0.04	0.04
1200	0.03	0.02	0.02	0.02	0.03	0.04
1800	0.03	0.01	0.02	0.01	0.03	0.03
3000	0.02	0.01	0.01	0.01	0.02	0.03
6000	0.02	0.01	0.01	0.01	0.02	0.02
10000	0.02	0.01	0.01	0.01	0.02	0.02

Table 3 shows simulation results for the different circuits and methods presented based on the VFM. The aim of this series of simulations is to analyse the influence of non-scanned resistors, Rns, on the measurement of the resistor, R_{ij}. Again, it is an 8 × 8 array in which the resistance of the switches is set to 1 Ω. Different values of Rns are used for each value of the resistor to be measured, R_{ij}. Both Rns and R_{ij} vary in the selected range [100 Ω, 10 kΩ]. The first question to consider is that the errors for any method are very large if Rns < 600 Ω. This comes about since the parasitic current flowing through the Rns of row *i* is excessive for the OA (even though the feedback resistor, Rg, has been chosen so there is no saturation of voltages of the OA in the selected range of individual resistors). The situation gets worse with increased R_{ij}, since $Vo(i, j)$ decreases, and in consequence, the current flowing through all the Rns of row *i* increases. This situation does not occur in the ZPM circuits since the row selection switches are responsible for providing this current. The range limitation of Rns is therefore a drawback of VFM-based circuits compared to those based on the ZPM. It is worth to note that situations with all Rns very low are uncommon in arrays of sensors. For example, in [33], a 16 × 16 tactile sensor array designed with a continuous electroactive material using a piezoresistive sheet of capLINQ (code MVCF-40012BT50KS/2A) has 10 kΩ for no-pressed tactels; however, only the few pressed ones might have resistances around 300 Ω.

If this saturation does not come about (due to the current to be provided by the OA), the proposed method outperforms the other methods in both the first and the second steps for all combinations of R_{ij} and Rns. Finally, the results obtained by the proposed method are very similar in the VFM and ZPM circuits (outside of the saturation range of the VFM circuit).

Table 4 shows the influence of the resistances of the switches, Rs, on the error in estimating R_{ij} for the methods based on the VFM. Once again, the error increases with the value of Rs and, except for VFM-Wu, also increases with decreasing R_{ij}. The biggest errors occur, as in all previous cases, for the classical approach, while the proposed method

TABLE 4: Results obtained for methods based on the VFM in an 8 × 8 array with different values of R_{ij} if Rns = 5 kΩ and the resistances of the switches take different values. (a) Classical approach. (b) VFM circuit proposed by Wu et al. (c) Results for the proposed method in the first step. (d) Results for the proposed method in the second step.

(a) VFM classical approach

Rns = 5 kΩ	Error (%), R_{ij} (Ω)					
Rs (Ω)	100	500	2000	5000	7500	10000
1	2.31	0.69	0.39	0.34	0.33	0.32
2	4.59	1.37	0.77	0.66	0.63	0.62
4	9.19	2.74	1.53	1.30	1.24	1.22
6	13.81	4.12	2.30	1.94	1.86	1.82
9	20.78	6.19	3.45	2.90	2.77	2.71
15	34.88	10.36	5.76	4.82	4.59	4.48
20	46.79	13.88	7.69	6.42	6.11	5.94
25	58.85	17.43	9.64	8.02	7.62	7.39

(b) VFM-Wu

Rns = 5 kΩ	Error (%), R_{ij} (Ω)					
Rs (Ω)	100	500	2000	5000	7500	10000
1	0.74	0.07	0.22	0.25	0.25	0.25
2	0.16	0.64	0.79	0.81	0.82	0.82
4	0.98	1.76	1.90	1.93	1.93	1.93
6	2.12	2.85	2.99	3.02	3.02	3.03
9	3.83	4.47	4.59	4.62	4.63	4.63
15	7.23	7.58	7.66	7.69	7.71	7.72
20	10.05	10.07	10.09	10.12	10.15	10.18
25	12.86	12.46	12.41	12.45	12.49	12.54

(c) VFM proposed method, 1st step

Rns = 5 kΩ	Error (%), R_{ij} (Ω)					
Rs (Ω)	100	500	2000	5000	7500	10000
1	0.05	0.01	0.01	0.02	0.02	0.02
2	0.12	0.02	0.01	0.02	0.02	0.03
4	0.40	0.05	0.03	0.03	0.03	0.04
6	0.82	0.11	0.05	0.05	0.05	0.05
9	1.70	0.22	0.10	0.09	0.09	0.09
15	4.09	0.56	0.25	0.21	0.21	0.21
20	6.59	0.96	0.44	0.36	0.34	0.34
25	9.42	1.45	0.65	0.54	0.51	0.50

(d) VFM proposed method, 2nd step

Rns = 5 kΩ	Error (%), R_{ij} (Ω)					
Rs (Ω)	100	500	2000	5000	7500	10000
1	0.02	0.01	0.01	0.02	0.02	0.02
2	0.02	0.01	0.01	0.02	0.02	0.02
4	0.04	0.01	0.01	0.02	0.02	0.02
6	0.08	0.01	0.01	0.02	0.02	0.02
9	0.19	0.02	0.01	0.02	0.02	0.03
15	0.66	0.05	0.03	0.02	0.03	0.03
20	1.29	0.09	0.04	0.04	0.04	0.04
25	2.15	0.16	0.06	0.05	0.05	0.06

presents the best results for all combinations, even in the first step. Comparing the errors obtained for ZPM-based methods to the errors of VFM-based methods, it is found that they are quite similar for the classical approach. The errors are similar (although slightly higher) in both steps for the proposed methods and a great deal higher in the case of VFM-Wu.

Two sets of simulations have been carried out for the circuits of the ZPM of Table 5 and the VFM of Table 6 in order to analyse the errors by changing the size of the array. In both simulations, Rs = 5 Ω and Rns = 2 kΩ. For all methods, errors increase for bigger arrays. The classical approach shows higher errors than Wu proposals, while the proposed methods give the lowest errors. On the other hand, very similar errors appear for the classical approach in both circuits while errors in ZPM-Wu are lower than those obtained in VFM-Wu. It is worth noting that the proposed method shows the best results in all cases and, when the size increases, the advantage grows. So when the size varies from 4 × 4 to 16 × 16, errors for the proposed methods in the biggest array are less than the error for the smallest array with the other approaches.

5. Conclusions

The main problem in determining sensor resistances in a resistive array is the appearance of crosstalk. This phenomenon is due mainly to the resistance of the sensor selection switches to be measured. Crosstalk appears in the two types of circuits known in the literature for reading the array: the zero potential method (ZPM) and the voltage feedback method (VFM). This article presents a recursive method to obtain the individual values of the resistors in an array of resistive sensors that reduces the influence of crosstalk in determining the resistances of the sensors. The new method proposed is similar for both types of circuits and is based on the recursive calculation of the row and column voltages of the array.

The proposed method converges very quickly, outperforming, in its first or second iteration, the classical equations of the ZPM and VFM and also the circuits proposed by Wu to reduce errors. The method does not require any additional hardware and is applied directly to the classic ZPM and VFM circuits. The errors found in the calculations using the proposed method are similar for both types of circuit and increase slowly as the array size and selection switch resistance increase.

As a further conclusion, it should be noted that a decrease has been observed in the range of resistance values that can be measured in VFM circuits compared to ZPM circuits for equal output voltage ranges.

TABLE 5: Results obtained for methods based on the ZPM by changing the size of the array and maintaining Rns = 2 kΩ and Rs = 5 Ω. (a) Classical approach. (b) ZPM circuit proposed by Wu et al. (c) Results for the proposed method in the first step. (d) Results for the proposed method in the second step.

(a) ZPM classical approach

Rns = 2 kΩ	Error (%), R_{ij} (Ω)					
Size	100	500	2000	5000	7500	10000
4 × 4	11.58	3.51	1.99	1.66	1.57	1.51
6 × 6	12.63	4.53	2.99	2.64	2.53	2.46
8 × 8	13.69	5.55	3.98	3.61	3.49	3.39
10 × 10	14.76	6.57	4.98	4.58	4.43	4.31
12 × 12	15.83	7.59	5.98	5.54	5.35	5.20
14 × 14	16.91	8.62	6.98	6.49	6.26	6.07
16 × 16	17.99	9.66	7.98	7.43	7.16	6.93

(b) ZPM-Wu

Rns = 2 kΩ	Error (%), R_{ij} (Ω)					
Size	100	500	2000	5000	7500	10000
4 × 4	0.75	0.74	0.73	0.70	0.68	0.66
6 × 6	1.25	1.24	1.22	1.18	1.14	1.11
8 × 8	1.74	1.74	1.70	1.64	1.58	1.53
10 × 10	2.24	2.23	2.18	2.09	2.01	1.93
12 × 12	2.74	2.72	2.65	2.52	2.41	2.30
14 × 14	3.24	3.21	3.12	2.95	2.80	2.65
16 × 16	3.73	3.70	3.59	3.36	3.17	2.98

(c) ZPM proposed method, 1st step

Rns = 2 kΩ	Error (%), R_{ij} (Ω)					
Size	100	500	2000	5000	7500	10000
4 × 4	0.58	0.07	0.02	0.00	0.02	0.04
6 × 6	0.70	0.13	0.06	0.03	0.01	0.01
8 × 8	0.84	0.20	0.12	0.09	0.06	0.04
10 × 10	0.99	0.29	0.20	0.16	0.13	0.10
12 × 12	1.15	0.40	0.29	0.24	0.21	0.17
14 × 14	1.33	0.52	0.40	0.33	0.29	0.25
16 × 16	1.52	0.66	0.52	0.44	0.39	0.34

(d) ZPM proposed method, 2nd step

Rns = 2 kΩ	Error (%), R_{ij} (Ω)					
Size	100	500	2000	5000	7500	10000
4 × 4	0.03	0.00	0.01	0.03	0.05	0.06
6 × 6	0.05	0.00	0.01	0.03	0.05	0.07
8 × 8	0.06	0.00	0.01	0.03	0.05	0.07
10 × 10	0.08	0.01	0.01	0.03	0.05	0.07
12 × 12	0.10	0.02	0.00	0.03	0.05	0.07
14 × 14	0.13	0.03	0.01	0.02	0.05	0.07
16 × 16	0.16	0.04	0.02	0.01	0.04	0.06

TABLE 6: Results obtained for methods based on the VFM by changing the size of the array and maintaining Rns = 2 kΩ and Rs = 5 Ω. (a) Classical approach. (b) VFM circuit proposed by Wu et al. (c) Results for the proposed method in the first step. (d) Results for the proposed method in the second step.

(a) VFM classical approach

Rns = 2 kΩ	Error (%), R_{ij} (Ω)					
Size	100	500	2000	5000	7500	10000
4 × 4	11.60	3.53	2.01	1.71	1.64	1.60
6 × 6	12.66	4.54	3.01	2.69	2.61	2.56
8 × 8	13.72	5.56	4.02	3.67	3.57	3.50
10 × 10	14.79	6.59	5.02	4.64	4.52	4.43
12 × 12	15.86	7.62	6.02	5.61	5.45	5.33
14 × 14	16.94	8.65	7.02	6.56	6.37	6.21
16 × 16	18.02	9.68	8.02	7.51	7.27	7.07

(b) VFM-Wu

Rns = 2 kΩ	Error (%), R_{ij} (Ω)					
Size	100	500	2000	5000	7500	10000
4 × 4	10.77	2.76	1.25	0.95	0.88	0.84
6 × 6	11.27	3.26	1.75	1.43	1.35	1.30
8 × 8	11.77	3.75	2.23	1.89	1.79	1.73
10 × 10	12.27	4.25	2.71	2.35	2.22	2.13
12 × 12	12.77	4.74	3.19	2.79	2.63	2.52
14 × 14	13.27	5.24	3.66	3.22	3.03	2.88
16 × 16	13.77	5.73	4.13	3.63	3.41	3.21

(c) VFM proposed method, 1st step

Rns = 2 kΩ	Error (%), R_{ij} (Ω)					
Size	100	500	2000	5000	7500	10000
4 × 4	0.61	0.08	0.05	0.04	0.05	0.05
6 × 6	0.73	0.14	0.09	0.09	0.09	0.09
8 × 8	0.87	0.22	0.15	0.15	0.15	0.15
10 × 10	1.02	0.31	0.24	0.22	0.22	0.22
12 × 12	1.18	0.42	0.33	0.31	0.30	0.30
14 × 14	1.15	0.55	0.44	0.41	0.39	0.38
16 × 16	0.07	0.68	0.56	0.52	0.50	0.48

(d) VFM proposed method, 2nd step

Rns = 2 kΩ	Error (%), R_{ij} (Ω)					
Size	100	500	2000	5000	7500	10000
4 × 4	0.06	0.01	0.01	0.02	0.02	0.02
6 × 6	0.07	0.02	0.01	0.02	0.03	0.03
8 × 8	0.09	0.02	0.02	0.03	0.03	0.04
10 × 10	0.11	0.03	0.03	0.03	0.04	0.04
12 × 12	0.13	0.04	0.04	0.04	0.05	0.05
14 × 14	0.08	0.06	0.05	0.05	0.06	0.06
16 × 16	1.56	0.07	0.06	0.06	0.07	0.07

Conflicts of Interest

The authors declare that they have no conflicts of interest.

References

[1] I. Fratoddi, I. Venditti, C. Cametti, and M. V. Russo, "Chemiresistive polyaniline-based gas sensors: a mini review," *Sensors and Actuators B: Chemical*, vol. 220, pp. 534–548, 2015.

[2] J. Wang, S. Chan, R. R. Carlson et al., "Electrochemically fabricated polyaniline nanoframework electrode junctions that function as resistive sensors," *Nano Letters*, vol. 4, no. 9, pp. 1693–1697, 2004.

[3] A. Depari, M. Falasconi, A. Flammini et al., "A new low-cost electronic system to manage resistive sensors for gas detection," *IEEE Sensors Journal*, vol. 7, no. 7, pp. 1073–1077, 2007.

[4] D. Stratos, G. Maria, F. Eleftherios, and L. George, "Comparison of three resistor network division circuits for the readout of 4×4 pixel SiPM arrays," *Nuclear Instruments and Methods in Physics Research Section A: Accelerators, Spectrometers, Detectors and Associated Equipment*, vol. 702, pp. 121–125, 2013.

[5] J. F. Wu, L. Wang, J. Q. Li, and Z. Z. Yu, "A small size device using temperature sensor array," *Chinese Journal of Sensors and Actuators*, vol. 24, pp. 1649–1652, 2011.

[6] Y.-J. Yang, M.-Y. Cheng, S.-C. Shih et al., "A 32 × 32 temperature and tactile sensing array using PI-copper films," *The International Journal of Advanced Manufacturing Technology*, vol. 46, no. 9-12, pp. 945–956, 2010.

[7] M. Arlit, E. Schleicher, and U. Hampel, "Thermal anemometry grid sensor," *Sensors*, vol. 17, no. 7, p. 1663, 2017.

[8] W. Z. Wan Hasan, F. R. M. Rashidi, M. N. Hamidon, and Y. Wahab, "Design of readout circuit for piezoresistive pressure sensor using nodal array approach reading technique," *Pertanika Journal of Science & Technology*, vol. 25, pp. 215–224, 2017.

[9] W. E. Snyder and J. St. Clair, "Conductive elastomers as sensor for industrial parts handling equipment," *IEEE Transactions on Instrumentation and Measurement*, vol. 27, no. 1, pp. 94–99, 1978.

[10] L. Shu, X. Tao, and D. D. Feng, "A new approach for readout of resistive sensor arrays for wearable electronic applications," *IEEE Sensors Journal*, vol. 15, no. 1, pp. 442–452, 2015.

[11] F. Vidal-Verdú, M. J. Barquero, J. Castellanos-Ramos et al., "A large area tactile sensor patch based on commercial force sensors," *Sensors*, vol. 11, no. 5, pp. 5489–5507, 2011.

[12] Y. Wang, A. X. Wang, Y. Wang, M. K. Chyu, and Q.-M. Wang, "Fabrication and characterization of carbon nanotube-polyimide composite based high temperature flexible thin film piezoresistive strain sensor," *Sensors and Actuators A: Physical*, vol. 199, pp. 265–271, 2013.

[13] O. Oballe-Peinado, J. A. Hidalgo-Lopez, J. Castellanos-Ramos et al., "FPGA-based tactile sensor suite electronics for real-time embedded processing," *IEEE Transactions on Industrial Electronics*, vol. 64, no. 12, pp. 9657–9665, 2017.

[14] M.-S. Suen, Y.-C. Lin, and R. Chen, "A flexible multifunctional tactile sensor using interlocked zinc oxide nanorod arrays for artificial electronic skin," *Sensors and Actuators A: Physical*, vol. 269, pp. 574–584, 2018.

[15] R. S. Saxena, N. K. Saini, and R. K. Bhan, "Analysis of crosstalk in networked arrays of resistive sensors," *IEEE Sensors Journal*, vol. 11, no. 4, pp. 920–924, 2011.

[16] H. Liu, Y.-F. Zhang, Y.-W. Liu, and M.-H. Jin, "Measurement errors in the scanning of resistive sensor arrays," *Sensors and Actuators A: Physical*, vol. 163, no. 1, pp. 198–204, 2010.

[17] T. D'Alessio, "Measurement errors in the scanning of piezoresistive sensors arrays," *Sensors and Actuators A: Physical*, vol. 72, no. 1, pp. 71–76, 1999.

[18] J. Wu, L. Wang, and J. Li, "VF-NSE method measurement error analysis of networked resistive sensor array," *Sensors and Actuators A: Physical*, vol. 211, pp. 45–50, 2014.

[19] J. A. Hidalgo-López, J. Romero-Sánchez, and R. Fernández-Ramos, "New approaches for increasing accuracy in readout of resistive sensor arrays," *IEEE Sensors Journal*, vol. 17, no. 7, pp. 2154–2164, 2017.

[20] Ó. Oballe-Peinado, F. Vidal-Verdú, J. Sánchez-Durán, J. Castellanos-Ramos, and J. Hidalgo-López, "Improved circuits with capacitive feedback for readout resistive sensor arrays," *Sensors*, vol. 16, no. 2, p. 149, 2016.

[21] J. S. Kim, D. Y. Kwon, and B. D. Choi, "High-accuracy, compact scanning method and circuit for resistive sensor arrays," *Sensors*, vol. 16, no. 2, p. 155, 2016.

[22] R. Yarahmadi, A. Safarpour, and R. Lotfi, "An improved-accuracy approach for readout of large-array resistive sensors," *IEEE Sensors Journal*, vol. 16, no. 1, pp. 210–215, 2016.

[23] F. Reverter, F. Vidal-Verdú, and J. A. Hidalgo-Lopez, "Advanced techniques for directly interfacing resistive sensors to digital systems," in *Advanced Interfacing Techniques for Sensors: Measurement Circuits and Systems for Intelligent Sensors*, B. George, J. K. Roy, V. J. Kumar, and S. C. Mukhopadhyay, Eds., pp. 139–165, Springer International Publishing, Cham, Switzerland, 2017.

[24] J. F. Wu, "Scanning approaches of 2-D resistive sensor arrays: a review," *IEEE Sensors Journal*, vol. 17, no. 4, pp. 914–925, 2017.

[25] L. Wang, X.-L. Wen, J.-J. Pan, and L. Yang, "Improved FRZPC for the two-dimensional resistive sensor array," *IET Science, Measurement and Technology*, vol. 12, no. 2, pp. 278–282, 2018.

[26] J. Wu, L. Wang, and J. Li, "Design and crosstalk error analysis of the circuit for the 2-D networked resistive sensor array," *IEEE Sensors Journal*, vol. 15, no. 2, pp. 1020–1026, 2015.

[27] X. Zhang and X. Ye, "Zero potential method measurement error analysis for networked resistive sensor arrays," *IET Science, Measurement and Technology*, vol. 11, no. 3, pp. 235–240, 2017.

[28] J. Wu, L. Wang, J. Li, and A. Song, "A novel crosstalk suppression method of the 2-D networked resistive sensor array," *Sensors*, vol. 14, no. 7, pp. 12816–12827, 2014.

[29] J.-F. Wu, J.-Q. Li, and A.-G. Song, "Readout circuit based on double voltage feedback loops in the two-dimensional resistive sensor array: design, modelling and simulation evaluation," *IET Science, Measurement and Technology*, vol. 11, no. 3, pp. 288–296, 2017.

[30] F. Castro, T. Pentiado, J. Blanco, R. Xavier, M. Sanches, and A. de Carvalho, "Crosstalk error analysis in IIDFC readout

circuit for use in piezoresistive composite," *IEEE Sensors Journal*, vol. 18, no. 1, pp. 382–389, 2018.

[31] J.-F. Wu, F. Wang, Q. Wang, J.-Q. Li, and A.-G. Song, "An improved zero potential circuit for readout of a two-dimensional resistive sensor array," *Sensors*, vol. 16, no. 12, p. 2070, 2016.

[32] J. Millman, "A useful network theorem," *Proceedings of the IRE*, vol. 28, no. 9, pp. 413–417, 1940.

[33] J. A. Hidalgo-López, Ó. Oballe-Peinado, J. Castellanos-Ramos, J. A. Sánchez-Durán, R. Fernández-Ramos, and F. Vidal-Verdú, "High-accuracy readout electronics for piezoresistive tactile sensors," *Sensors*, vol. 17, no. 11, 2017.

4

Evaluation of Low-Cost Sensors for Ambient $PM_{2.5}$ Monitoring

Marek Badura[1] **Piotr Batog,**[2] **Anetta Drzeniecka-Osiadacz,**[3] **and Piotr Modzel**[3]

[1]*Department of Air Conditioning, Heating, Gas Engineering and Air Protection, Faculty of Environmental Engineering, Wrocław University of Science and Technology, 50-373 Wrocław, Poland*
[2]*INSYSPOM, 54-427 Wrocław, Poland*
[3]*Department of Climatology and Atmosphere Protection, Institute of Geography and Regional Development, Faculty of Earth Science and Environmental Management, University of Wrocław, 51-621 Wrocław, Poland*

Correspondence should be addressed to Marek Badura; marek.badura@pwr.edu.pl

Academic Editor: Michele Penza

Low-cost sensors are an opportunity to improve the spatial and temporal resolution of particulate matter data. However, such sensors should be calibrated under conditions close to the final ones before any monitoring actions. The paper presents the results of a collocated comparison of four models of low-cost optical sensors with a TEOM 1400a analyser. SDS011 (Nova Fitness), ZH03A (Winsen), PMS7003 (Plantower), and OPC-N2 (Alphasense) sensors were used in this research. Three copies of each sensor model were placed in a common box to compare the sensor performance under the same measurement conditions. Monitoring of the $PM_{2.5}$ fraction was conducted for almost half a year from 21 August 2017 to 19 February 2018 in Wrocław (Poland). Reproducibility between sensor units was assessed on the basis of coefficient of variation (CV). CV values were lower than 7% in the case of SDS011 and PMS7003 sensors and equal to 20% for OPC-N2 units. CV was higher than 50% for ZH03A, mainly due to malfunctions. During the measurements, the trends of outputs from sensors were generally similar to TEOM data, but significant overestimation of $PM_{2.5}$ concentrations was observed for the sensor raw data. A high linear relationship between TEOM and sensors was noticed for 1 min, 15 min, and 1-hour averaged data for PMS7003 sensors ($R^2 \approx 0.83$–0.89), for SDS011 units ($R^2 \approx 0.79$–0.86), and for one unit of ZH03A ($R^2 \approx 0.74$–0.81). R^2 values for daily averages were at the level 0.91–0.93 for PMS7003, 0.87–0.90 for SDS011, and 0.89 for ZH03A. OPC-N2 had only a moderate linear relationship with TEOM ($R^2 \approx 0.53$–0.69 for daily data and 0.43–0.61 for shorter time averages). Quite large dispersion of data and high relative errors of $PM_{2.5}$ estimation were observed for concentration ranges below 20–30 $\mu g/m^3$. The impact of high relative humidity level was observed for SDS011 and OPC-N2 devices—clear overestimation of outputs was observed above 80% RH.

1. Introduction

Particulate matter (PM) is presently considered as one of the most serious air pollutants. PM could originate from natural sources (e.g., volcanic dust or desert dust particles and sea spray aerosols), but anthropogenic activities are also responsible for its emission. Transport, agriculture, industrial processes, and household fuel combustion can be considered the main sources of that contaminant [1].

The adverse health effects of particulates are nowadays well documented. Exposure to PM is linked mainly to respiratory and cardiovascular health effects [2–5]. Moreover, the International Agency for Research on Cancer (IARC) classified particulates as carcinogenic to humans [6].

Despite the fact that particulates may vary in origin, chemical composition, shape, or surface area, the harmful effects of PM are usually related to the size of particles. The particles with smaller aerodynamic diameters (d_a) have greater ability to penetrate into the lower parts of the respiratory tract. Coarse PM (particles with d_a in the range 2.5–10 μm) deposit mainly in the upper respiratory tract, and fine PM (particles with d_a below 2.5 μm) deposit throughout the respiratory tract and can penetrate to the lower parts and alveoli, whereas ultrafine particles (particles with d_a below

Table 1: Characteristic of PM sensors used in the research.

Sensor model	SDS011	ZH03A	PMS7003	OPC-N2
Manufacturer	Nova Fitness	Winsen	Plantower	Alphasense
Approximate price ($)	20	20	20	500
Dimensions (mm)	71 × 70 × 23	50 × 32.4 × 21	48 × 37 × 12	75 × 63.5 × 60
Approximate weight (g)	50	30	30	105
Power supply voltage (V)	5	4.5–5.5	4.5–5.5	4.8–5.2
Working current (mA)	220	70–140	≤100	175
Detectable size range (μm)	0.3–10	0.3–10	0.3–10	0.38–17
Size bins	Not available	Not available	6 size bins	16 size bins
Estimated PM_x concentration	$PM_{2.5}$, PM_{10}	PM_1, $PM_{2.5}$, and PM_{10}	PM_1, $PM_{2.5}$, and PM_{10}	PM_1, $PM_{2.5}$, and PM_{10}
Concentration range (μg/m^3)	0–999.9	0–1000 (for $PM_{2.5}$)	Effective range: 0–500 Maximum range: above 1000 (for $PM_{2.5}$)	0.01–1500·10^3 (for PM_{10})

0.1 μm) may cross the endothelial barrier and enter the blood [2–5]. The mass concentrations of PM_{10} (d_a below 10 μm) and $PM_{2.5}$ (d_a below 2.5 μm) are broadly used for air quality standard establishment and for assessing the air pollution levels.

Today, information about the current air quality is available to the public through websites or mobile apps (for example, [7]). High-quality data for such portals comes from the monitoring stations of governmental agencies, where automated measuring systems (AMS) are used. The most commonly used instruments are β-ray attenuation monitors (BAMs) and tapered element oscillating microbalances (TEOMs). These instruments are capable of measuring daily and hourly PM variation. However, the spatial coverage of monitoring stations is not very dense and nowadays new techniques are sought to improve the spatial and temporal resolution of PM data. Technological progress leads, among other things, to the development of miniaturized low-cost sensor devices and makes them promising tools for air quality monitoring application [8].

Currently, low-cost pollution monitoring is possible via different commercial sensors and a growth in the popularity of the use of such devices is observed worldwide [8–10]. Sensors for particulate matter measurements are also available in many types. The common feature of all commercial PM sensors is the principle of operation—they measure light scattered by particles carried in an air stream through a light beam. Prices of such optical sensors range from tens to hundreds of US dollars, because they are generally cheap to manufacture. In addition, PM sensors are easy to use and in many cases ready to connect to microcomputers. Because of this, they are often adopted for use by citizen scientists [9, 11, 12].

Another advantage of optical PM sensors is the low energy consumption—they require power supply voltage at the level of 5 V, and the working current is usually lower than 250 mA (see Table 1 for examples). They are also relatively small and light, and the output data could be collected with high frequency. For these reasons, they are good candidates for creating widely dispersed sensor networks [12–15]. That kind of monitoring may supplement the existing AMS infrastructure [9, 12].

Optical sensors have, however, a major drawback—the amount of scattered light is dependent on particle parameters: size, shape, density, and refractive index [16]. In other words, calibration factors obtained for one type of particulate matter (in most cases in laboratory conditions) may not be appropriate in the measurement environment. Therefore, PM sensors should be calibrated or recalibrated under conditions close to the final ones [17]. In the case of ambient air monitoring, the practical way to calibrate low-cost sensors is using collocated data from more recognized, higher-class, instruments [18]. Devices that meet regulatory requirements might be used for that purpose—for example, those that fulfil the criteria of European Standards EN 12341 [19] or EN 16450 [20]. Other types of control instruments, commonly used in low-cost sensor testing, are research-grade devices that were previously strictly tested, and their performance was reported in peer-review journals [21].

The article presents the results of a collocated comparison of four models of low-cost optical particulate matter sensors and a TEOM 1400a analyser. A great advantage of the TEOM device is the possibility of near real-time monitoring (at the level of minutes) [22]. This type of equipment is widely used in regulatory monitoring stations and for scientific research, especially for PM_{10} monitoring [23, 24]. Furthermore, TEOM 1400a proved to be also useful in $PM_{2.5}$ measurements [25] and in low-cost sensor testing campaigns [26, 27].

PM sensors were evaluated in several respects. Firstly, the operational stability of sensors was checked. This property is especially important for measurement systems dedicated to long-term monitoring. In case of low-cost PM sensors, accumulation of particles inside the measuring chamber could be the main cause of sensor aging. Moreover, severe meteorological conditions (e.g., high humidity or extreme temperatures) can affect the functioning of electronics. In fact, many sensor models are dedicated strictly to indoor environments and special attention should be paid when they are used outdoors [28].

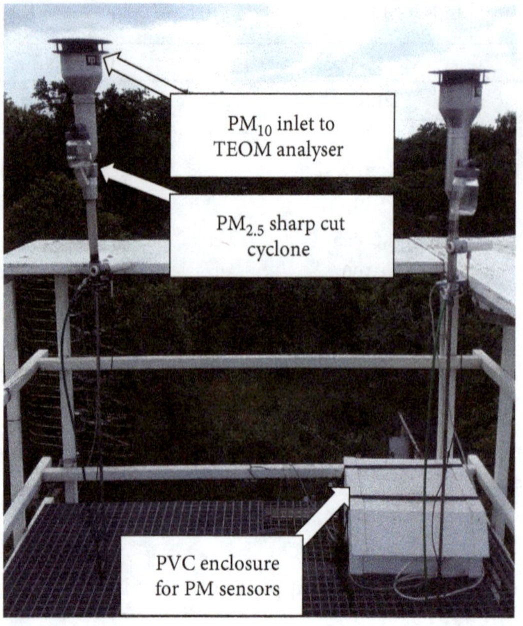

FIGURE 1: The placement of the PM sensor measurement box and TEOM inlet.

Secondly, the precision of sensors was tested in terms of reproducibility between units of the same sensor model (so-called intramodel variability). Calibration efforts of devices with high reproducibility are considered smaller, because they do not have to be calibrated individually.

Thirdly, the relationship to TEOM instrument and linearity of sensor responses were assessed. Linear responses are favourable in case of many measurement devices, because they simplify calibration procedures.

Additionally, the performance of sensor operation in air with high relative humidity (RH) was examined. The impact of humidity is particularly important for monitoring of ambient air, where large variations of that parameter occur. Some sensors seem affected by humidity, and their performance may deteriorate in very humid air [21].

All these features were considered as crucial for the end-users and for the creators of sensor systems and distributed sensor nets.

2. Materials and Methods

2.1. Measurement Site Description. Measurements were carried out at the Meteorological Observatory of Department of Climatology and Atmosphere Protection of University of Wrocław (the location is presented in Figure S1 in the Supplementary Material). The observatory is located in an area of detached houses and allotments, close to the large city park (about 100 ha of land). Emission sources of particulate matter are related mainly to individual heating systems in households and, to a lesser extent, city road transport. There are no large point sources of particulate emission nearby.

The Observatory is equipped with PM_{10} and $PM_{2.5}$ measuring instruments (TEOMs); however, PM_{10} instruments were out of service during this campaign.

2.2. Measurements of $PM_{2.5}$ Fraction. $PM_{2.5}$ was measured by means of a TEOM 1400a analyser (Rupprecht & Patashnick Co., USA) and low-cost PM sensors, placed in a measurement box. TEOM was equipped with a PM_{10} inlet with $PM_{2.5}$ sharp cut cyclone. The instrument also used a heated flow tube (40°C) to lower the RH and remove moisture droplets from the sampled air. In this type of device, sampled air is drawn through a filter, attached to the end of a hollow tapered glass tube. The tube is maintained in oscillation by an electronic feedback system. Particulates deposited on the filter decrease the frequency of oscillation of the tube, and this change is used to determine the PM mass increase. The device calculates PM mass concentration by dividing the obtained mass by the air flow rate [23]. TEOM provided 1-minute averaged data, stored in the database. To maintain the accuracy of the device, the service and calibration were performed annually by authorised service personnel.

PM sensors were enclosed in a PVC box ($56 \times 50 \times 26$ cm) with a rainproof lid. The airflow through the box was possible by means of two air inlets (13×10 cm), located at one wall of the housing, and a fan, mounted at the opposite wall. Air inlets were fitted out with mesh filters to provide protection against insects. The box was fastened to the measurement platform, below TEOM intake (see Figure 1). The distance between the inlet of the measurement box and the inlet of TEOM was approximately 1.5–1.8 m.

Four commercially available PM sensors were chosen for this research: SDS011 (Nova Fitness Co. Ltd., China), ZH03A (Zhengzhou Winsen Electronics Technology Co. Ltd., China), PMS7003 (Beijing Plantower Co. Ltd., China), and OPC-N2 (Alphasense, UK). A short characteristic of those devices is shown in Table 1.

SDS011, ZH03A, and PMS7003 sensors were chosen mainly because of their low price (≤20$) and small size. The earlier versions of Plantower sensors (i.e., PMS1003, PMS3003, and PMS5003) were evaluated as promising tools for $PM_{2.5}$ monitoring [29–31]; however, high air humidity influenced the sensor outputs [32, 33]. The SDS011 sensor from Nova Fitness also demonstrated its usefulness for particulates measurements, but in a portable version [34]. Based on the knowledge of the authors, the performance of ZH03A sensors (Winsen) has not been tested in scientific literature so far.

SDS011, ZH03A, and PMS7003 sensors compose of a small measurement chamber with light source (light emitting diode), light receptor (photodiode detector), and a set of focusing lenses. The airflow through every sensor is possible by means of a microfan. The output signals are digital and have a form of mass concentration, mainly for $PM_{2.5}$ and PM_{10} fractions. All output values are determined on the basis of light scattering intensity by algorithms implemented in sensors, but the details of the calculations are not provided by the manufacturers. PM mass concentration for Plantower PMS7003 is also available with so-called atmospheric environment correction factor ("AE"). Moreover, this sensor provides the number of particles per unit volume for 6 size bins. In case of all the tested low-cost sensors, the factory calibration procedures were not specified in the data sheets.

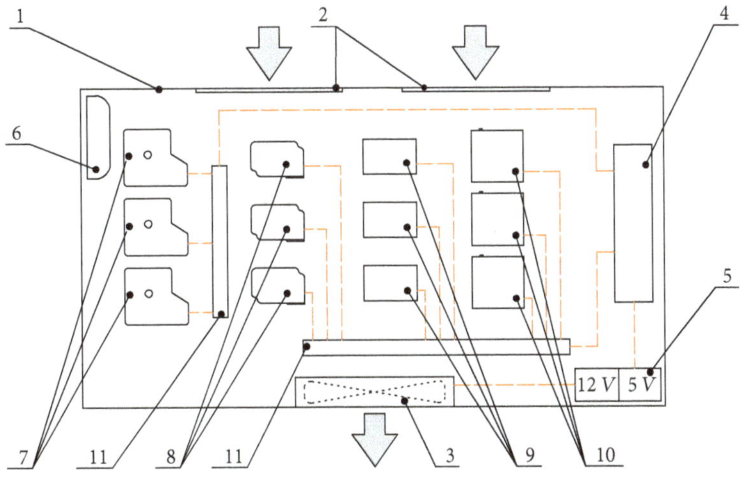

FIGURE 2: Schematic of PM sensor measurement setup: 1—PVC enclosure, 2—air inlets with mesh filters, 3—air outlet with fan, 4—Raspberry Pi microcomputer, 5—power suppliers, 6—temperature and humidity datalogger, 7—OPC-N2 sensors (Alphasense), 8—ZH03A sensors (Winsen), 9—PMS7003 sensors (Plantower), 10—SDS011 sensors (Nova Fitness), and 11—USB hubs. Figure 2 is reproduced from Badura et al. [38].

OPC-N2 devices were selected as representatives of the higher-price segment (~500$). Their performance was evaluated both in the laboratory [35] and in the field conditions [36, 37]. The results presented in the literature were promising, although high ambient RH affected sensor performance [36].

OPC-N2 enclosure is a little bigger compared to previously described sensors, and they use an embedded diode laser light source that operates at a wavelength of 658 nm. Every OPC-N2 uses an elliptical mirror and dual-element photodetector, and the airflow is produced also by a small fan. Those devices are referred to as optical particle counters, and they classify particles to one of 16 size bins. The resulting particle size histograms are used to determine the mass concentration for PM_1, $PM_{2.5}$, and PM_{10} fractions. The calculations are made with assumptions of particle density (1.65 g/cm^3 by default) and refractive index (1.5 by default). OPC-N2 sensors are laboratory-calibrated using Polystyrene Spherical Latex Particles of a known diameter and known refractive index. Correction factors can be applied by the user when properties of particulates are known. In this study, the default values from the manufacturer were adopted.

The measurement setup consisted of three copies of each sensor model. Sensor units were mounted to a PVC board at a distance of a few centimetres from each other. Multiport USB hubs were used to connect the sensors with a Raspberry Pi microcomputer. All electronic connectors were coated with an anticorrosion agent. Sensor signals were read with a 1- or 2-second resolution and averaged in 1-minute intervals. Averaged data was stored in the database for further analysis. Two power supplies were utilized in this setup: 5 V for the microcomputer and 12 V to power the fan in the box. Temperature and relative humidity were measured inside the box by means of AR235 datalogger (APAR, Poland). A schematic of the described arrangement is shown in Figure 2 [38]. The applied setup allowed the comparison of the sensor performance under the same measurement conditions.

3. Data Analysis

3.1. Data Preparation. For the purpose of this research, data registered from 21 August 2017 to 19 February 2018 was used. To examine the relationship between TEOM and sensors, only sensor outputs related to the $PM_{2.5}$ mass concentration were taken into account. It was assumed that factory $PM_{2.5}$ outputs, determined by the sensor manufacturers, should reflect the $PM_{2.5}$ concentration in the best way.

TEOM and sensor signals were analysed in different time scales: (1) 1 minute, (2) 15 minutes, (3) 1 hour, and (4) 24 hours. The first two data types represent the short-time measurements, especially useful for indication of elevated PM concentration events or "hot-spots." The latter two are the most popular in informing the public about air quality. In every case, the averaging was made only for data sets with at least 75% completeness. Data analysis was performed in a MATLAB environment.

3.2. Reproducibility between Units. Precision of sensors could be assessed in terms of repeatability of measurements of a particular sensor unit and reproducibility of units of the same sensor model. Estimating the repeatability of PM sensors is very difficult in nonlaboratory conditions due to problems in maintaining constant particle type and concentration [17]. For this reason, only the variability of output signals from copies of the same sensor model (intramodel variability) was evaluated in this study.

Firstly, to visualize the dispersion of the data, scatterplots of sensor unit outputs versus the mean value of the 1-minute averaged data were presented. Secondly, the percentage coefficient of variation (CV) was calculated. For every sensor model, the temporary coefficient of variation was calculated as

$$CV_{(t)} = 100 \cdot \frac{\sigma}{\mu}, \%, \tag{1}$$

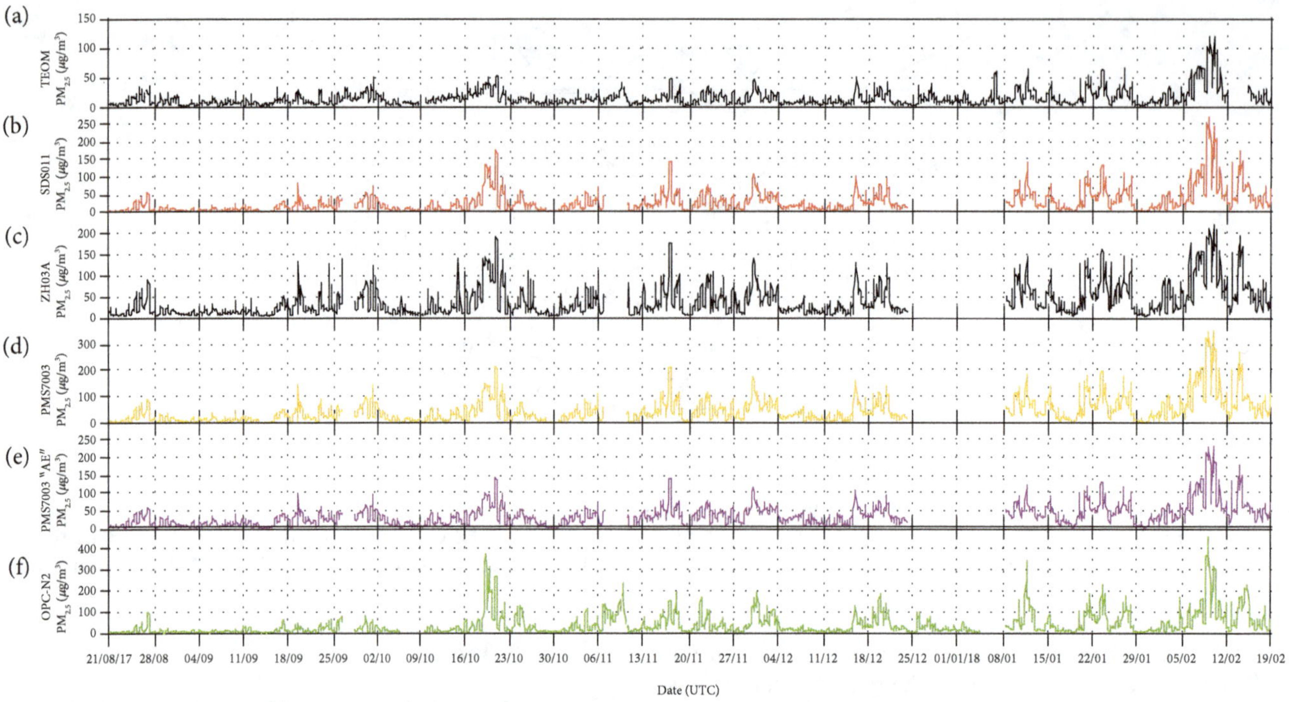

FIGURE 3: Results of $PM_{2.5}$ measuring campaign: (a) TEOM 1400a analyser, (b) SDS011 (Nova Fitness), (c) ZH03A (Winsen), (d) PMS7003 (Plantower), (e) PMS7003 "AE" (Plantower), and (f) OPC-N2 (Alphasense). 1-hour averaged data for selected sensors was plotted for clarity. Please note the different scales on the y-axes.

where σ is the standard deviation and μ is the mean of the 1-minute averaged data from units of the same sensor model. The final CV value for each model was determined as the average of all temporary $CV_{(t)}$ values. Sensors with low CV values are characterized with high reproducibility. The CV value below 10% is considered acceptable in the low-cost sensor testing studies [35, 39, 40].

3.3. Relationship between Sensors and TEOM Control Measurements. The relationship to TEOM control measurements was assessed on the basis of coefficient of determination (R^2) from ordinary least-squares regression fittings. Linear regression models were assumed with TEOM readouts as dependent variable and PM sensor data as predictor variables. The R^2 value near 1 reflects very good agreement with control measurements and linearity of responses. Small R^2 values indicate poor relationship.

Calculations between sensors and TEOM were made for all considered time scales. Pairwise matched data was used in every case.

3.4. Influence of $PM_{2.5}$ Concentration Range on Sensor Performance. Parameters of linear functions (slopes and intercepts) from regression fittings were used to calibrate the low-cost sensors. Afterwards, the distribution of relative error of $PM_{2.5}$ estimation over the measured mass concentrations was examined. On the basis of obtained results, the relationship between TEOM and PM sensors was considered for different $PM_{2.5}$ concentration ranges: low ($\leq 20\,\mu g/m^3$), medium (20–$60\,\mu g/m^3$), and high ($>60\,\mu g/m^3$). Combined ranges were taken into account too. Calculations for 15 min and 1-hour time scales were made as an example.

3.5. Influence of Air Humidity Level on Sensor Performance. Conditions of a high air humidity can affect the performance of optical PM sensors in several ways. The first group of problems is related to a failure of the electronic circuits in a very moist environment. This can lead to biased measurement results or damage of sensors [41]. The second group of problems is associated with particle properties and the principle of operation of light-scattering devices. The hygroscopic growth of some particles (e.g., hygroscopic salts) is the most discussed cause of overestimated outputs and large positive artefacts in sensor signals [36, 42]. Moreover, in conditions close to the 100% RH, the formation of fog or mist may occur and the water droplets may be detected as particles [33]. The absorption of infrared radiation by water was described as responsible for deterioration of sensor performance too [41].

The purpose of this research was the evaluation of sensor performance in ambient air with variable humidity. The mechanisms that affect the operation of PM sensors in humid air were beyond the scope of this study. The RH level, above which the overestimation of particles concentration appears, was previously reported as 70%–75% [33, 42], in some cases even 60% [43], but the level of 85%–95% was also very often documented [28, 36].

In this study, sensor performance was tested in three ranges of relative humidity: (1) below 80%, (2) between 80% and 90%, and (3) above 90%. The comparison of

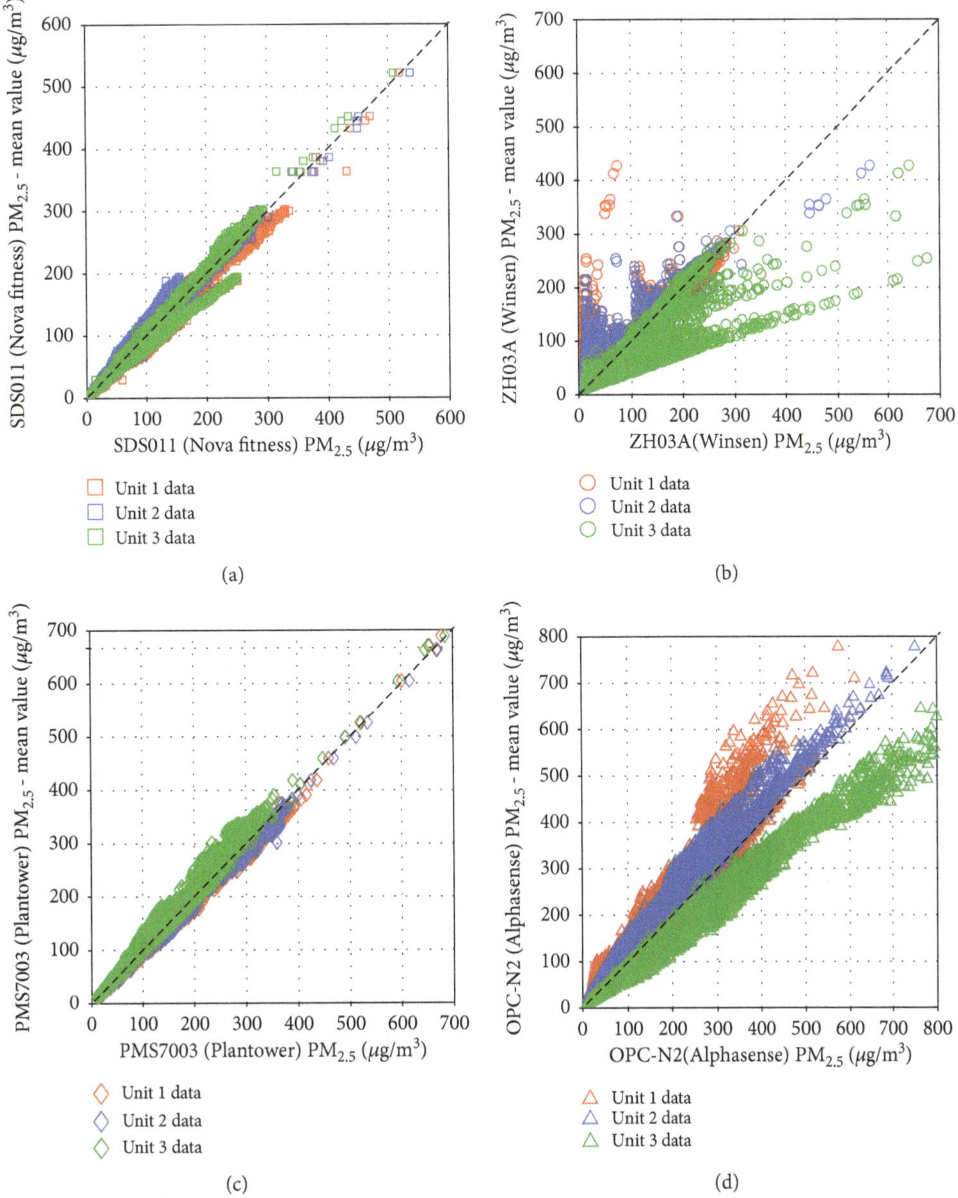

FIGURE 4: Scatterplots of sensor unit outputs versus the mean values of the 1-minute averaged data: (a) SDS011 (Nova Fitness), (b) ZH03A (Winsen), (c) PMS7003 (Plantower), and (d) OPC-N2 (Alphasense). Dashed lines denote the ideal relationship.

parameters of linear functions (slope and intercept) was made to assess the impact of RH level. Calculations for 15 min and 1-hour averages were made as an example.

4. Results and Discussion

4.1. Evaluation of PM Sensor Operation Stability. Figure 3 presents the results of $PM_{2.5}$ measurements for the period of 26 weeks (from 21 August 2017 to 19 February 2018). For the sake of clarity, 1-hour averaged data was presented. In case of PM sensors, outputs from only one selected copy of each model were plotted. Data gaps for TEOM device (Figure 3(a)) were related to power outages or maintenance activities. Power failures were also partly responsible for data losses from PM sensors (Figures 3(b)–3(f)). In addition to this, problems with USB hubs and data transfer to the database were observed.

TABLE 2: Coefficients of variation (CV, %) for tested PM sensors.

Sensor model	SDS011	ZH03A	PMS7003	PMS7003 "AE"	OPC-N2
CV, %	6.53	54.1* /10.8**	6.74	5.94	20.0

*CV calculated only for two units for the period 21 August 2017–24 December 2017; **CV calculated for three units after replacement of unit no. 1 and no. 2 on 08 January 2018.

All measurements were conducted under varying meteorological conditions. Sensors were exposed to temperatures in the range from −8°C to +36°C and relative humidity in

Table 3: Coefficients of determination (R^2) for tested PM sensors at different time scales.

Sensor model	SDS011			ZH03A			PMS7003			PMS7003 "AE"			OPC-N2		
Unit	1	2	3	1*	2**	3	1	2	3	1	2	3	1	2	3
1 min	0.79	0.83	0.80	—	0.70	0.74	0.85	0.85	0.83	0.83	0.83	0.80	0.59	0.43	0.46
15 min	0.80	0.85	0.81	—	0.71	0.77	0.87	0.87	0.85	0.85	0.85	0.81	0.60	0.44	0.47
1 hour	0.82	0.86	0.83	—	0.73	0.81	0.88	0.89	0.86	0.87	0.86	0.83	0.61	0.44	0.48
24 hours	0.88	0.90	0.87	—	0.78	0.89	0.93	0.93	0.91	0.91	0.91	0.88	0.69	0.53	0.59

*Unit no. 1 was excluded from calculations due to malfunction; **R^2 calculated for the period 21 August 2017–24 December 2017, before sensor replacement.

the range of about 27%–94% (parameters measured inside the measurement box). These conditions have not affected the data availability or electronic parts of sensors.

The maximum value of 1 min averaged outputs from TEOM was 145 $\mu g/m^3$, and for 1-hour data it was at the level of 120 $\mu g/m^3$. No saturation effect was observed for the tested PM sensors in the range of registered data. In general, the operation of tested PM sensors was stable throughout this 6-month study. Problems were observed only for Winsen ZH03A sensors. Unit no. 1 of that device was stable only for about three weeks, and after that time, the outputs did not correlate with other units. Failure of unit no. 2 was also noticed—this copy was stable for about eleven weeks. Units 1 and 2 were replaced by new specimens on 08 January 2018.

4.2. Reproducibility between Units. Figure 4 presents the relationships between signals of sensors units and the mean values of their 1-minute data. The smallest scatter between units was observed for the PMS7003 sensor from Plantower (Figure 4(c)). Slightly smaller results were recorded only for unit no. 3.

Units of SDS011 (Nova Fitness) were also characterized with high reproducibility (Figure 4(a)), but some part of the data was distant from the ideal relationship. The reason for this may be the high humidity of air (see Figure S2 in the Supplementary Material and further discussion).

OPC-N2 units were also characterized with high variability (see Figure 4(d)). Outputs from unit no. 3 were especially distant from other unit signals. The impact of humidity level was not so clearly observable in this case, due to the overall large data scatter.

In case of Winsen ZH03A units, high dispersion of data (Figure 4(b)) resulted from malfunctions of two units. Presumably, the accumulation of particles could be responsible for that situation.

These observations were confirmed by the values of coefficients of variation (see Table 2). The best reproducibility in the full measurement period was observed for Nova Fitness SDS011 and Plantower PMS7003 units. In both cases, the mean CV values were below 7%, so they met the acceptable level of sensor variations (10% [35, 39, 40]). In case of Plantower PMS7003, slightly better results were obtained for the signals with "AE" correction factor (CV below 6%). Good repeatability between Plantower sensors was also previously reported for air quality monitors with earlier versions of that sensor (PMS1003 [29], PMS3003 [30], and PMS5003 [31]).

Very high CV value was calculated for Winsen units—54.1%. This value represents only the reproducibility for two copies of that sensor (unit no. 2 and no. 3), before the replacement action on 08 January 2018. Unit no. 1 was excluded from the first calculations, because of its malfunction. CV at the level of 11% was computed for unit no. 3 and new units (no. 1 and no. 2) after the replacement. This result can be considered decent, but it represents only the 6-week measurement period.

In this campaign, Alphasense OPC-N2 counters were characterized with only moderate precision (CV value equal 20%). A previously reported laboratory evaluation of OPC-N2 sensors [35] revealed quite good reproducibility between units (CV between 4.2% and 16%). On the other hand, field evaluations of OPC-N2 [36, 37] demonstrated modest intramodel variability, with CV values higher than 20% [36]. Taking into account the cost of OPC-N2 (several times higher than other sensors), its usefulness in widely dispersed sensor networks for $PM_{2.5}$ monitoring may be considered as limited.

4.3. Relationship between Sensors and TEOM Control Measurements. Taking into account the whole measurement period, the trends of outputs from PM sensors were generally similar to TEOM data. What is important, the occurrence of episodes of elevated $PM_{2.5}$ concentrations could be readily observed from signals of each tested sensor (e.g., episode from 05 February 2018 to 12 February 2018 in Figure 3). Nevertheless, raw outputs from sensors overestimated TEOM data by a factor of about 2.5 in case of Nova Fitness devices, 3 for Winsen units, 3.5 for Plantower, and 4.5–5 for OPC-N2 sensors.

On this occasion, it should be mentioned that TEOM can underestimate PM concentrations, because of the use of the heated inlet [41]. This situation could lead to the loss of the semivolatile constituents of ambient particles [42, 43]. Generally, TEOM 1400a tracked well the variations of $PM_{2.5}$ mass concentration observed by means of a BAM analyser from the nearest governmental air quality station (Figure S3 in the Supplementary Material). The correlation between data sets was very high for nonheating periods (Figure S3a), but in heating periods, the outputs from TEOM were around 30%–40% lower than results from BAM (Figure S3b). However, it has to be taken into account that the governmental urban background station was located within 5 km from the TEOM measuring point (see Figure S1) and some different sources of $PM_{2.5}$ emission occurred in its

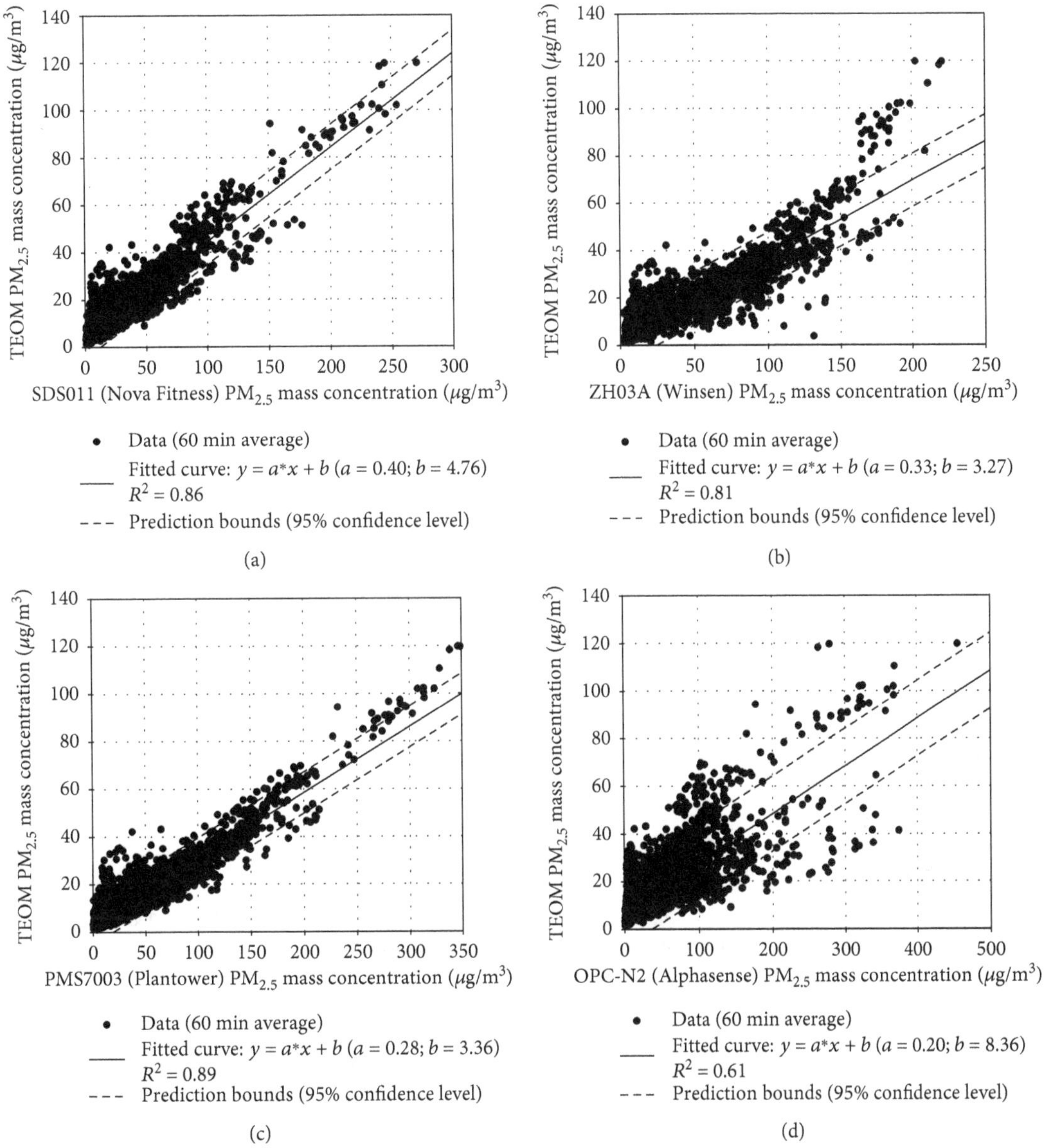

FIGURE 5: Results of linear fittings for 1-hour averaged data: (a) SDS011 (Nova Fitness), (b) ZH03A (Winsen), (c) PMS7003 (Plantower), and (d) OPC-N2 (Alphasense).

vicinity. During this research campaign, TEOM was highly linearly correlated with this measuring station: Pearson's correlation coefficient was at the level of 0.91 and coefficient of determination at the level of 0.81 for 1-hour averages (see Figure S4). For this reason, TEOM was considered a sufficient source of information about $PM_{2.5}$ changes and sufficient to assess the linearity of sensor outputs.

Table 3 presents the coefficients of determination (R^2) from linear regression fittings between TEOM and sensor outputs. The best results were obtained for Plantower sensors, for all adopted averaging times. The level of 0.83 to 0.87 was achieved for short-time averages (1 min and 15 min). R^2 values nearly 0.9 were calculated for 1-hour means, and the level of 0.9 was exceeded for daily averages. It should be noted that Plantower outputs with "AE" correction factors were highly correlated to data without that factor (Pearson's correlation coefficients at the level of 0.99), but their relationship to TEOM was somewhat lower.

Previously reported evaluations of earlier versions of Plantower sensors showed also quite high linear correlations with comparison instruments: $R^2 > 0.57$ for 1-hour data and $R^2 > 0.66$ for 24-hour data from air quality monitor with PMS3003 [30], $R^2 > 0.71$ for 1-hour data and $R^2 > 0.77$ for daily data for instrument with PMS5003 sensor [31], and $R^2 > 0.8$ for 1-hour averages and $R^2 > 0.88$ for 24-hour data, when PMS1003 sensors were used [29].

Nova Fitness SDS011 sensors showed also high linearity to TEOM data. Coefficients of determination for 1 min, 15 min, and 1-hour data were at the level of about 0.8. The R^2 value around 0.9 was obtained for 24-hour averages.

In case of Winsen ZH03A sensors, only unit no. 3 was observed to be stable during the whole measurement period.

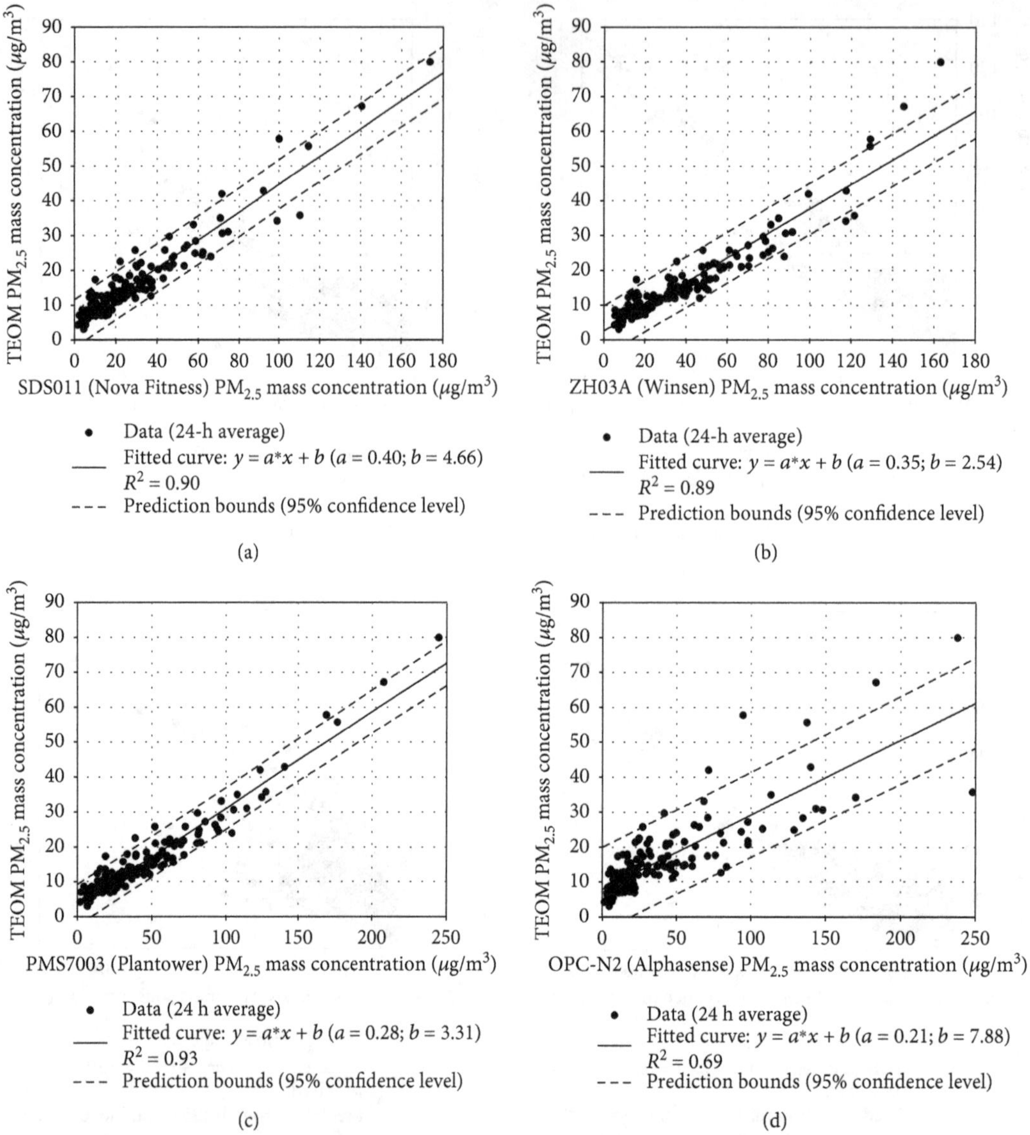

FIGURE 6: Results of linear fittings for 24-hour averaged data: (a) SDS011 (Nova Fitness), (b) ZH03A (Winsen), (c) PMS7003 (Plantower), and (d) OPC-N2 (Alphasense).

R^2 values were equal to 0.74 and 0.77 for 1 min and 15 min mean, respectively. The R^2 level of 0.8 was achieved for 1-hour time scales and nearly 0.9 for 24-hour averages. Unit no. 2 was used only to 24 December 2017, and the fitting results were a little bit lower than in the previous case. Overall, fitting results for ZH03A were satisfactory, but above the level of about 60 $\mu g/m^3$ data significantly deviated from the linear model (see Figure 5(b)).

Alphasense OPC-N2 sensors demonstrated only moderate linear relationship with TEOM. R^2 values reached the level of 0.5–0.7 for 24-hour averages and 0.4–0.6 for shorter time scales. OPC-N2 sensors were characterized with some variability between units, and quality of fittings for sensor no. 1 was slightly better than for other units. The relationship between TEOM and OPC-N2 devices was more similar to power-law function, but the quality of such fittings was also moderate because of large dispersion of data. Nonlinear characteristics of OPC-N2 sensors and moderate correlation with control measurements were also reported in [37].

In case of all PM sensors, characteristic temporary peaks were observed for data averaged in short-time scales. For that reason, it was assumed that sensor outputs should be smoothed by averaging for at least 15 minutes when they are used for ambient air monitoring. This period of time should be sufficient to detect short-term pollution events. Averaging in longer time periods (1 hour, 24 hours) could be also useful, particularly to preserve consistency with data from typical governmental measurement stations. Examples of relationships between TEOM and PM sensors are presented in Figures 5 and 6 for 1-hour and 24-hour averaging, respectively. Only results from units with the highest R^2 values were presented.

Tables S1–S4 in the Supplementary Material present the computed parameters of linear models for all averaging

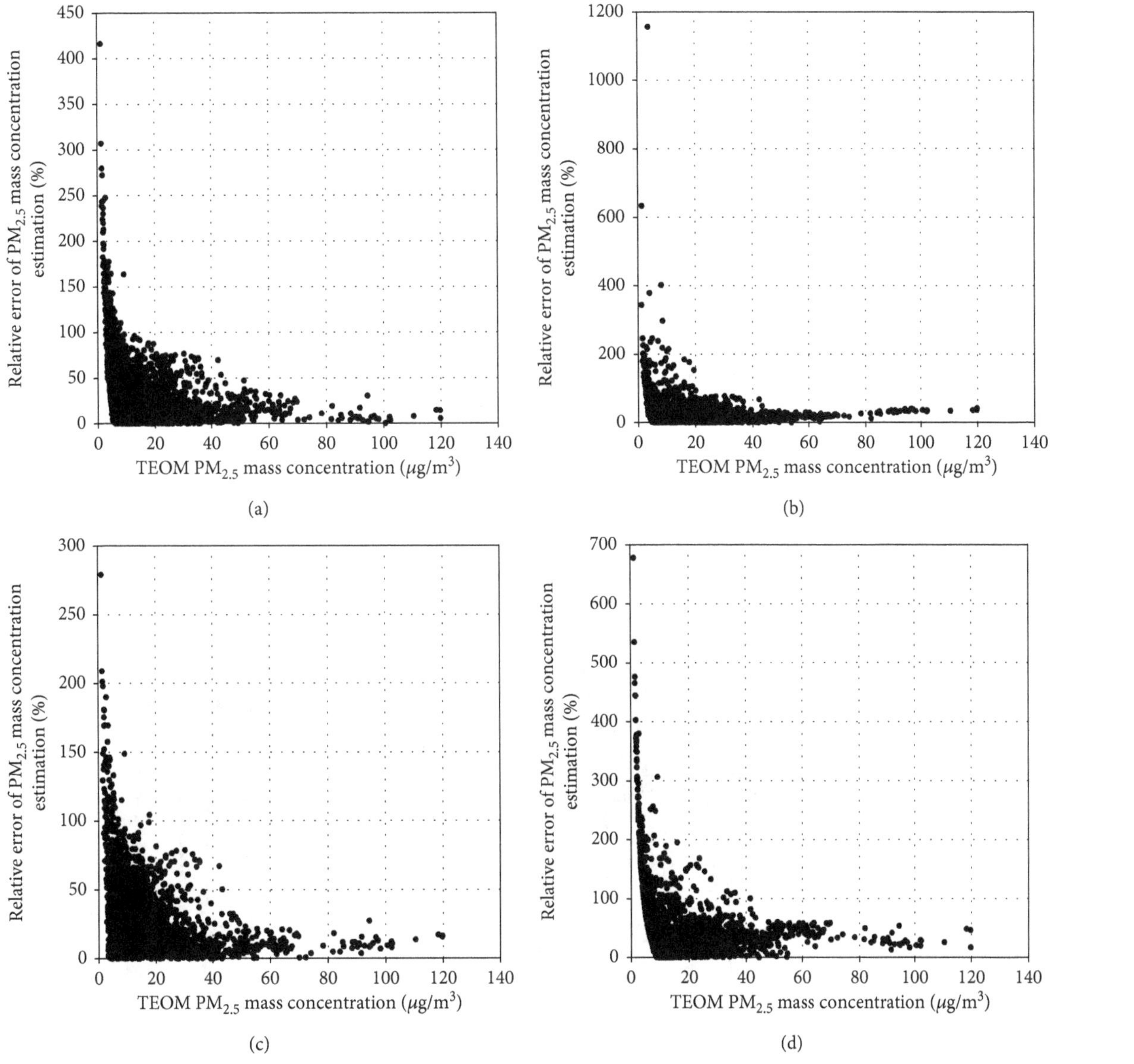

Figure 7: The distribution of relative error of $PM_{2.5}$ estimation over the mass concentrations measured by TEOM in case of (a) SDS011 (Nova Fitness), (b) ZH03A (Winsen), (c) PMS7003 (Plantower), and (d) OPC-N2 (Alphasense). Please note the different scales on the y-axes.

periods. Slopes of regression and intercepts were very similar for all units of sensors with the lowest intramodel variability: PMS7003 and SD011. In case of the ZH03A sensor, some differences in intercepts were observed between units no. 2 and no. 3. Two copies of OPC-N2 (no. 1 and no. 2) were characterized with very similar parameters of regression, but unit no. 3 had value of slope quite distinct from others. Additionally, the slopes for most sensors were the same for short-term averaging (1 min/15 min), hourly averaging, and daily averaging. Only small differences in intercepts were noticed for different averaging times.

4.4. Influence of $PM_{2.5}$ Concentration Range on Sensor Performance. Figure 7 presents the distribution of the relative error of $PM_{2.5}$ estimation by means of calibrated low-cost sensors. 1-hour averaged data for selected units was plotted for clarity. Generally, the biggest relative errors were observed for $PM_{2.5}$ concentrations below 20–30 $\mu g/m^3$—the error values were up to a few hundred percent. Between 20 $\mu g/m^3$ and 60 $\mu g/m^3$, the errors were mostly below 90%–100% (only OPC-N2 showed errors up to 170%). Above the level of 60 $\mu g/m^3$, relative errors were between ~0% and 31% in case of SDS011 (12% on average) and PMS7003 (11% on average). Some higher errors were observed for the stable unit of Winsen ZH03A (3%–41%; 25% on average). The OPC-N2 counter was characterized with relative errors between 13% and 58% (37% on average).

Figure 8 presents the comparisons of R^2 values from regression fittings for adopted ranges of $PM_{2.5}$ concentration, low (≤20 $\mu g/m^3$), medium (20–60 $\mu g/m^3$), and high (>60 $\mu g/m^3$), and also for combined ranges. Only sensors with low intramodel variability and highly

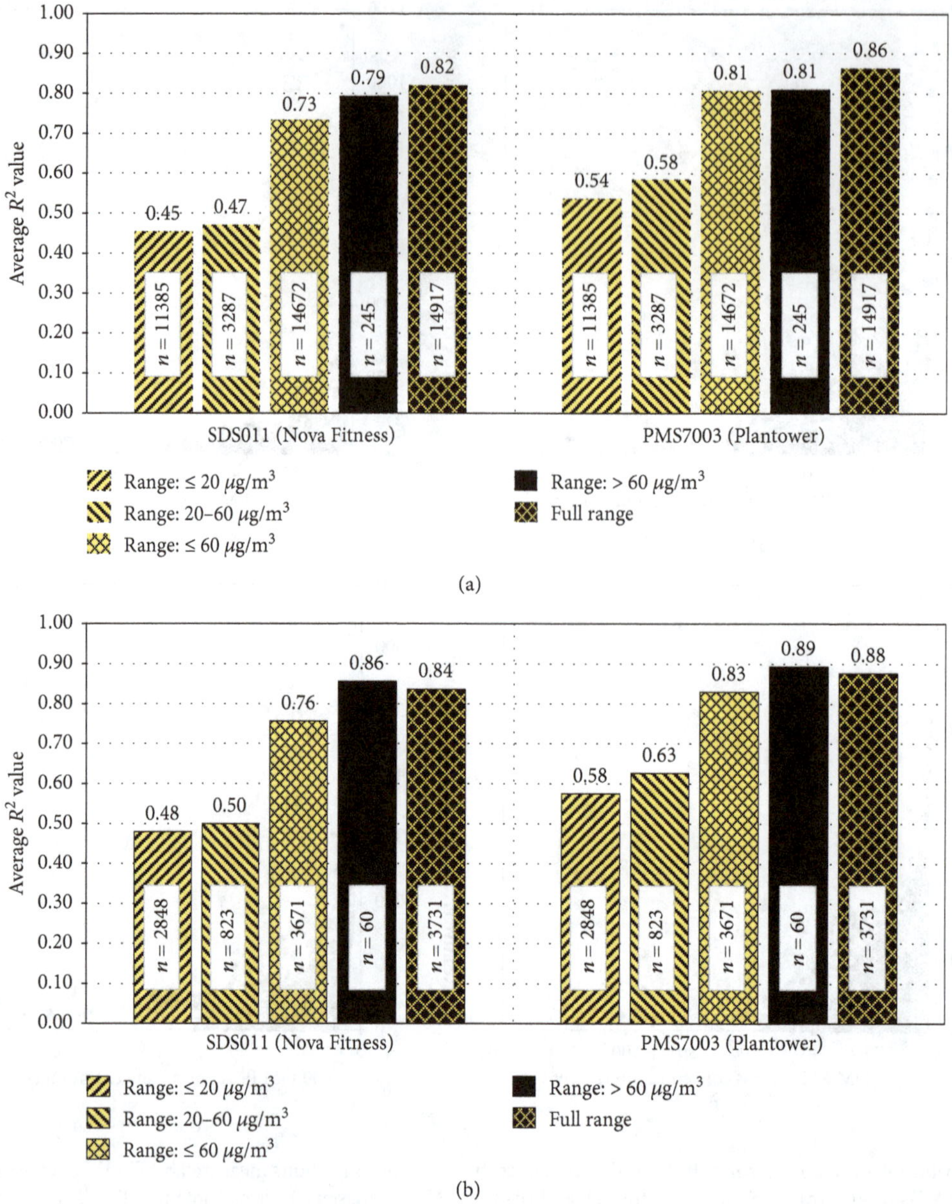

FIGURE 8: Averaged coefficients of determination (R^2) for SDS011 and PMS7003 sensors in different concentration ranges for (a) 15 min and (b) 1-hour averages. *n* in the text box inside the bar indicates the number of samples used for fitting.

correlated to TEOM were taken into that analysis (i.e., SDS011 and PMS7003). Averaged R^2 values from three units are presented for readability.

In case of both tested sensors, irrespective of averaging time, the quality of fittings was only moderate when sensors were measuring low levels of $PM_{2.5}$ (R^2 at the level of 0.4–0.5 for SDS011 and R^2 around 0.5–0.6 for PMS7003). This range of concentrations was characterized with a large dispersion of registered data. Only slightly better results were reached for the range of medium concentrations. However, the increase in R^2 parameter was noticeable when data from low and medium ranges were used together in regression analysis—R^2 reached the level of 0.7 for SDS011 and 0.8 in case of PMS7003. Relatively high R^2 values were also calculated for a range with higher $PM_{2.5}$ concentrations—R^2 around 0.8 for 15 min data and near 0.9 for 1 h averages. This range of data was mainly responsible for the overall quality of sensors fittings, but it should be noted that in the whole study more than 75% of TEOM data was below 20 $\mu g/m^3$ and only 1.5% was bigger than 60 $\mu g/m^3$ (mainly after 01 January 2018). Therefore, further measurements are necessary to confirm the results of the conducted analysis.

The dependency of fitting quality on the studied range of $PM_{2.5}$ concentrations was also discussed in [27, 32]. Poor agreement of low-cost sensors with collocated instruments was observed at low ambient $PM_{2.5}$ concentration levels (<20/40 $\mu g/m^3$), and the improvement of fitting was noticed for wider ranges of concentrations (up to 250 $\mu g/m^3$) [27]. The fitting quality depends also on the precision and accuracy of the control instrument used for comparison. This

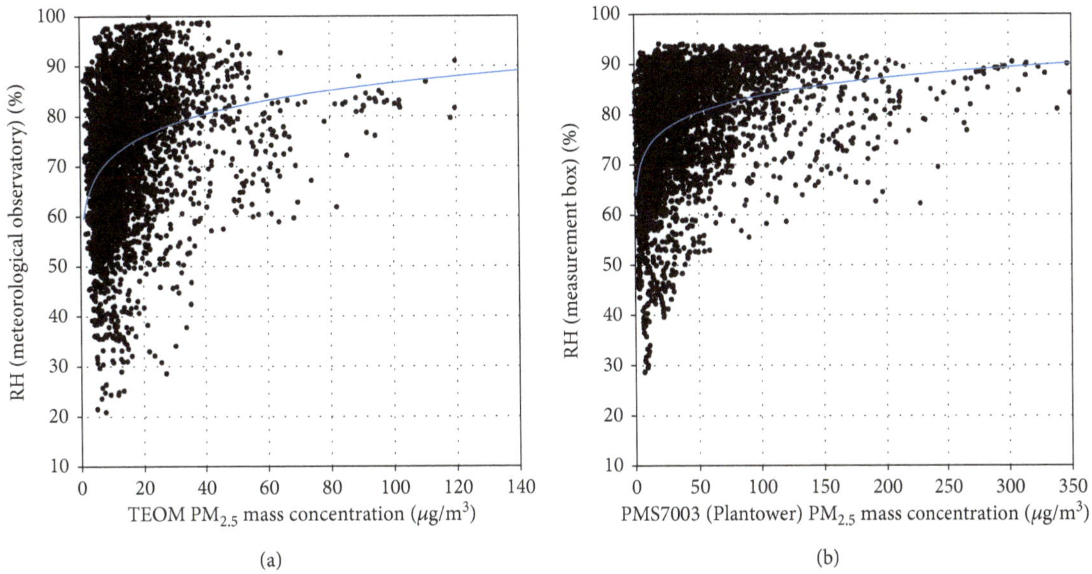

FIGURE 9: Distribution of $PM_{2.5}$ outputs on relative humidity (RH): (a) TEOM 1400a vs. RH at the meteorological observatory and (b) PMS7003 sensor vs. RH in the measurement box. 1-hour averaged data was plotted for clarity.

issue is particularly important in case of low-level concentrations and short-time sampling. In case of this measuring campaign, TEOM might be responsible for some of the inaccuracies. In this type of device, large fluctuations of signal may occasionally occur [42], which can lead to erroneous measurements.

Generally, the results of analysis showed that calibration by collocation is not a simple task. A long period of time might be necessary for acquisition of a sufficiently wide range of particulate matter concentrations. In some circumstances, calibration actions can take weeks or months for ambient air monitoring systems. Measuring properties of the control instrument also affect the results of such comparison.

Regardless of the fact that correlations with the TEOM instrument in the range of low $PM_{2.5}$ concentrations were only modest, the analysis showed that PM sensors could be reasonably good at indicating high levels of $PM_{2.5}$. Thus, they may become a valuable tool to inform the public about increased air pollution events.

4.5. Influence of Air Humidity Level on Sensor Performance. During the measuring campaign, strong diurnal variations of relative humidity and temperature were observed. Generally, lower temperatures and higher RH levels occurred at night. Many elevated $PM_{2.5}$ concentrations happened also at night, and household heating systems could have made a big contribution to that situation. Figure S5 in the Supplementary Material presents examples of such episodes.

The daily distribution of measured $PM_{2.5}$ concentrations (see Figure S6 in the Supplementary Material) showed that the highest values of $PM_{2.5}$ occurred in the mornings (6–9 UTC) and in the evenings (18–20 UTC), when RH values were also usually high (compare with Figure S7). This co-occurrence of elevated RH and $PM_{2.5}$ concentrations may explain the relatively high outputs from low-cost PM sensors at high RH levels. An example of the relationship between RH values and $PM_{2.5}$ data from TEOM and low-cost sensor is presented in Figure 9. Similar relationships were observed for all tested sensors. It should be noted that RH registered inside the measurement box did not reach the highest values registered at the meteorological observatory. This is due to slightly higher temperature inside the box, which results from the heat emitted by electronic circuits during their work.

High levels of RH had, however, an impact on performance of some sensors. In the case of Nova Fitness SDS011, higher dispersion of data between units was observed already above 80%, and beyond 90% differences in data sets were even more visible. Figure S2 in the Supplementary Material presents those differences.

The other aspect of assessment of the humidity impact was based on a comparison of parameters of linear functions for all considered RH ranges. Examples of such comparison are presented in Figure 10. It was noticed that in some cases slopes of linear equations can vary significantly when the RH level is elevated.

For SDS011 (Figure 10(a)), the value of slope changed from a level of 0.47 for RH below 80% to 0.41 for RH range 80%–90% and around 0.37 when RH was higher than 90%. Therefore, the overestimation of outputs was observed when measurements were conducted in very humid air.

A similar situation was even more visible in the case of OPC-N2 (see Figure 10(d)). Slope changed from 0.38, for RH under 80%, to 0.14 when RH was higher than 90%. The impact of high ambient RH was also noticed in [36]—OPC-N2 demonstrated significant positive artefacts during measurements in humid air (>85%).

In the case of ZH03A, changes of slope values were less significant (see Figure 10(b)). The slope was equal around 0.35–0.36 up to 90% RH, and above that level the value of 0.33 was computed. PMS7003 was characterized with stable slope value for all ranges of RH (see Figure 10(c)). For both

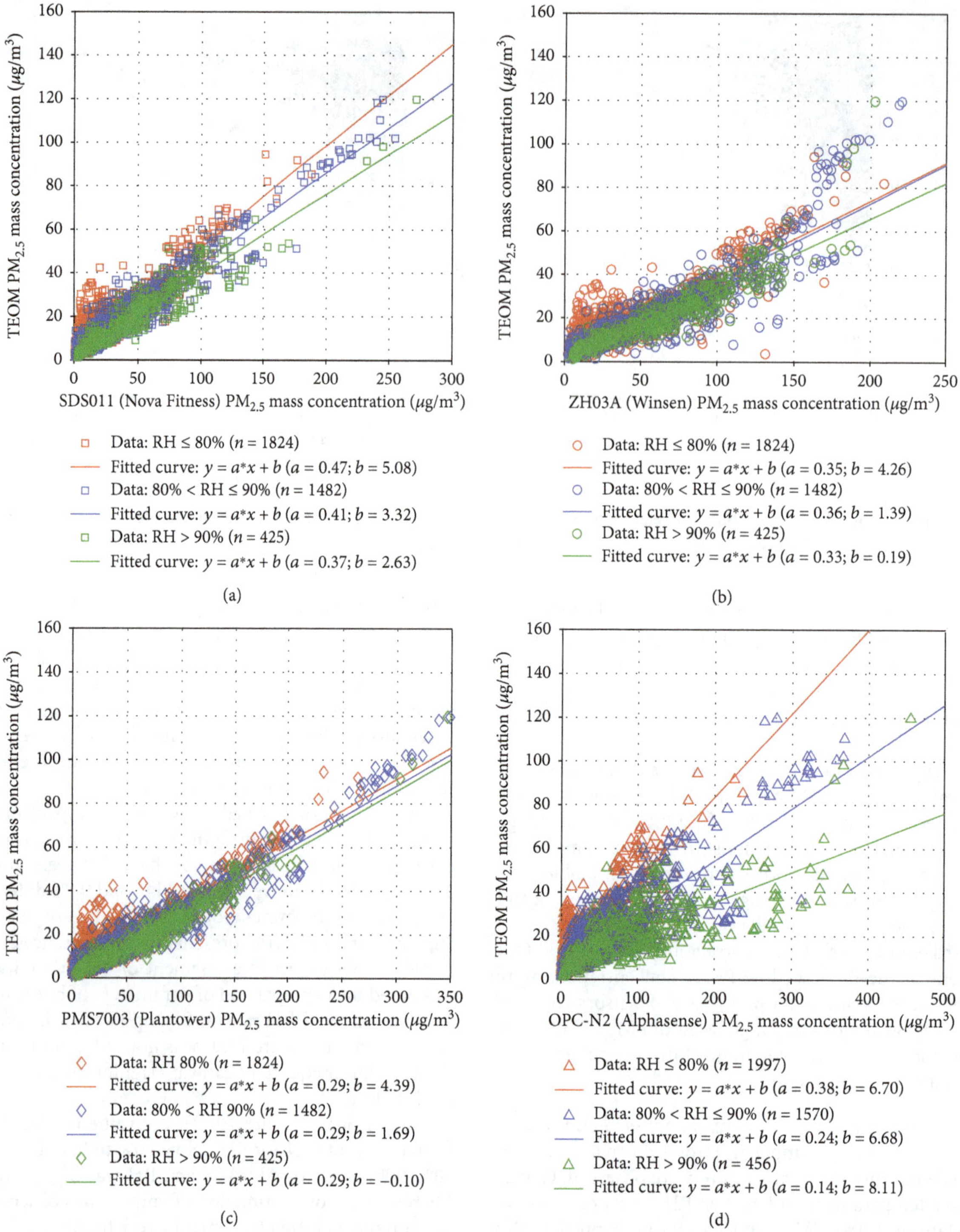

FIGURE 10: Results of linear fittings in different relative humidity (RH) ranges for 1-hour averaged data: (a) SDS011 (Nova Fitness), (b) ZH03A (Winsen), (c) PMS7003 (Plantower), and (d) OPC-N2 (Alphasense). *n* in the brackets indicates the number of samples used for fitting.

sensor models, small alterations of intercepts were observed when the RH level was changing.

To sum up, the analysis showed the necessity of the use of correction factors for high humidity levels in the case of some PM sensors. Such action has been proposed for OPC-N2 in [36]. Similar corrections could be also beneficial for Nova Fitness SDS011 devices.

5. Conclusions

Many models of low-cost PM sensors are available these days on the market. The conducted comparison of four models of PM sensors and TEOM analyser demonstrated that optical PM sensors generally follow the trend of $PM_{2.5}$ changes in atmospheric air. However, some temporary peaks were observed for the raw short-term data in the case of all tested sensors. Data filtering or smoothing for at least 15 minutes may be advantageous in that situation and should not affect the ability to detect short-term pollution events.

An important thing that should be noted is that the raw outputs from the off-the-shelf devices may significantly overestimate the $PM_{2.5}$ concentrations (a factor of 2.5–5 might be observed, as the results of this study showed). It may be due to the fact that sensors are calibrated by the manufacturer using particles with properties completely different than particulate matter in the monitored air. For this reason, calibration (or recalibration) of PM sensors should be made in the final environment of measurements. The most common method of such calibration is based on the use of the data from the collocated higher-class instrument.

As the results of this study showed, that kind of action could be in some ways be complicated. Sensors like OPC-N2 (Alphasense) or ZH03A (Winsen) were characterized with relatively high intramodel variability. On the other hand, SDS011 (Nova Fitness) and PMS7003 (Plantower) were precise in terms of reproducibility between units. This feature could be very useful when a large amount of sensors is to be used, e.g., in dense monitoring networks. Calibration factors from one unit can be used for others in such case.

The other aspect of calibration is related to linearity of sensor responses. OPC-N2 units exposed only a moderate linear relationship with TEOM (the average R^2 value at the level of 0.5–0.6). The results for linear fittings were much better for PMS7003, SDS011, and ZH03A sensors. In particular, PMS7003 sensors reached R^2 higher than 0.8 for short-time averaging times and exceeded 0.9 for daily averages.

The conducted data analysis showed also that a wide range of $PM_{2.5}$ concentrations is desirable to properly describe sensor characteristics. The situation when the particulate's concentration range is narrow could lower the quality of fittings of sensor data and increase the calibration efforts. The duration of collocated calibration is site-specific, but it should last at least a few weeks to acquire a sufficient amount of data.

This research was conducted under different meteorological conditions. PM sensors worked both at temperatures below 0°C and at temperatures above 30°C, and no problems with functioning were noticed. Malfunction was only observed for two copies of ZH03A sensors, and particle accumulation could be responsible for that situation.

During the measurement period, the relative humidity varied from 27% to the level of above 90%. Some impact of high relative humidity (RH > 80%) was observed in the case of SDS011 and OPC-N2 sensors. Larger dispersion of SDS011 outputs was noticed in very humid air, and substantial change of slopes of linear models occurred for both devices. The use of correction factors for high humidity levels should be advantageous for those sensors.

Overall, the study showed that low-cost optical PM sensors could be effective tools for ambient air quality monitoring. The results of this research can help the authors of measuring systems and users of low-cost PM sensors in choosing the appropriate conditions for sensor work and calibration.

Conflicts of Interest

The authors declare that there is no conflict of interest regarding the publication of this paper.

Acknowledgments

This work was cofinanced within the specific subsidy granted for the Faculty of Environmental Engineering, Wroclaw University of Science and Technology, by the Ministry of Science and Higher Education (statutory project No. 0402/0130/17). The work was partially financed by the European Union and cofinanced by NFOSiGW within LIFE+ Program, Project LIFE-APIS/PL-Air Pollution and biometeorological forecast and Information System (LIFE12 ENV/PL/000056).

References

[1] European Environment Agency, "Air quality in Europe – 2017 report," Publications Office of the European Union. EEA Report no. 13/2017, 2017.

[2] J. O. Anderson, J. G. Thundiyil, and A. Stolbach, "Clearing the air: a review of the effects of particulate matter air pollution on human health," *Journal of Medical Toxicology*, vol. 8, no. 2, pp. 166–175, 2012.

[3] K. H. Kim, E. Kabir, and S. Kabir, "A review on the human health impact of airborne particulate matter," *Environment International*, vol. 74, pp. 136–143, 2015.

[4] M. Guarnieri and J. R. Balmes, "Outdoor air pollution and asthma," *Lancet*, vol. 383, no. 9928, pp. 1581–1592, 2014.

[5] F. R. Cassee, M.-E. Héroux, M. E. Gerlofs-Nijland, and F. J. Kelly, "Particulate matter beyond mass: recent health evidence on the role of fractions, chemical constituents and sources of emission," *Inhalation Toxicology*, vol. 25, no. 14, pp. 802–812, 2013.

[6] International Agency for Research on Cancer, "Outdoor air pollution a leading environmental cause of cancer deaths," *Press Release No. 221*, IARC Communications Group, 2013.

[7] European Environment Agency, "European Air Quality Index," May 2018 http://airindex.eea.europa.eu/.

[8] P. Kumar, L. Morawska, C. Martani et al., "The rise of low-cost sensing for managing air pollution in cities," *Environment International*, vol. 75, pp. 199–205, 2015.

[9] E. G. Snyder, T. H. Watkins, P. A. Solomon et al., "The changing paradigm of air pollution monitoring," *Environmental Science & Technology*, vol. 47, no. 20, pp. 11369–11377, 2013.

[10] N. Castell, F. R. Dauge, P. Schneider et al., "Can commercial low-cost sensor platforms contribute to air quality monitoring

and exposure estimates?," *Environment International*, vol. 99, pp. 293–302, 2017.

[11] M. Budde, M. Köpke, and M. Beigl, "Design of a light-scattering particle sensor for citizen science air quality monitoring with smartphones: tradeoffs and experiences," in *Pro Science 3: Conference Proceedings: 2nd International Conference on Atmospheric Dust - DUST 2016*, pp. 13–20, Castellaneta Marina, Italy, 2016.

[12] M. Budde, L. Zhang, and M. Beigl, "Distributed, low-cost particulate matter sensing: scenarios, challenges, approaches," in *Proceedings of the 1st International Conference on Atmospheric Dust*, pp. 230–236, Castellaneta Marina, Italy, June 2014.

[13] M. Gao, J. Cao, and E. Seto, "A distributed network of low-cost continuous reading sensors to measure spatiotemporal variations of PM2.5 in Xi'an, China," *Environmental Pollution*, vol. 199, pp. 56–65, 2015.

[14] N. Zikova, M. Masiol, D. Chalupa, D. Rich, A. Ferro, and P. Hopke, "Estimating hourly concentrations of $PM_{2.5}$ across a metropolitan area using low-cost particle monitors," *Sensors*, vol. 17, no. 8, p. 1922, 2017.

[15] S. Patel, J. Li, A. Pandey, S. Pervez, R. K. Chakrabarty, and P. Biswas, "Spatio-temporal measurement of indoor particulate matter concentrations using a wireless network of low-cost sensors in households using solid fuels," *Environmental Research*, vol. 152, pp. 59–65, 2017.

[16] A. Schmidt-Ott and Z. D. Ristovski, "Measurement of airborne particles," in *Indoor Environment: Airborne Particles and Settled Dust*, pp. 56–81, Wiley-VCH Verlag GmbH & Co. KGaA, 2003.

[17] A. C. Rai, P. Kumar, F. Pilla et al., "End-user perspective of low-cost sensors for outdoor air pollution monitoring," *Science of the Total Environment*, vol. 607-608, pp. 691–705, 2017.

[18] R. Williams, T. Watkins, and R. Long, "Low cost sensor calibration options," in *EM: Air and Waste Management Association's Magazine for Environmental Managers*, pp. 10–15, Air & Waste Management Association, 2014.

[19] EN 12341:2014, *Ambient Air - Standard Gravimetric Measurement Method for the Determination of the PM10 or PM2,5 Mass Concentration of Suspended Particulate Matter*, European Committee for Standardization, 2014.

[20] EN 16450:2017, *Ambient Air - Automated Measuring Systems for the Measurement of the Concentration of Particulate Matter (PM10; PM2,5)*, European Committee for Standardization, 2017.

[21] World Meteorological Organization, "Low-cost sensors for the measurement of atmospheric composition: overview of topic and future applications," in *WMO-No. 1215*, A. C. Lewis, E. Schneidemesser, and R. E. Peltier, Eds., World Meteorological Organization, 2018.

[22] R. McDonald, S. Hu, D. Martuzevicius, S. A. Grinshpun, G. LeMasters, and P. Biswas, "Intensive short term measurements of the ambient aerosol in the greater Cincinnati airshed," *Aerosol Science and Technology*, vol. 38, Supplement 2, pp. 70–79, 2004.

[23] H. Patashnick and E. G. Rupprecht, "Continuous PM-10 measurements using the tapered element oscillating microbalance," *Journal of the Air & Waste Management Association*, vol. 41, no. 8, pp. 1079–1083, 1991.

[24] US Environmental Protection Agency (EPA), *List of Designated Reference and Equivalent Methods*, United States Environmental Protection Agency, 2017.

[25] K. Zhu, J. (. J.). Zhang, and P. J. Lioy, "Evaluation and comparison of continuous fine particulate matter monitors for measurement of ambient aerosols," *Journal of the Air & Waste Management Association*, vol. 57, no. 12, pp. 1499–1506, 2007.

[26] K. K. Johnson, M. H. Bergin, A. G. Russell, and G. S. W. Hagler, "Using low cost sensors to measure ambient particulate matter concentrations and on-road emissions factors," *Atmospheric Measurement Techniques*, pp. 1–22, 2016.

[27] K. K. Johnson, M. H. Bergin, A. G. Russell, and G. S. W. Hagler, "Field test of several low-cost particulate matter sensors in high and low concentration urban environments," *Aerosol and Air Quality Research*, vol. 18, no. 3, pp. 565–578, 2018.

[28] R. Williams, A. Kaufman, T. Hanley, J. Rice, and S. Garvey, *Evaluation of Field-Deployed Low Cost PM Sensors*, U.S. Environmental Protection Agency, Washington, DC, 2014, EPA/600/R-14/464 (NTIS PB 2015–102104).

[29] K. E. Kelly, J. Whitaker, A. Petty et al., "Ambient and laboratory evaluation of a low-cost particulate matter sensor," *Environmental Pollution*, vol. 221, pp. 491–500, 2017.

[30] South Coast Air Quality Management District (SCAQMD), "Field evaluation laser egg PM sensor," May 2018 http://www.aqmd.gov/docs/default-source/aq-spec/field-evaluations/laser-egg—field-evaluation.pdf.

[31] South Coast Air Quality Management District (SCAQMD), "Field evaluation of SainSmart," May 2018 http://www.aqmd.gov/docs/default-source/aq-spec/field-evaluations/sainsmart—field-evaluation.pdf.

[32] T. Zheng, M. H. Bergin, K. K. Johnson et al., "Field evaluation of low-cost particulate matter sensors in high- and low-concentration environments," *Atmospheric Measurement Techniques*, vol. 11, no. 8, pp. 4823–4846, 2018.

[33] R. Jayaratne, X. Liu, P. Thai, M. Dunbabin, and L. Morawska, "The influence of humidity on the performance of a low-cost air particle mass sensor and the effect of atmospheric fog," *Atmospheric Measurement Techniques*, vol. 11, no. 8, pp. 4883–4890, 2018.

[34] K. N. Genikomsakis, N.-F. Galatoulas, P. I. Dallas, L. C. Ibarra, D. Margaritis, and C. S. Ioakimidis, "Development and on-field testing of low-cost portable system for monitoring PM2.5 concentrations," *Sensors*, vol. 18, no. 4, p. 1056, 2018.

[35] S. Sousan, K. Koehler, L. Hallett, and T. M. Peters, "Evaluation of the Alphasense optical particle counter (OPC-N2) and the Grimm portable aerosol spectrometer (PAS-1.108)," *Aerosol Science and Technology*, vol. 50, no. 12, pp. 1352–1365, 2016.

[36] L. R. Crilley, M. Shaw, R. Pound et al., "Evaluation of a low-cost optical particle counter (Alphasense OPC-N2) for ambient air monitoring," *Atmospheric Measurement Techniques*, vol. 11, no. 2, pp. 709–720, 2018.

[37] South Coast Air Quality Management District (SCAQMD), "Field evaluation AlphaSense OPC-N2 sensor," May 2018 http://www.aqmd.gov/docs/default-source/aq-spec/field-evaluations/alphasense-opc-n2—field-evaluation.pdf.

[38] M. Badura, P. Batog, A. Drzeniecka-Osiadacz, and P. Modzel, "Optical particulate matter sensors in $PM_{2.5}$ measurements in atmospheric air," *E3S Web of Conferences*, vol. 44, article 00006, 2018.

[39] S. Sousan, K. Koehler, G. Thomas et al., "Inter-comparison of low-cost sensors for measuring the mass concentration of occupational aerosols," *Aerosol Science and Technology*, vol. 50, no. 5, pp. 462–473, 2016.

[40] S. Sousan, K. Koehler, L. Hallett, and T. M. Peters, "Evaluation of consumer monitors to measure particulate matter," *Journal of Aerosol Science*, vol. 107, pp. 123–133, 2017.

[41] J. D. Wanjura, B. W. Shaw, C. B. Parnell Jr, R. E. Lacey, and S. C. Capareda, "Comparison of continuous monitor (TEOM) and gravimetric sampler particulate matter concentrations," *Transactions of the ASABE*, vol. 51, no. 1, pp. 251–257, 2008.

[42] G. Allen, C. Sioutas, P. Koutrakis, R. Reiss, F. W. Lurmann, and P. T. Roberts, "Evaluation of the TEOM® method for measurement of ambient particulate mass in urban areas," *Journal of the Air & Waste Management Association*, vol. 47, no. 6, pp. 682–689, 1997.

[43] L.-A. Tortajada-Genaro and E. Borrás, "Temperature effect of tapered element oscillating microbalance (TEOM) system measuring semi-volatile organic particulate matter," *Journal of Environmental Monitoring*, vol. 13, no. 4, pp. 1017–1026, 2011.

5

Stainless Steel Microfibers for Strain-Sensing Smart Clay Bricks

Antonella D'Alessandro, Andrea Meoni, and Filippo Ubertini

Department of Civil and Environmental Engineering, Via G. Duranti 93, 06125 Perugia, Italy

Correspondence should be addressed to Antonella D'Alessandro; antonella.dalessandro@unipg.it

Academic Editor: Fanli Meng

Life cycle monitoring of structural health of civil constructions is crucial to guarantee users' safety. An optimal structural health monitoring system allows to automatically detect, locate, and quantify any damage in structural elements, thus anticipating major risks of local or global failures. Critical issues affecting traditional monitoring systems are sensors' placement, hardware durability, and long-term reliability of the measurements. Indeed, sensors' deployment is crucial for an effective investigation of the static and dynamic characteristics of the structural system, whereby durability and long-term stability of sensing systems are necessary for long-term monitoring. A very attractive solution to some of these challenges is developing sensors made of the same, or similar, material of the structure being monitored, allowing a spatially distributed and long-term reliable monitoring system, by the use of self-sensing construction materials. Within this context, the authors have recently proposed new "smart clay bricks" that are strain-sensing clay bricks aimed at embedding intelligent monitoring capabilities within structural masonry buildings. While previous work focused on smart bricks doped with titanium dioxide and using embedded point electrodes, this work proposes an enhanced version of smart bricks based on the addition of conductive micro stainless steel fibers that possess higher electrical conductivity and a more suitable fiber-like aspect ratio for the intended application, as well as plate copper electrodes deployed on top and bottom surfaces of the bricks. The paper thus presents preparation and experimental characterization of the new smart bricks. The influence of different amounts of fibers is investigated, allowing the identification of their optimal content to maximize the gauge factor of the bricks. Both electrical and electromechanical experimental tests were performed. Overall, the presented results demonstrate that the new smart bricks proposed in this paper possess enhanced strain-sensing capabilities and could be effectively utilized as sensors within structural masonry buildings.

1. Introduction

Structural health monitoring (SHM) of buildings is a very attractive technology for enhancing safety of users and occupants [1–3]. Among different construction types, traditional masonry structures, including historic and monumental buildings, entail the greatest monitoring challenges due to the peculiar heterogeneity and complexity of their components, to material nonlinearity and to the possible activation of both local and global failure mechanisms. Traditional sensing systems for SHM are made of materials that are very different from those of the monitored constructions. Furthermore, they are usually applied on the external surface of the elements and have durability drawbacks. The high cost of such sensing systems is also a quite often limiting issue. In the last few years, much scientific attention has been devoted to smart construction materials exhibiting self-sensing capabilities [4–6]. Nevertheless, only a few of those works have been focused on smart materials suitable for structural applications, with the majority of the contributions devoted to cement-based materials and concretes [7–9]. The sensing ability of smart materials is typically obtained by doping traditional matrices with conductive nano- or microfillers. Most of the researches concern carbon-based inclusions, such as carbon nanotubes and carbon fibers [10]. The authors have studied the behavior of cementitious materials with carbon additions with a specific attention to dynamic strain sensing [11] and recently presented a new structural multifunctional

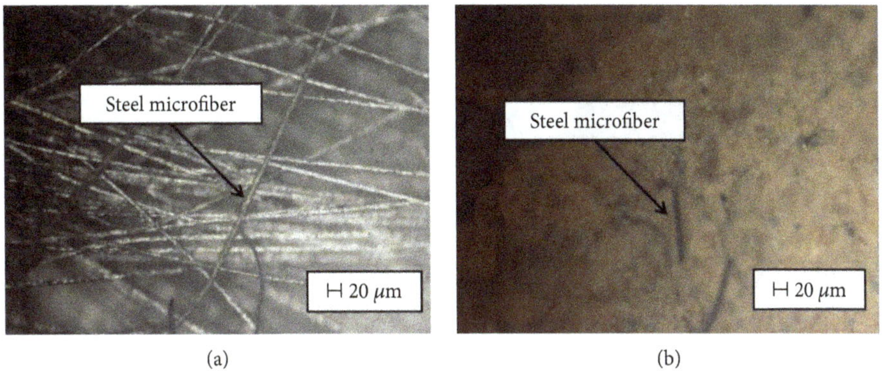

FIGURE 1: Micrographs of (a) stainless steel microfibers used as conductive filler for smart clay bricks and (b) a fragment of burned brick doped with stainless steel microfibers.

clay-based brick termed "smart brick" [12] with piezoresistive properties. Compared to a previous work, with this paper the authors propose a new fabrication methodology for smart bricks using ultrathin stainless steel microfibers (diameter of 12 μ) with a high level of conductivity and thermal resistance as reinforcement. It is also shown that plate electrodes made of copper to be attached on the top and the bottom surfaces of the bricks are more effective than previously adopted embedded wire stainless steel electrodes. After defining the new smart bricks and their fabrication procedure, laboratory tests are described and obtained results are discussed in order to demonstrate the smart bricks' strain-sensing properties.

2. Materials and Preparation of Smart Bricks with Stainless Steel Microfibers

Novel clay bricks were manufactured with different amounts of conductive micro inclusions consisting of stainless steel fibers: from 0 to 1% with steps of 0.25%, 1.5, and 2%, with respect to the total weight of the fresh clay. The fibers, model R.STAT/S, had a diameter of 12 μ and length of 5 mm and were made of a special alloy (AISI 316L with Fe 68%, Cr 18%, Ni 12%, and Mo 2%) keeping low electrical resistivity values, of the order of 60–80 Ω/cm, even after exposure to high temperatures, such as those reached during baking of clay around 1000°C. Figure 1 shows a micrograph of the stainless steel microfibers (Figure 1(a)) and the magnification of a fragment of brick doped with stainless steel microfibers (Figure 1(b)) obtained using an inverted microscope, model Nikon Epiphot 300. The pictures clearly show the dimensional characteristics of the metallic fibers utilized in the experimentation and the acceptable dispersion obtained through the mechanical addition to the clay matrix.

The performance of the fibers at high temperature was investigated through thermogravimetric tests. Small amounts of fibers, kept in closed cups (Figure 2), were heated in an oven up to 900°C, which is the temperature reached by the bricks during their preparation. Four subsequent thermal cycles were performed at increasing maximum temperature, from 300 to 900°C (Figure 2(a)). The fibers did not show any weight loss (Figure 2(b)). Before and after any thermal cycle, the electrical resistance of the fibers was measured using a multimeter instrumented with two clamps: no relevant decrease was observed (Figure 2(c)). The thermogravimetric tests confirmed that the stainless steel fibers were valid candidates as inclusions for conductive clay bricks.

The samples fabricated for the electromechanical tests were prisms with a square base of 5 cm sides and height of 7 cm. The preparation process is sketched in Figure 3. Fibers and fresh clay were mixed for 10 minutes through an electric mixer for consistent mixtures (Figures 3(a) and 3(b)). The doped clay was poured into oiled molds sprinkled with sand (Figures 3(c) and 3(d)). Four stainless steel wires were embedded symmetrically in the central axis of the greater side of the brick (Figure 3(e)). Then, the bricks were dried at a temperature of 90°C for four hours and then burned at 900°C (Figures 3(f) and 3(g)) for six hours. This procedure was the same adopted by the authors for the preparation of bricks with nanotitania described in [12]. Particles of titania had a spherical shape, so they did not possess a prevalent dimension as the stainless steel fibers. Their dispersion was successfully achieved through a mechanical mixer. Each brick sample was instrumented with two 2 cm-long electrical strain gauges, with a gauge factor of 2.1, applied on opposite lateral sides. Figure 3 refers to the case of embedded point stainless steel electrodes. In this paper, also horizontal copper plates are considered. In this case, copper plates of 50 μ thickness were cut one cm larger than two opposite sides of the samples and applied after baking in order to be used as external conductive electrodes (Figure 4).

3. Methodology

Both electrical (with no mechanical loading) and electromechanical (under mechanical loading) tests were carried out. The samples were placed horizontally along the larger dimension. The two-probe method was adopted to measure bricks' electrical properties. Two different configurations were implemented: the first using the embedded stainless steel electrodes placed at a distance of 4 cm and the second with two copper plates externally applied at a distance of 5 cm. Figure 4 shows the sketch of both setups and the photographs of the samples instrumented for electrical and electromechanical tests.

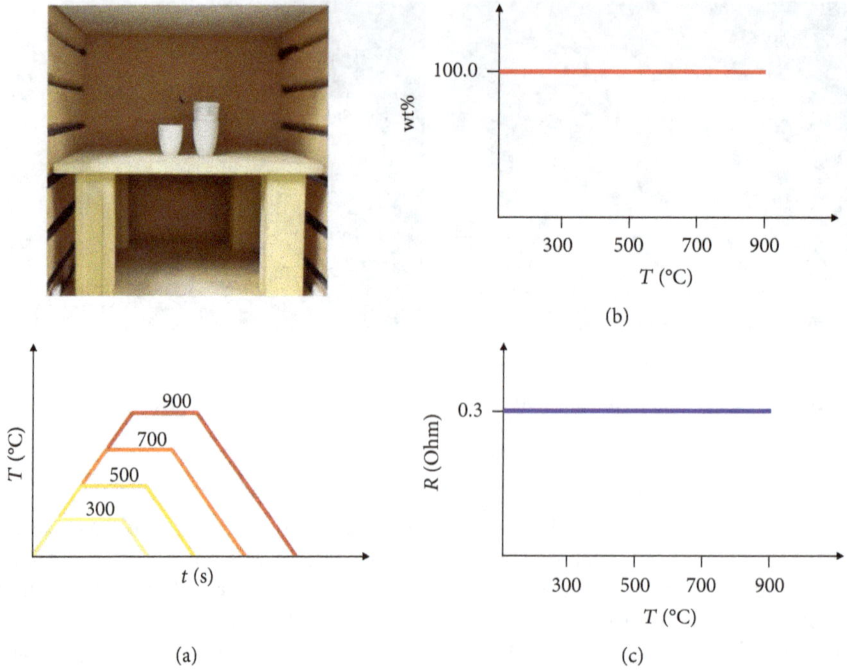

FIGURE 2: Thermogravimetric tests on stainless steel fibers: (a) waveforms of imposed thermal cycles; (b) residual weight after thermal cycles at increasing temperatures; (c) electrical resistance of the fibers after the thermal cycles at increasing temperatures.

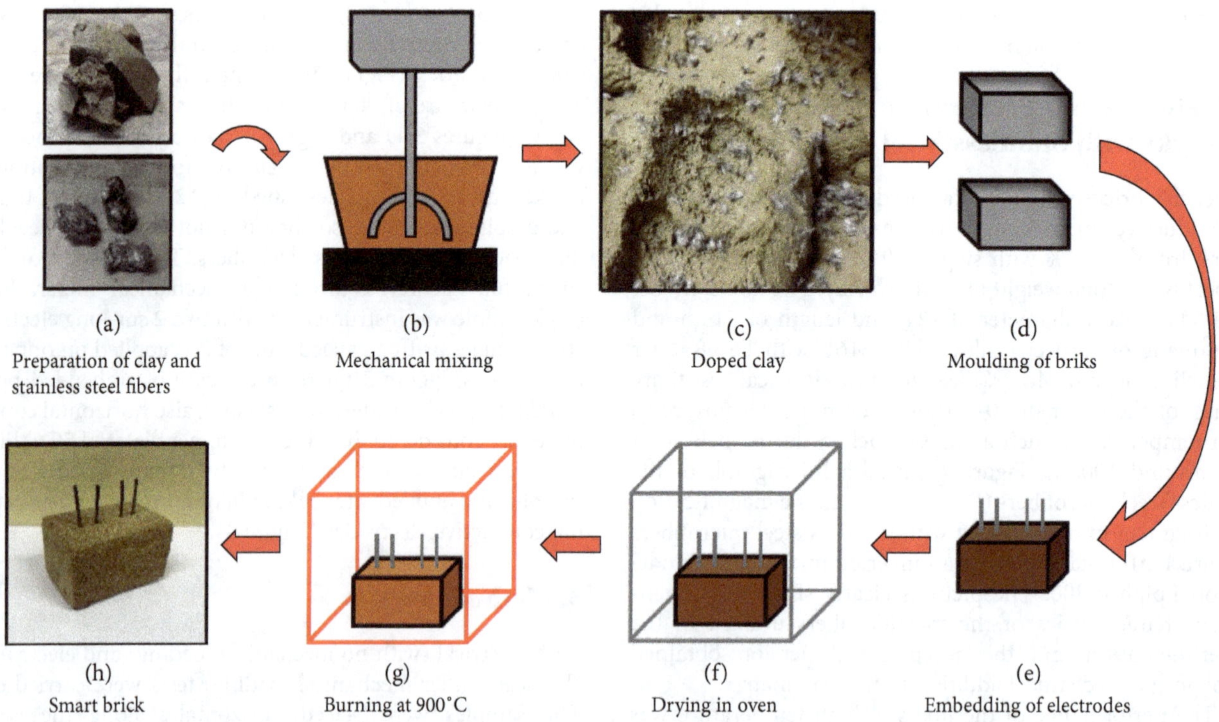

FIGURE 3: Preparation process of the clay brick samples with stainless steel microfibers.

3.1. Electrical Tests. Electrical tests were aimed at investigating the variation of electrical conductivity of the bricks with increasing amount of fillers and the effect of electrodes' configuration on the measurements. For this purpose, both embedded and external configurations of the electrodes were investigated. The tests were also aimed at verifying the repeatability of the measurements. The data acquisition system consisted of an NI PXIe-1073 chassis with a NI PXI4071 digital multimeter. A voltage square wave with an amplitude of ±10 V (20 V peak-to-peak), 50% of duty cycle, and

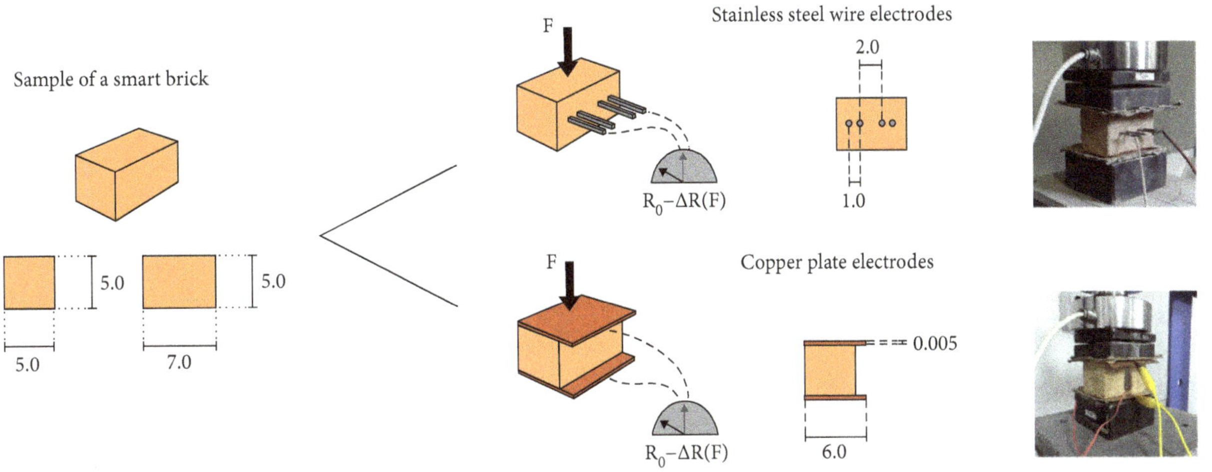

FIGURE 4: Bricks' dimensions and electrodes' configurations for electrical and electromechanical measurements (dimensions in cm).

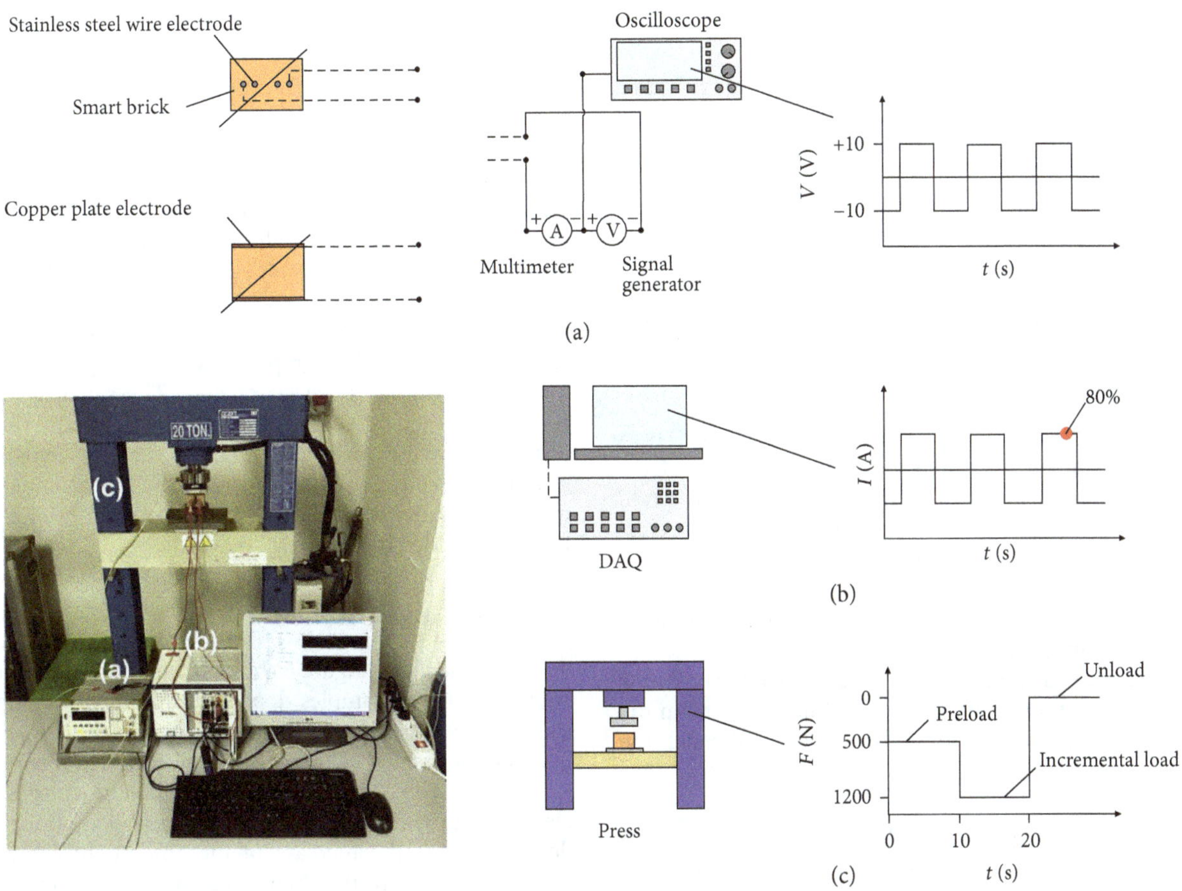

FIGURE 5: Photo evidence of the setup for the electrical and electromechanical tests with key components annotated (a) signal generator of the voltage square wave, (b) data acquisition system of the current (I), (c) applied square load.

frequency of 1 Hz (Figure 5(a)) was applied through a function generator, model Rigol DG1022, in order to produce the cycles of charge and discharge on the samples and eliminate the polarization effect. This effect is typical of dielectric materials, such as clay-based ones, and affects the electric measurements causing a time drift of the signal due to the generation of an electrical field opposite to the applied one. The adopted electrical configuration produces a low noise level, as demonstrated in previous works [13]. The current intensity measurements were taken in the positive portion of the biphasic signal, approximately at 80% of its constant part (Figure 5(b)), following the procedure defined in [13].

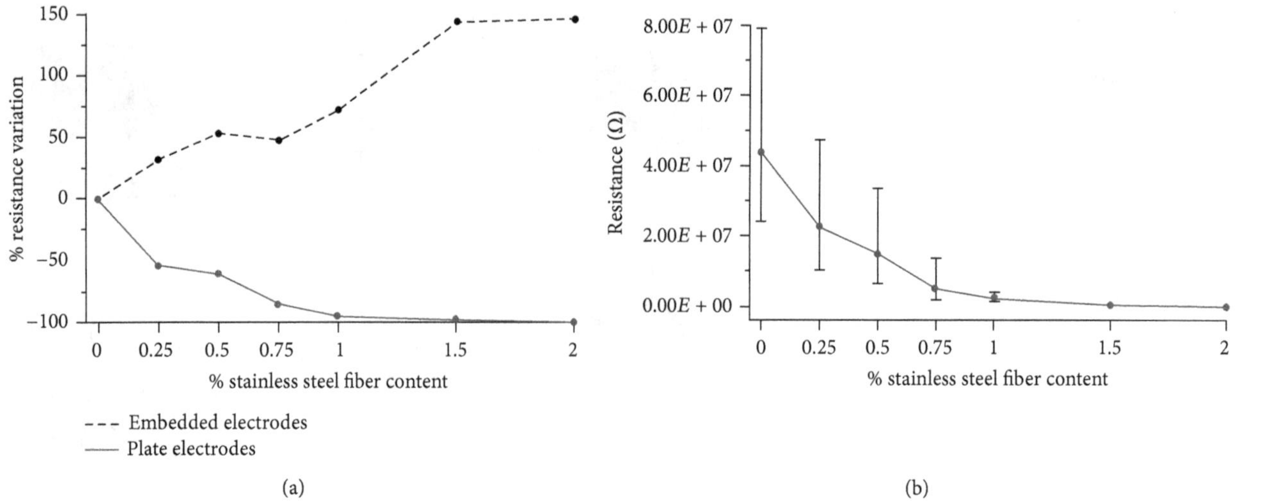

FIGURE 6: (a) Comparison between electrical resistance measured with embedded point electrodes and with copper plate electrodes for different stainless steel fibers contents; (b) average electrical resistance for different stainless steel fiber contents, with error bars representing the maximum and the minimum value of the electrical resistance variation obtained during electrical tests (10 repetitions of measurement).

The sampling rate in samples per second (S/s) was the same of the frequency of the applied voltage square wave.

The electrical resistance R was calculated following Ohm's law:

$$R|_{t=\bar{t}} = \frac{V}{I_{t=\bar{t}}}, \tag{1}$$

where V was the applied constant voltage and I the measured current intensity, $\bar{t}$ being the specific time instant when the sample was taken.

3.2. Electromechanical Tests. The electromechanical tests were carried out for analyzing the strain sensitivity of the bricks and evaluate their monitoring performance as strain sensors. All the samples were subjected to step loads: the electrical resistance was obtained through the electrical measurements between two conductive electrodes (both stainless steel and copper ones), while the actual strain was measured using traditional strain gauges, by averaging their results. The calculation of the gauge factor (GF) of all the samples permits to identify the optimal amount of added fillers to obtain the most performing brick sensor. It was obtained through the following equation:

$$\mathrm{GF} = -\frac{\Delta R/R_0}{\Delta \varepsilon}, \tag{2}$$

where ΔR is the incremental variation in electrical resistance, R_0 the unstrained electrical resistance, and ε the axial strain (positive in compression). In plate electrodes, part of the change in electrical resistance with increasing compression load could be attributable to a change in contact resistance, at low values of compression. The detailed analysis of this aspect goes however beyond the purposes of the present study where a preload is applied to mitigate this phenomenon. The data acquisition system and the voltage applying methodology were the same with the electrical tests. The data acquisition system for the strain gauges was model PXIe-4330, 8 channels, 24-bit resolution, 25 kHz of maximum sampling rate, and antialiasing filters. The step loads were applied using a LAUMAS CL press with a 2000 kg load cell. The load step was between 500 and 1200 N. Both clay bricks with stainless steel fibers and titania particles were investigated, in order to compare their electrical behavior and their sensitivity with different amounts of inclusions.

Figure 5 shows the setup for the electrical and electromechanical tests. The photograph shows the complete measurement setup during an electrical test. The signal generator (Figure 5(a)) provided the voltage square wave, and the data acquisition system recorded the current flowing between the electrodes. Figure 5(b) also illustrates the biphasic measurement procedure, while Figure 5(c) shows the cyclical load history provided by the press.

4. Results and Discussion

4.1. Electrical Tests. Figure 6 shows the outputs of the electrical tests on brick samples doped with different amounts of stainless steel microfibers. Figure 6(a) depicts the comparison between the results obtained using embedded steel wire electrodes and external copper plate electrodes: the external plate electrodes provided an electrical resistance variation which decreased with the increase of the amount of conductive fillers. On the contrary, when using embedded stainless steel electrodes, an unexpected increase in electrical resistance variation with increasing amount of steel microfibers was evidenced. Such effect was probably due to the contact resistance between the wire electrodes and the brick matrix with steel fibers. Indeed, the internal electrodes represented points of discontinuity in the material, and during the burning, such discontinuity originated internal local macro- and microporosity which affected the contact between clay and metallic electrodes. So, the increase of the electrical resistance variation with the increase of the amount of added conductive

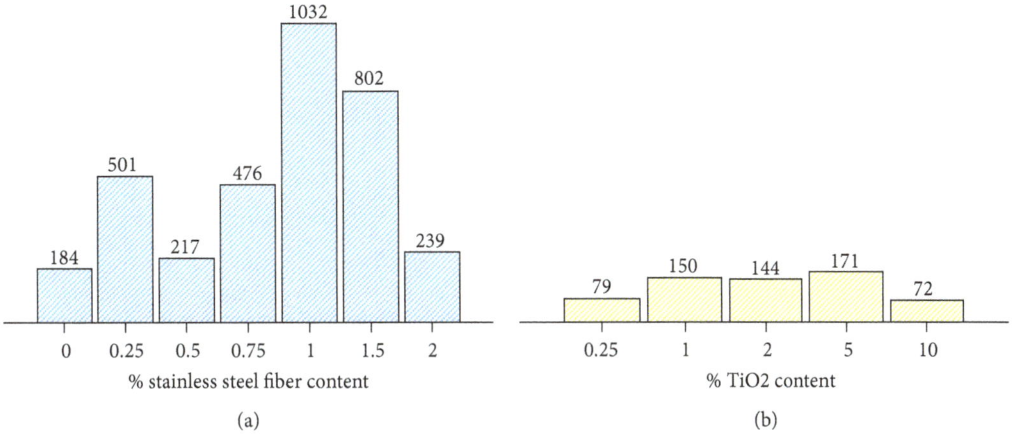

FIGURE 7: Gauge factor values of smart bricks as a function of (a) stainless steel fiber content (with copper plate electrodes) and (b) content of titania (with stainless steel embedded point electrodes).

fillers was probably due to the higher interference effect of the embedded electrodes on the composite with higher presence of solid inclusions. Figure 6(b) shows the average electrical resistance in bricks with increasing amounts of conductive fillers and the observed variability for the different measurements, in the form of error bars. The results show that the presence of the conductive fillers in the clay determines a smaller variability of the results, that is, a higher repeatability of the measurements.

4.2. Electromechanical Tests. The gauge factor was calculated from results obtained with bricks lying in the horizontal configuration, for both typologies of inclusions and by increasing their amounts. Figure 7(a) shows that, in the case of bricks with stainless steel microfibers with copper plate electrodes, the sample with the highest sensitivity is the one with 1% of conductive additions, with a notable GF of 1032. The variability of the measured GF with different amounts of fiber contents is probably due to the possible occurrence of local concentrations of fibers due to the bond among them. For comparative purposes, Figure 7(b) reports the gauge factors of bricks doped with titania and embedded electrodes as used in the previous work [12]: the brick with 5% of filler shows the highest value of the gauge factor. The comparison of such results allows to conclude that the bricks with steel microfibers manifest the highest sensitivity with a much lower amount of filler in comparison to smart bricks doped with titania. This is conceivably due to the peculiar aspect ratio of the inclusions and to the higher electrical conductivity of stainless steel microfibers with respect to titania. Gauge factors of bricks proposed in this work are also seen to be much higher than those of bricks with titania. However, this result cannot be generalized due to the different configurations of electrodes (copper plate electrodes for bricks made with stainless steel microfibers and stainless steel point electrodes for bricks doped with titania and proposed in the previous work). As a final investigation on the strain-sensing properties of smart bricks doped with stainless steel microfibers, Figure 8 depicts the time histories of measured strain and of corresponding normalized variation of electrical resistance, during an electromechanical test on samples with plate electrodes. Particularly, Figure 8(a) refers to a normal brick, while Figure 8(b) refers to the brick with 1% of steel microfibers. The graphs show that the presence of the fillers increases the stiffness of the bricks, producing a lower strain under the same applied load. Furthermore, the sensitivity of the doped brick is notably higher than that of the normal one, and also the quality of the output signal appears to be much improved with the addition of microfibers.

5. Conclusions

This paper presented a new formulation for smart clay bricks, a novel sensing technology recently proposed by the authors for structural health monitoring of masonry structures, based on the use of stainless steel microfibers as conductive fillers and superficial horizontal copper plates (applied postbaking of clay) as electrodes. Traditional clay bricks were doped with different amounts of fillers, from 0% to 2%. Such peculiar fillers possessed a great resistance to high temperature, as demonstrated through the thermogravimetric tests. This characteristic was essential to ensure brick's conductivity and piezoresistivity after heating at 900°C that is reached during their production process. Electrical tests carried out with different electrodes' configurations showed that horizontal plate copper electrodes, placed on the top and bottom external surfaces of the bricks, exhibited more reliable and repeatable results in comparison to those obtained using embedded electrodes. Furthermore, higher amounts of fillers dispersed in the clay matrix produced samples with a better homogeneity. Electromechanical tests were aimed at investigating the sensing capabilities of the doped bricks. A biphasic methodology was used to apply the voltage to the samples, removing the polarization effect which usually affects traditional DC electrical measurements. The sample with 1% of stainless steel microfibers was the most sensitive sample among those with different contents of microfibers or titania. Overall, the presented results demonstrate that the new typology of clay-based smart bricks is promising for structural health monitoring application to

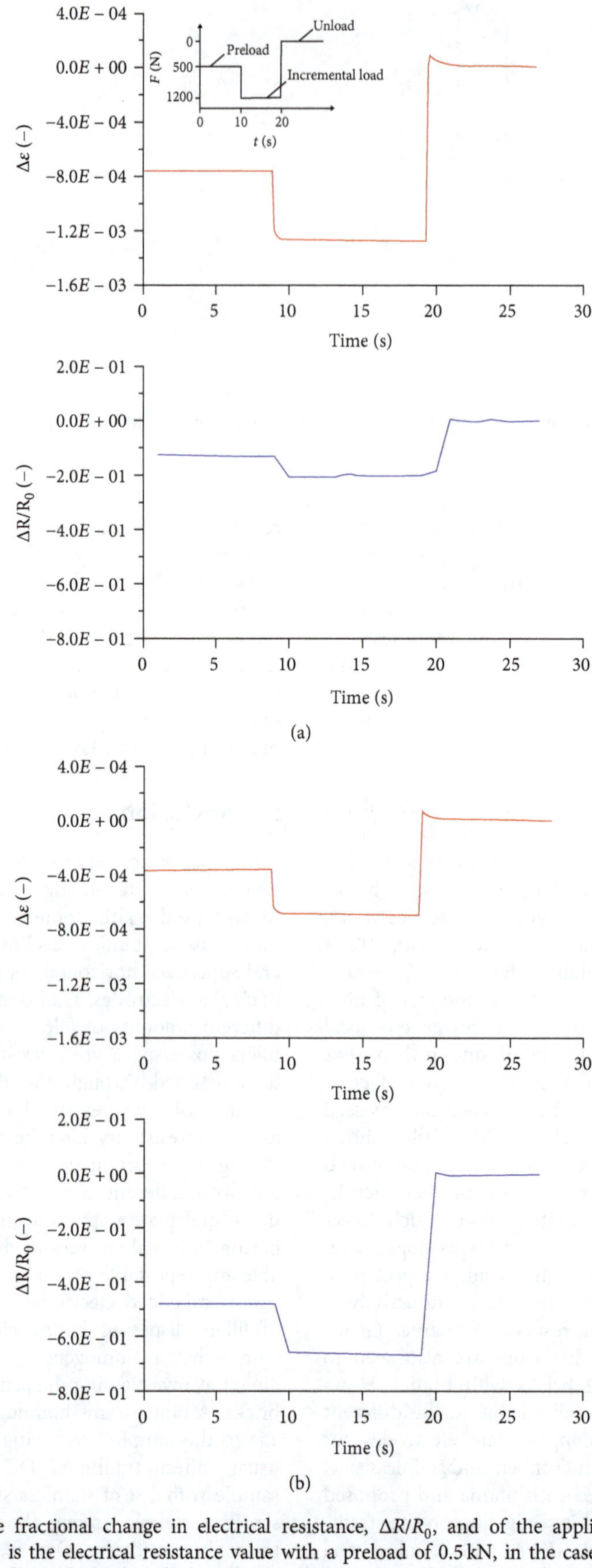

Figure 8: Time histories of the fractional change in electrical resistance, $\Delta R/R_0$, and of the applied strain, $\Delta\varepsilon$, obtained from the electromechanical tests where R_0 is the electrical resistance value with a preload of 0.5 kN, in the case of (a) load applied on clay brick with 0.00% of stainless steel microfibers; (b) load applied on clay brick with 1.00% of stainless steel microfibers.

masonry structures, outperforming the previously proposed smart bricks incorporating titanium dioxide particles and using embedded point electrodes. Their applications may potentially include structural health monitoring of new and existing masonry structures, including historical and monumental ones, which is also made possible by the circumstance that the proposed smart bricks are aesthetically similar to normal ones.

Conflicts of Interest

The authors declare that there is no conflict of interest regarding the publication of this paper.

Acknowledgments

This work was supported by the Italian Ministry of Education, University and Research (MIUR) through the funded project of Relevant National Interest "SMART-BRICK: Novel strain-sensing nanocomposite clay brick enabling self-monitoring masonry structures" (Protocol no. 2015MS5L27).

References

[1] J. M. W. Brownjohn, "Structural health monitoring of civil infrastructure," *Philosophical Transactions of the Royal Society A: Mathematical, Physical and Engineering Sciences*, vol. 365, no. 1851, pp. 589–622, 2007.

[2] S. Laflamme, L. Cao, E. Chatzi, and F. Ubertini, "Damage detection and localization from dense network of strain sensors," *Shock and Vibration*, vol. 2016, Article ID 2562949, 13 pages, 2016.

[3] F. Magalhaes, A. Cunha, and E. Caetano, "Vibration based structural health monitoring of an arch bridge: from automated OMA to damage detection," *Mechanical Systems and Signal Processing*, vol. 28, pp. 212–228, 2012.

[4] B. Han, X. Yu, and J. Ou, "Multifunctional and smart nanotube reinforced cement-based materials," in *Nanotechnology in Civil Infrastructure. A Paradigm shift*, K. Gipalakrishnan, B. Birgisson, P. Taylor, and N. Attoh-Okine, Eds., pp. 1–48, Springer, 2011.

[5] K. J. Loh, T.-C. Hou, J. P. Lynch, and N. A. Kotov, "Carbon nanotube sensing skins for spatial strain and impact damage identification," *Journal of Nondestructive Evaluation*, vol. 28, no. 1, pp. 9–25, 2009.

[6] J. Wu, C. Song, H. S. Saleem, A. Downey, and S. Laflamme, "Network of flexible capacitive strain gauges for the reconstruction of surface strain," *Measurement Science and Technology*, vol. 26, no. 5, article 055103, 2015.

[7] D. D. L. Chung, "Piezoresistive cement-based materials for strain sensing," *Journal of Intelligent Material Systems and Structures*, vol. 13, no. 9, pp. 599–609, 2002.

[8] Y. Ding, Z. Chen, Z. Han, Y. Zhang, and F. Pacheco-Torgal, "Nano-carbon black and carbon fiber as conductive materials for the diagnosing of the damage of concrete beam," *Construction and Building Materials*, vol. 43, pp. 233–241, 2013.

[9] B. Han, S. Ding, and X. Yu, "Intrinsic self-sensing concrete and structures: a review," *Measurement*, vol. 59, pp. 110–128, 2015.

[10] L. Coppola, A. Buoso, and F. Corazza, "Electrical properties of carbon nanotubes cement composites for monitoring stress conditions in concrete structures," *Applied Mechanics and Materials*, vol. 82, pp. 118–123, 2011.

[11] A. D'Alessandro, F. Ubertini, E. García-Macías et al., "Static and dynamic strain monitoring of reinforced concrete components through embedded carbon nanotube cement-based sensors," *Shock and Vibration*, vol. 2017, Article ID 3648403, 11 pages, 2017.

[12] A. Downey, A. D'Alessandro, S. Laflamme, and F. Ubertini, "Smart bricks for strain sensing and crack detection in masonry structures," *Smart Materials and Structures*, vol. 27, no. 1, article 015009, 2018.

[13] A. Downey, A. D'Alessandro, F. Ubertini, S. Laflamme, and R. Geiger, "Biphasic DC measurement approach for enhanced measurement stability and multi-channel sampling of self-sensing multi-functional structural materials doped with carbon-based additives," *Smart Materials and Structures*, vol. 26, no. 6, p. 065008, 2017.

6

Design, Development, and Deployment of Real-Time Sensor Fusion (CnW+EKF) for a Linux-Based Embedded System using Qt-Anywhere

Felipe P. Vista IV and Kil To Chong

Electronic Engineering Department, Chonbuk National University, Jeonju 54896, Republic of Korea

Correspondence should be addressed to Kil To Chong; kitchong@jbnu.ac.kr

Academic Editor: Hyung-Sup Jung

This paper describes the design, development, and implementation of a real-time sensor fusion system that utilizes the classification and weighing plus extended Kalman filter algorithm to derive heading for navigation using inexpensive sensors. This algorithm was previously tested only through postprocessing using MATLAB and is now reprogrammed using Qt and deployed on a Linux-based embedded board for real-time operation. Various data from inexpensive sensors such as global positioning system devices, an electronic compass, and an inertial measurement unit were utilized to ultimately derive a more reliable and accurate heading value. The algorithm flow can be described with the GPS values first being evaluated and classified which are then fused with the EC heading using classification and weighing, whose result is then passed through an EKF to fuse with the IMU data. Real-time tests and trials were done to prove the operational capability of the developed process. The complete setup and configuration processes of the systems for development and deployment via Qt are also provided for those interested to replicate the process.

1. Introduction

The fusion of cheap sensor devices to generate information that would have performance similar to that of more expensive systems is a continuing and exciting research field. Existing works utilized GPS (global positioning system), EC (electronic compass), IMU (inertial measure unit), or a combination of any of them with varying fusion methods utilized such as numerical discretization [1], Kalman filter or its variations [2, 3], fuzzy logic [4], timing synchronization [5], dead reckoning [6], or ad hoc [7].

Sensor fusion is also utilized in various other sectors such as increasing the reliability of quality assessment and authentication of food and beverages [8] as well as a globally updated map of the plant and the dynamic information about velocities and positions of all automatic guided vehicles (AGVs) [9]. One other factory application that utilized sensor fusion is the monitoring of machining operations dependent on rotary cutters [10].

It is also used for the real-time recognition of human action that relied on diverse modality sensors (inertial and depth of vision) [11] while another work presented a technique for indoor position tracking and localization of pedestrians [12]. There is also the often-researched application of sensor data fusion for the pose estimation of a 3D mobile robot in indoor applications [13]. There is also research that looked on the performance of sensor fusion systems [14]; on the other hand, there was another study that designed and developed an open-source tool that can be used for evaluating data fusion systems which are primarily focused on maritime surveillance systems [15].

In the field of maritime and ships, there are systems designed to assist the ship captain in entering or leaving a harbor [16] including the Advanced Sensor Module of the MUNIN project for autonomous and unmanned shipping [17]. There are also related works on collision avoidance such as those of Flåten and Brekke [18] and Chen et al. [19]. Works similar to ours that used sensor fusion in deriving a

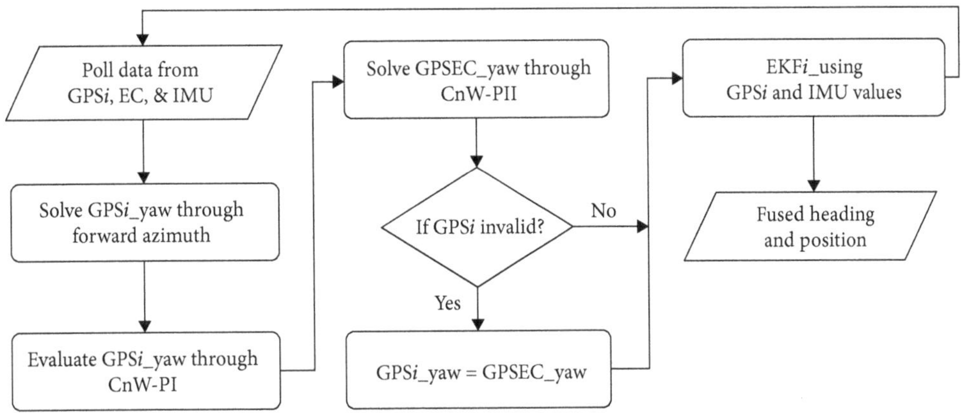

Figure 1: Simplified algorithm.

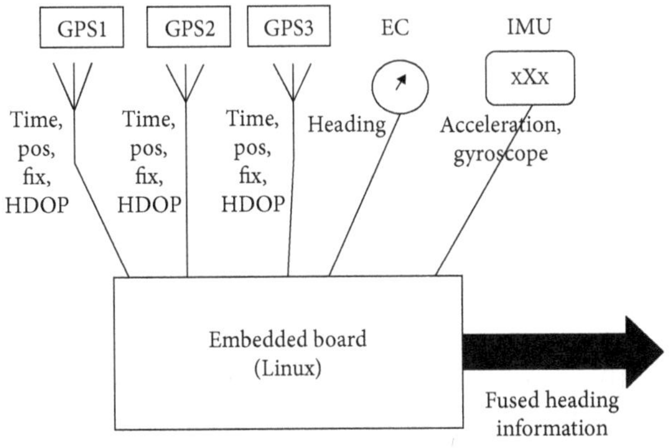

Figure 2: System design.

more accurate heading for navigation were by Hu and Huang [20] and Juang and Lin [21] while there were others that focused on the improved position and attitude aside from the heading such as those by Jaroś et al. [22], Bryne [23], Núñez et al. [24], and Feng-de et al. [25].

Cappello et al. applied sensor fusion to unmanned vehicles such as their work on integrated navigation and guidance systems (NGS) for small-sized UVs using low-cost off-the-shelf sensors [26]. They also worked on NGS for small- and medium-sized Remotely Piloted Aircraft Systems (RPAS) utilizing the GNSS- (global navigation satellite system-) and microelectromechanical system- (MEMS-) based IMU and vision-based navigation (VBN) sensor and the Aircraft Dynamics Model (ADM) that was treated as a virtual sensor [27, 28]. Other research works were implemented in an underwater environment that focused on robot pose estimation [29], towed array shape estimation [30], and passive target tracking [31].

These sensor fusion systems, especially the navigation systems, are usually deployed on embedded systems similar to ones that utilized field-programmable gate arrays (FPGA) with digital signal processors (DSP) [32]. The work presented in this paper details the development and deployment of a real-time sensor fusion system on an embedded board based on a previously proposed system of ours that was just done postprocessing [33]. Previously related works or our works are the real-time fusion of several GPS devices with an electronic compass initially on a notebook through classification and simplified weighing which was then followed by the first phase of real-time sensor fusion of three GPS devices and an electronic compass on an embedded board. There were also preceding studies done on postprocessing sensor fusion such as fuzzy logic and simplified classification and weighing. The current system was programmed using Qt-anywhere on a Linux desktop system and then deployed on an embedded board with the Linux system. This paper is arranged with the theoretical background of the implemented sensor fusion algorithm given in Section 2. The system design and implementation are then detailed in Section 3 while the developed system is presented in Section 4 followed by the concluding remarks.

2. Theoretical Background

The simplified algorithm for the real-time implementation of the previous theoretical work done via postprocessing is given in Figure 1. The currently implemented system on an embedded board utilized forward azimuth (FAz), classification and weighing (CnW), and extended Kalman filter (EKF) in order to derive the fused heading value from multiple inexpensive GPS, an EC, and an IMU.

The overall system design is in Figure 2 showing the inputs from various devices and the corresponding data that are utilized in the process with a 64-bit Ubuntu Linux desktop system acting as the platform for the design and development process of the real-time sensor fusion system. The developed program was then deployed on an embedded board with 32-bit Ubuntu Linux system (both BeagleBone Black and FreeScale). The graphical representation of the operational steps is shown in Figure 3.

2.1. Proposed Method. The GPS and EC both were sampled at 10 Hz while the IMU was sampled at 100 Mz. The data polled from these sensors were utilized for calculation of the heading through FAz (GPS data), classification and weighing-I (FAz data), classification and weighing-II (Faz, EC, and CnW-I data), and extended Kalman filter (FAz, CnW-I, and CnW-II data). The EKF prediction and update process involves a set of data each from GPS (FAz, GPSEC_yaw)

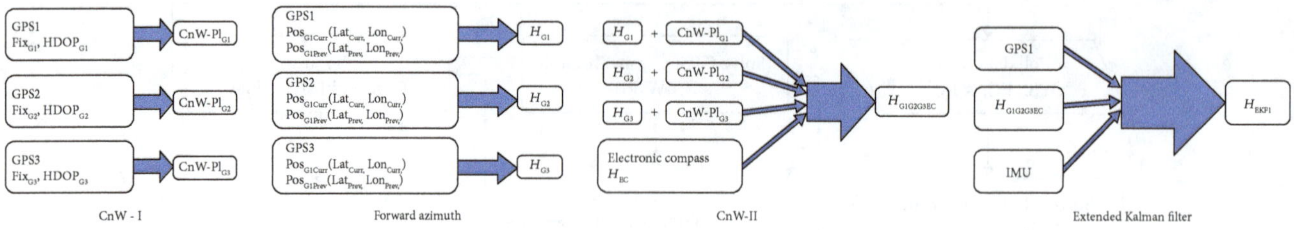

FIGURE 3: Graphical representation of the operational steps.

and IMU. The steps of the implemented algorithm are given as follows.

(1) Solve for individual heading values of the GPS devices (GPS*i*_yaw) through FAz, if there are valid GPS position values.

(2) Evaluate the derived individual GPS heading values through CnW-I.

(3) Fuse the solved individual GPS heading values with the EC heading value through CnW-II to derive GPSEC_yaw.

(4) Assign GPSEC_yaw as the heading value for any GPS device that does not have a valid position value (GPS*i*_yaw = GPSEC_yaw).

(5) Fuse the GPS and IMU values through EKF with the IMU accelerometer value treated as the local gravity vector measurement [9]. The inputs are lat_i, $long_i$, and GPS*i*_yaw while psi_i_EKF, lat_i_EKF, and $long_i$_EKF are the output values.

(6) Repeat steps 1 to 5 for the next set of IMU, EC, and GPS values.

2.2. Heading Derivation. Forward azimuth is described by the US Army as the angular measurement in clockwise direction of the line created by two points with the north set as the reference [34]. The equation for calculating the heading through forward azimuth is given in (1), taking note that the initial (lat_2, $long_2$) and succeeding (lat_2, $long_2$) position values are set so as to derive the value for initial heading. Other heading derivations were also tried such as centroid and direct arc tangent, but their performance was not as good as forward azimuth.

$$\begin{aligned} \arg_i &= \sin \Delta\mathrm{long}^* \cos \mathrm{lat}_2, \\ \arg_2 &= \cos \mathrm{lat}_1{}^* \sin \mathrm{lat}_2 - \sin \mathrm{lat}_1 \cos \mathrm{lat}_2 \Delta\mathrm{long}, \\ \psi &= \tan^{-1}(\arg_1, \arg_2) + 180. \end{aligned} \tag{1}$$

2.3. Classification and Weighing. The classification and weighing process evaluates the quality of the polled GPS data by checking the values of the FIX type and the HDOP (Horizontal Dilution of Precision). The descriptive classification for the type of FIX that was adapted from http://www.gpsinformation.org/dale/nmea.html is given as follows:

(i) Invalid value is "0."

(ii) Standard Positioning Service or GPS mode is "1."

(iii) Differential GPS mode is "2."

(iv) Precise Positioning System is "3."

(v) Real-time kinematics is "4."

(vi) Float real-time kinematics is "5."

(vii) Dead reckoning or estimated fix is "6."

(viii) Manual input mode is "7."

(ix) Simulation mode is "8."

In the case of HDOP values, we utilized our own classification as follows:

(i) greater than 0 and less than or equal to 1 ($0 < \mathrm{HDOP} \le 1$), it is "IDEAL."

(ii) greater than 1 and less than or equal to 2 ($1 < \mathrm{HDOP} \le 2$), it is "EXCELLENT."

(iii) greater than 2 and less than or equal to 5 ($2 < \mathrm{HDOP} \le 5$), it is "GOOD."

(iv) greater than 5 and less than or equal to 10 ($5 < \mathrm{HDOP} \le 10$), it is "MODERATE."

(v) greater than 10 and less than or equal to 20 ($10 < \mathrm{HDOP} \le 20$), it is "FAIR."

(vi) greater than 20 and less than or equal to 1 ($20 < \mathrm{HDOP} \le 1$), it is "POOR."

The combined FIX and HDOP classification is now given as follows:

(i) If FIX is 2 to 5 and HDOP is 1 to 2, then it is IDEAL, with weight of "3."

(ii) If FIX is 1 and HDOP is 1 to 2, then it is EXCELLENT, with weight of "2."

(iii) If FIX is 2 to 5 and HDOP is 3 to 5, then it is EXCELLENT, with weight of "2."

(iv) If FIX is 1 and HDOP is 3 to 5, then it is GOOD, with weight of "1."

(v) If FIX is 2 to 5 and HDOP is 6 to 10, then it is GOOD, with weight of "1."

(vi) Else, it is BAD, with weight of "0."

The resulting weights are then used as inputs along with the GPS and EC heading values into (2) in order to solve for the CnW-II heading.

$$\begin{aligned} h_{\text{fused}} &= (h_{\text{allGPS}})(w_{\text{allGPS}}) + (h_{\text{EC}})(w_{\text{EC}}), \\ h_{\text{allGPS}} &= \frac{\sum_1^n (w_i h_i)}{\sum_1^n w_i}, \\ w_{\text{allGPS}} &= \frac{\sum_1^n w_i}{n^* \text{idealValue}}, \\ w_{\text{EC}} &= (1 - w_{\text{allGPS}}), \end{aligned} \quad (2)$$

where n is the number of GPS devices, idealValue is the weight value given for IDEAL, h_{fused} is the fused heading from electronic compass and GPS, w_{allGPS} is the weight assigned for the calculated GPS heading, w_{EC} is the weight assigned to the electronic compass, h_{allGPS} is the fused calculated heading for all the "n" GPS, w_i is the weight value given to the "ith" GPS, h_i is the derived heading of the "ith" GPS, and h_{fused} is assigned as the "GPSEC_yaw."

2.4. Extended Kalman Filter. The extended Kalman filter is utilized to derive the state estimate through the following general steps: (i) predicting the state and error covariance, (ii) deriving the Kalman gain, (iii) finding the time update of estimate, and then (iv) solving for the time update of error covariance. The observability and controllability of the Kalman filter or extended Kalman filter have been extensively studied and proved as can be read in the works of Trzuskowsky et al. [16], Bustamante et al. [15], and Simonetti et al. [14].

The nonlinear measurement and dynamic models are given as follows:

$$\begin{aligned} x_{k+1} &= f(x_k) + w_k, \\ z_k &= h(x_k) + v_k, \end{aligned} \quad (3)$$

where x_k and z_k are the state estimate and measurement, respectively. $\mathbf{R}_k$ and $\mathbf{Q}_k$ are the measurement noise and process noise covariance matrix and are assumed as positive definite. It is the linearity of the system which is why EKF was chosen for this work. The predicted state $\hat{x}_{k+1}^-$ and error covariance of the predicted state P_{k+1}^- are derived as follows:

$$\begin{aligned} \hat{x}_{k+1}^- &= f(\hat{x}_k^+), \\ P_{k+1}^- &= \mathbf{F}_k \mathbf{P}_k F_k^T + \mathbf{Q}_k, \end{aligned} \quad (4)$$

where F_k is the nonlinear dynamic model Jacobian matrix and P_k is the updated covariance matrix of the state for the previous time step k. The predicted measurement $\hat{z}_{k+1}^-$ is calculated as follows:

$$\hat{z}_{k+1}^- = h(\hat{x}_{k+1}^-). \quad (5)$$

The innovation covariance P_{k+1}^{vv} of the residual error between predicted and observed measurements is derived through

$$P_{k+1}^{vv} = P_{k+1}^{yy} + R_{k+1} = H_{k+1} P_{k+1}^- H_{k+1}^T + R_{k+1}, \quad (6)$$

where $P_{k+1}^{yy} = H_{k+1} P_{k+1}^- H_{k+1}^T$ is the output covariance, H_{k+1} is the Jacobian of the measurement function $h(\hat{x}_{k+1}^-)$ evaluated about the state prediction $\hat{x}_{k+1}^-$, and R_{k+1} is the measurement noise covariance of the sensor at time $k + 1$.

The Kalman gain is now derived through

$$K_{k+1} = P_{k+1}^{xy} (P_{k+1}^{vv})^{-1} = P_{k+1}^- H_{k+1}^T (P_{k+1}^{vv})^{-1}. \quad (7)$$

The Kalman gain is computed by deriving the product of the predicted cross-correlation matrix with inverse of the innovation covariance matrix. The state distribution in the EKF algorithm is approximated by a Gaussian random variable that is propagated by first-order linearization of nonlinear functions.

The state vector $\mathbf{x}_k$ of the model is composed of the latitude, longitude, and heading as elements, $\mathbf{x}_k = [x_N, y_E, \psi]^T$, which is described as

$$\begin{aligned} \dot{x} = \begin{Bmatrix} \dot{x}_N \\ \dot{y}_E \\ \dot{\psi} \end{Bmatrix} &= \begin{bmatrix} \cos\psi & -\sin\psi & 0 \\ \sin\psi & \cos\psi & 0 \\ 0 & 0 & 1 \end{bmatrix} \begin{Bmatrix} u \\ v \\ r \end{Bmatrix} + w \\ &= \begin{bmatrix} u\cos\psi - v\sin\psi \\ u\sin\psi + v\cos\psi \\ r \end{bmatrix} + w = f(x) + w. \end{aligned} \quad (8)$$

On the other hand, the model for measurement $h(x_k)$ is given as

$$y_k = \begin{Bmatrix} x_{N,k} \\ y_{E,k} \\ \psi_k \end{Bmatrix} = \begin{bmatrix} 1 & 0 & 0 \\ 0 & 1 & 0 \\ 0 & 0 & 1 \end{bmatrix} \begin{pmatrix} x_{N,k} \\ y_{E,k} \\ \psi_k \end{pmatrix} = H_k x_k + v_k. \quad (9)$$

The measured values for estimating the state (latitude, longitude, and heading angle) are from three individual GPS devices and an EC. Process noise covariance (Q) and measurement noise covariance (R) values are set depending on the measured values as follows:

$$\begin{aligned} Q &= \begin{bmatrix} 0.01 & 0 & 0 \\ 0 & 0.01 & 0 \\ 0 & 0 & 10 \end{bmatrix}, \\ R_k &= \begin{bmatrix} 0.3 & 0 & 0 \\ 0 & 0.07 & 0 \\ 0 & 0 & 0.03 \end{bmatrix}. \end{aligned} \quad (10)$$

3. Design and Implementation

The system was designed and programmed using RBCDWBPA (Rapid By-Customer Demand with Business Process Approach) in tandem with the SFA (Systems

(a) GPS (b) EC (c) IMU

(d) FreeScale (e) BBB (f) LCD with cape

FIGURE 4: Devices utilized.

Features Analysis) development method on a 64-bit desktop computer with the Ubuntu Linux operating system using the Qt-anywhere 4.8-5 open-source version. It was initially developed on the desktop Linux box and then cross-compiled on the same Linux box for deployment to an ARM-based embedded board. It is not possible to directly connect the EC we are using to the serial port of both the Linux box and the embedded board due to the high-power output of their serial ports which would overload the serial port of the board. The EC therefore was connected to a simple Arduino-based system that was in turn polled by the Linux box or the embedded board.

3.1. Data and Devices. Three UIGGUB02-R001 from u-blox were utilized as inexpensive GPS receivers and were polled using the NMEA protocol. GPS data utilized in the sensor fusion process are timestamp (time when fix was taken), latitude, longitude, and COG (course over ground) which are all extracted from the RMC sentence (Recommended Minimum). On the other hand, the HDOP (Horizontal Dilution of Precision) and FIX (type of fix) are extracted from the GGA sentence which is the source for essential fix information. The EC used in this work was a Devantech Ltd. SEN0183 CMPS11 tilt-compensated model while the IMU was an EBIMU-9DOFV3 heading and attitude referencing system. Two embedded boards were utilized to demonstrate the generic design of the real-time sensor fusion system: namely, a FreeScale i-MX6 and a BeagleBone Black (BBB). A touchscreen with cape for a BeagleBone Black from Waveshare was used to serve as display while the FreeScale was connected to a HDMI-compatible display device. All the devices used in the implementation of the real-time sensor fusion system are shown in Figure 4.

3.2. Setting Up the System for Programming and Deployment. There are several steps that need to be done in setting up and configuring the desktop system for designing and programming the sensor fusion system as well as for cross-compiling and then deploying it on an embedded board such as the BBB. These important configurations and settings are broken down into four major substeps as follows:

(i) Configure TSLIB for Qt-desktop and Qt-arm on freshly installed Ubuntu 15.04-LTS.

(ii) Configure Qt-anywhere for the Qt-arm setup (cross-compile).

(iii) Configure fresh BBB for deployment.

(iv) Configure Qt-Creator for writing and deploying the program (both desktop and cross-compile versions).

The Qt utilized in this work is the Qt-4.8.5-anywhere open-source version while the toolchain used for compiling the fusion algorithm into an ARM-compatible system is the gcc-linaro-arm-linux-gnueabihf. TSLIB was also utilized to enable the touchscreen mode of the LCD display connected to a BBB.

A detailed step-by-step procedure for the given configuration settings is given in the Appendix. These steps are the result of various experimentations and tests that were done in order to be able to easily replicate the system configuration and settings needed to design, develop, and deploy a system onto an embedded system.

3.3. TinyEKF Library. The TinyEKF library by Simon D. Levy, which is available from GitHub (https://github.com/simondlevy/TinyEKF), was chosen to implement the EKF on an embedded board system. It is a straightforward C/C++ execution of the EKF that is generally enough for the implementation on STM32, Arduinos, and other microcontrollers. It is very suitable for utilization especially due to its utilization of static or compile-time memory allocation instead of "`malloc`" or "`new`."

There were customizations done on the TinyEKF library for it to be implemented with our sensor fusion system. The first thing was to make sure that the library would be recognized and compiled on a Qt environment

(a) SUBSYSTEM=="tty" value

(b) KERNELS=="1-4.1" value

FIGURE 5: A part of the result for "udevadm info –a –n /dev/tty/ACM0."

wherein they were initially converted to cpp files and then additional Qt-related header files were included especially "math.h" and "qmath.h."

For the tinyekf.cpp source file, the following changes were implemented:

(i) Comment out the line typedef struct ekf_t declaration.

(ii) Insert the line "mulmat(ekf.H, ekf.fx, ekf.hx, n, n, 1)" after the "/* \hat{x}_k = \hat{x_k} + G_k(z_k - h(\hat{x}_k)) */" line of the ekf_step function since the current implementation starts from "H" which is multiplied with "fx" to derive "hx."

For the Tinyekfconfig.h header file, the number of states ("Nsta") is set as "3" while the number of observables ("Mobs") is also set into "3."

3.4. Operating System Configuration. It is recommended to assign persistent names to all sensor devices so that the compiled system can be uploaded to similar embedded boards without the need for unique setting up and configuration of each fusion system as long as the same embedded board systems are utilized. This is to avoid the problem of setting the correct serial port to which a specific sensor is connected, that is, GPS, EC, and IMU. The Ubuntu Linux operating system usually assigns the next available serial port device name to the currently connected serial device, such as "/dev/ttyACM0," "/dev/ttyACM1," "/dev/ttyACM2," "/dev/ttyACM3," and "/dev/ttyUSB0" for the GPS1, GPS2, GPS3, EC, and IMU devices, respectively. A problem arises when the serial device gets disconnected and gets reconnected again since then there is a distinct probability that it will now be assigned a different available serial device name such as "/dev/ttyACM4," "/dev/ttyACM5," and "/dev/ttyACM6." When this happens, the correct information from the specified sensor cannot be retrieved due to naming convention in the fusion system, that is, if the serial port device name in the fusion system for GPS1 is set as "/dev/ttyACM0" but the device is actually now recognized as "/dev/ttyACM4" by the embedded board system.

This situation is addressed by using persistent names to the sensor devices connected via the serial port so that even if the sensors get disconnected and connected, the necessary information can be extracted from the correct specified sensor. Given as follows are the steps in setting up persistent names in the Ubuntu Linux system which is applicable for both desktop and embedded board systems.

1. Open a terminal and run the command "udevadm info -a -n /dev/tty/ACM0."

   *In our system, ACM0 is for GPS1, ACM1 is for GPS2, ACM2 is for GPS3, ACM3 is for EC, and ttyUSB0 is for IMUc.

   *Search for the needed values that will be encoded into the rule file (Figure 5).

2. Create the /etc/udev/rules.d/99-usb-serial.rules file.

3. Insert the following into the newly created file:

   SUBSYSTEM=="tty", ATTRS{idVendor}=="2a03", ATTRS{idProduct}=="0043", ATTRS{product}=="Arduino Uno", ATTRS{serial}=="85439303033351819221", SYMLINK+="fjEC"

   SUBSYSTEM=="tty", ATTRS{idVendor}=="10c4", ATTRS{idProduct}=="ea60", ATTRS{serial}=="0001", SYMLINK+="fjIMU"

   SUBSYSTEM=="tty", KERNELS=="1-4.1", ATTRS{idVendor}=="1546", ATTRS{idProduct}=="01a5", ATTRS{product}=="u-blox 5 - GPS Receiver", SYMLINK+="fjGPS1"

   SUBSYSTEM=="tty", KERNELS=="1-4.2", ATTRS{idVendor}=="1546", ATTRS{idProduct}=="01a5", ATTRS{product}=="u-blox 5 - GPS Receiver", SYMLINK+="fjGPS2"

   SUBSYSTEM=="tty", KERNELS=="1-4.3", ATTRS{idVendor}=="1546", ATTRS{idProduct}=="01a5", ATTRS{product}=="u-blox 5 - GPS Receiver", SYMLINK+="fjGPS3"

   *Note that the "KERNELS" information is especially necessary when setting up a BBB embedded board. It is used to uniquely identify each of the connected similar model GPS devices.

   *The SYMLINK+= fjGPS3 value will be the one utilized in the program instead of the default device name to identify the sensor device, that is, "/dev/fjGPS3" instead of "/dev/ttyACM2."

3.5. Multithreading. The multithreading capability of the Ubuntu system and the embedded boards was harnessed to simultaneously retrieve data from the various sensor devices as well as implement the various individual processes such as CnW-I, FAz, CnW-II, and EKF. There are a total of eight (8) threads running simultaneously. These threads and their specific jobs are described as follows:

(1) Thread 1 is the main thread for the fusion system.

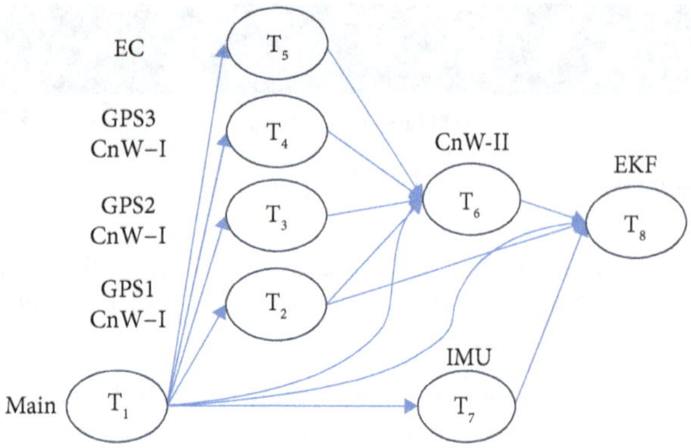

FIGURE 6: Multithreading design.

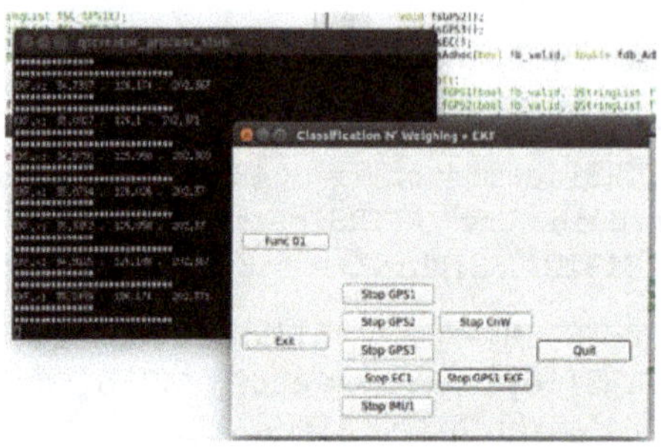

FIGURE 7: Real-time on a desktop.

(2) Thread 2 polls the data and performs CnW-I and then FAz on GPS1.

(3) Thread 3 polls the data and performs CnW-I and then FAz on GPS2.

(4) Thread 4 polls the data and performs CnW-I and then FAz on GPS3.

(5) Thread 5 polls the data from EC.

(6) Thread 6 performs CnW-II.

(7) Thread 7 polls the data from IMU.

(8) Thread 8 performs EKF on the data from GPS1 (those from GPS2 or GPS3 can also be used).

This multithreading design and the interconnection between the threads are graphically represented in Figure 6.

4. The Resulting Real-Time Sensor Fusion System

The sensor fusion was successfully implemented in real time on a Linux system desktop whose result is given in Figure 7. The same system was then cross-compiled in Qt for deployment on an ARM-based FreeScale board as well as on a BBB. The fusion system was able to perform as desired with the display shown in the LCD (Figure 8) and the output monitored through a remote login console on the embedded board (Figure 9). The main button for starting and stopping the whole sensor fusion process is labeled as "Start Sensor Fusion" which switches to "Stop Sensor Fusion" when the fusion process is running.

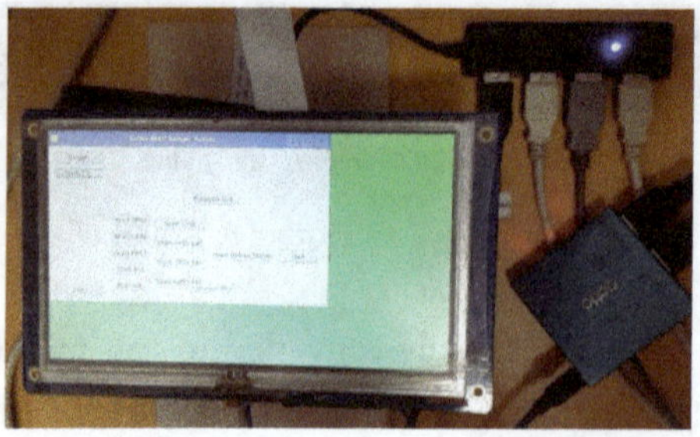

FIGURE 8: Program display on the embedded board.

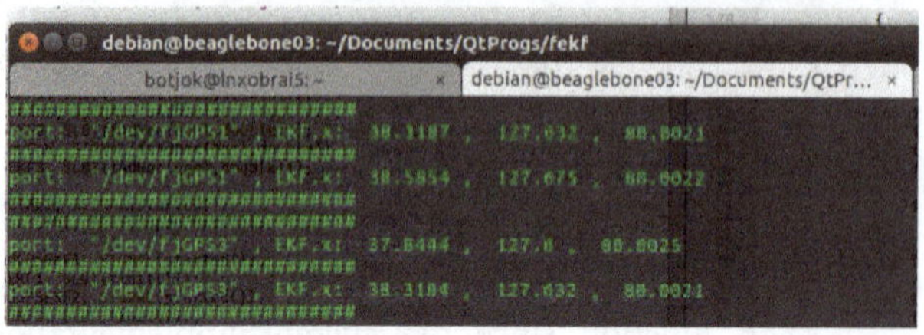

FIGURE 9: Fused heading and position on the embedded board.

The sensor fusion system developed also allowed the user to be able to control the polling of data from each of the sensors (start/stop GPS1, start/stop GPS2, start/stop GPS3, start/stop EC1, and start/stop IMU1). This was done so that it is possible to test the performance of the system by artificially introducing loss of data from either one or more of the sensors. It is also possible to start/stop the CnW-I and CnW-II processes as well as start the EKF process.

The system was able to run and still perform sensor fusion correctly when the sensor devices are disconnected and then reconnected or even switching the serial ports they are attached to. This is the main reason why it is important to implement the assigning of persistent names in the Ubuntu Linux systems of the embedded boards. This was tested for both the FreeScale and BBB with the same program deployed into it.

5. Conclusion

The real-time implementation of classification and weighing plus extended Kalman filter sensor fusion for the derivation of more accurate heading has been described. The theoretical background of the sensor fusion algorithm to the setting up and configuration of the desktop system and then the design and development using Qt-anywhere were described. The target embedded boards are then set up with the Ubuntu Linux system and then configured with the necessary libraries and drivers to be able to run the fusion system. The real-time sensor fusion system performed as designed, and the same configuration steps given in this paper allowed for the ease of deployment of the same system on an embedded board with very minimal changes that need to be done. Future work should focus on the addition of the covariance intersection algorithm into the real-time system to further improve the accuracy of the designed system.

Appendix

A. Configuration of TSLIB for Qt-Desktop and Qt-Arm on Freshly Installed Ubuntu 15.04-LTS

(1) Download toolchain (we use linaro)

∗ For gcc-linaro-arm-linux-gnueabihf-4.8-2013.10

$ wget -c https://launchpadlibrarian.net/155358238/gcc-linaro-arm-linux-gnueabihf-4.8-2013.10_linux.tar.xz

$ tar xJf gcc-linaro-arm-linux-gnueabihf-4.8-2013.10_linux.tar.xz

$ chmod 777 - R gcc-linaro-arm-linux-gnueabihf-4.8-2013.10_linux

∗ For gcc-linaro-4.9-2015.02-3-x86_64_arm-linux-gnueabihf

$ wget -c https://releases.linaro.org/archive/15.02/components/toolchain/binaries/arm-linux-gnueabihf/gcc-linaro-4.9-2015.02-3-x86_64_arm-linux-gnueabihf.tar.xzz

$ tar xJf gcc-linaro-4.9-2015.02-3-x86_64_arm-linux-gnueabihf.tar.xz

$ chmod 777 - R gcc-linaro-4.9-2015.02-3-x86_64_arm-linux-gnueabihf

(2) Download tslib

∗ For tslib-1.2

$ wget -c https://github.com/kergoth/tslib/releases/download/1.2/tslib-1.2.tar.gz

$ tar xvf tslib-1.2.tar.gz

∗ For tslib-1.3 (with multitouch support)

$ wget -c https://github.com/kergoth/tslib/releases/download/1.3/tslib-1.3.tar.gz

$ tar xvf tslib-1.3.tar.gz

(3) Configure environment

∗ For gcc-linaro-arm-linux-gnueabihf-4.8-2013.10

$ export CC=/home/botjok/Documents/Qt/gcc-linaro-arm-linux-gnueabihf-4.8-2013.10_linux/bin/arm-linux-gnueabihf-gcc

$ export CXX=/home/botjok/Documents/Qt/gcc-linaro-arm-linux-gnueabihf-4.8-2013.10_linux/bin/arm-linux-gnueabihf-g++

$ export CROSS_COMPILE=/home/botjok/Documents/Qt/gcc-linaro-arm-linux-gnueabihf-4.8-2013.10_linux/bin/arm-linux-gnueabihf-

∗ For gcc-linaro-4.9-2015.02-3-x86_64_arm-linux-gnueabihf

$ export CC=/home/botjok/Documents/Qt/gcc-linaro-4.9-2015.02-3-x86_64_arm-linux-gnueabihf/bin/arm-linux-gnueabihf-gcc

$ export CXX=/home/botjok/Documents/Qt/gcc-linaro-4.9-2015.02-3-x86_64_arm-linux-gnueabihf/bin/arm-linux-gnueabihf-g++

$ export CROSS_COMPILE=/home/botjok/Documents/Qt/gcc-linaro-4.9-2015.02-3-x86_64_arm-linux-gnueabihf/bin/arm-linux-gnueabihf-

(4) Install necessary packages for compiling tslib

$ sudo apt-get install autoconf autogen intltool libtool

$ sudo apt-get install libc6-i386 lib32stdc++6

$ sudo apt-get install lib32ncurses5 lib32z1

(5) Cross-compile tslib

∗ For tslib-1.2 and 4.8-2013.10 toolchain

$ cd tslib-1.2

$ ./autogen.sh

$ sudo ./configure CC=/home/botjok/Documents/Qt/tmp/gcc-linaro-arm-linux-gnueabihf-4.8-2013.10_linux/bin/arm-linux-gnueabihf-gcc CXX=/home/botjok/Documents/Qt/tmp/gcc-linaro-arm-linux-gnueabihf-4.8-2013.10_linux/bin/arm-linux-gnueabihf-g++ –host=arm-linux –prefix=/usr/local/Cross tslibSE

$ sudo make

$ sudo make install

∗ For tslib-1.2 and 4.9-2015.02-3 toolchain

$ cd tslib-1.2

$ ./autogen.sh

$ sudo ./configure CC=/home/botjok/Documents/Qt/tmp/gcc-linaro-4.9-2015.02-3-x86_64_arm-linux-gnueabihf/bin/arm-linux-gnueabihf-gcc CXX=/home/botjok/Documents/Qt/tmp/gcc-linaro-4.9-2015.02-3-x86_64_arm-linux-gnueabihf/bin/arm-linux-gnueabihf-g++ –host=arm-linux –prefix=/usr/local/Cross tslibSE4.9

$ sudo make

$ sudo make install

∗ For tslib-1.3 and 4.8-2013.10 toolchain

$ cd tslib-1.3

$ ./autogen.sh

$ sudo ./configure CC=/home/botjok/Documents/Qt/tmp/gcc-linaro-arm-linux-gnueabihf-4.8-2013.10_linux/bin/arm-linux-gnueabihf-gcc CXX=/home/botjok/Documents/Qt/tmp/gcc-linaro-arm-linux-gnueabihf-4.8-2013.10_linux/bin/arm-linux-gnueabihf-g++ –host=arm-linux –prefix=/usr/local/Crosstslib1.3bSE

$ sudo make

$ sudo make install

∗ For tslib-1.3 and 4.9-2015.02-3 toolchain

$ cd tslib-1.3

$ ./autogen.sh

$ sudo ./configure CC=/home/botjok/Documents/Qt/tmp/gcc-linaro-4.9-2015.02-3-x86_64_arm-linux-gnueabihf/bin/arm-linux-gnueabihf-gcc CXX=/home/botjok/Documents/Qt/tmp/gcc-linaro-4.9-2015.02-3-x86_64_arm-linux-gnueabihf/bin/arm-linux-gnuea bihf-g++ –host=arm-linux –prefix=/usr/local/Cross tslib1.3SE4.9

$ sudo make

$ sudo make install

(6) Check if the compiled library is ARM-compatible

$ file /usr/local/Crosstslib1.3SE4.9/bin/ts_calibrate

– result should be something similar as follows, especially the ", ARM,"

/usr/local/Crosstslib1.3SE4.9/bin/ts_calibrate: ELF 32-bit LSB executable, ARM, EABI5 version 1 (SYSV), dynamically linked, interpreter /lib/ld-linux-armhf.so.3, for GNU/Linux 2.6.16, BuildID[sha1]=2dc789dcb60d6d9758c4d90b4c92c722bc661a34, not stripped

∗ The generated library will be used when compiling Qt and will also be copied to the BBB or other boards

B. Configuration of Qt-Anywhere for Qt-Arm Setup (Cross-Compile)

(1) Download Qt-4.8.5-everywhere open-source version

$ wget -c http://download.qt.io/archive/qt/4.8/4.8.5/qt-everywhere-opensource-src-4.8.5.tar.gz

$ tar xzvf qt-everywhere-opensource-src-4.8.5.tar.gz

(2) Download toolchain (we use linaro)

∗ For gcc-linaro-arm-linux-gnueabihf-4.8-2013.10

$ wget -c https://launchpadlibrarian.net/155358238/gcc-linaro-arm-linux-gnueabihf-4.8-2013.10_linux.tar.xz

$ tar xJf gcc-linaro-arm-linux-gnueabihf-4.8-2013.10_linux.tar.xz

$ chmod 777 - R gcc-linaro-arm-linux-gnueabihf-4.8-2013.10_linux

∗ For gcc-linaro-4.9-2015.02-3-x86_64_arm-linux-gnueabihf

$ wget -c https://releases.linaro.org/archive/15.02/components/toolchain/binaries/arm-linux-gnueabihf/gcc-linaro-4.9-2015.02-3-x86_64_arm-linux-gnueabi hf.tar.xzz

$ tar xJf gcc-linaro-4.9-2015.02-3-x86_64_arm-lin ux-gnueabihf.tar.xz

$ chmod 777 - R gcc-linaro-4.9-2015.02-3-x86_64_arm-linux-gnueabihf

(3) Configure mkspec file for ARM of Qt-everywhere

$ vi qt-everywhere-opensource-src-4.8.5/mksepcs/qws/linux-arm-gnueabi-g++/qmake.conf {use gedit instead of vi for graphical method}

∗ This is the original file content

```
######################QMAKE.CONF###################

#

# qmake configuration for building with arm-none-linux-gnueabi-g++

#

include(../../common/linux.conf)

include(../../common/gcc-base-unix.conf)

include(../../common/g++-unix.conf)

include(../../common/qws.conf)

# modifications to g++.conf

QMAKE_CC = arm-none-linux-gnueabihf-gcc

QMAKE_CXX = arm-none-linux-gnueabihf-g++

QMAKE_LINK = arm-none-linux-gnueabihf-g++

QMAKE_LINK_SHLIB = arm-none-inux-gnueabihf-g++

# modifications to linux.conf

QMAKE_AR = arm-linux-gnueabihf-ar cqs

QMAKE_OBJCOPY = arm-linux-gnueabihf-objcopy*

QMAKE_STRIP = arm-linux-gnueabihf-strip*

load(qt_config)

################END_OF_QMAKE.CONF#################
```

∗ This will be the updated file content for gcc-linaro-4.9-2015.02-3-x86_64_arm-linux-gnueabihf

```
######################QMAKE.CONF###################

#

# qmake configuration for building with arm-none-linux-gnueabi-g++
```

```
#
include(../../common/linux.conf)
include(../../common/gcc-base-unix.conf)
include(../../common/g++-unix.conf)
include(../../common/qws.conf)
# modifications to g++.conf
QMAKE_CC = /home/botjok/Documents/Qt/tmp/gcc-linaro-4.9-2015.02-3-x86_64_arm-linux-gnueabihf/bin/arm-linux-gnueabihf-gcc
QMAKE_CXX = /home/botjok/Documents/Qt/tmp/gcc-linaro-4.9-2015.02-3-x86_64_arm-linux-gnueabihf/bin/arm-linux-gnueabihf-g++
QMAKE_LINK = /home/botjok/Documents/Qt/tmp/gcc-linaro-4.9-2015.02-3-x86_64_arm-linux-gnueabihf/bin/arm-linux-gnueabihf-g++
QMAKE_LINK_SHLIB = /home/botjok/Documents/Qt/tmp/gcc-linaro-4.9-2015.02-3-x86_64_arm-linux-gnueabihf/bin/arm-linux-gnueabihf-g++
# modifications to linux.conf
QMAKE_AR = /home/botjok/Documents/Qt/tmp/gcc-linaro-4.9-2015.02-3-x86_64_arm-linux-gnueabihf/bin/arm-linux-gnueabihf-ar cqs
QMAKE_OBJCOPY = /home/botjok/Documents/Qt/tmp/gcc-linaro-4.9-2015.02-3-x86_64_arm-linux-gnueabihf/bin/arm-linux-gnueabihf-objcopy
QMAKE_STRIP = /home/botjok/Documents/Qt/tmp/gcc-linaro-4.9-2015.02-3-x86_64_arm-linux-gnueabihf/bin/arm-linux-gnueabihf-strip
# include the compiled TSLIB library
QMAKE_INCDIR +=/usr/local/Crosstslib1.3SE4.9/include
QMAKE_LIBDIR +=/usr/local/Crosstslib1.3SE4.9/lib
# flags to successfully compile Qt4.8.5 linked with TSLIB
QMAKE_LFLAGS += -Wl,-rpath-link=/usr/local/Crosstslib1.3SE4.9/lib
load(qt_config)
################END_OF_QMAKE.CONF#####
############
```

(4) Install packages needed to compile Qt-4.8.5-everywhere

∗ Follow http://doc.qt.io/qt-4.8/requirements-x11.html

$ sudo apt-get update

$ sudo apt-get install libxrender-dev libxrandr-dev libxcursor-dev libxfixes-dev libxinerama-dev libfontconfig-dev libfreetype6-dev libxi-dev libxt-dev libxext-dev libx11-dev libsm-dev libice-dev libglib2.0–0 libc6-amd64 libxft-dev

∗ Install package to utilize virtual framebuffer for Qt for 32-bit ARM

$ sudo apt-get install libxtst-dev

∗ Install additional packages to configure Qt for 32-bit ARM on a 64-bit Ubuntu 16.04 LTS

$ sudo apt-get install lib32stdc++6 zlib1g:i386

(5) Compile Qt-4.8.5-everywhere for Qt-4.8.5-arm

∗ Configure Qt-4.8.5-everywhere using the following command

$ cd qt-everywhere-opensource-src-4.8.5

$ ./configure -opensource -confirm-license -prefix /usr/local/Qt-4.8.5-arm -embedded arm -little-endian -no-pch -no-webkit -xplatform qws/linux-arm-gnueabi-g++ -qt-mouse-tslib -qt-kbd-tty -qt-kbd-linuxinput -qt-mouse-linuxinput

∗ Compile and install Qt-4.8.5-arm

∗∗ check the number of processors of your linux box

$ cat /proc/cupinfo | grep processor | wc -l

∗∗ use option "-j<n+1>", where "n" is the number of processors

$ make -j5 ARCH=arm CROSS_COMPILE=/home/botjok/Documents/Qt/tmp/gcc-linaro-4.9-2015.02-3-x86_64_arm-linux-gnueabihf/bin/arm-linux-gnueabihf-

$ make install

C. Configuration of Fresh BBB for Deployment

Using Windows system:

(1) Prepare the SD card

∗ Use Debian-sdcard-beaglebone-2015.02.27-LCD-cape-LCD7-v2.0.img file

∗ Follow http://www.waveshare.com/wiki/LCD_CAPE_(7inch)#Introduction

(2) Extract the system image file

(3) Format the SD card

∗ Use HPUSBDisk.exe to format the TF card. Note: the TF card capacity should be 4 GB or above!

∗ Choose the device as your TF card and File system as FAT32. Then, click "Start".

(4) Write the system image file to the SD card

∗ Use Win32DiskImager.exe to write the system image

∗ Launch Win32DiskImager.exe and select the extracted system image. Then, click “Write”.

(5) Insert the SD card into BeagleBone and boot

(6) Configure network properties

∗ Connect debugger cable to the board and use PUTTY.EXE to remotely login to BeagleBone Board or SSH to the BeagleBone Board using the IP address assigned from the router

On BeagleBone Black board:

(1) Expand the file system partition on a microSD

∗ Follow http://elinux.org/Beagleboard:Expanding_File_System_Partition_On_A_microSD

$ sudo su

fdisk /dev/mmcblk0

d

2

n

p

2

w

∗ Reboot -> (after reboot)

$ sudo resize2fs /dev/mmcblk0p2

(2) Disable boot to GUI

∗ Follow http://unix.stackexchange.com/questions/143855/disable-gui-on-beaglebone-black-running-debian

$ sudo vi/boot/uboot/uEnv.txt {or you can use gedit instead of vi}

∗ Update /boot/uboot/uEnv.txt file by appending the following lines

for BBB debian OS, this disables lightdm run from “/etc/init.d/lightdm”

commwent it to enable GUI

optargs=text

(3) Setup TSLIB

∗ Follow http://wiki.mentorel.com/doku.php/tslib

∗ Copy compiled TSLIB(CrosstslibSE) folder to host

∗ Copy contents of TSLIB/lib/∗.∗ to /usr/lib/

cp –R /home/debian/Qt/CrosstslibSE/lib/∗ /usr/lib/

∗ Copy contents of TSLIB/bin/∗.∗ to /usr/bin/

cp –R /home/debian/Qt/CrosstslibSE/bin/∗ /usr/bin/

∗ Update following environment settings

export TSLIB_CONSOLEDEVICE=none

export TSLIB_FBDEVICE=/dev/fb0 {the display device}

export TSLIB_TSDEVICE=/dev/input/event1 {the touchscreen device}

export TSLIB_CALIBFILE=/etc/pointercal

export TSLIB_CONFFILE=/etc/ts.conf

export TSLIB_PLUGINDIR=/usr/lib/ts

export QWS_MOUSE_PROTO = tslib:/dev/input/event1 {the mouse/touchscreen device}

∗ Run the ts_calibrate program to test TSLIB setup and calibrate the touch screen; it can only be run as ROOT

/usr/bin/ts_calibrate

(4) Setup Qt-4.8.5 library and plugins on the BBB

∗ Follow https://sites.google.com/site/timpicuc/Downhome/beaglebone-black/setting-up-qt-for-bbb-with-debian

∗ create /opt/Qt-arm folder

mkdir /opt/Qt-arm

∗ copy Qt-4.8.5-arm/lib and Qt-4.8.5-arm/plugins folders to /opt/Qt-arm/

cp –R /home/debian/Qt/Qt-4.8.5-arm/lib /home/debian/Qt/Qt-4.8.5-arm/plugins/ /opt/Qt-arm/

∗ update LD_LIBRARY_PATH

export LD_LIBRARY_PATH=/opt/Qt-arm/lib

∗ export QT_QWS_FONTDIR path to solve font problem

export QT_QWS_FONTDIR=/opt/Qt-arm/lib/fonts

D. Configuration of Qt-Creator for Writing and Deploying the Program (Both Desktop and Cross-Compile Versions)

(1) Install Qt-Creator

∗ We use Qt-Creator-3.4.2 since it works for Qt-4.8.5

$ wget -c https://download.qt.io/official_releases/qtcreator/3.4/3.4.2/qt-creator-opensource-linux-x86_64-3.4.2.run

$ chmod +x qt-creator-opensource-linux-x86_64-3.4.2.run

$ sudo qt-creator-opensource-linux-x86_64-3.4.2.run

∗ Run Qt-Creator from the launcher or from the terminal

$/home/botjok/qtcreator-3.4.2/bin/qtcreator

(2) Configure Qt-Creator

∗ Configure "Build & Run"—"Qt Versions" option

(i) Go to "Tools"->"Options" menu.

(ii) Click "Build & Run" on the left window pane.

(iii) Select "Qt Versions" tab of the "Build & Run" window plane.

(iv) Select "Manual" option then click then "Add" button on the right. Browse to the location of the compiled Qt-for-embedded earlier, in our case it was "/usr/local/Qt-4.8.5-arm/bin/qmake." Press "Open" button or double-click on "qmake" file in the browser window.

∗∗ there will be warning: "No compiler can produce code for this Qt version. Please define one or more compilers."

(v) Leave "Version Name" as it is and then Click "Apply" button.

∗ Configure "Build & Run"—"Compilers" option

(i) Go to "Tools"->"Options" menu.

(ii) Click "Build & Run" on the left window pane.

(iii) Select "Compilers" tab of the "Build & Run" window plane.

(iv) Click "Add"->"GCC."

(v) Select "Browse" and then go to the location of "/home/botjok/Documents/Qt/gcc-linaro-4.9-2015.02-3-x86_64_arm-linux-gnueabihf/bin/arm-linux-gnueabihf-g++". Press "Open" button or double-click the "arm-linux-gnueabihf-g++" file.

(vi) Update "Name" to "armGCC".

(vii) Click the "Apply" or "Ok" button.

∗ Configure "Build & Run"—"Debuggers" option

(i) Go to "Tools"->"Options" menu.

(ii) Click "Build & Run" on the left window pane.

(iii) Select "Debuggers" tab of the "Build & Run" window plane.

(iv) Click "Add."

(v) Select "Browse" and then go to the location of "/home/botjok/Documents/Qt/gcc-linaro-4.9-2015.02-3-x86_64_arm-linux-gnueabihf/bin/arm-linux-gnueabihf-gdb." Press "Open" button or double-click the "arm-linux-gnueabihf-gdb" file.

(vi) Update "Name" to "armGDB."

(vii) Click the "Apply" or "Ok" button.

∗ Configure "Devices" option

(i) Go to "Tools"->"Options" menu.

(ii) Click "Devices" on the left window pane.

(iii) Click "Add", select "Generic Linux Device", and then click "Start Wizard."

(iv) Fill up the information in the "New Generic Linux Device Configuration Setup" window.

∗∗ Name: "BeagleBoneBlack03"

∗∗ Device hostname/IP Address: "beaglebone03" or "192.168.0.93" {hostname set or IP address assigned)

∗∗ Username: "debian" {userID to login to the device}

∗∗ Authentication type:

∗∗∗∗ Choose "Password" if using password authentication, encode Password: "∗∗∗∗∗∗∗∗

" {password for userID}

∗∗∗∗ Choose "Key" if choosing token key, then select "Browse" button to select the private key file

(v) Click "Next."

(vi) Click the "Finish" button.

∗∗ The connection to the device will be tested; press the "Close" button.

(i) Click "Apply" or "Ok."

∗ Configure "Build & Run"—"Kits" option

(i) Go to "Tools"->"Options" menu.

(ii) Click "Build & Run" on the left window pane.

(iii) Select "Kits" tab of the "Build & Run" window plane.

(iv) Click "Add."

(v) Fill up the following information

∗∗ Name: "BeagleBoneBlack"

∗∗ File system name: "" {leave blank}

∗∗ Device type: "Generic Linux Type"

∗∗ Device: "BeagleBoneBlack03" {select the one we created earlier, Click "Manage" to create/update device}

∗∗ Sysroot: "" {leave blank}

∗∗ Compiler: "armGCC" {select the one we created earlier, click "Manage" to create/update compiler}

∗∗ Environment: {leave it as it is}

∗∗ Debugger: "armGDB" {select the one we created earlier, click "Manage" to create/update debugger}

∗∗ Qt version: "Qt 4.8.5 (Qt-4.8.5-arm)" {select the one we created earlier, click "Manage" to create/update Qt version}

∗∗ Qt mkspec: "" {leave it as it is}

(vi) Press the "Apply" or "Ok" button.

Conflicts of Interest

The authors declare that they have no conflicts of interest.

Acknowledgments

This research is supported by the Brain Research Program through the National Research Foundation of Korea (NRF) funded by the Ministry of Science, ICT, & Future Planning (Grant no. NRF-2017M3C7A1044815). This was also supported by the EUREKA project (Korea Institute for Advancement of Technology and Ministry of Trade, Industry & Energy with EUREKA Project ID 10 735).

References

[1] T. Yairi, K. Hori, and S. Nakasuka, "Sensor space discretization in autonomous agent based on entropy minimization of behavior outcomes," in *Proceedings. 1999 IEEE/SICE/RSJ. International Conference on Multisensor Fusion and Integration for Intelligent Systems. MFI'99 (Cat. No.99TH8480)*, pp. 111–116, Taipei, Taiwan, August 1999.

[2] M. Joerger and B. Pervan, "Measurement-level integration of carrier-phase GPS and laser-scanner for outdoor ground vehicle navigation," *Journal of Dynamic Systems, Measurement, and Control*, vol. 131, no. 2, article 021004, 2009.

[3] F. Samadzadegan and G. Abdi, "Autonomous navigation of unmanned aerial vehicles based on multi-sensor data fusion," in *2012 20th Iranian Conference on Electrical Engineering (ICEE)*, pp. 868–873, Tehran, Iran, May 2012.

[4] M. A. A. Akhoundi and E. Valavi, *Multi-Sensor Fuzzy Data Fusion Using Sensors with Different Characteristics*, ArXiv Prepr. ArXiv10106096, 2010.

[5] B. Li, C. Rizos, H. K. Lee, and H. K. Lee, "A GPS-slaved time synchronization system for hybrid navigation," *GPS Solutions*, vol. 10, no. 3, pp. 207–217, 2006.

[6] Y. L. Zhang, J. H. Park, N. O. Sel, and K. T. Chong, "Robot navigation using a DR/GPS data fusion," *Applied Mechanics and Materials*, vol. 392, pp. 261–266, 2013.

[7] Z. Ercan, V. Sezer, H. Heceoglu et al., "Multi-sensor data fusion of DCM based orientation estimation for land vehicles," in *2011 IEEE International Conference on Mechatronics (ICM)*, pp. 672–677, Istanbul, Turkey, April 2011.

[8] E. Borràs, J. Ferré, R. Boqué, M. Mestres, L. Aceña, and O. Busto, "Data fusion methodologies for food and beverage authentication and quality assessment – a review," *Analytica Chimica Acta*, vol. 891, pp. 1–14, 2015.

[9] E. Cardarelli, L. Sabattini, C. Secchi, and C. Fantuzzi, "Multisensor data fusion for obstacle detection in automated factory logistics," in *2014 IEEE 10th International Conference on Intelligent Computer Communication and Processing (ICCP)*, pp. 221–226, Cluj Napoca, Romania, September 2014.

[10] J. A. Duro, J. A. Padget, C. R. Bowen, H. A. Kim, and A. Nassehi, "Multi-sensor data fusion framework for CNC machining monitoring," *Mechanical Systems and Signal Processing*, vol. 66-67, pp. 505–520, 2016.

[11] C. Chen, R. Jafari, and N. Kehtarnavaz, "A real-time human action recognition system using depth and inertial sensor fusion," *IEEE Sensors Journal*, vol. 16, no. 3, pp. 773–781, 2016.

[12] A. Colombo, D. Fontanelli, D. Macii, and L. Palopoli, "Flexible indoor localization and tracking based on a wearable platform and sensor data fusion," *IEEE Transactions on Instrumentation and Measurement*, vol. 63, no. 4, pp. 864–876, 2014.

[13] Y. Dobrev, S. Flores, and M. Vossiek, "Multi-modal sensor fusion for indoor mobile robot pose estimation," in *2016 IEEE/ION Position, Location and Navigation Symposium (PLANS)*, pp. 553–556, Savannah, GA, USA, April 2016.

[14] A. Simonetti, D. Accardo, D. S. Domenico, and P. Calcagni, *Developing an Attitude and Heading Reference System Based on Advanced MEMS Gyros*, AIAA Infotech@ Aerospace, 2015.

[15] A. L. Bustamante, J. M. M. López, and J. G. Herrero, "Player: an open source tool to simulate complex maritime environments to evaluate data fusion performance," *Simulation Modelling Practice and Theory*, vol. 76, pp. 3–21, 2017.

[16] A. Trzuskowsky, C. Hoelper, and D. Abel, "ANCHOR: navigation, routing and collision warning during operations in harbors," *IFAC-PapersOnLine*, vol. 49, no. 23, pp. 220–225, 2016.

[17] W. C. Bruhn, H.-C. Burmeister, M. T. Long, and J. A. Moræus, "Conducting look-out on an unmanned vessel: Introduction to the advanced sensor module for MUNIN's autonomous dry bulk carrier," in *The 10th International Symposium ISIS 2014 "Integrated Ship's Information Systems"*, Hamburg, Germany, September 2014.

[18] A. L. Flåten and E. F. Brekke, "Performance prediction of tracking sensors for surface vehicle collision avoidance," in *2016 19th International Conference on Information Fusion (FUSION)*, pp. 1894–1900, Heidelberg, Germany, July 2016.

[19] D. Chen, C. Dai, X. Wan, and J. Mou, "A research on AIS-based embedded system for ship collision avoidance," in *2015 International Conference on Transportation Information and Safety (ICTIS)*, pp. 512–517, Wuhan, China, June 2015.

[20] P. Hu and C. Huang, "Shipborne high-accuracy heading determination method using INS- and GPS-based heading determination system," *GPS Solutions*, vol. 21, no. 3, pp. 1059–1068, 2017.

[21] J. C. Juang and C. F. Lin, "A sensor fusion scheme for the estimation of vehicular speed and heading angle," *IEEE*

Transactions on Vehicular Technology, vol. 64, no. 7, pp. 2773–2782, 2015.

[22] K. Jaroś, A. Witkowska, and R. Śmierzchalski, "Data fusion of GPS sensors using particle Kalman filter for ship dynamic positioning system," in *2017 22nd International Conference on Methods and Models in Automation and Robotics (MMAR)*, pp. 89–94, Miedzyzdroje, Poland, August 2017.

[23] T. H. Bryne, *Nonlinear Observer Design for Aided Inertial Navigation of Ships*, Norges Teknisk-Naturvitenskapelige Universitet, 2017.

[24] J. M. Núñez, M. G. Araújo, and I. García-Tuñón, "Real-time telemetry system for monitoring motion of ships based on inertial sensors," *Sensors*, vol. 17, no. 5, p. 948, 2017.

[25] Q. Feng-de, W. Feng-wu, L. Jiang, and T. Guan-jun, "A method for the measurement of ship attitude based on multi-sensor data fusion," in *2015 Ninth International Conference on Frontier of Computer Science and Technology (FCST)*, pp. 196–199, Dalian, China, August 2015.

[26] F. Cappello, S. Ramasamy, R. Sabatini, and J. Liu, "Low-cost sensors based multi-sensor data fusion techniques for RPAS navigation and guidance," in *2015 International Conference on Unmanned Aircraft Systems (ICUAS)*, pp. 714–722, Denver, CO, USA, June 2015.

[27] F. Cappello, R. Sabatini, S. Ramasamy, and M. Marino, "Particle filter based multi-sensor data fusion techniques for RPAS navigation and guidance," in *2015 IEEE Metrology for Aerospace (MetroAeroSpace)*, pp. 395–400, Benevento, Italy, June 2015.

[28] F. Cappello, S. Ramasamy, and R. Sabatini, "Multi-sensor data fusion techniques for RPAS navigation and guidance," in *AIAC16: 16th Australian International Aerospace Congress*, pp. 64–72, Barton, ACT: Engineers Australia, May 2015.

[29] J. Choi, S. Kim, Y. Lee, T.-J. Kim, and H.-T. Choi, "Relative pose estimation of underwater robot by fusing inertial sensors and optical image," in *2014 11th International Conference on Ubiquitous Robots and Ambient Intelligence (URAI)*, pp. 204–208, Kuala Lumpur, Malaysia, November 2014.

[30] J. L. Odom and J. L. Krolik, "Passive towed array shape estimation using heading and acoustic data," *IEEE Journal of Oceanic Engineering*, vol. 40, no. 2, pp. 465–474, 2015.

[31] K. L. Prasanna, S. K. Rao, B. O. L. Jagan, A. Jawahar, and S. B. Karishma, "Data fusion in underwater environment," *International Journal of Engineering and Technology*, vol. 8, no. 1, pp. 225–234, 2016.

[32] P. Scherz, A. Haderer, K. Pourvoyeur, and A. Stelzer, "Embedded sensor fusion system for unmanned vehicle navigation," in *2008 IEEE/ASME International Conference on Mechtronic and Embedded Systems and Applications*, pp. 192–197, Beijing, China, October 2008.

[33] F. P. Vista IV, D.-J. Lee, and K. T. Chong, "Design of an EKF-CI based sensor fusion for robust heading estimation of marine vehicle," *International Journal of Precision Engineering and Manufacturing*, vol. 16, no. 2, pp. 403–407, 2015.

[34] US Army, *Advanced Map and Aerial Photograph Reading*, U.S. Army Headquarters, War Department, 1941.

7

Sequential Human Activity Recognition based on Deep Convolutional Network and Extreme Learning Machine using Wearable Sensors

Jian Sun,[1,2] Yongling Fu,[1] Shengguang Li,[2] Jie He,[3] Cheng Xu,[3] and Lin Tan[2]

[1]*School of Mechanical Engineering & Automation, Beihang University, Beijing 100191, China*
[2]*The F.R.I. of Ministry of Public Security, Beijing 100048, China*
[3]*School of Computer and Communication Engineering, University of Science and Technology Beijing, Beijing 100083, China*

Correspondence should be addressed to Shengguang Li; shijsun@163.com

Academic Editor: Jaime Lloret

Human activity recognition (HAR) problems have traditionally been solved by using engineered features obtained by heuristic methods. These methods ignore the time information of the streaming sensor data and cannot achieve sequential human activity recognition. With the use of traditional statistical learning methods, results could easily plunge into the local minimum other than the global optimal and also face the problem of low efficiency. Therefore, we propose a hybrid deep framework based on convolution operations, LSTM recurrent units, and ELM classifier; the advantages are as follows: (1) does not require expert knowledge in extracting features; (2) models temporal dynamics of features; and (3) is more suitable to classify the extracted features and shortens the runtime. All of these unique advantages make it superior to other HAR algorithms. We evaluate our framework on OPPORTUNITY dataset which has been used in OPPORTUNITY challenge. Results show that our proposed method outperforms deep nonrecurrent networks by 6%, outperforming the previous reported best result by 8%. When compared with neural network using BP algorithm, testing time reduced by 38%.

1. Introduction

Human activity recognition (HAR) is a new technology that can recognize human activities or gestures through computer system. Identified signals can be obtained from different types of detectors, such as audio sensors, image sensors, barometers, and accelerometers. With the rapid development of human-computer interaction (HCI) and wireless body area networks (WBANs), more and more technologies and methods have been applied to the sensor-based human activity recognition. Meanwhile, the growing maturity of ubiquitous computing [1] and machine learning algorithms has made human activity recognition widely used in athletic competition [2], medical care [3], smart home [4], and health care for the old people [5].

There are two methods of human activity recognition: human activity recognition based on visual images [6, 7] and based on wearable sensors [8]. Human motion analysis in computer vision involves object detection, tracking, and human motion recognition [6]. Computer vision-based human activity recognition method has many limitations. For example, the difficulty of motion detection will be greatly improved under unconstrained conditions, occlusion of the object, and video data acquisition problems for a long time. In addition, the camera needs to be deployed in advance, which cannot be used in some special scenarios, such as emergency rescue. Compared with computer vision, it is more advantageous to obtain signals from wearable sensors than video cameras, due to the following reasons: (1) wearable sensors alleviate the limitations of environmental constraints and fixed scenes that cameras often suffer from [9, 10]; (2) wearable sensors can better protect the privacy of users, as they can acquire signals for a specific target; and (3) multiple sensors can be deployed more accurately and efficiently on the body for signal acquisition.

In this paper, we study activity recognition based on wearable sensors. This work is motivated by requirements of activity recognition: decreasing dependence on engineered features to address increasingly complex recognition problems, improving recognition accuracy, and improving recognition efficiency. Human activity recognition is challenging due to the large variability of the given action. In order to obtain high accuracy, a large number of data are required. For example, the OPPORTUNITY Activity Recognition Challenge that was organized in 2011, which aims at recognizing activities and gestures in a complex home environment, showed that recognition accuracy of 17 gestures could not exceed 88% [11]. Therefore, addressing the recognition problem in complex scenes will require further improving recognition performance to face a wider set of activities.

Statistical learning methods have been widely used to solve activity recognition problems [12]. Gupta and Dallas used a Naïve Bayes (NB) and a K-nearest neighbor (KNN) classifier to recognize seven motions, such as walking, running, and jumping [13]. However, they relied on handcrafted features and could not find discriminative features to accurately distinguish different activities. The feature extraction methods such as symbolic representation [14], statistics of raw data [15], and transform coding [16] are widely applied in human activity recognition, but they are heuristic and require expert knowledge to design features.

Some methods can automatically extract features without requiring expert knowledge such as convolutional neural network (CNN). Yang et al. [17] used deep convolutional neural networks to automatically learn features from the original inputs. Through the deep structure, the learning features are considered as higher-level abstract representations of low-level raw time series signals. However, this structure ignored the temporal dependencies on the features and was not suitable to recognize real-time sensor signals. Applying the time dependence to the features obtained from the original sensors is a key factor for the success of sequential human activity recognition.

HAR also faces many challenges, such as large variability of a given action, similarity between classes, time consumption, and the high proportion of Null class. All of these challenges have led researchers to develop representation methods of systematic features and efficient recognition methods to effectively solve these problems. Ordóñez and Roggen [18] proposed deep convolutional network with utilization of CNN and LSTM. This paper took advantage of LSTM to solve sequential human activity recognition problem and achieved a good precision. But the complex network framework suffered from low efficiency and can hardly meet real-time requirements in practice applications.

For the above reasons, we creatively integrate extreme learning machine (ELM) [19] into deep convolutional network. Different from [18], we adopt ELM to improve the real-time performance of our framework. In ELM, the parameters of the hidden-layer nodes were chosen randomly and the output weights were solved by the least squares method. It was training-efficient and could be having a very good classification performance.

The contributions of this paper are as follows:

(i) We propose a hybrid deep structure called CNN-LSTM-ELM, which solves the problem of sequential activity recognition

(ii) The framework is composed of convolutional layers, LSTM recurrent layers, and ELM classifier, which can automatically learn feature representations and model the temporal dependencies between features, and improved the real-time capability with ELM

(iii) We show that the proposed deep framework using the ELM classifier is superior to the one using fully connected layer in running time, namely, with high efficiency

The rest of this paper are organized as follows. In Section 2, we provide a primer on the relevant background in deep learning for human activity recognition. A detailed description is presented in Section 3, illustrating the structure of our proposed CnvLSTM-ELM model. Experiments are demonstrated in Section 4, and the results show the superiorities of our proposed model. Then, conclusions come in Section 5.

2. Related Work

Continuous human action recognition is a challenging issue in machine learning with some difficulties to determine the parameters and sizing required because it highly depended upon some issues like feature selection in continuously training data streaming and the typical of classifier methods and we have less prior knowledge to determine the final size of training data and the size of machine learning architectures. In human action data features, we have to deal with a variance problem as explained in the survey paper [6, 7, 15]. In the typical of classifier methods, it makes uneasy requirements and conditions for any machine learning methods [16], such as neural networks [17], dynamic Bayesian networks [4], extreme learning machine [19], and deep learning [12] that may not give good generalization accuracy and processing speed for all human action recognition cases. Here we, summarize that state of the art for e-health monitoring and other proposals for human activity recognition.

2.1. Convolutional Neural Network. The input of neural network is generally the original signal; however, applying features extracted from the original signal to the neural network tends to improve performance. Extracting more useful features from the original signal requires sufficient expert knowledge, which inevitably limits a systematic exploration of the feature space [20]. Convolutional neural networks have been proposed to address this problem. Generally, CNN can be considered to comprise two parts. The first part is the hierarchical feature extractor, which contains convolutional layers and pooling layers. The input of each layer is the output of its previous layer. As a result, the original signal is mapped into feature vectors. The second part is a fully connected layer, and the feature vectors are classified by the fully connected layer.

The most widely used deep learning approach in the ubiquitous computing field in general and in human activity recognition using wearables in particular employs CNNs. CNNs typically contain multiple hidden layers that implement convolutional filters that extract abstract representations of input data. Combined with pooling and/or subsampling layers and fully connected special layers, CNNs are able to learn hierarchical data representations and classifiers that lead to extremely effective analysis systems. A multitude of applications are based on CNNs, including but not limited to [21–24].

Recently, sophisticated model optimization techniques have been introduced that actually allow for the implementation of deep CNNs in resource-constrained scenarios, most prominently for real-time sensor data analysis on smartphones and even smart watches [25].

2.2. Long Short-Term Memory (LSTM) Network. The de facto standard workflow for activity recognition in ubiquitous and wearable computing [26] treats individual frames of sensor data as statistically independent, that is, isolated portions of data are converted into feature vectors that are then presented to a classifier without further temporal context. However, ignoring the temporal context beyond frame boundaries during modelling may limit the recognition performance for more challenging tasks. Instead, approaches that specifically incorporate temporal dependencies of sensor data streams seem more appropriate for human activity recognition. In response to this, recurrent deep learning methods have now gained popularity in the field. Most prominent models based on so-called LSTM units [27] have been used very successfully. In [18], deep recurrent neural networks have been used for activity recognition on the OPPORTUNITY benchmark dataset. The LSTM model was combined with a number of preceding CNN layers in a deep network that learned rich, abstract sensor representations and very effectively could cope with the nontrivial recognition task. Through large-scale experimentation in [28], appropriate training procedures have been analyzed for a number of deep learning approaches to HAR including deep LSTM networks. In all of the previous work, single LSTM models have been used and standard training procedures have been employed for parameter estimation. The majority of existing methods [18, 27, 28] is based on (variants of) sliding-window procedures for frame extraction. The focus of this paper is on capturing diversity of the data during training and to incorporate diverse models into ensemble classifiers.

2.3. Extreme Learning Machine (ELM). The sequential concept in neural networks was mainly introduced by Yingwei et al. [29] for function approximation and time series prediction using a minimal radial basis function neural network (RBFNN), named minimal resource allocation neural network (MRAN). MRAN with sequential learning algorithms then become popular for feedforward networks with RBF nodes. Not like MRAN, OS-ELM (online sequential ELM) can handle both additive and RBF nodes in a unified framework and can learn the training data not only one-by-one but also chunk-by-chunk (with fixed or varying length). OS-ELM originates from the batch learning extreme learning machine (ELM) that was developed by Huang et al. [19].

OS-ELM problem cannot automatically determine the optimal number of hidden nodes in a hidden layer. In most of the cases, searching for the suitable number of hidden nodes relies on the trial and error method. There are two disadvantages: (1) it wastes lots of time and (2) it cannot always guarantee to get the optimal solution, because too small a network cannot learn the problem well and too large a network may lead to overfitting and poor generalization performance. According to Jun and Er [30], OSELM as well as the ELM requires much more hidden nodes compared with MRAN that would increase the processing time in real-time applications. The EOS-ELM algorithm adapts the node location, adjustment, and pruning method of the minimal resource allocation neural network (MRAN), so that the number of hidden nodes used in the OS-ELM can be modified. As a further improvement by Lan et al. [31], CEOS-ELM was proposed with searching capability for the optimal network architecture, and it can handle both additive and radial basis function (RBF) hidden nodes. The optimal number of hidden nodes can be obtained automatically in the sequential training process.

3. The Proposed Architecture

Current popular algorithms mainly use CNN for feature extraction and classification. But these methods are often not ideal in dealing with time series problems. We propose a hybrid activity recognition architecture; the flow chart of the architecture is depicted in Figure 1. We introduce the LSTM units to model temporal dependencies on the features extracted by CNN, which can be used to deal with time series problems. In addition, we use the ELM classifier with better generalization performance to classify the features that contain time information, which can improve the classification performance and shorten the running time. The structure combines convolutional layers, LSTM layers, and ELM classifier.

3.1. Feature Extraction Using CNN-LSTM. The convolutional neural network has great potential to identify various salient features in human activity recognition signals. In particular, the processing unit of the lower layers obtains local features of the signal (to represent the nature of each elementary motion in human activity). Higher-layer processing units represent the data in a more abstract way (to characterize the saliency of several basic motion combinations). Note that, as described below, each layer can have multiple convolution operations and pooling operations (specified by different parameters), so the CNN takes the features contained in the data into consideration from different aspects, and the

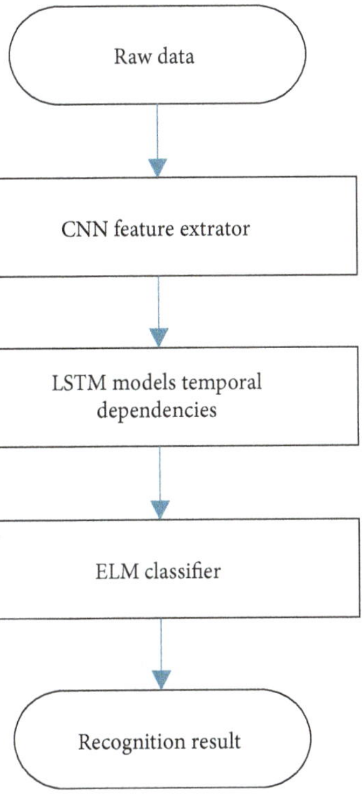

FIGURE 1: Hybrid activity recognition architecture: the structure combines convolutional layers, LSTM layers, and ELM classifier.

learned features are more comprehensive than those of artificial extraction.

When these operations with the same parameters are applied to sequence signals (or their mappings) at different time periods, the property of translation invariance is obtained. Therefore, what matters is the salient mode of the signal, not its position or scale. However, in human activity recognition, we are faced with multiple time series signal channels, and traditional CNN cannot be used directly. Our challenges include that: (1) the processing unit in the CNN needs to be applied along the time dimension, (2) sharing or unifying the units in the CNN between multiple sensors. Next, we will define the convolution operation and the pooling operation along the time dimension, and then present the entire architecture of the CNN used in human activity recognition.

We use a sliding window strategy to decompose the time series signal into a collection of short signals. Specifically, an example used by CNN is a two-dimensional matrix containing r original samples (each sample contains D attributes). Here, r is selected as the sampling rate (for example, 30 is used in the experiment), and the sliding step size of the window is selected as $r/2$. We can choose a smaller step size to increase the number of instances, but it may result in higher computational cost. For training data, the actual label of the matrix instance is determined by the most frequently occurring labels of the r original records. For the jth feature map in the ith layer of the CNN, it is also a matrix. For convenience, the value of the Xth row of the sensor D is expressed as $v_{ij}^{x,d}$.

The convolutional neural network assumes that the inputs and outputs of the model are independent from each other. However, the collected data is time dependent; thus, some time information must be included in the input data. Long short-term memory (LSTM) network has been proposed to solve this problem. The LSTM is an extension of the recurrent neural network that uses memory cells instead of loop units to store and output information. In this study, we used LSTM to model the features extracted by CNN and output the feature vectors containing time relationships.

The feature extraction network proposed in this paper combines four convolutional layers, two LSTM layers, and a fully connected (FC) layer, as shown in Figure 2. The convolution layer acts as a feature extractor and provides an abstract representation of the original data in the form of a feature graph. The LSTM layer builds the time dynamics for the feature graph. In order to illustrate the advantages of the proposed ConvLSTM-FC network, we compare it with the nonrecurrent deep convolution neural network (namely, baseline CNN). They all use four convolutional layers. The input is processed in a hierarchical manner, with each layer handling the input and then passing it to the next layer. Under these two frameworks, the number of convolution kernels in the convolution layer is the same as that in the dense layer. The only difference is that the ConvLSTM-FC uses a loop unit, while the baseline CNN is nonrecurrent and fully connected.

The input of the network is sequential data. These sequences are extracted from the sensor data by using a fixed-length sliding window. The convolution layer performs convolution operations on these data sequences and represents the data in forms of feature graphs. The LSTM layer constructs the time correlation of the feature graph and extracts the features containing time information. When the whole ConvLSTM-FC training is completed, we only reserve the convolution layer and the LSTM layer.

3.2. Classification Using ELM. As we all know, the more the discriminative features are and the more the powerful classifier is, the higher the recognition rate will be. The full-connected layer is equivalent to a general single hidden-layer feedforward neural network classifier, which is trained by backpropagation (BP). On one hand, the BP algorithm can make the weight converge to a certain value, but it is not guaranteed to be the global minimum of the error surface [32]. On the other hand, the network may be overtrained and obtain nonideal generalization performance when BP learning is used. In other words, the full-connected layer is not suitable for classifying the discriminative deep convolutional features.

Given a training set of instance-label pairs (X_i, y_i) (X_i stands for features, and y_i is labels), where $X_i = [x_{i1}, x_{i2}, \cdots, x_{in}]^T \in R^n$ and $y_i = [y_{i1}, y_{i2}, \cdots, y_{in}] \in R^m$, for a single hidden-layer neural network with M hidden nodes, the output of the network (denoted as o_j) is calculated by the following equation:

$$\sum_{i=1}^{M} \beta_i f(W_i \cdot X_j + b_i) = o_j, \quad j = 1, \ldots, N, \tag{1}$$

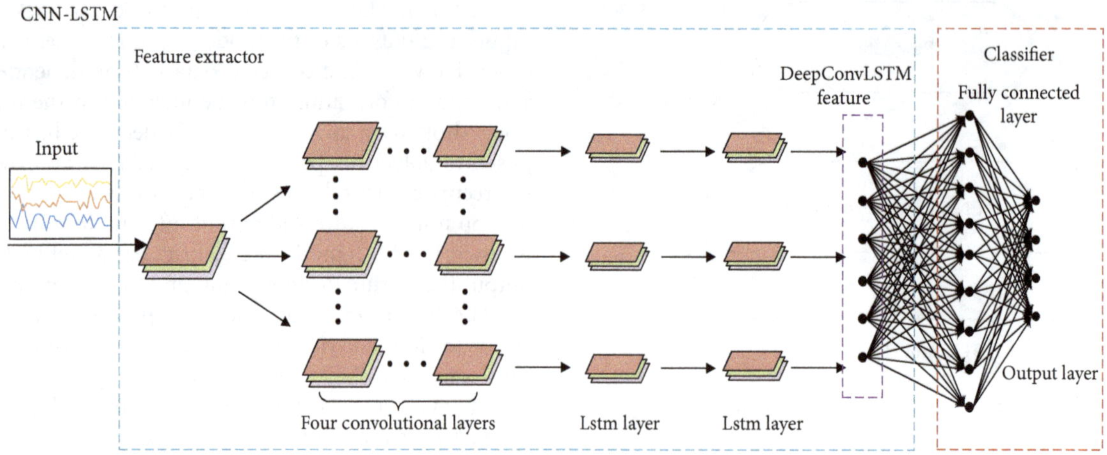

Figure 2: Our proposed CnvLSTM-FC model, combining convolutional layers, LSTM layers, and ELM classifier.

where $f(x)$ is an activation function, $W_i = [w_{i1}, w_{i2}, \cdots, w_{in}]$ is the input weight matrix, β_i is the output weight vector, and b_i is the bias of the ith hidden node.

As shown in Figure 3, the learning goal of ELM is to minimize the output error (see (2)), that is, there exist β_i, W_i, and b_i that make (3) hold true.

$$\sum_{J=1}^{M} \left\| o_j - y_j \right\| = 0, \tag{2}$$

$$\sum_{i=1}^{M} \beta_i f\left(W_i \cdot X_j + b_i\right) = y_j, \quad j = 1, \ldots, N. \tag{3}$$

Equation (4) is the compact version of (3), where H is the hidden-layer output matrix (see (4)), β is the output weight vector, and Y is the desired output.

$$\begin{gathered} H\beta = Y, \\ \beta = \begin{bmatrix} \beta_1^T \\ \vdots \\ \beta_M^T \end{bmatrix}_{M\times m}, \\ Y = \begin{bmatrix} y_1^T \\ \vdots \\ y_N^T \end{bmatrix}_{N\times m} \end{gathered} \tag{4}$$

$$H_{w,b,X} = \begin{bmatrix} f(W_1 \cdot X_1 + b_1) & \cdots & f(W_M \cdot X_1 + b_M) \\ \vdots & \cdots & \vdots \\ f(W_1 \cdot X_N + b_1) & \cdots & f(W_M \cdot X_N + b_M) \end{bmatrix}_{N\times M}. \tag{5}$$

The traditional gradient-based learning algorithm requires that all parameters be adjusted during the iteration. However, the output weight of ELM is solved in a noniterative way and there is no dependency between the input weights and the output weights. A noniterative solution of

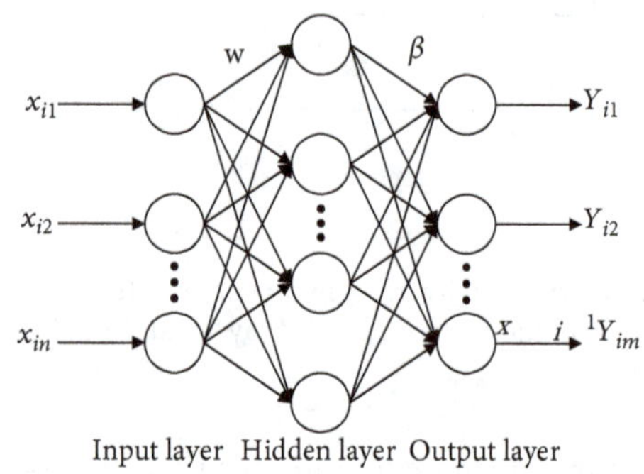

Figure 3: ELM classifier architecture.

ELM provides a speedup of 5 orders of magnitude compared to multilayer perceptron (MLP) [33] or 6 orders of magnitude compared to support vector machines (SVM) [34].

Using ELM to classify deep convolutional features can get a good recognition result and has a good generalization ability. Unlike BP, the weights between the input layer and the hidden layer of ELM are randomly set, and the output weights are solved through the method of least squares [19]. Therefore, the training speed of ELM is very fast.

3.3. Data Preprocessing. In this paper, we use the OPPORTUNITY dataset to evaluate the ConvLSTM-ELM model and compare the performance with baseline CNN and other literatures using some other machine learning algorithms used in the dataset experiment. The baseline CNN provides a performance reference for deep networks.

In this paper, we have used the OPPORTUNITY dataset to train and test our model. OPPORTUNITY activity recognition dataset is composed of a set of complex human natural activities collected in an environment where rich sensors are installed [35]. We only consider the on-body sensors, including inertial measurement units and 3-axis accelerometers. The wearing position of the sensors is shown in Figure 4. Each sensor channel is treated as an individual channel, a total of 113 channels. The sampling frequency of these sensors is 30 Hz. OPPORTUNITY dataset contains several

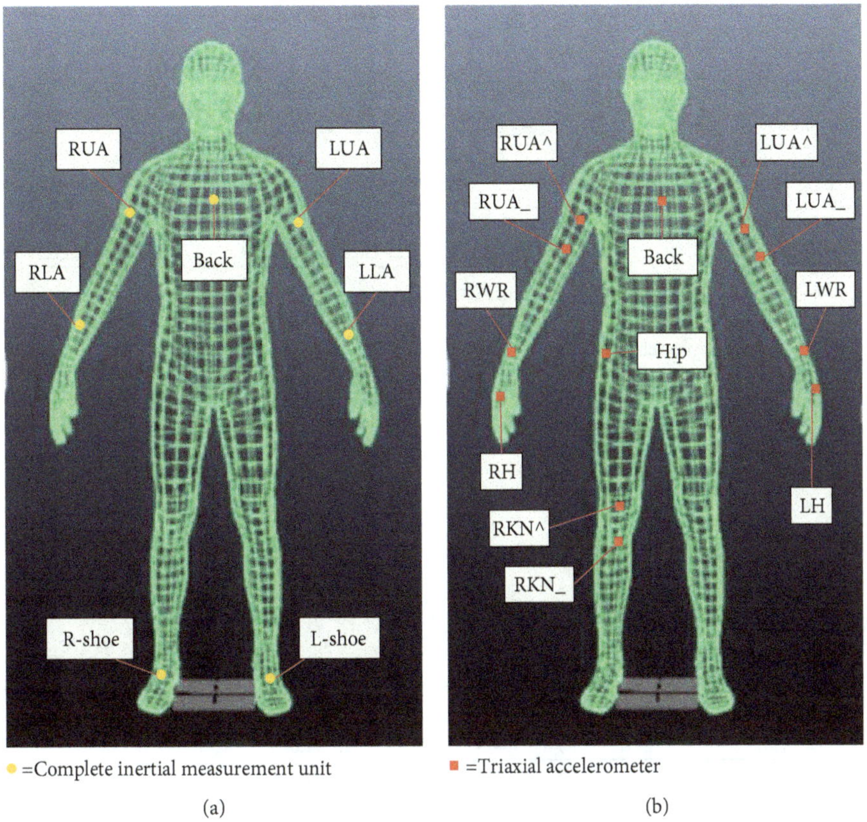

FIGURE 4: Position of on-body sensors used in the OPPORTUNITY dataset ((a) IMU sensors; (b) 3-axis accelerometers) [11].

gestures and postures, and we mainly realize the recognition of gestures either including or ignoring the Null class. This is an 18-class classification problem; the gestures in the dataset are summarized in Table 1.

We used 0 to fill in missing values of the sensor data, and each sensor channel was normalized to interval [0, 1]. We used a fixed-length sliding window to segment data; each segmentation of data was called a sequence. The length of the window is 500 ms and the step size is 250 ms. The model is trained with a learning rate of 0.001 and a decay factor of 0.9. CNN-LSTM works as the feature extractor. The whole setting of the CNN-LSTM architecture will be detailed later in Section 4, which is shown in Table 2. When the training is completed, parameters of convolutional layers and LSTM layers are reserved, and the full-connected layer is removed. Then, ELM is fed with features extracted by CNN-LSTM and trained as classifier.

Corresponding to the serialized window data is the pose observed during the time window. Given a sliding window with a length of T, we choose the label of the $t = T$ moment as the class of the sequence; in other words, we choose the label of the last data in the sequence as the label of the sequence, as shown in Figure 5.

3.4. Model Implementation. We build and train the neural network in Theano using Lasagne [36]. The model runs on a 1536 core, 1050 MHz clock speed, and 8 GB RAM GPU. The detailed software and hardware parameters are shown in Tables 3 and 4, respectively.

TABLE 1: Class labels for the mode of gesture recognition.

Gestures	
Open dishwasher	Close dishwasher
Open fridge	Close fridge
Open drawer 1	Close drawer 1
Open drawer 2	Close drawer 2
Open drawer 3	Close drawer 3
Open door 1	Close door 1
Open door 2	Close door 2
Drink from cup	Clean table
Toggle switch	Null

TABLE 2: CNN-LSTM architecture parameters.

Layer	Type	Number of neurons	Kernel size	Stride
1	Input	24×113	—	—
2	Convolutional	20×113	5×1	1
3	Convolutional	16×113	5×1	1
4	Convolutional	12×113	5×1	1
5	Convolutional	8×113	5×1	1
6-7	LSTM	128	—	—
8	Fully connected	n	—	—

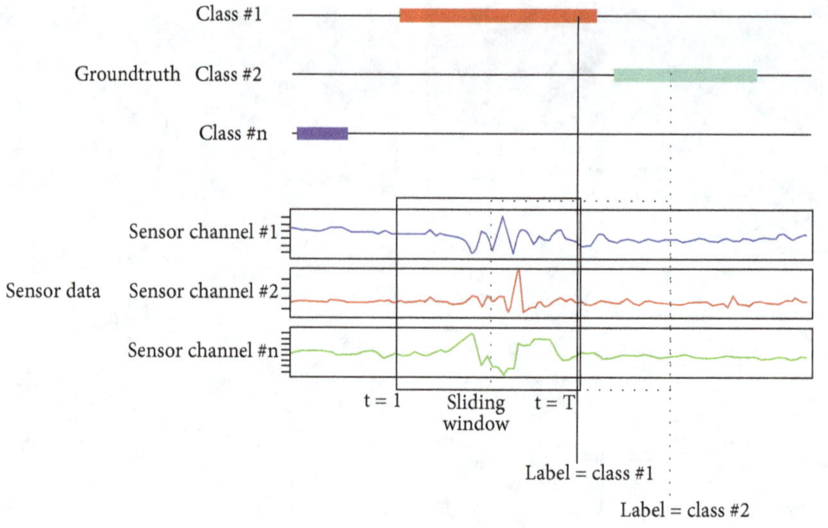

FIGURE 5: Sequence tags after splitting data with sliding windows.

TABLE 3: Software parameter configuration.

Operating system	Ubuntu 14.04
Programming language	Python 2.7.11
IDE	Spyder 3.1.4

TABLE 4: Hardware parameter configuration.

CPU	Intel Xeon E3-1505M v5
CPU dominant frequency	2.8 GHz
Core/thread count	4 cores/8 threads
RAM volume	64 GB
Graphics card	NVIDIA Quadro M5000M
Graphic memory	8 GB
Clock frequency	1050 MHz
CUDA cores	1536

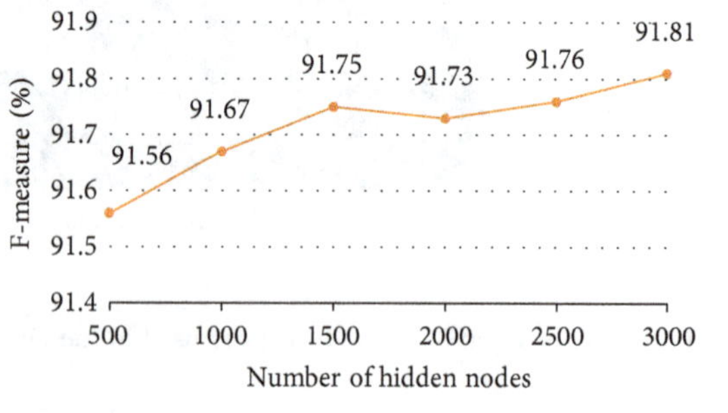

FIGURE 6: CNN-LSTM-ELM performance with different numbers of hidden nodes.

TABLE 5: The comparison of runtime.

	Training time (s)	Testing time (s)
CNN-LSTM [18]	12154.5	8.025
CNN-LSTM-ELM	11456.124	4.901

4. Result and Analysis

In this section, we test our model on the public OPPORTUNITY dataset. We will give a brief introduction to the dataset used in the experiment and the indicators to evaluate the performance of proposed models. Discussion on the experimental results is detailed. We demonstrated the performance of the proposed method and evaluated the impact of some key parameters on the methods.

4.1. Performance Measure. Human activity datasets collected in natural scenes are often imbalanced between classes [37]. Some classes may contain a large number of samples while other classes have only a few samples. The gestures of OPPORTUNITY dataset are extremely imbalanced; the Null class accounts for more than 70% of all the data. The classifier predicts the classification accuracy of each class; the Null class can achieve very high accuracy. The overall classification accuracy is not an appropriate index for performance evaluation. *F*-measure (F_1) considers the correct classification of each class as equally important. It takes into account both the precision and the recall of each class to compute the score and can evaluate the model better than the precision. Precision is defined as $P-(\mathrm{TP}/(\mathrm{TP}+\mathrm{FP}))$, and recall corresponds to $R-(\mathrm{TP}/(\mathrm{TP}+\mathrm{FN}))$, where TP and FP are the number of true and false positives, respectively, and FN corresponds to the number of false negatives. Class imbalance is countered by weighting classes according to their sample proportion:

$$F_1 = \sum_i 2^* w_i \frac{\mathrm{precision}_i \cdot \mathrm{recall}_i}{\mathrm{precision}_i + \mathrm{recall}_i}, \tag{6}$$

where $w_i = n_i/N$ is the proportion of samples of the ith class, with n_i being the number of samples of the ith class and N being the total number of samples.

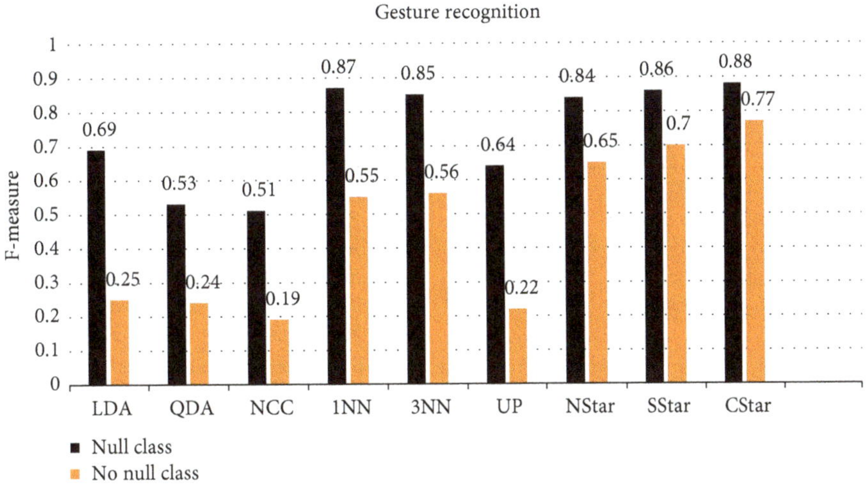

FIGURE 7: F1 score comparison of the proposed architecture and the OPPORTUNITY challenge models, either including or ignoring the Null class.

4.2. Parameter Evaluation. We describe the impact of key parameters of the architecture. We evaluate the influence of ELM hidden-layer nodes. For ELM, it is fed with discriminative features and a different number of hidden nodes are chosen, and the corresponding recognition performance is recorded in Figure 6.

Apparently, the recognition performance increases as the number of hidden nodes increases. When there are 500 hidden-layer nodes, the recognition performance is up to 91.56%, which is already better than that of many methods. It reaches 91.81% with 3000 hidden-layer nodes. The results show that the performance may be further improved if the number of hidden nodes is increased.

4.3. Runtime Analysis of CNN-LSTM-ELM. The runtime analysis of the CNN-LSTM-ELM is done on OPPORTUNITY dataset. It has 46495 training samples and 9894 testing samples, and each sample is a sequence. The model training and testing are run on a GPU with 1536 cores, 1050 MHz clock speed, and 8 GB RAM.

Because the traditional feedforward neural network adopts the iterative algorithm of gradient descent to update the weight parameters, it has obvious defects: the learning speed is slow, so the time cost is unacceptable; the learning rate is difficult to be determined and the network is easy to fall into the local minimum; it would probably be overtrained; and its generalization performance is not guaranteed to be optimal. These defects become a bottleneck for the wide application of feedforward neural network using iterative algorithms. In order to solve these problems, ELM is utilized instead of the traditional algorithm with gradient descent. The algorithm can calculate the output weights of the learning network by one step. Compared with the iterative algorithm, the ELM greatly improves generalization ability and the learning speed of the network. Therefore, for our network compared with the traditional network, the runtime mainly depends on the use of ELM classifier.

The CNN-LSTM with the fully connected layer is used to compare the runtime with CNN-LSTM-LSTM; training time and testing time are shown in Table 5.

As can be seen from Table 4, when compared to the CNN-LSTM, training time is reduced by 5% and testing time is reduced by 38%. The length of the sliding window is 500 ms and the testing samples are composed of 9894 windows, so the testing samples contain the original data of 4947 s. CNN-LSTM-ELM recognizes the test samples and only takes 4.901 s, indicating that our framework can recognize real-time sensor data.

4.4. Performance Comparison. The classification results of the proposed deep methods on the OPPORTUNITY dataset are shown in Figure 7 and Table 2. We report classification performance either including or ignoring the Null class. Including the Null class may lead to a much higher recognition rates for Null class than rare classes. In other words, some samples belonging to rare classes are often wrongly recognized as the Null class. By providing both results, we can better understand the type of errors caused by the model.

Figure 7 includes the classification performance of past published classification techniques implemented on OPPORTUNITY dataset. All of these methods are based on sliding window and only the feature extraction and classifier are different. Compared to the best approach of the OPPORTUNITY challenge, our method improves the performance by 8% on average. There is more than 13% improvement in the gesture recognition task including the Null class when compared to the OPPORTUNITY challenge methods.

From the results in Table 6, CNN-LSTM-ELM has better performance than other deep models either including or ignoring the Null class. CNN-LSTM-ELM improves by 6% compared to CNN used by Yang et al. [17]. The Baseline CNN has the same number of convolutional layers as the CNN-LSTM-ELM, but it uses the traditional classifier based on the gradient descent algorithm. When compared with

TABLE 6: F_1 score comparison of the proposed architecture and deep architecture, either including or ignoring the Null class.

	Deep architectures	
	Gesture recognition	Gesture recognition (no Null class)
CNN [17]	0.851	
Baseline CNN	0.883	0.783
CNN-LSTM [18]	0.897	0.802
CNN-LSTM-ELM	**0.918**	**0.906**

baseline CNN, CNN-LSTM-ELM improves by 7.5% on average. For the recognition of these similar gestures, CNN-LSTM-ELM can get better classification performance because LSTM cells have the ability to model time dynamics of the data sequences. However, the baseline CNN ignores the temporal dependencies between the data sequences.

5. Conclusions

In this article, we demonstrated the advantages of a deep architecture based on the combination of convolutional layers, LSTM recurrent layers, and ELM classifier for wearable activity recognition. This combined structure is used to study multichannel time series data. It mainly utilizes convolution operations and LSTM cells to capture significant features of sensor signals with different time scales. Then, all the extracted features are classified by the ELM classifier. The main advantages of this framework are as follows: (i) feature extraction is automatically performed without manual intervention; (ii) LSTM cells can capture the temporal dependencies on features extracted by convolution operations; and (iii) ELM classifier has outstanding generalization ability and fast learning speed. This framework outperformed best result of the OPPORTUNITY challenge by 13% on average in an 18-class gesture recognition task. In the experiment, we illustrate that the proposed method is superior to other best methods, so we believe that the proposed method can be used as a powerful tool for human activity recognition problems.

As for future work, we aim to improve sequential learning adaptive capability; thus, the machine learning has capability to continuously learn while at the same time doing verification (learning by doing). Transfer learning approach based on existing models to perform activity recognition on large-scale data may be a potential working direction.

Conflicts of Interest

The authors declare that they have no conflicts of interest.

Acknowledgments

This work is supported by Projects of International Cooperation and Exchanges NSFC (Key Program) no. 61327807.

References

[1] T. Plötz, N. Y. Hammerla, and P. Olivier, "Feature learning for activity recognition in ubiquitous computing," *Proceedings of the Twenty-Second International Joint Conference on Artificial Intelligence*, 2011, pp. 1729–1734, Barcelona, Catalonia, Spain, July 2011.

[2] J. Margarito, R. Helaoui, A. M. Bianchi, F. Sartor, and A. Bonomi, "Userindependent recognition of sports activities from a single wrist-worn accelerometer: a template-matching-based approach," *IEEE transactions on bio-medical engineering*, vol. 63, no. 4, pp. 1–796, 2015.

[3] S. Sendra, L. Parra, J. Lloret, and J. Tomás, "Smart system for children's chronic illness monitoring," *Information Fusion*, vol. 40, pp. 76–86, 2018.

[4] P. C. Roy, S. Giroux, B. Bouchard et al.L. Chen, C. Nugent, J. Biswas et al., "A possibilistic approach for activity recognition in smart homes for cognitive assistance to Alzheimer's patients," in *Activity Recognition in Pervasive Intelligent Environments*, Atlantis Ambient and Pervasive Intelligence, pp. 33–58, Atlantis Press, 2011.

[5] C. Xu, J. He, X. Zhang, C. Wang, and S. Duan, "Detection of freezing of gait using template-matching-based approaches," *Journal of Sensors*, vol. 2017, Article ID 1260734, 8 pages, 2017.

[6] A. Bux, P. Angelov, and Z. Habib, "Vision based human activity recognition: a review," in *Advances in Computational Intelligence Systems*, P. Angelov, A. Gegov, C. Jayne, and Q. Shen, Eds., vol. 513 of Advances in Intelligent Systems and Computing, Springer, Cham, 2017.

[7] S. R. Ke, H. L. U. Thuc, Y. J. Lee, J.-N. Hwang, J.-H. Yoo, and K.-H. Choi, "A review on video-based human activity recognition," *Computers*, vol. 2, no. 2, pp. 88–131, 2013.

[8] C. Xu, J. He, X. Zhang, C. Yao, and P.-H. Tseng, "Geometrical kinematic modeling on human motion using method of multi-sensor fusion," *Information Fusion*, vol. 41, 2017.

[9] Q. V. Le, W. Y. Zou, S. Y. Yeung, and A. Y. Ng, "Learning hierarchical invariant spatio-temporal features for action recognition with independent subspace analysis," in *CVPR 2011*, pp. 3361–3368, Providence, RI, USA, June 2011.

[10] S. Ji, M. Yang, and K. Yu, "3D convolutional neural networks for human action recognition," *IEEE Transactions on Pattern Analysis and Machine Intelligence*, vol. 35, no. 1, pp. 221–231, 2013.

[11] R. Chavarriaga, H. Sagha, A. Calatroni et al., "The opportunity challenge: a benchmark database for on-body sensor-based activity recognition," *Pattern Recognition Letters*, vol. 34, no. 15, pp. 2033–2042, 2013.

[12] W. Liu, J. Yang, L. Wang, C. Wu, and R. Zhang, "Movement behavior recognition based on statistical mobility sensing," *Adhoc & Sensor Wireless Networks*, vol. 25, no. 3-4, pp. 323–340, 2015.

[13] P. Gupta and T. Dallas, "Feature selection and activity recognition system using a single triaxial accelerometer," *IEEE transactions on bio-medical engineering*, vol. 61, no. 6, pp. 1780–1786, 2014.

[14] J. Lin, E. Keogh, S. Lonardi, and B. Chiu, "A symbolic representation of time series, with implications for streaming algorithms," in *DMKD '03 Proceedings of the 8th ACM SIGMOD workshop on Research issues in data mining and knowledge discovery*, pp. 2–11, San Diego, CA, USA, June 2003.

[15] C. Xu, J. He, X. Zhang, P. H. Tseng, and S. Duan, "Toward near-ground localization: modeling and applications for TOA ranging error," *IEEE Transactions on Antennas and Propagation*, vol. 65, no. 10, pp. 5658–5662, 2017.

[16] A. Rghioui, S. Sendra, J. Lloret, and A. Oumnad, "Internet of things for measuring human activities in ambient assisted living and e-health," *Network Protocols and Algorithms*, vol. 8, no. 3, pp. 15–28, 2016.

[17] M. Kahng, P. Y. Andrews, A. Kalro, and D. H. Chau, "ActiVis: visual exploration of industry-scale deep neural network models," *IEEE Transactions on Visualization and Computer Graphics*, vol. 24, no. 1, pp. 88–97, 2018.

[18] F. J. Ordóñez and D. Roggen, "Deep convolutional and LSTM recurrent neural networks for multimodal wearable activity recognition," *Sensors*, vol. 16, no. 1, p. 115, 2016.

[19] G. B. Huang, H. Zhou, X. Ding, and R. Zhang, "Extreme learning machine for regression and multiclass classification," *IEEE Transactions on Systems, Man, and Cybernetics, Part B (Cybernetics)*, vol. 42, no. 2, pp. 513–529, 2012.

[20] C. Xu, J. He, and X. Zhang, "DFSA: a classification capability quantification method for human action recognition," in *2017 IEEE SmartWorld, Ubiquitous Intelligence & Computing, Advanced & Trusted Computed, Scalable Computing & Communications, Cloud & Big Data Computing, Internet of People and Smart City Innovation (SmartWorld/SCALCOM/UIC/ATC/CBDCom/IOP/SCI)*, San Francisco, CA, USA, August 2017.

[21] C. Xu, J. He, X. Zhang et al., "Recurrent transformation of prior knowledge based model for human motion recognition," *Computational Intelligence and Neuroscience*, vol. 2018, Article ID 4160652, 12 pages, 2018.

[22] C. A. Ronao and S. B. Cho, "Deep convolutional neural networks for human activity recognition with smartphone sensors," in *Neural Information Processing*, pp. 46–53, Springer International Publishing, 2015.

[23] J. B. Yang, M. N. Nguyen, P. P. San, X. L. Li, and S. Krishnaswamy, "Deep convolutional neural networks on multichannel time series for human activity recognition," in *IJCAI'15 Proceedings of the 24th International Conference on Artificial Intelligence*, pp. 3995–4001, Buenos Aires, Argentina, July 2015.

[24] N. M. Rad, A. Bizzego, S. M. Kia, G. Jurman, P. Venuti, and C. Furlanello, "Convolutional neural network for stereotypical motor movement detection in autism," 2015, https://arxiv.org/abs/1511.01865.

[25] S. Bhattacharya and N. D. Lane, "Sparsification and separation of deep learning layers for constrained resource inference on wearables," in *Proceedings of the 14th ACM Conference on Embedded Network Sensor Systems CD-ROM - SenSys '16*, pp. 176–189, Stanford, CA, USA, November 2016.

[26] A. Bulling, U. Blanke, and B. Schiele, "A tutorial on human activity recognition using body-worn inertial sensors," *ACM Computing Surveys (CSUR)*, vol. 46, no. 3, pp. 1–33, 2014.

[27] S. Hochreiter and J. Schmidhuber, "Long short-term memory," *Neural computation*, vol. 9, no. 8, pp. 1735–1780, 1997.

[28] N. Y. Hammerla, J. M. Fisher, P. Andras, L. Rochester, R. Walker, and T. Plotz, "PD disease state assessment in naturalistic environments using deep learning," in *Proceedings of the Twenty-Ninth AAAI Conference on Artificial Intelligence*, pp. 1742–1748, Austin, TX, USA, January 2015.

[29] L. Yingwei, N. Sundararajan, and P. Saratchandran, "A sequential learning scheme for function approximation using minimal radial basis function neural networks," *Neural Computation*, vol. 9, no. 2, pp. 461–478, 1997.

[30] Y. Jun and M.-J. Er, "An enhanced online sequential extreme learning machine algorithm," in *2008 Chinese Control and Decision Conference*, pp. 2902–2907, Yantai, Shandong, China, July 2008.

[31] Y. Lan, Y. C. Soh, and G.-B. Huang, "A constructive enhancement for online sequential extreme learning machine," in *2009 International Joint Conference on Neural Networks*, pp. 1708–1713, Atlanta, GA, USA, June 2009.

[32] B. F. Books and S. Haykin, *Neural Networks a Comprehensive Foundation*, Pearson Education, Singapore, 2010.

[33] F. Rosenblatt, "The perceptron: a probabilistic model for information storage and organization in the brain," *Psychological Review*, vol. 65, no. 6, pp. 386–408, 1958.

[34] C. Cortes and V. Vapnik, "Support-vector networks," *Machine Learning*, vol. 20, no. 3, pp. 273–297, 1995.

[35] D. Roggen, A. Calatroni, M. Rossi et al., "Collecting complex activity datasets in highly rich networked sensor environments," in *2010 Seventh International Conference on Networked Sensing Systems (INSS)*, pp. 233–240, Kassel, Germany, June 2010.

[36] S. Dieleman, J. Schlüter, C. Raffel et al., *Lasagne: First Release*, Zenodo, Geneva, Switzerland, 2015.

[37] C. X. Ling and V. S. Sheng, *Class Imbalance Problem*, Springer US, 2011.

8

Human Emotional Care Purposed Automatic Remote Portrait Drawing Generation and Display System using Wearable Heart Rate Sensor and Smartphone Camera with Depth Perception

Gyeoungrok Lee,[1] Dongwann Kang,[2] and Kyunghyun Yoon[1]

[1]*School of Computer Science and Engineering, Chung-Ang University, Seoul, Republic of Korea*
[2]*Faculty of Science and Technology, Bournemouth University, Poole, UK*

Correspondence should be addressed to Kyunghyun Yoon; khyoon@cglab.cau.ac.kr

Academic Editor: Ka L. Man

We propose a system that automatically generates portrait drawings for the purpose of human emotional care. Our system comprises two parts: a smartphone application and a server. The smartphone application enables the user to take photographs throughout the day while acquiring heart rates from the smartwatch worn by the user. The server collects the photographs and heart rates and displays portrait drawings automatically stylized from the photograph for the most exciting moment of the day. In the system, the user can recall the exciting and happy moment of the day through admiring the drawings and heal the emotion accordingly. To stylize photographs as portrait drawings, we employ nonphotorealistic rendering (NPR) methods, including a portrait etude stylization proposed in this paper. Finally, the effectiveness of our system is demonstrated through user studies.

1. Introduction

Internet of Things (IoT) has undergone rapid advances following the popularization of smartphones and wearable sensor devices. Owing to these circumstances, IoT has become nearly ubiquitous in our lives, widely supporting facets of our well-being as well. Especially, in healthcare front, many applications (e.g., Apple's HealthKit) based on IoT have been developed and employed. However, such applications generally focus on physical healthcare only. Consequently, the needs for emotional and mental health-care applications are gradually increasing.

We present a system developed with the aim of caring for the user's emotions; the proposed system automatically generates and displays the users' portrait drawings by utilizing IoT. In this system, when a user takes photographs throughout the day using a smartphone, the most exciting portrait is selected automatically based on the user's heart rate taken from a smartwatch equipped with the heart rate sensor. The selected photo is then rendered as portrait drawings with various styles by using the nonphotorealistic rendering (NPR) techniques at the server located at the user's home. Finally, when the user returns home, the stylized portrait drawings are displayed in a digital photo frame that is connected to the server. Consequently, the user can recall the most exciting and also the happiest moment of the day through admiring the drawings and heal their emotions accordingly.

The major contributions of this study are as follows. First, we present an IoT-based application for human emotional care as mentioned earlier. This is achieved by employing the NPR techniques, which aim to acquire aesthetic images. The study not only shows that the NPR can be utilized for emotion-aware applications but also that using IoT technique can further expand it. Second, we propose a novel portrait stylization method inspired by Henri de Toulouse-Lautrec's etude, which mostly stylizes the facial region while abstractly representing the others. In this study, our system cares for human emotions using automatically stylized portrait drawings. Selective focusing and expressing the facial region during stylization process, therefore, play a key role in enabling the resulting drawings to convey facial expression representing the most exciting moment of the day. Finally, by conducting user studies, we show that our system makes users feel better after recalling the most exciting moments of their days by admiring the drawings our system generates. Through the user study, we also confirm that the stylization method proposed in this paper generates results that reasonably resemble Henri de Toulouse-Lautrec's etude.

The remainder of this paper is organized as follows. In Section 2, we provide an overview of related works on IoT applications and artistic stylization methods. Next, we describe our system which automatically and remotely generates portrait drawings using IoT technique in Section 3. We then present our portrait stylization approaches in Section 4. In Section 5, we present experimental results and their evaluation. Finally, in Section 6, we conclude with a summary of our method and the scope for future development.

2. Related Work

Due to the popularization of mobile devices, the range of Internet of Things (IoT) technology has been rapidly broadening. This has made IoT technology nearly ubiquitous in our lives, and it widely supports facets of our well-being. In this regard, many studies focused on various IoT applications. For healthcare, [1, 2] developed applications which monitor heart rate measured by heart rate sensor of smartwatch and provide personalized information for the user's heart care without going to hospital or pharmacy to get doctor's diagnosis. For fitness, applications where activity tracker communicates with fitness facilities like running machine and cycling fitness to obtain precise fitness data, such as exercise distance, speed, and slopeness, and tracks these on smartphone to suggest personalized exercise plan are widely used [3, 4]. However, despite increasing interest in physical heath, studies for emotional care have rarely conducted. In this paper, we aim at developing a system for human emotional care. To do this, we utilize stylization techniques.

Artistic image stylization is a technique that aims at converting an original image into an artistic image with certain style [5]. Early studies on artistic image stylization [6, 7] focused on expressing painterly style by generating brush strokes. In the studies, brush stroke's properties such as direction, size, and texture were mainly managed to stylize original image. Meanwhile, several studies have focused on expressing specific artistic style. Van Gogh's painting styles were imitated by manipulating brush stroke's properties [8, 9], and Seurat's pointillism was simulated as well by employing color juxtaposition theory [10]. These studies were conducted while targeting on mimicking the impressionists' painting style, because the brush strokes they used were quite unique. In this paper, we aim at generating Toulouse-Lautrec's portrait etude who is one of the famous impressionists in 18 centuries.

3. Remote Portrait Generation and Display System for Human Emotional Care

3.1. System Overview. The system proposed in this paper consists of a mobile application part including smartphone application and smartwatch and a server part including server and digital photo frame. In our framework, the user wears a smartwatch and takes portrait photographs throughout the day by using our smartphone application. While taking a photograph, the application also acquires the heart rate from the smartwatch and sends it to the server located at the user's home. The server then stylizes the photographs as portrait drawings with given styles if they are judged as the most exciting moment of the day based on the heart rate. When the user comes back home, the server displays the drawings by using a digital photo frame, and through admiring these drawings and recalling the moments, the user experiences an emotional healing. We present the detail of each part in the next two sections.

FIGURE 1: Our smartphone application.

3.2. Mobile Application Part for Acquiring Photograph and Heart Rate. User's photographs are acquired throughout the day by using a smartphone and smartwatch equipped with heart rate sensor, built-in camera, and depth perception sensor. When the user takes a photograph using a smartphone, our smartphone application (Figure 1) simultaneously obtains user's heart rate information from the smartwatch and sends the photograph with heart rate to the server. In this study, we used Samsung Gear S which is a smartwatch equipped with heart rate sensor and integrated it with our smartphone application developed on Android environment.

Our system mainly aims at generating portrait drawings. Consequently, dividing input photograph into various

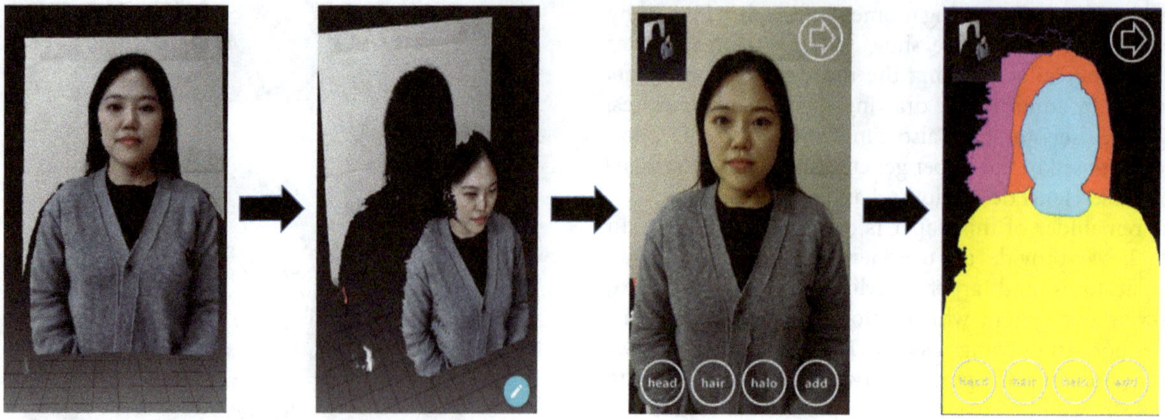

Photograph taken with depth perception | Object and background separation | User-assisted segmentation UI | Segmented result

FIGURE 2: Segmentation process using depth perception.

regions such as facial area and background enables the system to obtain a better quality of the drawings. To do this, we segment the input photograph by using the depth perception of Google's Tango, which is an augmented reality computing platform. Figure 2 shows the process that is performed on Lenovo Phab 2 Pro, which is a smartphone employed in this study and supports the Tango. The Tango reconstructs a 3-dimensional surface from the photograph taken by using depth sensor and motion sensor. Starting from the surface, we search for the point of the greatest depth and find the center of the bounding box, which is detected as the face from the photograph. By using these two points as seeds, we then perform watershed segmentation on depth domain to separate object from the background. Optionally, we also provide semiautomatic segmentation using the seeds mentioned above and user-selected seeds for hair and body. Finally, the segmentation result is incorporated along with the input photograph and is sent to the server to be used in portrait drawing generation process described in the following section.

3.3. Server Part for Generating and Displaying Portrait Drawings. When a photograph is received with its segmentation result and heart rate, the server which is located at the user's home judges whether the photograph is the most exciting moment of the day. In this study, we regard the photograph incorporated with the highest heart rate as the most exciting moment. Therefore, the server renders portrait drawings only if the heart rate of the received photograph is higher than that of previously received photographs. For the style of portrait drawings, we provide oil paintings, pastel drawing, pencil drawings, and portrait etudes, as described in Section 4. Using these four styles, the server generates portrait drawings and preserves them until a new photograph corresponding to a more exciting moment is received. The server monitors the user's location, which is computed on the smartphone using GPS. When the user is within a predefined distance from the server, the server finally displays the portrait drawings on the digital photo frame connected to it.

This paper supposes the system to be used for home use. Therefore, it is recommended to use a low-power device as the server of our system for a practical use. To this end, we used Raspberry Pi 2 B+ and Camel PF1710IPS as the server and digital photo frame, respectively, as shown in Figure 3. When a user admires our resulting drawings, interactions which enable the user to browse and select preferred drawings are provided by an infrared remote controller. To do this, we capture infrared remote signals from the GPIO 22 pin of Raspberry Pi.

4. Portrait Stylization Methods

This study aims at enabling users to admire portrait drawings with various styles to arouse their interests. To achieve this, algorithms for stylizing various styles are required. As mentioned in Section 2, lots of stylization methods have been proposed. Among them, we employ Seo et al.'s painterly rendering algorithm [11] using a stroke texture database. In their study, each individual brush stroke texture, which is captured from a glass sheet where an artist paints on, is stored in their stroke texture database and is utilized for generating brush strokes in the rendering process. At this time, if the database is replaced with another one which consists of brush strokes of different medium, the style of resulting drawing is consequently changed. In this manner, we used three mediums (i.e., oil paint, pastel, and pencil) to provide each style (Figure 4).

In this study, we focus on generating portrait drawings using the most exciting moments taken by users throughout the day. Therefore, conveying facial expression with abstract representation of the others is sometimes more effective for giving a strong impression to the users. In addition to the styles mentioned above, we propose a portrait stylization inspired by Henri de Toulouse-Lautrec's etude, which mostly stylizes the facial region while abstractly representing the others. As shown in Figure 5, Toulouse-Lautrec left many portrait etudes. The main feature of his etude style is represented as simplification except for facial region. In his etudes,

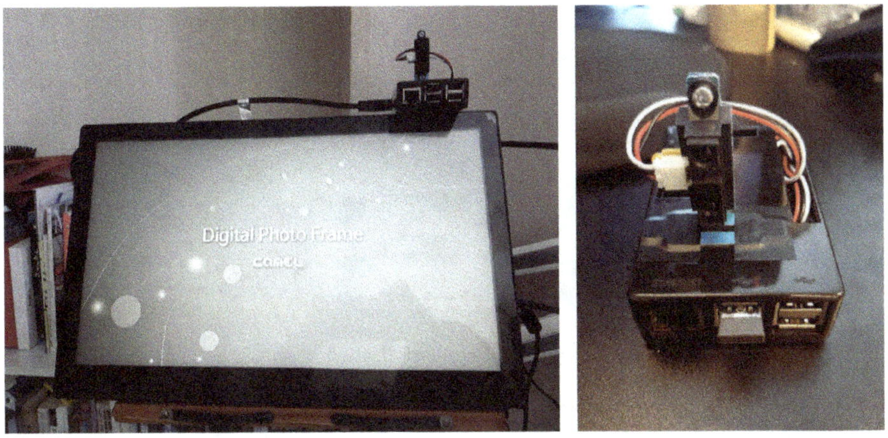

FIGURE 3: Digital photo frame and server with IR receiver.

(a) (b) (c)

FIGURE 4: Brush stroke textures: oil paint (a), pastel (b), and pencil (c).

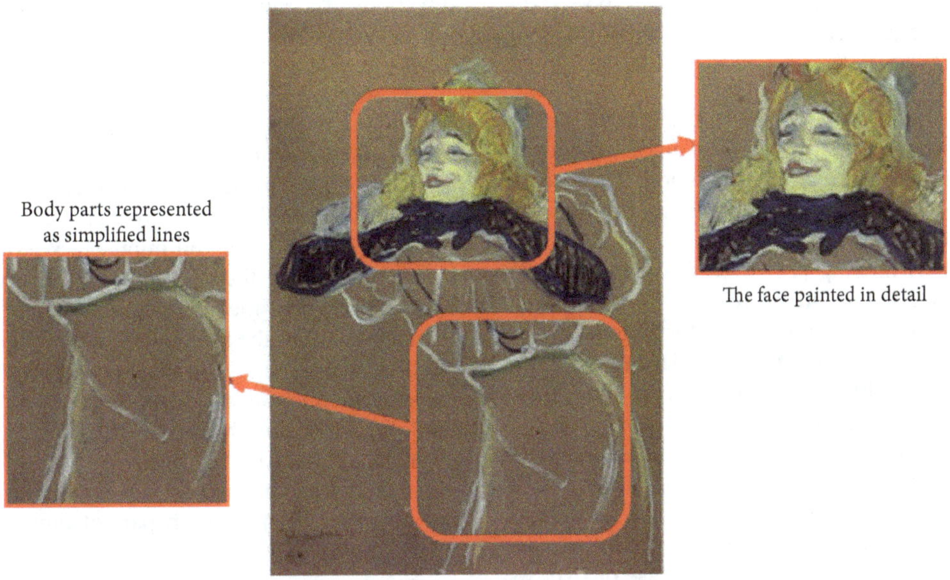

FIGURE 5: The main feature of Toulouse-Lautrec's portrait etude.

the background is generally eliminated. Only face part is depicted in detail while most of body parts are simplified as a few lines. In this manner, we extract simplified edges for representing body and manipulate them to mimic the lines he used. Then, we paint brush strokes on facial region mostly to make user concentrate on the facial expression. We present the detail of proposed algorithm in the following section.

4.1. Portrait Etude Stylization. Figure 6 shows our portrait etude stylization process. The process is divided into three steps. The first step is line processing. In this step, we extract

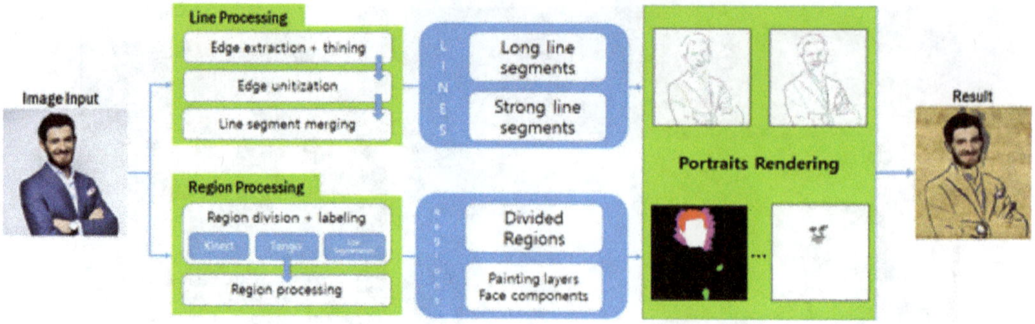

FIGURE 6: Portrait etude stylization process.

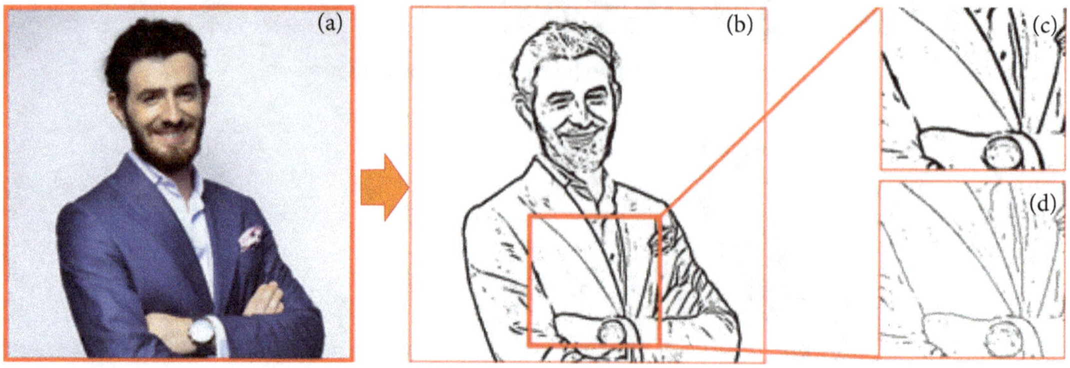

FIGURE 7: Edges extracted through FDoG (c) and thinned edges (d).

edges from input image and merge them into long simple lines. Next step is region processing; we divide the input image into several regions such as face, hair, body, and background. Final step is rendering processing. In this step, we render line and brush strokes.

4.1.1. Line Processing. In this step, we focus on the expression of simple lines shown throughout Toulouse-Lautrec's portrait etudes, especially for body expression. To generate simplified lines, we first extract edges from input image and vectorize them. In this study, flow-based difference of Gaussian (FDoG) filtering [12] is used to extract edges. As shown in Figure 7, the filtered result is a raster image including thick edges. Therefore, vectorizing edges is necessary to obtain simplified lines. To achieve this, we first use the thinning algorithm to make 1-pixel wide edges (Figure 7(d)). We then use the tracing algorithm to detect and trace each point on edges by using 3×3 kernels as shown in Figure 8. In the figure, the tracing starts from the end point of the edge (a) and proceeds along the edge (b–d). In cases of (a), (b), and (c), there is only one candidate for next edge point which is not traced yet. However, in case of (d), two candidates are given, so choosing one which makes the edge simpler is required. To solve this, we select the point where the angle difference is minimized through the vector calculation (inner product) of the tracked edge points and candidate points.

If more than one pixel is cut off on the edge (Figure 9(a)), it will not be traced to single line (Figure 9(b)) because the kernel covers only adjacent neighbor pixels. To solve this, we merge nearby lines considering the distance and direction between them (Figure 9(c)). If the distance between each end point of two lines and the difference between their directions are below predefined threshold, we merge them into the single line. In this paper, experimentally, thresholds for distance and direction are set to a default value of 5 and 15°, respectively.

Among the separated line segments, we select long and strong line segments only by using their lengths and average gradient values to represent overall structure abstractly. Figure 10 shows the line segments and a histogram which represents their length and numbers. We obtain a threshold value for dividing them into long and short line segments by applying the k-means clustering algorithm with $k = 2$on the histogram. In the figure, long lines (blue) and short lines are divided by using the threshold whose value is set to 32. In the same manner, we divide the line segments into strong and weak lines by using their gradient intensity. In our method, strong lines tend to be detected more in the areas where detailed expression such as face is needed. These two lines are utilized to express each part of input image separately in Section 4.1.3.

4.1.2. Region Processing. To express each region separately according to what the region is, it is needed to divide the input image into regions and label them. As shown in Figure 5, Toulouse-Lautrec's etude can be mainly divided into face, hair, body, and halo in background. As mentioned in Section 3.2, we obtain such regions by using depth perception and user assistance.

In Toulouse-Lautrec's etudes, he briefly depicted facial components such as eyebrows, eyes, nose, and mouth by

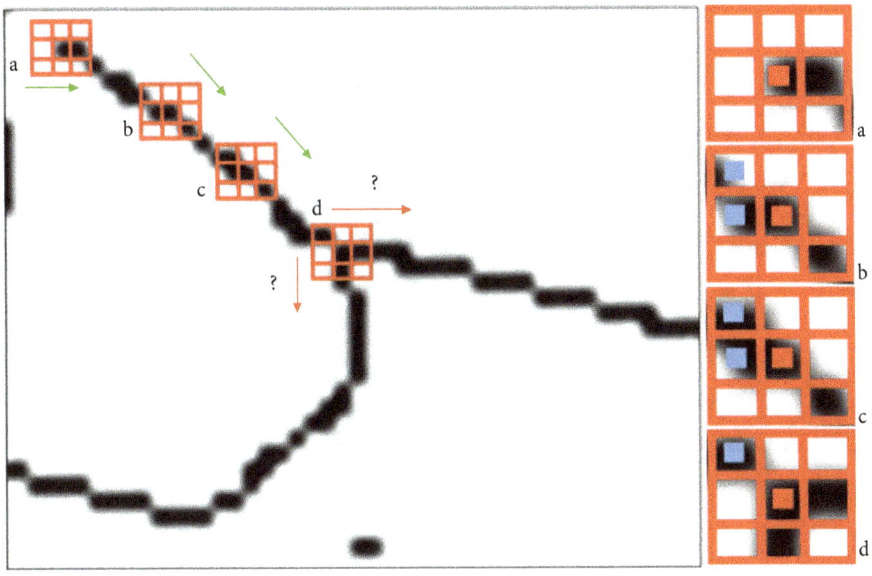

FIGURE 8: Edge tracing algorithm.

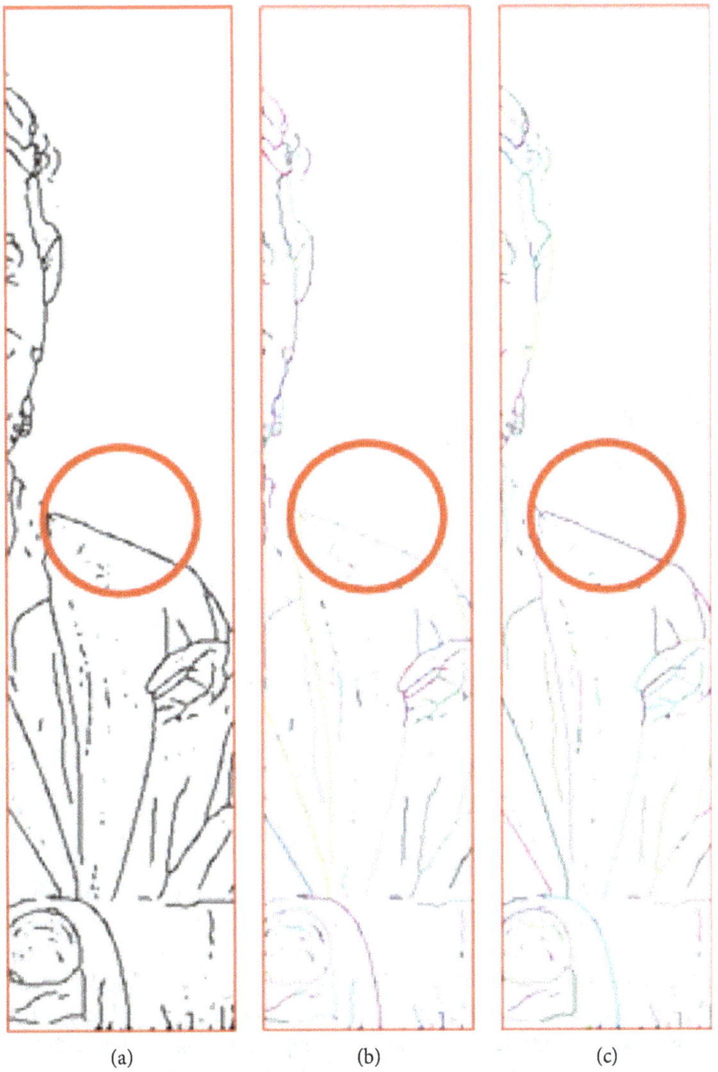

FIGURE 9: Line merging: (a) original edge, (b) tracked line, and (c) merged line.

Figure 10: Line selection: (a) original edges, (b) line length histogram, (c) separated line segments, (d) long lines, and (e) strong lines.

Figure 11: Face edge extraction: (a) facial component distance and (b) face edges.

using simple lines rather than colored brush strokes. To express this, we only draw the lines which correspond to these components. To achieve this, we extract 77 face landmarks from the input image by using active shape model (ASM). We then construct a distance map around facial components which are generated by connecting the landmarks as shown in Figure 11. Finally, we remove out-of-range lines by using the distance map.

For halo region in background, we represent the backlight behind the face. In Toulouse-Lautrec's etudes, the halo is often represented as straight hatching lines with arbitrary direction or following the silhouette of the head. To express this, we calculate modified flow map where the direction is generated by mixing the flow direction obtained by ETF [12] and arbitrary direction as shown in Figure 12.

4.1.3. Rendering Lines and Brush Strokes. Before rendering lines and brush strokes, we create a rectangular background using groundwood paper texture which is shown in Toulouse-Lautrec's etudes. We then render in the order of long lines, halo, face, hair, and facial components. As shown in Figure 13, we render each line through following steps: we first create a triangle strip mesh following the gradient direction; we then color it with the color sampled from input image

FIGURE 12: Flow direction modification for halo: (a) original painting gradient, (b) gradient correction map, and (c) modified gradient.

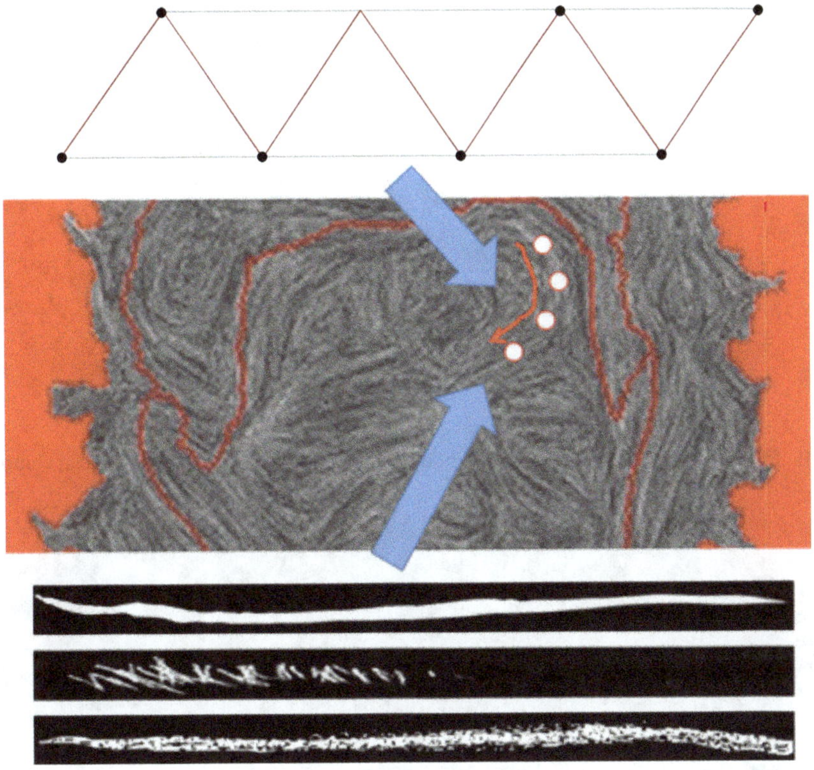

FIGURE 13: Line rendering using a triangle strip mesh and line texture.

where the mesh is located on; and we finally map predefined line texture captured from Toulouse-Lautrec's etudes.

Our line drawing is performed using the long and strong line segments obtained previously. To render lines with various width, we adjust the width by using the following equation:

$$\text{Thickness} = \frac{\alpha \times \max(w, h)}{\text{mag}}, \quad (1)$$

where w and h are the width and height of the input image, mag is the average gradient strength of the area through which the line segment passes, and α controls the maximum width of lines. Figure 14 shows the lines resulting from adjusting α.

Rendering brush strokes is performed on each region separately. We depict face and hair in detail but body and halo roughly. To do this, we arrange the brush intervals narrow and short on face and hair. On the other hand, for body and halo, the spacing between brushes was widened, and the length was arranged long.

After rendering brush strokes on the entire facial region, we draw facial components defined in Section 4.1.2 to convey facial expression. To achieve this, we first render additional thin brushes on the facial region and then draw the edges corresponding to the facial components. Figure 15 shows the results without and with drawing facial components. As shown in the figure, drawing facial components convey facial expression better.

Figure 16 shows our portrait etude rendering results. As shown in the figure, our method mainly focused on conveying facial expression while representing the others abstractly. In Section 5.2, we evaluate our results by using user study in terms of similarity to Toulouse-Lautrec's etudes.

FIGURE 14: Lines rendered using different user control maximum width.

(a) (b)

FIGURE 15: Effect of facial components: (a) without facial components and (b) with facial components.

FIGURE 16: Our portrait etude results.

5. Experimental Results

Figure 17 shows the portrait drawings our system generated. As mentioned above, our system generates four styles of portrait drawings from the most exciting photograph of the day and allows users to select preferred results to display them on the digital photo frame of the system. In the figure, each marked result represents the drawing selected by the user who took the input photograph. In case of that users do not perform the optional semiautomatic segmentation described in Section 3.2, only facial region and background are distinguished; consequently, our portrait etude is not generated. We note that the results which do not contain the portrait etude result correspond to this case. As shown in the figure, the system generated the drawings of the most exciting moments of the day. We can observe that the system rendered facial region in detail compared to the others. It is because our smartphone application divided input photograph into face and the other parts semiautomatically and used them to render each part separately.

In the next two sections, we show that our system makes users feel better after recalling the most exciting moments of their days by admiring the drawings our system generates and also show that our portrait etude stylization

FIGURE 17: Input photographs and the portrait drawings our system generated. Marked result represents the photograph/drawing selected by the user.

method generates results that reasonably resemble Toulouse-Lautrec's etude.

5.1. Evaluation of the System. We suppose that a user takes photograph throughout the day and uses our system after coming back home where the system is installed. However, we installed the system at an office in the university and conducted a user study. We lent each participant a smartphone where our application was installed and a smartwatch, demonstrated how the devices worked to them, and let them freely take portrait photographs containing themselves throughout the day while wearing the watch. We then made them visit the office, demonstrated how the digital photo frame worked to them, and finally let them admire the original photograph they took and the drawings generated. Participants consisted of 19 undergraduate and graduate students who did not have any expertise in fine art. When they admired the photograph and drawings displayed on digital photo frame, we checked which styles they chose and also checked whether they had performed semiautomatic segmentation after taking the photograph. We then inquired them about the next three questions. (1) Do you think you felt better after admiring the drawings? (2) Do you agree with that the drawings correspond with the most

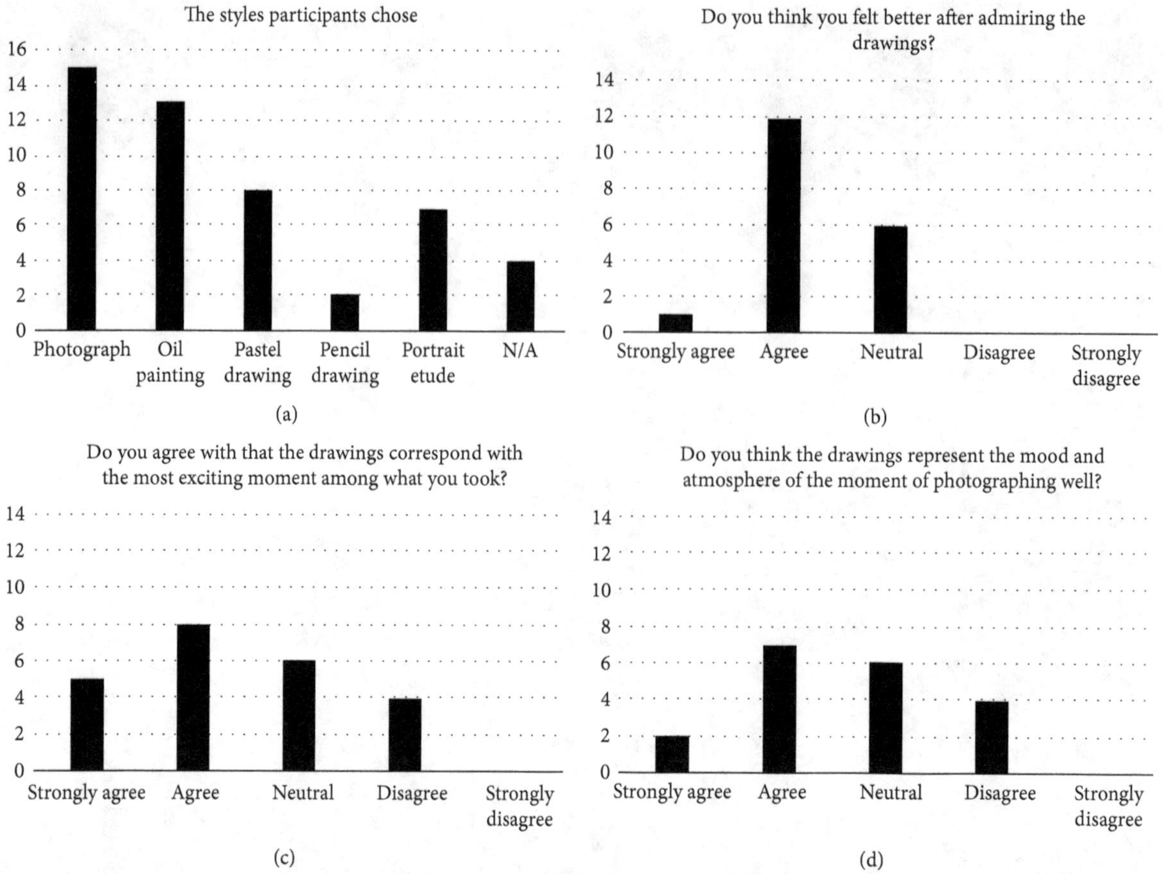

FIGURE 18: User study results.

exciting moment among what you took? (3) Do you think the drawings represent the mood and atmosphere of the moment of photographing well? For the answers, we use the following scale: 1 = strongly disagree, 2 = disagree, 3 = neutral, 4 = agree, and 5 = strongly agree. In addition, we asked them to give reasons and any feedback.

Figure 18(a) shows the styles participants chose. In the experiments, participants were able to choose multiple styles including original photograph to display them. Otherwise, they were also able to select none of the styles to display all of them. As shown in the figure, among 19 participants, 15 of them chose their preferred styles; the others chose nothing. Regarding the styles chosen by participants, photograph, oil painting, pastel drawing, pencil drawing, and portrait etude were chosen 15, 13, 8, 2, and 7 times, respectively. At this time, we note that portrait etudes were not generated for 7 cases where semiautomatic segmentation was not performed. The result shows that participants mostly preferred original photograph. Among drawings, oil painting was the most preferred style in our system. In this regard, participants responded that original photograph was easy to recognize. They also answered that oil painting was the most familiar style among what we provided. Meanwhile, pencil drawing was chosen only twice. They also mentioned that this was because facial detail was not depicted well in our pencil drawings.

Figure 18(b) shows the answers to question 1. The answers ranged from 3 to 5, and its average value was 3.74. From this result, we confirm that they mostly felt better after admiring the drawings our system generated.

Figure 18(c) shows the answers to question 2. The answers were distributed from 2 to 5, and the average value was 3.37. This was the worst score among our three questions. We assume that this is because our system depends on heart rate to select the most exciting moment. For instance, a participant answered 2 = disagree to Figure 19 of which input photograph was taken after exercise, and we can suppose that heart rate was affected by exercise rather than emotion. In this regard, we present a possible solution for this problem in Section 6.

Lastly, Figure 18(d) shows the answers to question 3. The answers ranged from 3 to 5. Its average value scored 3.95, and this was the best score in our experiments. Especially, the score of participants who chose our portrait etude as their one of preferred styles was 4.42. We analyze that this is because the algorithm effectively conveyed facial expression by focusing on mainly expressing facial region while representing the others abstractly.

Taking all together, we confirm that our system generated portrait drawings which effectively conveyed the mood of the moment of photographing and that users felt better after admiring the drawings, although the drawings did not

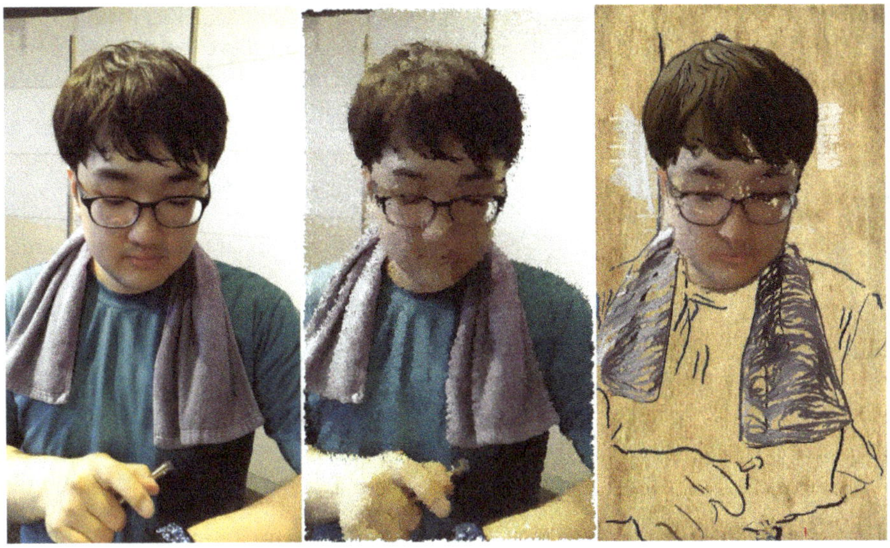

FIGURE 19: The photograph taken after exercise and its drawing results.

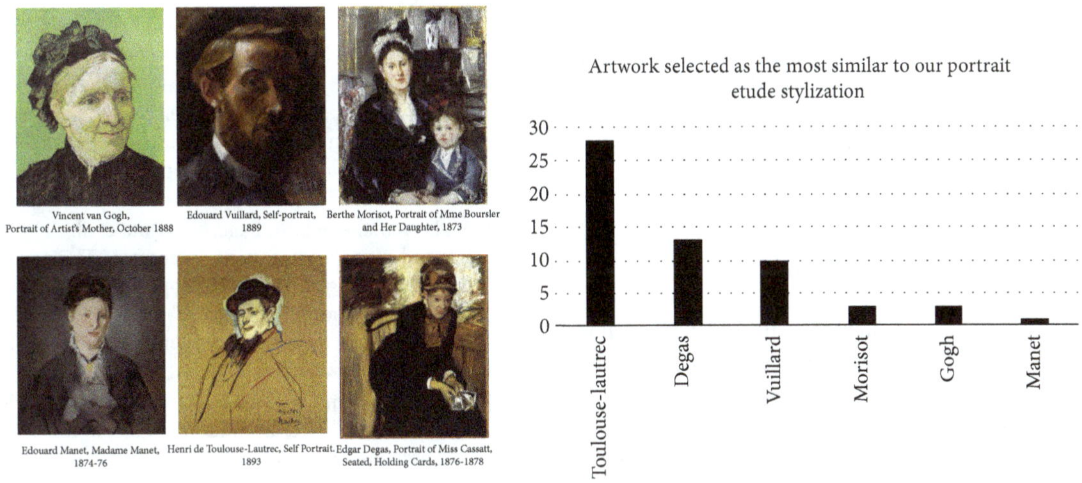

FIGURE 20: The artworks used for comparison and the number of participants' selections.

well correspond with the most exciting moment of the day. Besides multiple-choice answers, participants' feedbacks were as follows: the styles provided were too limited; portrait etude looked like incomplete; pencil drawing lacked detailed depiction; segmentation process was bothersome; not only the best moment but also the other moments could be displayed; smartphone application drained battery too fast. For these, we also suggest several solutions in Section 6.

5.2. Evaluation of Portrait Etude Rendering. For the style of portrait etude we proposed in Section 4.1, we conducted a user study as well. 31 participants including 19 who participated in the experiment in Section 5.1 consisted of undergraduate and graduate students who did not have expertise in fine art. We showed them 6 portrait artworks (Figure 20) including Toulouse-Lautrec's etude and let them choose the artworks whose styles were similar to our portrait etude drawing results. We allowed multiple-choice questions, consequently 58 selections were made. Figure 20 shows the artworks and the number of participants' selections. As shown in the figure, Toulouse-Lautrec's artwork was selected as the one which was the most similar to our portrait stylization by being chosen by 28 participants. Artworks of Edgar Degas, Eduard Vuillard, Berthe Morisot, Vincent Van Gogh, and Edouard Manet were chosen by 13, 10, 3, 3, and 1 participants, respectively. From this result, we confirm that our portrait etude stylization method generated resulting drawings quite similar to Toulouse-Lautrec's artwork which we targeted. Participants who mainly chose the artworks of Toulouse-Lautrec, Degas, and Vuillard said that line style, halo effect, selective abstraction, and incompleteness were similar to the characteristics found in our resulting drawings. On the other hand, the participants who chose Morisot and Gogh explained that the strokes used for facial area in our results were similar to those of artworks.

We also asked participants how similar Toulouse-Lautrec's artwork and our resulting drawings are. For

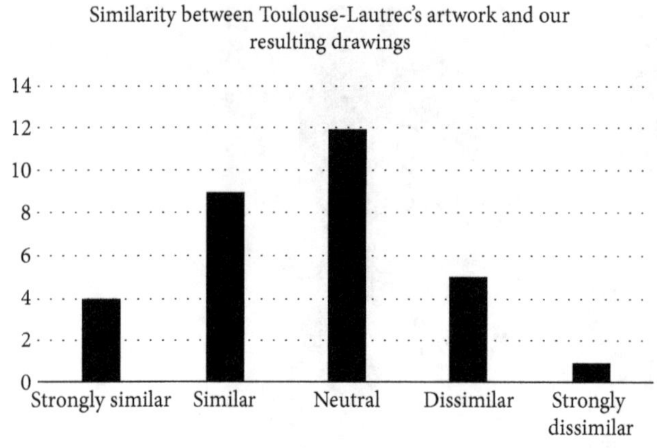

Figure 21: User test result for similarity between Toulouse-Lautrec's artwork and our results.

answers, we used the scale same with Section 5.1. Figure 21 shows the result. The answers were widely distributed from 1 to 5, and the average value was 3.32. Although participants chose Toulouse-Lautrec's artwork as the style most similar to ours, but the score they evaluated was not very high when the result was directly compared with his artwork. In this regard, they mentioned that painting nonfacial parts and using more abstract lines were required to mimic his artwork.

6. Conclusions and Future Work

In this study, we proposed a system that automatically generated portrait drawings for the purpose of human emotional care. In the system, our smartphone application enabled the user to take photographs throughout the day while acquiring heart rates from the smartwatch worn by the user. Meanwhile, the server collected the photographs and heart rates and displayed portrait drawings automatically stylized from the photograph for the most exciting moment of the day. Using the system, the user could recall the exciting and happy moment of the day through admiring the drawings and heal the emotion accordingly. To stylize photographs as portrait drawings, we employed NPR methods, including a portrait etude stylization proposed in this paper. Finally, our system and proposed stylization method were evaluated through user studies.

We collected some ideas for future work from the feedback of user study participants. As mentioned before, participants answered that the drawings our system generated occasionally did not correspond with the most exciting moment of the day. We analyzed that this was caused that our system entirely depended on heart rate to judge it. To solve this problem, we will employ emotion recognition from facial expression. In addition, participants also mentioned that semiautomatic segmentation was bothersome. To provide more better user experiences, we will employ automatic segmentation and labelling methods using semantic recognition. We also believe that this will be beneficial to represent more regions with their own styles to enhance our portrait etude stylization.

Conflicts of Interest

The authors declare that there is no conflict of interest regarding the publication of this paper.

Acknowledgments

The authors thank the participants of the user studies of the paper. This work was supported by the National Research Foundation (NRF) of Korea grant funded by the Korea government (MSIT) (no. NRF-2017R1A2B4007481) and Chung-Ang University research grant in 2017.

References

[1] "Hello Heart," http://helloheart.com.

[2] D. Phan, L. Y. Siong, P. N. Pathirana, and A. Seneviratne, "Smartwatch: performance evaluation for long-term heart rate monitoring," in *2015 International Symposium on Bioelectronics and Bioinformatics (ISBB)*, pp. 144–147, Beijing, China, October 2015.

[3] M. Hassanalieragh, A. Page, T. Soyata et al., "Health monitoring and management using Internet-of-Things (IoT) sensing with cloud-based processing: opportunities and challenges. In: services computing (SCC)," in *2015 IEEE International Conference on Services Computing*, pp. 285–292, New York, NY, USA, July 2015.

[4] S. M. R. Islam, D. Kwak, M. H. Kabir, M. Hossain, and K. S. Kwak, "The Internet of Things for health care: a comprehensive survey," *IEEE Access*, vol. 3, pp. 678–708, 2015.

[5] J. E. Kyprianidis, J. Collomosse, T. Wang, and T. Isenberg, "State of the "art": a taxonomy of artistic stylization techniques for images and video," *IEEE Transactions on Visualization and Computer Graphics*, vol. 19, no. 5, pp. 866–885, 2013.

[6] P. Haeberli, "Paint by numbers: abstract image representations," in *SIGGRAPH '90 Proceedings of the 17th Annual Conference on Computer Graphics and Interactive Techniques*, vol. 24, pp. 207–214, Dallas, TX, USA, August 1990.

[7] A. Hertzmann, "Painterly rendering with curved brush strokes of multiple sizes," in *SIGGRAPH '98 Proceedings of the 25th Annual Conference on Computer Graphics and Interactive Techniques*, pp. 453–460, New York, NY, USA, 1998.

[8] J. Hays and I. Essa, "Image and video based painterly animation," in *NPAR '04 Proceedings of the 3rd International Symposium on Non-Photorealistic Animation and Rendering*, pp. 113–120, Annecy, France, June 2004.

[9] S. C. Olsen, B. A. Maxwell, and B. Gooch, "Interactive vector fields for painterly rendering," in *GI '05 Proceedings of Graphics Interface 2005*, pp. 241–247, Victoria, British Columbia, May 2005.

[10] S. H. Seo and K. H. Yoon, "Color juxtaposition for pointillism based on an artistic color model and a statistical analysis," *The Visual Computer*, vol. 26, no. 6–8, pp. 421–431, 2010.

[11] S. Seo, S. Ryoo, and P. Jinwann, "Interactive painterly rendering with artistic error correction," *Multimedia Tools and Applications*, vol. 65, no. 2, pp. 221–237, 2013.

[12] H. Kang, S. Lee, and C. K. Chui, "Coherent line drawing," in *NPAR '07 Proceedings of the 5th International Symposium on Non-Photorealistic Animation and Rendering*, pp. 43–50, San Diego, CA, USA, August 2007.

9

Optimal Attitude Determination from Vector Sensors using Fast Analytical Singular Value Decomposition

Zhuohua Liu,[1] Wei Liu,[2,3] Xiangyang Gong,[1] and Jin Wu[4]

[1]State Key Laboratory of Networking and Switching Technology, Beijing University of Posts and Telecommunications, Beijing 100876, China
[2]Beijing University of Posts and Telecommunications, Beijing 100876, China
[3]State Key Laboratory of Information Security, Institute of Information Engineering, Chinese Academy of Sciences, Beijing 100093, China
[4]School of Automation, University of Electronic Science and Technology of China, Chengdu, China

Correspondence should be addressed to Xiangyang Gong; xygong@bupt.edu.cn and Jin Wu; jin_wu_uestc@hotmail.com

Academic Editor: Francesco Dell'Olio

A novel algorithm is proposed in this paper to solve the optimal attitude determination formulation from vector observation pairs, that is, the Wahba problem. We propose here a fast analytic singular value decomposition (SVD) approach to obtain the optimal attitude matrix. The derivations and mandatory proofs are presented to clarify the theory and support its feasibility. Through simulation experiments, the proposed algorithm is validated. The results show that it maintains the same attitude determination accuracy and robustness with conventional methodologies but significantly reduces the computation time.

1. Introduction

In aerial applications, strapdown vector sensors are often required to directly give the rotation matrix between the body frame and inertial frame [1]. With the development of modern global navigation satellite systems (GNSS), GNSS-based attitude determination has been a vital part in spacecraft instrumentation and measurement [2]. This leads to different algorithms of attitude determination on various platforms, for example, remote sensing and marine navigation [3, 4]. In such applications, the attitude determination from vector observation pairs is usually an effective solution [5, 6]. This approach is referred to as the Wahba problem, posed in 1965, which minimizes the following loss function

$$L(\mathbf{R}) = \frac{1}{2}\sum_{i=1}^{n} a_i \left\| \mathbf{D}_i^b - \mathbf{R}\mathbf{D}_i^r \right\|^2, \quad (1)$$

in which $\mathbf{D}_i^b$ and $\mathbf{D}_i^r$ are the ith pair of vector observation that was obtained in the body frame and reference frame, respectively [7]. a_i is the positive weight of the ith pair with the sum of 1. Here, $\mathbf{R}$ is the 3-dimensional direction cosine matrix (DCM) belonging to SO(3) [8]. Initial research on this problem mainly focuses on the exact solution rather than its robustness [9]. For instance, the above loss function can be explicitly given by the augmented rotation equations. Brute force calculation of such a system requires a large matrix memory and its sophisticated generalized inverse. Such solution is not only time and space consuming, but it will also cause nonorthogonality of the DCM. This in fact has a demand on later matrix correction that deduces the optimality of the solution. A robust framework was proposed by Davenport in 1968, who analytically changed the Wahba problem into an eigenvalue problem [9]. The target, that is, DCM, is shifted to an attitude quaternion accordingly. Following this formulation, many famous attitude estimators were then proposed in the next 30 years, including Shuster's QUaternion ESTimator (QUEST), Markley's Fast Optimal Attitude Matrix (FOAM), Mortari's ESOQ, and Wu's FLAE [10–13].

The robust solutions to Wahba's problem provides a possibility of accurate attitude determination from sensors like the sun sensor, nadir sensor, star tracker, magnetometer, and so on [14]. It can significantly improve the attitude estimation results combined with gyroscopes [15–17]. Apart from Davenport's method, it is also noticeable that the singular value decomposition (SVD) could be efficient. Markley proposed the SVD-based Wahba solution in 1988 where the DCM is composed of decomposition results from the SVD [18]. The SVD algorithm is proven to be very robust, hence facing some critical cases of sensor combinations, where its accuracy is always maintained. However, conventional SVD computation is very complicated. Could there be a fast analytical SVD algorithm that can boost its execution on embedded platforms?

The answer is positive. As a matter of fact, a 3-matrix SVD can be given with much faster speed. And in this paper, we focus on presenting a novel fast SVD algorithm. The derivations are given to show why it is fast. Yet simulations are carried out to compare the proposed algorithm with representative ones on accuracy, robustness, and time consumption, which reflect its superiority in real applications.

This paper is structured as follows: Section 2 contains our fast SVD theory and its relationship with optimal attitude determination. Section 3 includes the experimental results. Section 4 consists of concluding remarks.

2. Our Method

The introduced matrix theory in this paper is mainly referenced from [19]. Given an arbitrary nonsingular real squared matrix $\mathbf{A}$, the matrix $\mathbf{B} = \mathbf{A}^{\mathrm{T}}\mathbf{A}$ is real symmetry. One can compute $\mathbf{A}$'s singular values with $\mathbf{B}$'s eigenvalues $\lambda_1 \geq \lambda_2 \geq \lambda_3 > 0$ with $\sigma_i = \sqrt{\lambda_i}$, $i = 1, 2, 3$. The Householder transform and QR factorization are usually adopted for SVD. However, in real applications, they are proven to be time-consuming. Here, we define that matrix $\mathbf{A}$ is decomposed by

$$\mathbf{A} = \mathbf{UDV}, \tag{2}$$

where $\mathbf{D} = \mathrm{diag}\,(\sigma_1, \sigma_2, \sigma_3)$ and $\mathbf{U}$ and $\mathbf{V}$ are 3×3 orthonormal matrices. That is to say, the factorization process of $\mathbf{A}$ can be significantly simplified if the preliminary decomposition of $\mathbf{B}$ is completed. Actually, in scientific computations, such technique is very common which uses the QR decomposition or Jacobi method imposed on $\mathbf{B}$.

Now we propose the analytic eigenvalue decomposition of $\mathbf{B}$. The characteristic polynomial of $\mathbf{B}$ is given by

$$\begin{aligned}
f(\lambda) = \det\,(\lambda\mathbf{I} - \mathbf{B}) &= \begin{vmatrix} \lambda - \mathrm{B}_{11} & -B_{12} & -B_{13} \\ -B_{12} & \lambda - \mathrm{B}_{22} & -B_{23} \\ -B_{13} & -B_{23} & \lambda - \mathrm{B}_{33} \end{vmatrix} \\
&= (\lambda - \mathrm{B}_{11})\left[(\lambda - B_{22})(\lambda - B_{33}) - B_{23}^2\right] \\
&\quad + B_{12}[-B_{12}(\lambda - B_{33}) - B_{13}B_{23}] \\
&\quad - B_{13}[B_{12}B_{23} + (\lambda - B_{22})B_{13}] = \lambda^3 - (B_{11} + B_{22} + B_{33})\lambda^2 \\
&\quad + \left(B_{11}B_{22} + B_{11}B_{33} + B_{22}B_{33} - B_{23}^2 - B_{12}^2 - B_{13}^2\right)\lambda \\
&\quad - \left(B_{11}B_{22}B_{33} - B_{11}B_{23}^2 - B_{12}^2B_{33} + 2B_{12}B_{13}B_{23} - B_{13}^2B_{22}\right) \\
&= \lambda^3 + a\lambda^2 + b\lambda + c,
\end{aligned} \tag{3}$$

where

$$\begin{aligned}
a &= -(B_{11} + B_{22} + B_{33}), \\
b &= \left(B_{11}B_{22} + B_{11}B_{33} + B_{22}B_{33} - B_{23}^2 - B_{12}^2 - B_{13}^2\right), \\
c &= -\left(B_{11}B_{22}B_{33} - B_{11}B_{23}^2 - B_{12}^2B_{33} + 2B_{12}B_{13}B_{23} - B_{13}^2B_{22}\right),
\end{aligned} \tag{4}$$

with B_{ij} denoting the element of $\mathbf{B}$ in the ith row and jth column. Letting $\lambda = x - a/3$, $f(\lambda) = 0$ can be converted to

$$x^3 + px + q = 0, \tag{5}$$

where

$$\begin{aligned}
p &= \frac{3b - a^2}{3}, \\
q &= \frac{2a^3 - 9ab + 27c}{27}.
\end{aligned} \tag{6}$$

The discriminant is established by

$$\Delta = \left(\frac{p}{3}\right)^3 + \left(\frac{q}{2}\right)^2. \tag{7}$$

As far as the symbol of the discriminant is concerned, there are three specific situations:

(1) $\Delta > 0$: there are one real and a pair of conjugate complex roots.

$$\begin{aligned}
x_1 &= k_1 + k_2, \\
x_2, x_3 &= -\frac{1}{2}x_1 \pm \frac{\sqrt{3}j}{2}(k_1 - k_2),
\end{aligned} \tag{8}$$

where

$$\begin{aligned}
k_1 &= \frac{1}{2}\sqrt[3]{-4q + 8\sqrt{\Delta}}, \\
k_2 &= \frac{1}{2}\sqrt[3]{-4q - 8\sqrt{\Delta}},
\end{aligned} \tag{9}$$

in which j denotes the unitary imaginary number.

(2) $\Delta = 0$: there are three real roots in which two of them are equal to each other.

$$\begin{aligned}
x_1 &= \sqrt[3]{-4q} \\
\text{or } x_1 &= -\mathrm{sign}\,(q)\frac{2}{3}\sqrt{-3p}, \\
x_2 = x_3 &= -\frac{x_1}{2}.
\end{aligned} \tag{10}$$

Especially when $p = q = 0$, the roots are zeros.

(3) $\Delta < 0$: there are three different real roots.

$$x_1 = \frac{2}{3}\sqrt{-3p}\cos\frac{\theta}{3},$$
$$x_2, x_3 = \frac{2}{3}\sqrt{-3p}\cos\left(\frac{\theta \pm 2\pi}{3}\right), \quad (11)$$

where

$$\theta = \arccos\frac{9q}{2p\sqrt{-3p}}. \quad (12)$$

Based on the three conditions and inserting the roots into $\lambda = x - a/3$, **B**'s eigenvalues can be obtained. Here we assume that they are sorted in descending order described as before. Note that **B** is really symmetric. Then its eigenvalues are real numbers, that is, the roots of (5) are real as well. Hence only the abovementioned cases (2) and (3) would be adopted. Apart from this, according to the Vieta theorem [20], the eigenvalues are associated with

$$\lambda_1 + \lambda_2 + \lambda_3 = -a,$$
$$\lambda_1\lambda_2 + \lambda_2\lambda_3 + \lambda_3\lambda_1 = b \quad (13)$$
$$\lambda_1\lambda_2\lambda_3 = -c.$$

Since the eigenvalues are assumed to be real, their maximum condition number could be $-a$, while the minimum one is that of $-c/b$. Now we define the condition number of the characteristic polynomial equation $f(\lambda) = 0$ as

$$\mathrm{cond}(f(\lambda)) = \frac{ab}{c}. \quad (14)$$

Before computations, it is suggested that the users check the condition number first. If the condition number is too large, that means the numerical stability of the calculated results are not satisfactory. In such circumstance, we may resort to a more robust way, for example, the Jacobi method.

Based on the matrix theory, the orthogonal eigenvectors corresponding to the above-computed eigenvalues can be determined as follows:

(1) The eigenvalues have a multiplicity of 3 ($\lambda_1 = \lambda_2 = \lambda_3$). Three orthonormal eigenvectors can be immediately established by

$$\mathbf{v}_1 = \begin{bmatrix}1\\0\\0\end{bmatrix}, \quad \mathbf{v}_2 = \begin{bmatrix}0\\1\\0\end{bmatrix}, \quad \mathbf{v}_3 = \begin{bmatrix}0\\0\\1\end{bmatrix}. \quad (15)$$

(2) The eigenvalues have a multiplicity of 2 ($\lambda_1 = \lambda_2 > \lambda_3$ or $\lambda_1 > \lambda_2 = \lambda_3$). The eigenvector associated with the eigenvalue λ_2 meets

$$(\lambda_2\mathbf{I} - \mathbf{B})\mathbf{v}_2 = 0. \quad (16)$$

Letting $\mathbf{C} = \lambda_2\mathbf{I} - \mathbf{C}$, we have rank $(\mathbf{C}) = 1$, that is, the three row vectors of **R** are linearly correlated. Assuming that the maximum element of **R** is R_{ij}, it is obtained that

$$\mathbf{C}_i\mathbf{v} = [C_{i1}C_{i2}C_{i3}]\mathbf{v} = \mathbf{0}, \quad (17)$$

where $\mathbf{T}_i$ is the ith row vector of the matrix **C**. If $j = 1$ or $j = 2$, the eigenvector can be directly established by

$$\mathbf{v}_2 = [C_{i2} - C_{i1}0]^{\mathrm{T}}. \quad (18)$$

And if $j = 3$, we have

$$\mathbf{v}_2 = [0 - C_{i3}C_{i2}]^{\mathrm{T}}. \quad (19)$$

When $\lambda_1 = \lambda_2 > \lambda_3$, the eigenvectors are composed by

$$\mathbf{v}_1 = \mathbf{C}_i \times \mathbf{v}_2,$$
$$\mathbf{v}_3 = \mathbf{v}_1 \times \mathbf{v}_2, \quad (20)$$

while when $\lambda_1 > \lambda_2 = \lambda_3$, the eigenvectors are composed by

$$\mathbf{v}_3 = \mathbf{C}_i \times \mathbf{v}_2,$$
$$\mathbf{v}_1 = \mathbf{v}_2 \times \mathbf{v}_3. \quad (21)$$

(3) The eigenvalues are all different: the eigenvector associated with the eigenvalue λ_1 meets

$$(\lambda_1\mathbf{I} - \mathbf{B})\mathbf{v}_2 = \mathbf{0}. \quad (22)$$

Letting $\mathbf{C} = \lambda_1\mathbf{I} - \mathbf{B}$, we have rank $(\mathbf{C}) = 2$, which describes that the three row vectors of **C** should be on the same plane, with the preliminary definition that the zero vector can be on an arbitrary plane. Note that at least two of them are not parallel with each other. Here we define

$$\mathbf{v}_{t1} = \mathbf{C}_1 \times \mathbf{C}_2,$$
$$\mathbf{v}_{t2} = \mathbf{C}_2 \times \mathbf{C}_3, \quad (23)$$
$$\mathbf{v}_{t3} = \mathbf{C}_1 \times \mathbf{C}_3.$$

Theoretically, $\mathbf{v}_{t1}, \mathbf{v}_{t2}, \mathbf{v}_{t3}$ are parallel. Every one of them can be the eigenvector only if it is not a zero vector. In engineering practice, the item with the maximum norm is usually chosen for computing the eigenvector $\mathbf{v}_1$ belonging to the eigenvalue λ_1. Using the same technique, the eigenvector $\mathbf{v}_2$ can be computed. While for $\mathbf{v}_3$, it can be given by $\mathbf{v}_3 = \mathbf{v}_1 \times \mathbf{v}_2$.

Finally the eigenvectors $\mathbf{v}_1, \mathbf{v}_2, \mathbf{v}_3$ are normalized. Letting

$$\mathbf{V} = \begin{bmatrix} \frac{\mathbf{v}_1}{|\mathbf{v}_1|} & \frac{\mathbf{v}_2}{|\mathbf{v}_2|} & \frac{\mathbf{v}_3}{|\mathbf{v}_3|} \end{bmatrix}, \tag{24}$$

one can write

$$\mathbf{BV} = \mathbf{V\Lambda}, \tag{25}$$

in which

$$\Lambda = \text{diag}\,(\lambda_1 \quad \lambda_2 \quad \lambda_3). \tag{26}$$

Inserting $\mathbf{B} = \mathbf{A}^{\mathrm{T}}\mathbf{A}$ and $\Lambda = \mathbf{D}^2$ into (26), we obtain

$$\mathbf{A}^{\mathrm{T}}\mathbf{AV} = \mathbf{VD}^2, \tag{27}$$

which is transformed into

$$\mathbf{A} = \left(\mathbf{A}^{\mathrm{T}}\right)^{-1}\mathbf{VDDV}^{\mathrm{T}} = \mathbf{UDV}^{\mathrm{T}}. \tag{28}$$

Extracting $\mathbf{U} = (\mathbf{A}^{\mathrm{T}})^{-1}\mathbf{VD} = \mathbf{A}(\mathbf{A}^{\mathrm{T}}\mathbf{A})^{-1}\mathbf{VD} = \mathbf{A}(\mathbf{VD}^2\mathbf{V}^{\mathrm{T}})^{-1}$ $\mathbf{VD} = \mathbf{AVD}^{-1}$, we can easily verify that $\mathbf{U}$ is an orthogonal matrix, namely, $\mathbf{U}^{\mathrm{T}}\mathbf{U} = |(\mathbf{A}^{\mathrm{T}})^{-1}\mathbf{VD}|^{\mathrm{T}}(\mathbf{AVD}^{-1}) = \mathbf{DV}^{\mathrm{T}}\mathbf{A}^{-1}$ $\mathbf{AVD}^{-1} = \mathbf{I}$. Now, we have finished the presentation of the proposed SVD method.

The proposed SVD method can be applied to productions in which the computation resources are highly restricted. Relating to the attitude determination, we hereinafter introduce two propositions.

Proposition 1. *For the 3×3 nonsingular matrix $\mathbf{A}$, if its SVD is $\mathbf{A} = \mathbf{UDV}^{\mathrm{T}}$, $\mathbf{C} = \mathbf{UV}^{\mathrm{T}}$ is one of the closest orthonormal matrices to $\mathbf{A}$, such that*

$$\arg\min_{\mathbf{C}} J(\mathbf{C}) = \arg\min_{\mathbf{C}} \|\mathbf{C}-\mathbf{A}\|^2 = \arg\min_{\mathbf{C}} \sum_{i=1}^{3}\sum_{j=1}^{3}(C_{ij}-A_{ij})^2. \tag{29}$$

Proof 1. (46) is equivalent to

$$\arg\min_{\mathbf{C}} J(\mathbf{C}) = \arg\min_{\mathbf{C}} \text{tr}\left((\mathbf{C}-\mathbf{A})^{\mathrm{T}}(\mathbf{C}-\mathbf{A})\right). \tag{30}$$

With the consideration of $\mathbf{C}^{\mathrm{T}}\mathbf{C} = \mathbf{I}$, introducing Lagrange multipliers, the Lagrangian is developed by

$$J_o = \text{tr}\left((\mathbf{C}-\mathbf{A})^{\mathrm{T}}(\mathbf{C}-\mathbf{A}) + \Theta\left(\mathbf{C}^{\mathrm{T}}\mathbf{C}-\mathbf{I}\right)\right), \tag{31}$$

where Θ is the symmetric Lagrange multiplication matrix. Using differential rules of matrix

$$\begin{aligned} &\frac{\partial}{\partial \mathbf{X}}\text{tr}(\mathbf{A}^{\mathrm{T}}\mathbf{X}) = \frac{\partial}{\partial \mathbf{X}}\text{tr}(\mathbf{X}^{\mathrm{T}}\mathbf{A}) = \mathbf{A}^{\mathrm{T}}, \\ &\frac{\partial}{\partial \mathbf{X}}\text{tr}(\mathbf{AX}^{\mathrm{T}}\mathbf{X}) = 2\mathbf{AX}. \end{aligned} \tag{32}$$

The above Lagrangian is optimized by

$$\begin{aligned} \frac{\partial J_o}{\partial \mathbf{C}} &= \frac{\partial}{\partial \mathbf{C}}\text{tr}\left(\mathbf{C}^{\mathrm{T}}\mathbf{C} - \mathbf{C}^{\mathrm{T}}\mathbf{A} - \mathbf{A}^{\mathrm{T}}\mathbf{C} + \mathbf{A}^{\mathrm{T}}\mathbf{A} + \Theta\mathbf{C}^{\mathrm{T}}\mathbf{C} - \Theta\right) \\ &= \frac{\partial}{\partial \mathbf{C}}\text{tr}\left((\mathbf{I}+\Theta)\mathbf{C}^{\mathrm{T}}\mathbf{C} - \mathbf{C}^{\mathrm{T}}\mathbf{A} - \mathbf{A}^{\mathrm{T}}\mathbf{C} + \left(\mathbf{A}^{\mathrm{T}}\mathbf{A} - \Theta\right)\right) \\ &= 2\left[(\mathbf{I}+\Theta)\mathbf{C}^{\mathrm{T}} - \mathbf{A}^{\mathrm{T}}\right] = 0, \end{aligned} \tag{33}$$

producing

$$\mathbf{A} = \mathbf{C}(\mathbf{I}+\Theta). \tag{34}$$

Then we have

$$\begin{aligned} \mathbf{A}^{\mathrm{T}}\mathbf{A} &= [\mathbf{C}(\mathbf{I}+\Theta)]^{\mathrm{T}}[\mathbf{C}(\mathbf{I}+\Theta)] \\ &= (\mathbf{I}+\Theta)\mathbf{C}^{\mathrm{T}}\mathbf{C}(\mathbf{I}+\Theta) = (\mathbf{I}+\Theta)^2, \end{aligned} \tag{35}$$

which arrives at

$$\mathbf{I}+\Theta = \pm\left(\mathbf{A}^{\mathrm{T}}\mathbf{A}\right)^{1/2}, \tag{36}$$

Inserting (36) into (34), $\mathbf{C}$ is determined by

$$\mathbf{C} = \pm\mathbf{A}\left(\mathbf{A}^{\mathrm{T}}\mathbf{A}\right)^{-1/2}. \tag{37}$$

When $\mathbf{A}$ approximates an attitude matrix, that is, det $(\mathbf{A}) \approx +1$, we choose the positive symbol, which gives

$$\begin{aligned} \mathbf{C} = \mathbf{A}\left(\mathbf{A}^{\mathrm{T}}\mathbf{A}\right)^{-1/2} &= \mathbf{UDV}^{\mathrm{T}}\left[\left(\mathbf{UDV}^{\mathrm{T}}\right)^{\mathrm{T}}\left(\mathbf{UDV}^{\mathrm{T}}\right)\right]^{-1/2} \\ &= \mathbf{UDV}^{\mathrm{T}}\left(\mathbf{VDDV}^{\mathrm{T}}\right)^{-1/2} = \mathbf{UDV}^{\mathrm{T}}\left(\mathbf{UDV}^{\mathrm{T}}\mathbf{UDV}^{\mathrm{T}}\right)^{-1/2} \\ &= \mathbf{UDV}^{\mathrm{T}}\left(\mathbf{UDV}^{\mathrm{T}}\right)^{-1/2} = \mathbf{UV}^{\mathrm{T}}. \end{aligned} \tag{38}$$

If we rewrite the SVD into the following form

$$\mathbf{A} = \mathbf{UDV}^{\mathrm{T}} = \mathbf{UV}^{\mathrm{T}}\cdot\mathbf{VDV}^{\mathrm{T}} = \mathbf{CS}, \tag{39}$$

where $\mathbf{C} = \mathbf{UV}^{\mathrm{T}}$ and $\mathbf{S} = \mathbf{VDV}^{\mathrm{T}}$ are a symmetric positive definite matrix, that is, det $(\mathbf{S}) > 0$, then (37) represents the orthonormal symmetric decomposition.

Proposition 2. *For the 3×3 nonsingular matrix $\mathbf{A}$, if its SVD is $\mathbf{A} = \mathbf{UDV}^{\mathrm{T}}$, the choice $\mathbf{C} = \mathbf{UV}^{\mathrm{T}}$ makes the following function maximum:*

$$J(\mathbf{C}) = \text{tr}\left(\mathbf{CA}^{\mathrm{T}}\right) = \max. \tag{40}$$

Proof 2. (38) is equivalent to

$$\begin{aligned} J(\mathbf{C}) = \text{tr}\left(\mathbf{CA}^{\mathrm{T}}\right) &= \text{tr}\left(\mathbf{C}\left(\mathbf{UDV}^{\mathrm{T}}\right)^{\mathrm{T}}\right) = \text{tr}\left(\mathbf{CVDU}^{\mathrm{T}}\right) \\ &= \text{tr}\left(\mathbf{U}^{\mathrm{T}}\mathbf{CV}\cdot\mathbf{D}\right) = \text{tr}(\mathbf{C}^*\mathbf{D}), \end{aligned} \tag{41}$$

where

$$\mathbf{C}^* = \mathbf{U}^{\mathrm{T}}\mathbf{CV}. \tag{42}$$

Then we have

```
% Optimal Attitude Determination Algorithm Using Fast Singular Value Decomposition
% author: liuzhuohua@bupt.edu.cn; jin_wu_uestc@hotmail.com
function [u1,u2, u3, ...
        d1, d2, d3, ...
        v1, v2, v3] = svd3_fast (A, ...
                         A11, A12, A13, ...
                         A21, A22, A23, ...
                         A31, A32, A33)
    B11 = A11 * A11 + A21 * A21 + A31 * A31;
    B12 = A11 * A12 + A21 * A22 + A31 * A32;
    B13 = A11 * A13 + A21 * A23 + A31 * A33;
    B22 = A12 * A12 + A22 * A22 + A32 * A32;
    B23 = A12 * A13 + A22 * A23 + A32 * A33;
    B33 = A13 * A13 + A23 * A23 + A33 * A33;
    B12B23 = B12 * B23;
    B13B22 = B13 * B22;
    B13B12 = B13 * B12;
    B11B23 = B11 * B23;
    B12B12 = B12 * B12;
    B11B22 = B11 * B22;
    a = -(B11 + B22 + B33);
    b =  B11B22 + B11 * B33 + B22 * B33 - B23^2 - B12B12 - B13^2;
    c = -(B11B22 * B33 - B11B23 * B23 - B12B12 * B33 + 2 * B12B23 * B13 - B13 * B13B22);
    a3 = a/3;
    p = b - a * a3;
    q = (2 * a^3-9 * a * b + 27 * c)/27;
    p2 = p * p;
    T = -q * sqrt(-27 * p)/2/p2;
    theta3 = acos(T)/3;
    pi23 = 2.094395102393195;
    x0 = 2 * sqrt(-p/3);
    x1 = x0 * cos(theta3);
    x2 = x0 * cos(theta3 - pi23);
    x3 = x0 * cos(theta3 + pi23);
    rs1 = x1 - a3;
    rs2 = x2 - a3;
    rs3 = x3 - a3;
    if rs1 < rs2, tmp = rs2; rs2 = rs1; rs1 = tmp; end
    if rs1 < rs3, tmp = rs3; rs3 = rs1; rs1 = tmp; end
    if rs2 < rs3, tmp = rs2; rs2 = rs3; rs3 = tmp; end
    v1 = [B12B23 - B13B22 + rs1 * B13
          B13B12 - B11B23 + rs1 * B23
          (rs1 - B11) * (rs1 - B22) - B12B12];
    v2 = [B12B23 - B13B22 + rs2 * B13
          B13B12 - B11B23 + rs2 * B23
          (rs2 - B11) * (rs2 - B22) - B12B12];
    v1 = v1/norm(v1);
    v2 = v2/norm(v2);
    v3 = [v1 (2) * v2 (3) - v1 (3) * v2 (2);
          v1 (3) * v2 (1) - v1 (1) * v2 (3);
          v1 (1) * v2 (2) - v1 (2) * v2 (1)];
    d1 = sqrt(rs1);
    d2 = sqrt(rs2);
    d3 = sqrt(rs3);
    u1 = A * v1/d1;
    u2 = A * v2/d2;
    u3 = A * v3/d3;
end
```

ALGORITHM 1: MATLAB code of the proposed fast SVD algorithm.

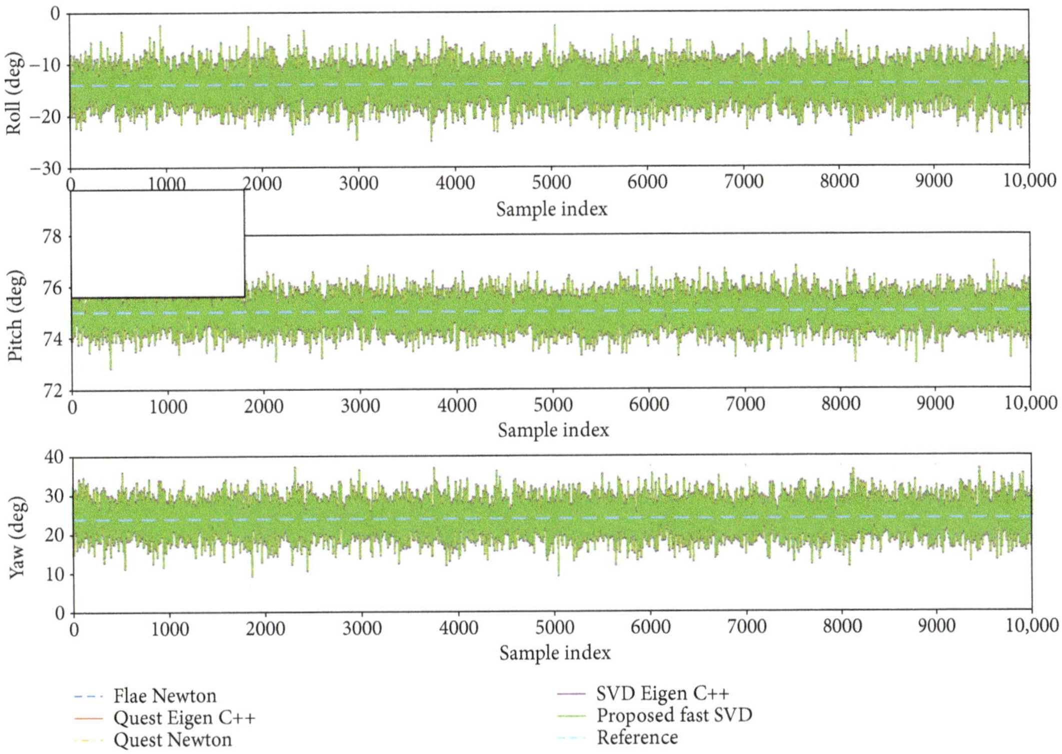

FIGURE 1: Static test results.

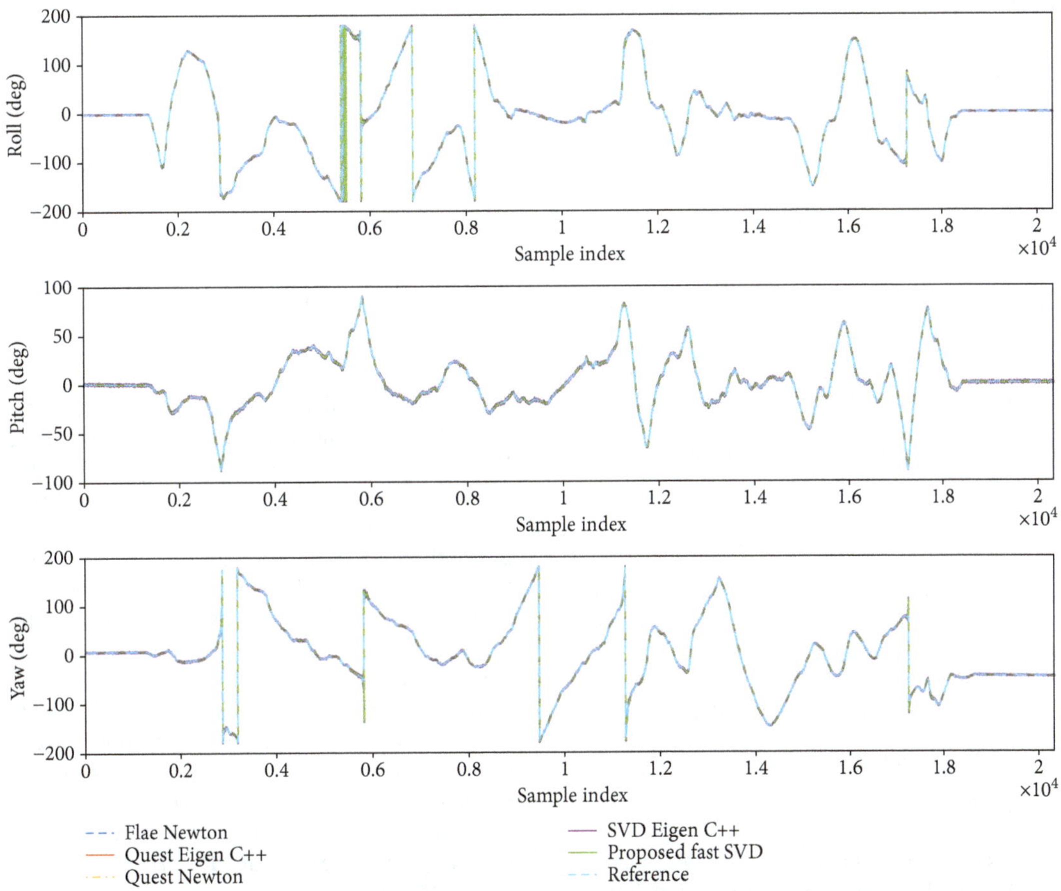

FIGURE 2: Dynamic test results.

$$J(\mathbf{C}) = \mathrm{tr}(\mathbf{C}^*\mathbf{D}) = \mathrm{tr}\left(\begin{bmatrix} C_{11}^* & C_{12}^* & C_{13}^* \\ C_{21}^* & C_{22}^* & C_{23}^* \\ C_{31}^* & C_{32}^* & C_{33}^* \end{bmatrix}\begin{bmatrix} \sigma_1 & 0 & 0 \\ 0 & \sigma_2 & 0 \\ 0 & 0 & \sigma_3 \end{bmatrix}\right) = (C_{11}^*\sigma_1 + C_{22}^*\sigma_2 + C_{33}^*\sigma_3) \le (\sigma_1 + \sigma_2 + \sigma_3). \quad (43)$$

Since $\sigma_1 \ge \sigma_2 \ge \sigma_3 > 0$, the above "equal" holds if and only if $C_{11}^* = C_{22}^* = C_{33}^* = 1$, which corresponds to $\mathbf{C}^* = \mathbf{I}$.

Finally (42) arrives at

$$\mathbf{C} = \mathbf{U}\mathbf{V}^{\mathrm{T}}. \quad (44)$$

It has been proposed by Arun et al. in 1987 [21] and Markley in 1988 [18], respectively, that for equal-weight and normal-weight cases the optimal DCM can be computed with

$$\mathbf{R} = \mathbf{V_R}\mathbf{M_R}\mathbf{U_R^T}, \mathbf{M_R} = \mathrm{diag}\,(1, 1, \det(\mathbf{U_R})\det(\mathbf{V_R})), \quad (45)$$

with the given

$$\mathbf{B_R} = \sum_{i=1}^{n} a_i \mathbf{D}_i^b (\mathbf{D}_i^r)^{\mathrm{T}} = \mathbf{U_R}\mathbf{D_R}\mathbf{V_R^T}. \quad (46)$$

The Wahba problem can be shifted to maximizing [22]

$$\underset{\mathbf{RR}^{\mathrm{T}}=\mathbf{I},\det(\mathbf{R})=+1}{\arg\min} J(\mathbf{R}) = \underset{\mathbf{RR}^{\mathrm{T}}=\mathbf{I},\det(\mathbf{R})=+1}{\arg\min} \mathrm{tr}\left(\mathbf{RB_R^T}\right). \quad (47)$$

According to the above propositions, the optimal DCM is calculated by

$$\mathbf{R} = \mathbf{U_R^T}\mathbf{V_R}. \quad (48)$$

The proposed fast SVD is now summarized in the MATLAB coding (see Algorithm 1).

3. Simulation Results

In this subsection, we conduct two simulations to verify the accuracy and efficiency of the proposed fast SVD method.

3.1. Static Simulation. The vector observation pairs are simulated with

$$\mathbf{D}_i^b = \mathbf{R}_{\text{true}}\mathbf{D}_i^r + \varepsilon_i, \quad i = 1, 2, \ldots n, \quad (49)$$

in which the noise items ε_i with different indexes are independent from each other and subject to Gaussian distribution. The reference DCM $\mathbf{R}_{\text{true}}$ here is chosen as

$$\mathbf{R}_{\text{true}} = \begin{pmatrix} 0.23644 & 0.18118 & -0.95460 \\ -0.10527 & 0.98145 & 0.16020 \\ 0.96592 & 0.06261 & 0.25113 \end{pmatrix}, \quad (50)$$

which corresponds to the Euler angles of roll: −14°, pitch: 75°, yaw: 24°. Here the standard deviation of noise is defined as 0.01 for each axis.

Table 1: Test cases.

Case	Reference vectors	Noise standard deviations
1	$\mathbf{D}_1^r = [1,0,0]^{\mathrm{T}}$, $\mathbf{D}_2^r = [0,1,0]^{\mathrm{T}}$, $\mathbf{D}_3^r = [0,0,1]^{\mathrm{T}}$	$\sigma_1 = 10^{-6}$, $\sigma_2 = 10^{-6}$, $\sigma_3 = 10^{-6}$
2	$\mathbf{D}_1^r = [1,0,0]^{\mathrm{T}}$, $\mathbf{D}_2^r = [0,1,0]^{\mathrm{T}}$	$\sigma_1 = 10^{-6}$, $\sigma_2 = 10^{-6}$
3	$\mathbf{D}_1^r = [1,0,0]^{\mathrm{T}}$, $\mathbf{D}_2^r = [0,1,0]^{\mathrm{T}}$, $\mathbf{D}_3^r = [0,0,1]^{\mathrm{T}}$	$\sigma_1 = 0.01$, $\sigma_2 = 0.01$, $\sigma_3 = 0.01$
4	$\mathbf{D}_1^r = [1,0,0]^{\mathrm{T}}$, $\mathbf{D}_2^r = [0,1,0]^{\mathrm{T}}$	$\sigma_1 = 0.01$, $\sigma_2 = 0.01$
5	$\mathbf{D}_1^r = [0.6,0.8,0]^{\mathrm{T}}$, $\mathbf{D}_2^r = [0.8,-0.6,0]^{\mathrm{T}}$	$\sigma_1 = 10^{-6}$, $\sigma_2 = 0.01$
6	$\mathbf{D}_1^r = [1,0,0]^{\mathrm{T}}$, $\mathbf{D}_2^r = [1,0.01,0]^{\mathrm{T}}$, $\mathbf{D}_3^r = [1,0,0.01]^{\mathrm{T}}$	$\sigma_1 = 10^{-6}$, $\sigma_2 = 10^{-6}$, $\sigma_3 = 10^{-6}$
7	$\mathbf{D}_1^r = [1,0,0]^{\mathrm{T}}$ $\mathbf{D}_2^r = [1,0.01,0]^{\mathrm{T}}$	$\sigma_1 = 10^{-6}$, $\sigma_2 = 10^{-6}$
8	$\mathbf{D}_1^r = [1,0,0]^{\mathrm{T}}$, $\mathbf{D}_2^r = [1,0.01,0]^{\mathrm{T}}$, $\mathbf{D}_3^r = [1,0,0.01]^{\mathrm{T}}$	$\sigma_1 = 0.01$, $\sigma_2 = 0.01$, $\sigma_3 = 0.01$
9	$\mathbf{D}_1^r = [1,0,0]^{\mathrm{T}}$, $\mathbf{D}_2^r = [1,0.01,0]^{\mathrm{T}}$	$\sigma_1 = 0.01$, $\sigma_2 = 0.01$
10	$\mathbf{D}_1^r = [1,0,0]^{\mathrm{T}}$, $\mathbf{D}_2^r = [0.96,0.28,0]^{\mathrm{T}}$, $\mathbf{D}_3^r = [0.96,0,0.28]^{\mathrm{T}}$	$\sigma_1 = 10^{-6}$, $\sigma_2 = 0.01$, $\sigma_3 = 0.01$
11	$\mathbf{D}_1^r = [1,0,0]^{\mathrm{T}}$, $\mathbf{D}_2^r = [0.96,0.28,0]^{\mathrm{T}}$	$\sigma_1 = 10^{-6}$, $\sigma_2 = 0.01$
12	$\mathbf{D}_1^r = [1,0,0]^{\mathrm{T}}$, $\mathbf{D}_2^r = [0.96,0.28,0]^{\mathrm{T}}$	$\sigma_1 = 0.01$, $\sigma_2 = 10^{-6}$

With this model, 10,000 groups of 3 pairs of vector observations are simulated for testing. The FLAE, QUEST, and Markley's SVD method are employed for comparison. For FLAE, we here use the Newton iteration to compute the eigenvalues of Davenport matrix **K**. While for QUEST, we use both Newton iteration and direct factorization with the Eigen math library. The Eigen library is also adopted for calculation of Markley's SVD method. We use the MATLAB software to evaluate the results, which are shown in Figure 1. We can see from the figure that the tested results of representative algorithms and the proposed method are all the same. Actually, since we have not broken the conventional framework of Wahba's problem, the final accuracy are all the same. However, as is known, SVD has been recognized as a robust computation tool. The numerical stability of SVD is far beyond the eigenvalue factorization.

By taking more cases into account, detailed time consumption results are tested. The designed cases from Markley

Table 2: Roll RMSE (deg).

Case	SVD Eigen	Fast SVD	QUEST Eigen	QUEST Newton	FLAE Newton
1	4.3120×10^{-05}	4.3120×10^{-05}	4.3120×10^{-05}	4.3120×10^{-05}	4.3120×10^{-05}
2	6.0836×10^{-05}	6.0836×10^{-05}	6.0836×10^{-05}	6.0836×10^{-05}	6.0836×10^{-05}
3	4.3242×10^{-01}	4.3242×10^{-01}	4.3242×10^{-01}	4.3242×10^{-01}	4.3242×10^{-01}
4	5.9031×10^{-01}	5.9031×10^{-01}	5.9031×10^{-01}	5.9031×10^{-01}	5.9031×10^{-01}
5	4.4283×10^{-01}	4.4283×10^{-01}	4.4283×10^{-01}	5.0414×10^{-01}	$2.4250 \times 10^{+01}$
6	4.9071×10^{-03}	4.9071×10^{-03}	4.9071×10^{-03}	4.9071×10^{-03}	4.9071×10^{-03}
7	8.0575×10^{-03}	8.0575×10^{-03}	8.0575×10^{-03}	8.0575×10^{-03}	8.0575×10^{-03}
8	$5.9527 \times 10^{+01}$	$5.9527 \times 10^{+01}$	$5.9527 \times 10^{+01}$	$5.9527 \times 10^{+01}$	$5.9527 \times 10^{+01}$
9	$7.7087 \times 10^{+01}$	$7.7087 \times 10^{+01}$	$7.7087 \times 10^{+01}$	$7.7087 \times 10^{+01}$	$7.7087 \times 10^{+01}$
10	$1.4617 \times 10^{+00}$	$1.4617 \times 10^{+00}$	$1.4617 \times 10^{+00}$	$4.7431 \times 10^{+01}$	$1.5650 \times 10^{+00}$
11	$2.0270 \times 10^{+00}$	$2.0270 \times 10^{+00}$	$2.0270 \times 10^{+00}$	$4.6515 \times 10^{+01}$	$2.9349 \times 10^{+00}$
12	$2.0594 \times 10^{+00}$	$2.0594 \times 10^{+00}$	$2.0594 \times 10^{+00}$	$3.6679 \times 10^{+01}$	$1.7381 \times 10^{+01}$

Table 3: Pitch RMSE (deg).

Case	SVD Eigen	Fast SVD	QUEST Eigen	QUEST Newton	FLAE Newton
1	4.0938×10^{-05}	4.0938×10^{-05}	4.0938×10^{-05}	4.0938×10^{-05}	4.0938×10^{-05}
2	5.3539×10^{-05}	5.3539×10^{-05}	5.3539×10^{-05}	5.3539×10^{-05}	5.3539×10^{-05}
3	4.0487×10^{-01}	4.0487×10^{-01}	4.0487×10^{-01}	4.0487×10^{-01}	4.0487×10^{-01}
4	5.3909×10^{-01}	5.3909×10^{-01}	5.3909×10^{-01}	5.3909×10^{-01}	5.3909×10^{-01}
5	4.0024×10^{-01}	4.0024×10^{-01}	4.0024×10^{-01}	4.5559×10^{-01}	$1.2596 \times 10^{+01}$
6	4.0103×10^{-05}	4.0103×10^{-05}	4.0103×10^{-05}	4.0103×10^{-05}	4.0103×10^{-05}
7	5.2893×10^{-05}	5.2893×10^{-05}	5.2893×10^{-05}	5.2893×10^{-05}	5.2893×10^{-05}
8	3.6948×10^{-01}	3.6948×10^{-01}	3.6948×10^{-01}	3.6948×10^{-01}	3.6948×10^{-01}
9	4.5164×10^{-01}	4.5164×10^{-01}	4.5164×10^{-01}	4.5164×10^{-01}	4.5164×10^{-01}
10	5.6876×10^{-05}	5.6876×10^{-05}	5.6876×10^{-05}	5.6877×10^{-05}	5.6876×10^{-05}
11	5.7268×10^{-05}	5.7268×10^{-05}	5.7268×10^{-05}	5.7270×10^{-05}	5.7265×10^{-05}
12	4.8652×10^{-01}	4.8652×10^{-01}	4.8652×10^{-01}	$7.9615 \times 10^{+00}$	$3.9358 \times 10^{+00}$

are employed (see Table 1, [5]). The results are categorized as follows (Tables 2–4). The presented materials show that the proposed algorithm has the same attitude accuracy with the compared representative ones.

3.2. Dynamic Simulation. In this subsection, the attitude is simulated using an attitude and heading reference system (AHRS) on which a MicroStrain 3DM-GX3-25 module is attached to the platform firmly. With a gimbal with angle encoders, the platform is rotated and the raw sensor data along with the reference angles are logged with 500 Hz. The reference angles are converted to DCMs that are inserted into (49) to simulate vector observations. With the abovementioned compared algorithms, the results are evaluated and plotted in Figure 2. The overall tests show that the proposed fast SVD method is not different compared with the representative methods. The superiority of the proposed method is going to be introduced in Subsection 3.3.

3.3. Computation Complexity. Several algorithms are tested with simulated pairs of vector observations. Here, in each round, there are 5 pairs of vector observations. The execution time consumption from different sources are collected with the internal timer of the MATLAB software and is shown in Figure 3. It is obvious that the proposed fast SVD is the best one compared with the other ones. This is because the proposed algorithm seeks the intelligent way by taking the simplest analytical form of SVD of the 3×3 matrix.

The low time consumption of the proposed method can significantly increase the CPU schedule of the attitude determination system especially on embedded platforms. The reserved time can be used for emergency backup to enhance the overall stability [23].

4. Conclusion

In this paper, an analytical singular value decomposition (SVD) of a 3×3 matrix is proposed. The algorithm is based on the SVD of a real symmetric matrix and the roots to its characteristic polynomial. Detailed derivations of roots to this cubic equation and further compositions of eigenvectors are introduced. With the presented theory, any 3×3 real matrix can be factorized with much faster speed. This is then applied to the optimal attitude determination problem. Simulations are carried out which contains static and dynamic

TABLE 4: Yaw RMSE (deg).

Case	SVD Eigen	Fast SVD	QUEST Eigen	QUEST Newton	FLAE Newton
1	4.3411×10^{-05}	4.3411×10^{-05}	4.3411×10^{-05}	4.3411×10^{-05}	4.3411×10^{-05}
2	4.8769×10^{-05}	4.8769×10^{-05}	4.8769×10^{-05}	4.8769×10^{-05}	4.8769×10^{-05}
3	4.3535×10^{-01}	4.3535×10^{-01}	4.3535×10^{-01}	4.3535×10^{-01}	4.3535×10^{-01}
4	4.8474×10^{-01}	4.8474×10^{-01}	4.8474×10^{-01}	4.8474×10^{-01}	4.8474×10^{-01}
5	2.5750×10^{-01}	2.5750×10^{-01}	2.5750×10^{-01}	2.9318×10^{-01}	$1.7425\times10^{+01}$
6	3.6295×10^{-05}	3.6295×10^{-05}	3.6295×10^{-05}	3.6295×10^{-05}	3.6295×10^{-05}
7	4.8501×10^{-05}	4.8501×10^{-05}	4.8501×10^{-05}	4.8501×10^{-05}	4.8501×10^{-05}
8	3.9564×10^{-01}	3.9564×10^{-01}	3.9564×10^{-01}	3.9564×10^{-01}	3.9564×10^{-01}
9	5.0110×10^{-01}	5.0110×10^{-01}	5.0110×10^{-01}	5.0110×10^{-01}	5.0110×10^{-01}
10	6.1477×10^{-05}	6.1477×10^{-05}	6.1477×10^{-05}	6.1480×10^{-05}	6.1482×10^{-05}
11	6.1345×10^{-05}	6.1345×10^{-05}	6.1345×10^{-05}	6.1343×10^{-05}	6.1348×10^{-05}
12	3.1358×10^{-01}	3.1358×10^{-01}	3.1358×10^{-01}	$5.1110\times10^{+00}$	$2.8665\times10^{+00}$

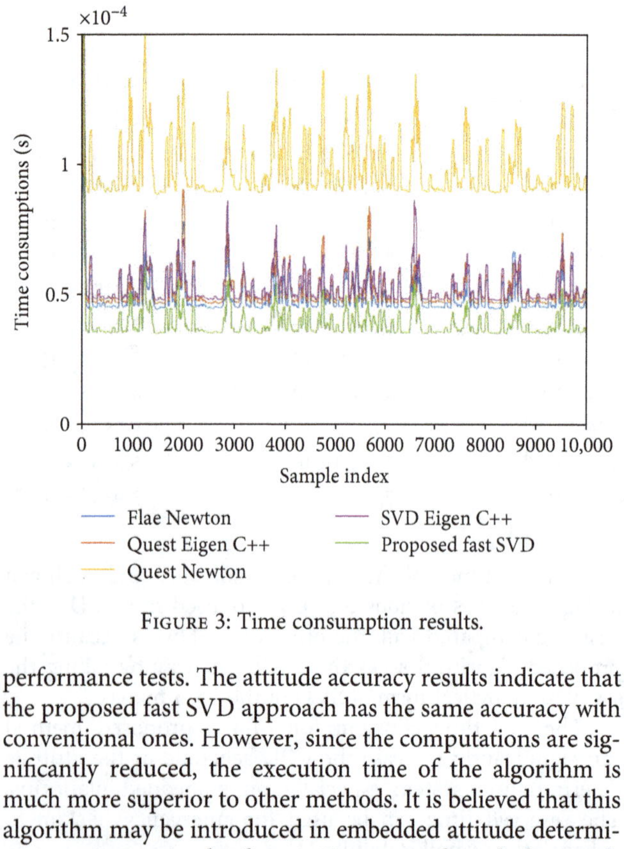

FIGURE 3: Time consumption results.

performance tests. The attitude accuracy results indicate that the proposed fast SVD approach has the same accuracy with conventional ones. However, since the computations are significantly reduced, the execution time of the algorithm is much more superior to other methods. It is believed that this algorithm may be introduced in embedded attitude determination systems in the future. However, when dealing with some critical cases, the proposed method may not be usable as the divisions may face NaN values.

Conflicts of Interest

The authors declare no conflict of interest regarding the content and publication of this paper.

Acknowledgments

The authors would like to thank Prof. Zebo Zhou for his constructive suggestions. This research is supported by the National Natural Science Foundation of China under Grant Nos. 61602053 and 41604025, the Fundamental Research Funds for Central Universities, and the Foundation of Key Laboratory (2017-MS-14).

References

[1] J. F. Vasconcelos, B. Cardeira, C. Silvestre, P. Oliveira, and P. Batista, "Discrete-time complementary filters for attitude and position estimation: design, analysis and experimental validation," *IEEE Transactions on Control Systems Technology*, vol. 19, no. 1, pp. 181–198, 2011.

[2] G. Giorgi and P. J. G. Teunissen, "Low-complexity instantaneous ambiguity resolution with the affine-constrained GNSS attitude model," *IEEE Transactions on Aerospace and Electronic Systems*, vol. 49, no. 3, pp. 1745–1759, 2013.

[3] G. Giorgi, P. J. G. Teunissen, S. Verhagen, and P. J. Buist, "Testing a new multivariate GNSS carrier phase attitude determination method for remote sensing platforms," *Advances in Space Research*, vol. 46, no. 2, pp. 118–129, 2010.

[4] G. Giorgi, P. J. G. Teunissen, and T. P. Gourlay, "Instantaneous global navigation satellite system (GNSS)-based attitude determination for maritime applications," *IEEE Journal of Oceanic Engineering*, vol. 37, no. 3, pp. 348–362, 2012.

[5] J. Wu, Z. Zhou, J. Chen, H. Fourati, and R. Li, "Fast complementary filter for attitude estimation using low-cost MARG sensors," *IEEE Sensors Journal*, vol. 16, no. 18, pp. 6997–7007, 2016.

[6] S. Guo, J. Wu, Z. Wang, and J. Qian, "Novel MARG-sensor orientation estimation algorithm using fast Kalman filter," *Journal of Sensors*, vol. 2017, Article ID 8542153, 12 pages, 2017.

[7] G. Wahba, "A least squares estimate of satellite attitude," *SIAM Review*, vol. 7, no. 3, p. 409, 1965.

[8] G. Chang, T. Xu, and Q. Wang, "Error analysis of Davenport's q method," *Automatica*, vol. 75, pp. 217–220, 2017.

[9] P. B. Davenport, "A vector approach to the algebra of rotations with applications," Tech. Rep., National Aeronautics and Space Administration, Washington, DC, USA, 1968.

[10] M. D. Shuster and S. D. Oh, "Three-axis attitude determination from vector observations," *Journal of Guidance, Control, and Dynamics*, vol. 4, no. 1, pp. 70–77, 1981.

[11] F. L. Markley, "Attitude determination using vector observations: a fast optimal matrix algorithm," *The Journal of the Astronautical Sciences*, vol. 41, no. 2, pp. 261–280, 1993.

[12] D. Mortari, "ESOQ: a closed-form solution to the Wahba problem," *The Journal of the Astronautical Sciences*, vol. 45, no. 2, pp. 195–204, 1997.

[13] J. Wu, Z. Zhou, B. Gao, R. Li, Y. Cheng, and H. Fourati, "Fast linear quaternion attitude estimator using vector observations," *IEEE Transactions on Automation Science and Engineering*, vol. 15, no. 1, pp. 307–319, 2018.

[14] Y. Yang, "Attitude determination using Newton's method on Riemannian manifold," *Proceedings of the Institution of Mechanical Engineers, Part G: Journal of Aerospace Engineering*, vol. 229, no. 14, pp. 2737–2742, 2015.

[15] H. Fourati, N. Manamanni, L. Afilal, and Y. Handrich, "Complementary observer for body segments motion capturing by inertial and magnetic sensors," *IEEE/ASME Transactions on Mechatronics*, vol. 19, no. 1, pp. 149–157, 2014.

[16] A. Makni, H. Fourati, and A. Y. Kibangou, "Energy-aware adaptive attitude estimation under external acceleration for pedestrian navigation," *IEEE/ASME Transactions on Mechatronics*, vol. 21, no. 3, pp. 1366–1375, 2016.

[17] A. P. Vinod, A. D. Mahindrakar, S. Bandyopadhyay, and V. Muralidharan, "A deterministic attitude estimation using a single vector information and rate gyros," *IEEE/ASME Transactions on Mechatronics*, vol. 20, no. 5, pp. 2630–2636, 2015.

[18] F. L. Markley, "Attitude determination using vector observations and the singular value decomposition," *The Journal of the Astronautical Sciences*, vol. 36, no. 3, pp. 245–258, 1988.

[19] R. A. Horn and C. R. Johnson, *Matrix Analysis*, Cambridge University Press, 2012.

[20] E. B. Vinberg, *A Course in Algebra*, vol. 56, American Mathematical Society, 2003.

[21] K. S. Arun, T. S. Huang, and S. D. Blostein, "Least-squares fitting of two 3-D point sets," *IEEE Transactions on Pattern Analysis and Machine Intelligence*, vol. PAMI-9, no. 5, pp. 698–700, 1987.

[22] J. L. Crassidis, F. L. Markley, and Y. Cheng, "Survey of nonlinear attitude estimation methods," *Journal of Guidance, Control, and Dynamics*, vol. 30, no. 1, pp. 12–28, 2007.

[23] Z. Zhou, Y. Li, J. Zhang, and C. Rizos, "Integrated navigation system for a low-cost quadrotor aerial vehicle in the presence of rotor influences," *Journal of Surveying Engineering*, vol. 143, no. 1, article 05016006, 2017.

10

Research on Real-Time Local Rainfall Prediction based on MEMS Sensors

Zeyi Chao,[1] Fangling Pu,[1,2] Yuke Yin,[1] Bin Han,[1] and Xiaoling Chen[3]

[1] *School of Electronic Information, Wuhan University, Wuhan, Hubei 430072, China*
[2] *Collaborative Innovation Center of Geospatial Technology, Wuhan University, Wuhan, Hubei 430079, China*
[3] *State Key Laboratory for Information Engineering in Surveying, Mapping and Remote Sensing (LIESMARS), Wuhan University, Wuhan, Hubei 430079, China*

Correspondence should be addressed to Fangling Pu; flpu@whu.edu.cn

Academic Editor: Fei Hao

A more accurate and timely rainfall prediction is needed for flood disaster reduction and prevention in Wuhan. The in situ microelectromechanical systems' (MEMS) sensors can provide high time and spatial resolution of weather parameter measurement, but they suffer from stochastic measurement error. In order to apply MEMS sensors in real-time rainfall prediction in Wuhan, firstly, seasonal trend decomposition using Loess (STL) algorithm is utilized to decompose the observed time series into trend, seasonal, and remainder components. The trend of the observed series is compared with the corresponding trend of the data downloaded from the authoritative website with the same weather parameter in terms of Euclidean distance and cosine similarity. The similarity demonstrates that the observation of MEMS sensors is believable. Secondly, the long short-term memory (LSTM) is used to predict the real-time rainfall based on the observed data. Compared with autoregressive and moving average (ARMA), random forest (RF), support vector machine (SVM), and back propagation neural networks (BPNNs), LSTM not only performs as well as ARMA in real-time rainfall prediction but also outperforms the other four models in seasonal rainfall pattern description and seasonal real-time rainfall prediction. Our experiment results show that more detailed, timely, and accurate rainfall prediction can be achieved by using LSTM on the MEMS weather sensors.

1. Introduction

Rainfall is one of the major severe phenomena within a climate system. It has a direct influence on water resource management, agriculture, and ecosystems [1]. The heavy rainfall mostly contributes to landslides, mudslides, floods, and other natural disasters. The rainfall-induced disasters in each year lead to serious damage and losses of both life and infrastructure [2]. Take the central large city, Wuhan, for instance, Wuhan locates at longitude between 113°41′ and 115°05′, latitude between 29°58′ and 31°22′. Wuhan belongs to subtropical moist monsoon climate [3]. In July 2016, affected by El Nino, Wuhan suffered heavy rainfall disaster. The rainfall from June 30 to July 6 reaches 582.5 mm, which is the highest weekly rainfall since the historical records [4]. It caused severe mass floods, 2.265 billion economic loss, and death of 14 citizens. Therefore, rainfall prediction is a problem that needs to be solved to avoid or minimize losses of both life and belongings.

According to the definition of World Meteorological Organization (WMO) (http://www.wmo.int/pages/prog/www/DPS/GDPS-Supplement5-AppI-4.html), the weather forecasting can be divided into real-time, short-term, middle-term, and long-term forecasting. Real time forecasting means 0–2 hours of weather prediction. Short-term forecasting ranges from 2 hours to 72 hours. Middle-term forecasting is beyond 72 hours and up to 240 hours. Long-term forecasting is beyond 10 days and up to 30 days and further extends to 2 years. Most weather forecasts provide today's and tomorrow's weather, an hour-by-hour forecast, and a 15-day forecast for a region. The hour level weather forecasting usually does not provide timely and accurate weather parameters for early warning, especially during rainstorm. The accurate early warning is difficult to extract from the current weather forecasting data to support the emergency response and disaster management. A shorter and more accurate weather prediction is needed for flood disaster reduction and prevention.

Weather parameters have been measured by gauges, weather radars, or satellites. The observation equipment evaluation index involves measurement precision, spatial range, time range, and deployment price. Satellite-based measurement has a wider spatial coverage but a longer repeated observation period. The common sampling interval for weather radar is 15 minutes [5]. The microelectromechanical systems' (MEMS) sensors for weather observation have the highest time resolution and spatial resolution. The local weather parameters around the sensors can be repeatedly measured within one minute or less. selected points are computed. Combined with the embedded technique and wireless communication interface, such as WiFi, Zigbee, or Bluetooth, an automatic meteorological station can be established. The price to build a meteorological network using MEMS sensors is cheaper than that paid on a radar or a satellite. However, the MEMS sensors suffer from deterministic and stochastic errors. The deterministic error can be removed by calibration techniques before sale. The stochastic error cannot be directly removed from the measurement. The stochastic error is modeled as additive noise [6]. The law of weather change is hidden in the sensor time series with error.

We have built several weather monitoring stations that utilized MEMS sensors to monitor local weather in China. Two stations located in Wuhan. Four stations were deployed around Erhai, Yunnan. Each station has been equipped with seven sensors, which are wind speed, wind direction, atmospheric temperature, atmospheric humidity, atmospheric pressure, rainfall sensor, and radiation sensor. The 7 weather parameters are sampled every 5 minutes. After obtaining weather sensor data for more than 3 years in Wuhan, we wish to demonstrate that the MEMS sensor observation is believable and the data are available to real-time rainfall prediction. The accurate and more timely rainfall prediction can strongly support flooding disaster relief decision.

We focus on the short-term weather variation of Wuhan. It is assumed that the observed sensor time series have trend, seasonal, and residual components. Trend component reflects the progression of the time series, while seasonal component represents the seasonal variation. The original observed data cannot exhibit the trend because of the stochastic error. According to Alexandrov et al. [7], Loess filter behaves well in capturing the details of the trends. Seasonal trend decomposition using Loess (STL) is a filtering procedure that was developed to decompose the time series into seasonal, trend, and remainder components [8]. STL has been demonstrated that it is a robust estimation of the trend and seasonal components. More importantly, STL is able to decompose the time series with missing values, which is the inevitable cases for the wireless sensor network (WSN) due to the interruption of communication link or power-off. Therefore, we applied STL to extract the trend of the observed data and that of the published weather data. The referenced weather data were downloaded from the website (https://rp5.ru/). The validity of the observed time series is proved by calculating the similarity parameters, Euclidean distance (ED), and cosine similarity (CS), between the two trends.

Rainfall is a complex meteorological phenomenon. Accurate and real-time rainfall prediction remains challenging for many decades because of its stochastic and nonlinear nature. The proposed methods for rainfall prediction can be roughly divided into two categories, classic algorithms and machine learning algorithms. The model-based approaches, singular spectrum analysis, and wavelet are the typical classic algorithms. Autoregressive and moving average (ARMA) model is the common time series model for rainfall prediction [9]. The wavelet decomposition uses a pair of filters to decompose the original time series iteratively. The trend is separated from noise at each scale domains. Because wavelet decomposition is usually dependent on other parameter estimation models, its application is less than the ARMA model. We select the ARMA model as the representative of classic algorithms.

In recent years, artificial intelligence (AI) develops rapidly. A great deal of machine learning and deep learning algorithms have been proposed for time series analysis. The models of machine learning algorithms have been applied to solve the real-time rainfall prediction problem. Support vector machine (SVM) shows better performance than random forest (RF) in 2 and 3 h ahead prediction [10]. Artificial neural networks (ANNs), such as radial basic function neural networks (RBFNNs) and back propagation neural networks (BPNNs), also perform satisfactorily in some cases. The best algorithm is difficult to choose, because the data used by the models are different. Recurrent neural networks (RNNs) have been proved well in time series prediction [11]. However, with the vanishing and exploding gradient problems, RNNs are not capable of solving the long-term dependency rainfall sequence. The long short-term memory (LSTM) was proposed in 1997 by Hochreiter and Schmidhuber [12] and was improved in 2000 by Gers et al. [13] to overcome the shortage of RNN. LSTM has been widely acknowledged as the current state-of-art prediction models for time series [14]. We choose the LSTM to predict real-time rainfall based on the observed sensor data at Wuhan University. The experiment results show that LSTM is better to describe the seasonal rainfall process than ARMA, RF, SVM, and BPNNs. LSTM is also found more suitable for the sensor time series that usually have missing data.

This paper has two main contributions. (i) The weather time series sampled by WSN with 5-minute interval have been proved to be effective and valuable in local meteorological monitoring. (ii) The real-time rainfall can be well predicted by LSTM.

The rest of this paper is organized as follows. In Section 2, the WSN for local weather parameter sampling is introduced. The validity of the observed time series is demonstrated in Section 3. Real-time rainfall prediction

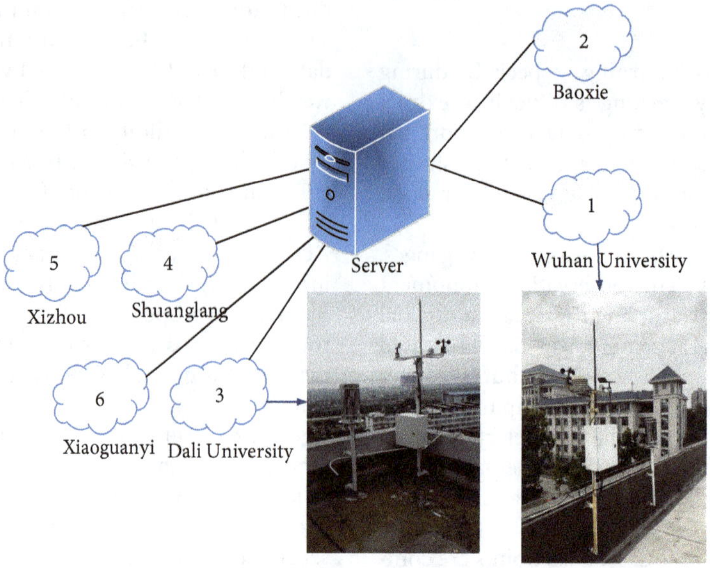

FIGURE 1: Star topology of a wireless sensor network.

武汉大学气象数据

日期	时间	风速 (m/s)	风向 (°)	大气温度 (°C)	大气湿度 (%RH)	气压 (hPa)	降雨量 (mm)	辐射 (W/平方米)	纬度	经度	高度 (m)
2017-12-02	16:45:58	0.6	166.0	12.4	57.6	1019.1	0.0	78.0	N 30°32.438606′	E 114°21.179687′	31.7
2017-12-02	16:40:58	0.0	166.0	12.5	57.8	1019.0	0.0	94.0	N 30°32.438582′	E 114°21.178710′	37.2
2017-12-02	16:35:58	0.6	166.0	12.6	56.9	1019.0	0.0	107.0	N 30°32.438511′	E 114°21.178710′	31.2
2017-12-02	16:30:58	0.6	166.0	12.5	55.9	1019.0	0.0	117.0	N 30°32.438546′	E 114°21.178710′	24.7
2017-12-02	16:25:58	0.6	166.0	13.1	56.6	1019.0	0.0	127.0	N 30°32.438523′	E 114°21.178710′	32.0

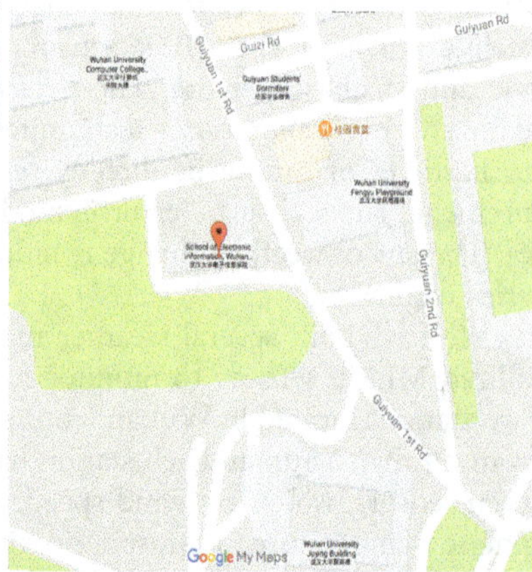

FIGURE 2: Webpage screenshot of the server.

using LSTM is presented in Section 4. The results of prediction are shown in Section 5. Conclusion is given in Section 6.

2. Weather Data Observed by a Wireless Sensor Network

2.1. WSN for Weather Observation. The WSN for local weather monitoring was built from December 2014 to July 2015. The star topology sensor network shown in Figure 1 is composed of one server and six automatic meteorological monitoring stations that are called sensor nodes. One node powered by electricity was set on the roof of Wuhan University. Another node powered by solar energy was set at Baoxie, suburb of Wuhan. The other four nodes that were deployed around Erhai Lake are all powered by electricity. All the nodes communicate with the server by GPRS (General Packet Radio Service) modules through the China Mobile's network. The server can send command to adjust the operation of each node. Weather data transmitted from nodes are all saved in the server. Besides, the server provides data services, such as publishing the real-time data on the website or providing data services to specific users. Figure 2 shows the webpage screenshot of our website.

Each sensor node is composed of seven sensors, one Globe Positioning Systems' (GPS) receiver module, and one GPRS module. Each measurement includes not only the values of the seven sensors but also the corresponding time and the geographical location information (longitude, latitude, and height). The default observation interval is 5 minutes. The repeating measurement cycle can be reset through the command issued from the server to the sensor node.

Table 1: Performance of weather sensors.

Sensors	Unit	Range	Resolution	Accuracy	Working condition
Wind speed	m/s	0~100	0.1	±0.3	−40°C~50°C; ≤100%RH
Wind direction	°	0~360	1	±3	−40°C~50°C; ≤100%RH
Temperature	°C	-50~70	0.1	±0.2	−50°C~80°C
Humidity	%RH	0~100	0.1	±3	−40°C~50°C; ≤100%RH
Pressure	hPa	500~1100	0.1	±0.3	−50°C~80°C
Rainfall	mm	0~4	0.1	±4%	0°C~50°C
Radiation	W/m^2	0~2000	0.1	±2%	−50°C~85°C; ≤100%RH

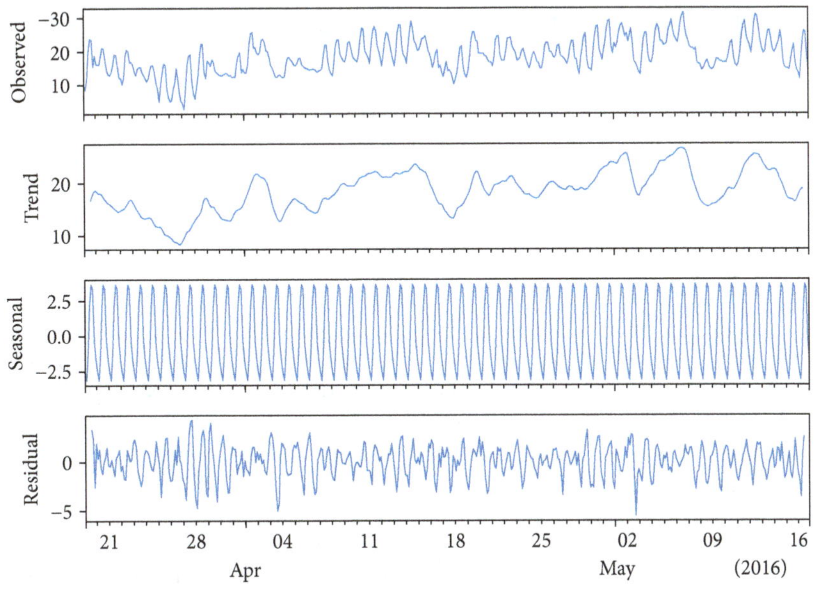

Figure 3: STL decomposition of observed atmospheric temperature data in spring 2016.

2.2. Performance of the MEMS Weather Sensors. The performance of each sensor is shown in Table 1. The measuring range and resolution can meet the requirement for weather observation, but the accuracy shows that the deviation is relatively high. Moreover, stochastic error exists in each measurement. Whether the time series observed by WSN is available needs to be proven.

2.3. Fault of the Time Series. There is another problem in the observed time series. The interval between consecutive measurements is not strictly in accordance with the set period. Missing data happened sometime. The reasons include out of power, interruption of communication connection, the system onboard rebooting, or attack on server by a hacker. Therefore, processing the observed weather data may encounter more difficulties than analyzing the authoritative released weather data.

3. Proof of Validity of the Observed Time Series

To prove the validity of the sensor data, we downloaded the weather data of Wuhan City from the website (https://rp5.ru/) as authoritative reference. The downloaded data are provided by the ground stations certified by National Oceanic and Atmospheric Administrator (NOAA). The station locates near Wuhan Tianhe International Airport. Although the station is about 35 kilometers away from our weather sensor node at Wuhan University, both time series reflect the Wuhan weather. Therefore, it is assumed that if the trend of the observed time series resembles closely the trend of the downloaded time series, the observation of the sensor nodes can be applied to weather research.

3.1. Weather Trend Extraction by STL. STL is a filtering procedure for decomposing a time series into three components: trend, seasonal, and remainder components. It is supposed that the weather time series D_v, such as atmospheric temperature, are composed of trend, seasonal, and remainder components, which are denoted by T_v, S_v, and R_v, respectively, for $v = 1, \ldots, N$. Then,

$$D_v = T_v + S_v + R_v. \quad (1)$$

T_v is the trend data that are extracted from the observed time series and the downloaded time series.

Key to the STL approach is the locally weighted scatterplot smoothing (Loess) process, in which the closest point is assigned to the biggest weight. Suppose y_i for $i = 1, \ldots, N$ is the observed time series, and the regression result $\widehat{y}_i$ is

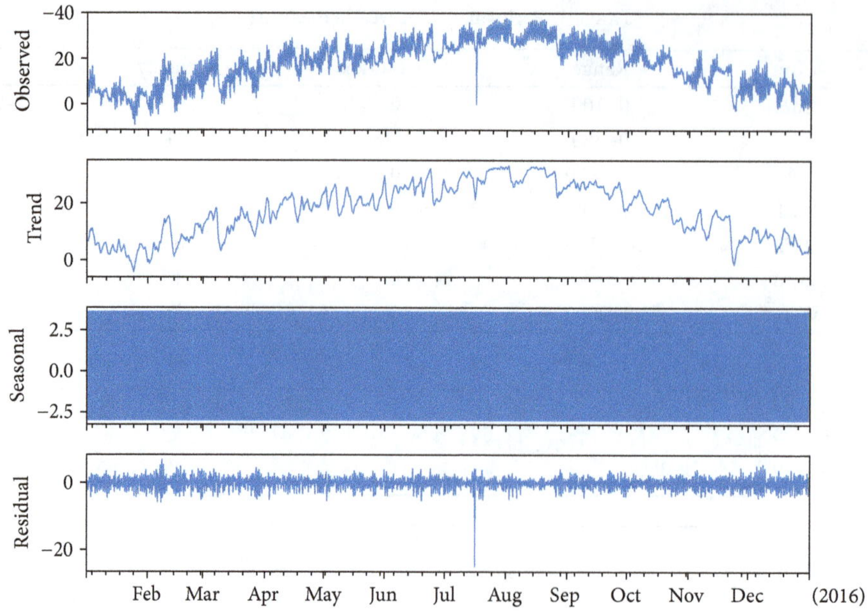

FIGURE 4: STL decomposition of downloaded atmospheric temperature data in 2016.

the smoothing y_i of given x_i. q points that are closest to x are used in the regression. The observed time series by WSN were resampled to match the 3-hour interval of the downloaded time series. Considering the 24 hours each day, we set q to be 8 with the interval of 3 hours. The weight function W is described in the following formula:

$$W_u = \begin{cases} \left(1 - u^3\right)^3 & \text{for } 0 \le u < 1, \\ 0 & \text{for } u \ge 1. \end{cases} \tag{2}$$

Thus, the weights of q selected points are computed according to the distance from x, as shown in the following formula:

$$W_i(x) = W\left(\frac{|x_i - x|}{d_q}\right), \tag{3}$$

where d_q is the distance between the qth farthest point x_i and x.

Weight of the point x_i decreases with the increasing distance between x_i and x. Weight of the qth farthest point comes to zero. Then, we can get the regression result $\hat{y}$ using polynomial regression with the weights based on least squares methods (LSM). The polynomial degree is usually set to be 1 or 2, corresponding to linear regression or quadratic regression separately. We chose the quadratic regression. With the smoother Loess, data is divided into trend component of low frequency, seasonal component of high frequency, and remainder component of random variation.

We applied STL to extract the trends of both observed weather data and downloaded weather data. The observed time series started in June 2015 at Wuhan University. Correspondingly, the downloaded weather data also started on the same date. Take atmospheric temperature data in spring 2016, for example, the result of STL decomposition on the observed time series is shown in Figure 3. The result of STL decomposition on the downloaded atmospheric temperature data in 2016 is shown in Figure 4. The falling peak in Figure 4 represents the missing data on 16 July 2016. Because of the robustness of STL, trend has not been affected.

3.2. Similarity Comparison. The paper [15] contributes to presenting an overview of nine similarity measurements for time series data. ED, dynamic time warping (DTW), and CS are widely used to measure the similarity of numerical time series. ED is straightforward for intuitive understanding and easy to implement. In particular, when the length of time series is relatively large, ED is an ideal choice [15]. DTW is designed to handle the time-warping series or the time series in which lengths do not matched [16]. CS is a similarity measure for orientation rather than magnitude. In CS, the cosine of the angle between two vectors is calculated. If the value of CS is equal to 1, the two vectors are in the same direction.

We chose ED and CS to measure the similarity of two trends, ED for magnitude and CS for orientation. Each year, the two types of dataset are divided into four subdatasets according to the seasonal partition of Wuhan [17]. The seasonal division is shown in Table 2. The results of ED and CS in 2016 are presented in Table 3. To make the similarity comparison more visible, two trends of temperature in spring 2016 and their subtraction are shown in Figure 5.

Table 3 exhibits that the trend of the observed time series is similar to that of the downloaded series in terms of atmospheric temperature, atmospheric humidity, atmospheric pressure, wind speed, and rainfall. The reason that the comparison of wind direction was not done is that the downloaded wind direction data is in word description, but the observed wind direction data is in degree number. Table 3 demonstrates that the weather parameters measured by the

Table 2: Seasonal partition of Wuhan.

Spring	Summer	Autumn	Winter
March 17–May 16	May 17–Sept 25	Sep 26–Nov 22	Nov 23–March 16

Table 3: Results of similarity comparison in four seasons.

(a) Spring

	Temperature	Humidity	Wind speed	Pressure	Rainfall
ED	0.055	0.831	0.059	0.087	0.054
CS	0.998	0.997	0.953	0.999	0.805

(b) Summer

	Temperature	Humidity	Wind speed	Pressure	Rainfall
ED	0.048	0.676	0.038	0.053	0.085
CS	0.999	0.998	0.948	0.999	0.814

(c) Autumn

	Temperature	Humidity	Wind speed	Pressure	Rainfall
ED	0.036	0.791	0.074	0.105	0.033
CS	0.999	0.999	0.970	0.999	0.843

(d) Winter

	Temperature	Humidity	Wind speed	Pressure	Rainfall
ED	0.052	0.662	0.035	0.419	0.007
CS	0.994	0.994	0.965	0.999	0.962

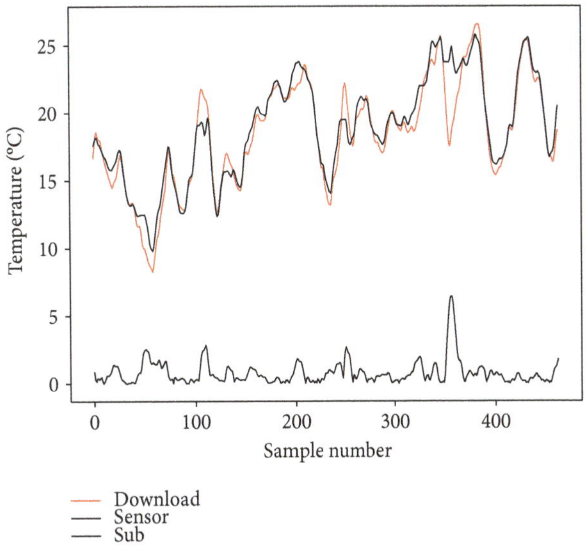

Figure 5: Comparison of the trends of the observed and downloaded temperature in spring 2016.

MEMS sensors can be used for local weather observation and climate research.

4. Local Real-Time Rainfall Prediction with LSTM

Our weather observation node may reflect the weather change of an area around Wuhan University. Real-time rainfall prediction information can attract students' more attention than the published one-hour forecasting, especially during a thunderstorm. They can decide whether to take an umbrella, whether to go outside, or whether their laboratory has the risk of inundation.

We set the observed time series $\mathbf{x} = [x_1, \ldots, x_T]$ as the input of LSTM. $\mathbf{x}_t \in R^d$ represents the weather vector observed at time step t. According to the correlation analysis of observed weather parameters, atmospheric temperature, atmospheric humidity, atmospheric pressure, and rainfall data are selected as the input vector $\mathbf{x}_t$. Then, the dimension of input data d equals to 4. T is the length of the observed time series. The output of LSTM is the predicted rainfall data at time step $t+1$, denoted as $\widehat{y}_t$. The dimension of output data equals to 1. The actual rainfall value is denoted as y_t. Each time step is set to 5 minutes. It means that LSTM can be used to predict next 5-minute rainfall. We also adjust the interval of each time step to predict 10- or 15-minute rainfall.

The basic unit in the hidden layer of an LSTM network is the memory block. Memory block contains one or more memory cells and three adaptive and multiplicative units [13]. These units are called gates, including input gate, forget gate, and output gate. The structure and data flow of memory block is shown in Figure 6. The symbols i, f, and o, respectively, refer to the input gate, forget gate, and output gate. The symbols g and h are the memory cell input and cell output activation function, respectively. At each time step t, input sequence $\mathbf{x}_t$ can be propagated to the memory cell, if the input gate i is activated. Forget gate f controls whether the past memory cell status $C_t - 1$ can be added as the input to the current memory cell. The updated status of memory cell will be propagated to the output based on the activation of the output gate o [18]. The overall data flow can be described with the following mathematical formulas:

$$
\begin{aligned}
i_t &= \sigma(\mathbf{W}_i \cdot [h_t - 1, \mathbf{x}_t] + \mathbf{b}_i),\\
f_t &= \sigma\left(\mathbf{W}_f \cdot [h_{t1}, \mathbf{x}_t] + \mathbf{b}_f\right),\\
g_t &= \tan h(\mathbf{W}_c \cdot [h_t - 1, \mathbf{x}_t] + \mathbf{b}_c),\\
C_t &= f_t * C_t - 1 + i_t * g_t,\\
o_t &= \sigma(\mathbf{W}_o \cdot [h_t - 1, \mathbf{x}_t] + \mathbf{b}_o),\\
h_t &= o_t * \tan h(C_t),
\end{aligned}
\tag{4}
$$

where $\mathbf{x}_t$ denote the input of memory block, and h_t denotes

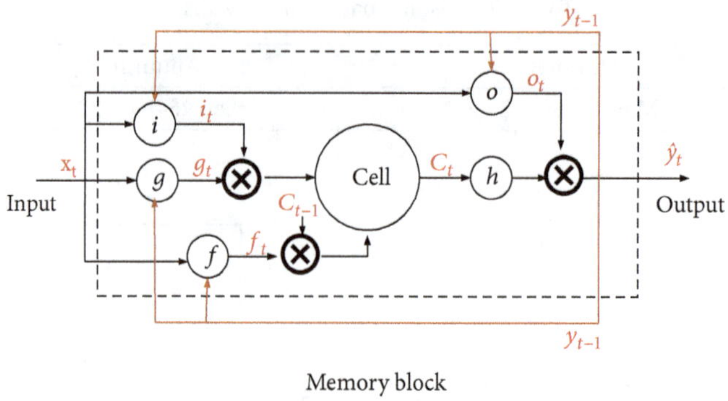

FIGURE 6: Structure and data flow of memory block.

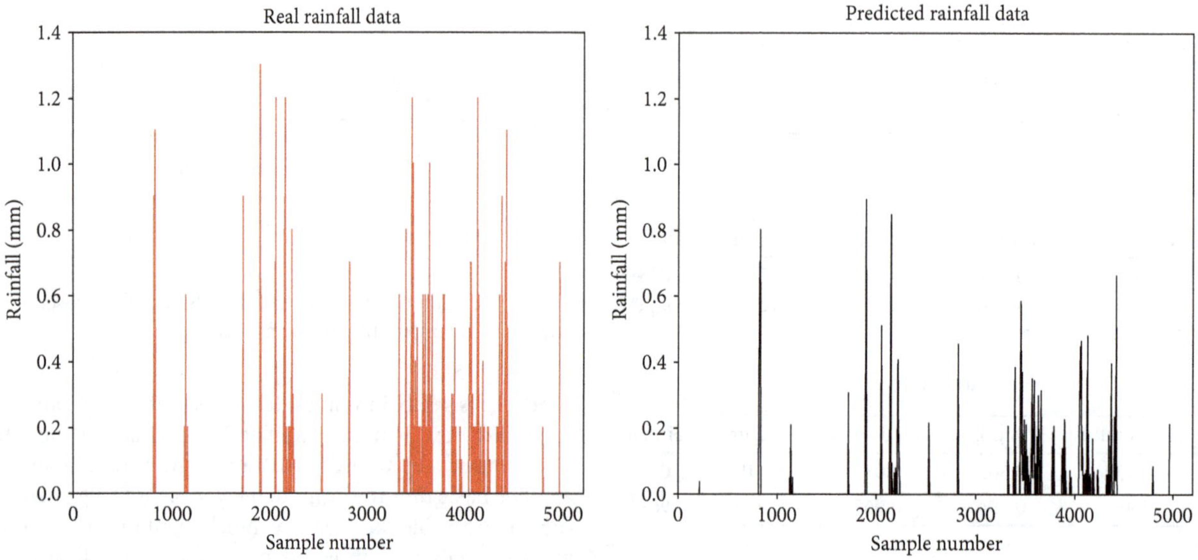

FIGURE 7: Five-minute rainfall prediction by LSTM.

the output of memory block. **W** denotes weight matrices and **b** denotes biases' vector, which are both shared in all time steps and learned during the phase of training. σ represents the sigmoid activation function [19]. At each time step t, the output of memory block h_t is updated by input sequence $\mathbf{x}_t$, input gate i_t, forget gate f_t, output gate o_t, memory cell status, output $h_t - 1$, and $C_t - 1$ at previous time step $t - 1$. The predicted rainfall data $\hat{y}_t$ is the output of LSTM output layer with the h_t as the input.

5. Results

The LSTM was used by two manners to predict rainfall. One is called real-time prediction. The other is called seasonal real-time prediction. Real-time prediction means to predict the rainfall based on the previous adjacent time sequence. Seasonal real-time prediction means to predict the rainfall based on the rules extracted by LSTM from the same season data of the previous year.

We have conducted real-time rainfall prediction by LSTM for 5 minutes, 10 minutes, and 15 minutes based on observations. Considering the weather characteristics of Wuhan, rainy days are much less than nonrainy days in the whole year. A great deal of observed rainfall is equal to zero. Assuming the observed rainfall as positive samples, to balance the positive and negative samples, we select the observations from March 2016 to July 2016 in Wuhan University. This period is called the rainy season of Wuhan. In addition, we select ARMA as the representative of classic algorithms and RF, SVM, and BPNNs as the representatives of machine learning algorithms for performance comparison.

TABLE 4: Prediction performance of the five models.

Model	5 min		10 min		15 min	
	MAE	RMSE	MAE	RMSE	MAE	RMSE
ARMA	0.0280	0.0936	0.0544	0.1677	0.0849	0.2469
RF	0.0330	0.1061	0.0634	0.1879	0.0984	0.2829
SVM	0.1262	0.1421	0.2452	0.2728	0.3548	0.3965
BPNNs	0.0351	0.1040	0.0655	0.1833	0.1023	0.2832
LSTM	0.0232	0.0964	0.0450	0.1756	0.0720	0.2525

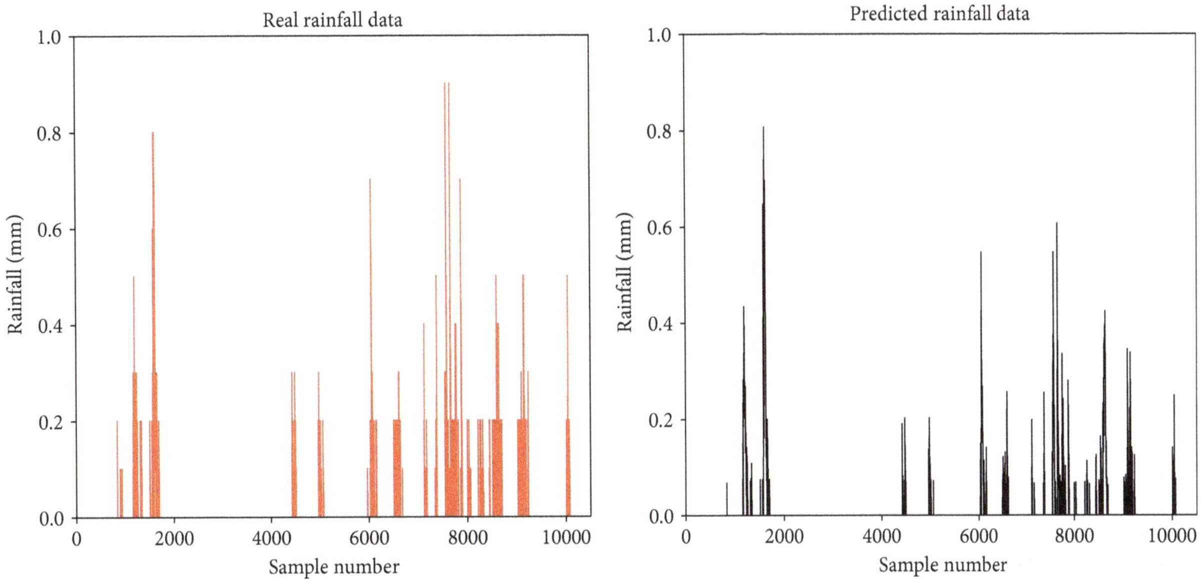

FIGURE 8: Five-minute rainfall prediction based on seasonal rules by LSTM.

Seasonal real-time rainfall was predicted for 5 minutes, 10 minutes, and 15 minutes in autumn 2016 based on the rules obtained in autumn 2015. ARMA is not suitable for seasonal real-time prediction because it predicts rainfall step by step. We used RF, SVM, BPNNs, and LSTM, respectively, to make seasonal real-time prediction.

5.1. Evaluation Metrics. For numeric weather prediction (NWP), several evaluation metrics have been defined to evaluate prediction performance in the paper [20]. We chose the mean absolute error (MAE) and root mean square error (RMSE) as the evaluation metrics, which are described in the following formulas:

$$\mathrm{MAE} = \frac{1}{T}\sum_{t=1}^{T}\left|\hat{y}_t - y_t\right|, \quad \mathrm{RMSE} = \sqrt{\frac{1}{T}\sum_{t=1}^{T}\left(\hat{y}_t - y_t\right)^2}, \tag{5}$$

where T is the length of test set, y_t denotes the actual rainfall data, and $\hat{y}_t$ is the predicted value at the given time t.

5.2. Experimental Results. The real-time rainfall prediction results by LSTM for 5 minutes is shown in Figure 7. Five minutes, 10 minutes, and 15 minutes real-time rainfall are all predicted by five models. The prediction performance are shown in Table 4. Except for SVM, the performances of other algorithms are quite near. LSTM has better performance than the other machine learning models. LSTM has similar performance with ARMA in real-time rainfall prediction. This phenomenon may be inconsistent with the conclusions that LSTM much outperforms classic models in time series prediction. However, it meets the conclusion that machine learning methods such as neural networks (NNs) do not often outperform simple methods for short individual series in the paper [21].

TABLE 5: Prediction performance based on seasonal rules by the four models.

Model	5 min		10 min		15 min	
	MAE	RMSE	MAE	RMSE	MAE	RMSE
RF	0.0127	0.0453	0.0239	0.0758	0.0376	0.1098
SVM	0.0776	0.0825	0.1222	0.1313	0.1619	0.1778
BPNNs	0.0146	0.0476	0.0285	0.0815	0.0418	0.1157
LSTM	0.0096	0.0430	0.0161	0.0725	0.0233	0.1027

The seasonal real-time rainfall prediction results for 5 minutes in autumn 2016 based on the previous seasonal rules with LSTM are shown in Figure 8. The prediction performance of the four models is shown in Table 5. LSTM outperforms the other models in seasonal rainfall prediction. If the seasonal prediction exhibits large deviation, it may mean that the abnormal rainfall happens. It can be concluded that LSTM not only performs well in real-time rainfall prediction but also works well in seasonal rainfall pattern detection as shown in Tables 4 and 5.

6. Conclusion

The developed automatic meteorological stations based on MEMS sensors can provide high time resolution and spatial resolution weather monitoring. However, MEMS sensors suffer from stochastic error. In order to demonstrate the observed time series by MEMS sensors, STL is utilized to extract trends from the time series observed at Wuhan University. The similarity between the trend of the observed series and the trend extracted from the downloaded authoritative time series with the same weather parameter is computed according to ED and CS. High similarity illustrates that the MEMS sensors of weather can be applied to meteorological research, and their measurement is believable. LSTM is used in two manners for rainfall prediction. Real-time prediction uses the previous sequence to predict the next 5-, 10-, and 15-minute rainfall. Seasonal real-time prediction applies the rule obtained from the previous season to predict the next same season rainfall. Evaluated by MAE and RMSE, LSTM is similar to ARMA but outperforms RF, SVM, and BPNNs in the real-time prediction. LSTM also outperforms RF, SVM, and BPNNs in the seasonal real-time prediction, while ARMA cannot work. Experiment results demonstrate that LSTM is powerful in the extraction of the change rules of seasonal rainfall. Application of LSTM to the time series observed by MEMS weather sensors will provide accurate and timely rainfall prediction to support local heavy rain emergency response.

Conflicts of Interest

The authors declare that there are no conflicts of interest.

Acknowledgments

This work was supported by the National Key Research and Development Program of China under Grant no. 2016YFB0502601. It was also supported by Major Science and Technology Program for Water Pollution Control and Treatment (2013ZX07105-005-003), and the Fundamental Research Funds for the Central Universities under Grant no. 2042017kf0211. Gratitude is due to Associate Professor Liqiong Chen from Wuhan University, who helped us to establish the WSN.

References

[1] S. Cramer, M. Kampouridis, A. A. Freitas, and A. K. Alexandridis, "An extensive evaluation of seven machine learning methods for rainfall prediction in weather derivatives," *Expert Systems with Applications*, vol. 85, pp. 169–181, 2017.

[2] J.-H. Seo, Y. H. Lee, and Y.-H. Kim, "Feature selection for very short-term heavy rainfall prediction using evolutionary computation," *Advances in Meteorology*, vol. 2014, 15 pages, 2014.

[3] L. Wu, Y. Fan, X. He, and G. Yu, "Change trend of extreme precipitation events in the Wuhan city (China) from 1951 to 2012," in *International Conference on Energy, Environment and Materials Engineering*, Shenzhen, Guangdong, China, February 2014.

[4] M. Li, C. Chen, and J. Jiang, "Features of flood risk map drawing for downtown areas of Wuhan city and its application," *China Water Resources*, vol. 5, pp. 49–51, 2017.

[5] H. Messer and O. Sendik, "A new approach to precipitation monitoring: a critical survey of existing technologies and challenges," *IEEE Signal Processing Magazine*, vol. 32, no. 3, pp. 110–122, 2015.

[6] P. Petkov and T. Slavov, "Stochastic modeling of mems inertial sensors," *Cybernetics and Information Technologies*, vol. 10, no. 2, pp. 31–40, 2010.

[7] T. Alexandrov, S. Bianconcini, E. B. Dagum, P. Maass, and T. S. McElroy, "A review of some modern approaches to the problem of trend extraction," *Econometric Reviews*, vol. 31, no. 6, pp. 593–624, 2012.

[8] R. B. Cleveland, W. S. Cleveland, and I. Terpenning, "STL: a seasonal-trend decomposition procedure based on loess," *Journal of Official Statistics*, vol. 6, no. 1, p. 3, 1990.

[9] L. Zhu, Y. Wang, and Q. Fan, "MODWT-ARMA model for time series prediction," *Applied Mathematical Modelling*, vol. 38, no. 5-6, pp. 1859–1865, 2014.

[10] P.-S. Yu, T.-C. Yang, S.-Y. Chen, C.-M. Kuo, and H.-W. Tseng, "Comparison of random forests and support vector machine for real-time radar-derived rainfall forecasting," *Journal of Hydrology*, vol. 552, pp. 92–104, 2017.

[11] T. Guo, Z. Xu, X. Yao, H. Chen, K. Aberer, and K. Funaya, "Robust online time series prediction with recurrent neural networks," in *2016 IEEE International Conference on Data Science and Advanced Analytics (DSAA)*, pp. 816–825, Montreal, QC, Canada, October 2016.

[12] S. Hochreiter and J. Schmidhuber, "Long short-term memory," *Neural Computation*, vol. 9, no. 8, pp. 1735–1780, 1997.

[13] F. A. Gers, J. Schmidhuber, and F. Cummins, "Learning to forget: continual prediction with LSTM," in *9th International Conference on Artificial Neural Networks: ICANN '99*, Edinburgh, UK, September 1999.

[14] Z. C. Lipton, D. C. Kale, C. Elkan, and R. Wetzel, "Learning to diagnose with LSTM recurrent neural networks," 2015, http://arxiv.org/abs/1511.03677.

[15] X. Wang, A. Mueen, H. Ding, G. Trajcevski, P. Scheuermann, and E. Keogh, "Experimental comparison of representation methods and distance measures for time series data," *Data Mining and Knowledge Discovery*, vol. 26, no. 2, pp. 275–309, 2013.

[16] H. Izakian, W. Pedrycz, and I. Jamal, "Fuzzy clustering of time series data using dynamic time warping distance," *Engineering Applications of Artificial Intelligence*, vol. 39, pp. 235–244, 2015.

[17] C. Chen, D. Gu, and Y. Lu, "Change of climatic seasons between urban area and suburb in Wuhan from 1960 to 2009," *Journal of Meteorology and Environment*, vol. 31, no. 1, pp. 96–100, 2015.

[18] N. Tax, I. Verenich, M. La Rosa, and M. Dumas, "Predictive business process monitoring with LSTM neural networks," in *Advanced Information Systems Engineering. CAiSE 2017*, Lecture Notes in Computer Science, E. Dubois and K. Pohl, Eds., pp. 477–492, Springer, Cham, 2017.

[19] S. Han, J. Kang, H. Mao et al., "ESE: efficient speech recognition engine with sparse LSTM on FPGA," in *FPGA '17 Proceedings of the 2017 ACM/SIGDA International Symposium on Field-Programmable Gate Arrays*, pp. 75–84, Monterey, CA, USA, February 2017.

[20] M. Shahrban, J. P. Walker, Q. J. Wang, A. Seed, and P. Steinle, "An evaluation of numerical weather prediction based rainfall forecasts," *Hydrological Sciences Journal*, vol. 61, no. 15, pp. 2704–2717, 2016.

[21] K. Bandara, C. Bergmeir, and S. Smyl, "Forecasting across time series databases using long short-term memory networks on groups of similar series," 2017, http://arxiv.org/abs/1710.03222.

11

Useful Piezoelectric Sensor to Detect False Liquor in Samples with Different Degrees of Adulteration

Luis Armando Carvajal Ahumada[1,2] **Andrés Felipe Sandoval Cruz**[1]
Mario Alejandro García Fox[1] **and Oscar Leonardo Herrera Sandoval**[1]

[1]*Facultad de Ingeniería y Ciencias Básicas, Universidad Central, Bogotá 110311, Colombia*
[2]*Servicio Nacional de Aprendizaje (SENA), TecnoParque nodo Bogotá, Bogotá, Colombia*

Correspondence should be addressed to Luis Armando Carvajal Ahumada; lcarvajala1@ucentral.edu.co

Academic Editor: Alfonso Salinas-Castillo

The main objective of this article is to demonstrate by experimental measurements the capability of the quartz crystal resonator (QCR) to characterize samples of liquor at different concentrations of adulteration with methanol, classify them according to their viscosity, and identify how to use this capability as a potential detection of fake liquor. According to the literature, the fake liquor is a common problem detected in several countries across the globe. The current strategy to detect the fake liquor is based on a piezoelectric biosensor that permits the classification of the samples according to its viscosity. The viscosity is a useful variable to determinate the degree of adulteration in the liquor samples because the viscosity value between ethanol (straight liquor) and methanol (most common fake liquor) is very different.

1. Introduction

The adulterated alcohol or "fake alcohol" is a common problem detected in some countries. For instance, in the UK, the illegal alcohol is distilled and sent without a license, and its sale costs the UK around £1.2 billion per year [1]. In Greece, the improvised production of beverages named "meth" exists, done in illegal laboratories in an empirical way. The production commonly includes illegal alcohol with high concentration of methyl alcohol [2]. In India, there is a similar situation related to the illicit liquor. In these cases, one dangerous adulterant is industrial methylated spirit, which causes mass poisoning of consumers [3]. Moreover, in Czech Republic, the Czech food safety authority informed the public of persons suffering from methanol poisoning after having consumed "on tap" spirits. The product was unlabeled and supplied "in bulk" in plastic barrels [4].

In particular, the Colombian case is not an exception. Studies performed by Colombian entities such as Externado University and the National Association of Industrialists (ANDI) reveal that the adulteration and falsification of locals have reached at 21% of the total offer. Javeriana University in another study concluded that one of each four bottles of liquors sold in Colombia has been falsified or adulterated. Moreover, Antioquia University from Colombia performed a study in which they observed that close to 50% of the liqueurs which are marketed in the Department of Antioquia, Colombia, are illegal, adulterated, or contraband [5].

In most cases, the commercial liquor is adulterated using methanol [1, 3–5], due to its low cost, and some physical characteristics are similar to those of the ethanol. Methanol (CH3OH) is an alcohol that is aliphatic, liquid, colorless, and volatile to room temperature. By itself, it is harmless, but its metabolites are extremely toxic. The methanol in industrial products, laboratories, and household products is commonly used. However, there is also a clandestine use like substituting of ethanol in alcoholic drinks. Although the intoxication may occur through the respiratory system or transdermally, in most cases, the exposition to methanol is through the oral way [6].

According to the "Report of the event notification for acute poisoning by chemical substances up to the epidemiological period XI" in 2010, out of a total of 23,844 reported intoxications, there were a total of 282 cases of methanol intoxication and 10 deaths, with a percentage of 1.18% and an incidence of 0.62 per one hundred thousand inhabitants. During 2011, 351 cases of methanol intoxication and 7 deaths were reported. Moreover, there is a high notification of consumption in the age ranging between 10 and 24 years [7]. A report of FND (National Federation of Departments) reveals that, of the percentage of illegal liquor (24%), the 70% correspond to adulterated drinks [4].

Although some cases of intoxication by methanol are due to suicide attempts, many times, the intoxication is caused by adulterated alcohol, which generally is distributed in bottles with labels very similar to those of the original drink but with methanol mixed to the original liquor. Therefore, the adulterated drink is not detectable to naked eye [2].

The symptoms, depending on the affected system, when ingesting adulterated alcohol are [8]

(i) nausea, vomiting, abdominal pain, and later-case affections in the pancreas (gastrointestinal system),

(ii) associated manifestations of neurologic type, which in cases of lack of treatment or late detection end in a coma state (central nervous system),

(iii) decreased visual acuity until irreversible blindness due to atrophy of the optic nerve, indicating ocular involvement and representing an advanced development of the intoxication.

According to the descriptive study of methanol intoxications reported in SIVIGILA 2010-2011 in Colombia [9], when the methanol is absorbed, it undergoes oxidation processes in the organ receptor which is the liver. The enzyme responsible for this transformation is the alcohol dehydrogenase, which oxidizes it to formaldehyde, and this in turn is oxidized to formic acid (responsible for blindness) by the aldehyde dehydrogenase. The treatment generally used to treat adulterated alcohol intoxication is based on the use of ethyl alcohol to prevent the formation of toxic methanol metabolites [9].

In accordance with the problematic presentation, several techniques have been developed with the aim of detecting the adulterated liquor. The more typical laboratory techniques are the nuclear magnetic resonance spectroscopy with multivariate analysis [4] and near-infrared spectrometry (NIR) [10]. However, these techniques do not have possibilities to be used outside of the laboratory.

In Colombia, different strategies have been generated to detect the illegal liquor, divided into contraband and fake liquor cases; several years ago, for the detection of adulterated liquor, the monitoring and control program of alcoholic beverages was implemented, led by the ANDI (National Association of Industrialists) and the Ministry of Health and Social Protection with the participation of the national police, making presence in factories, warehouses, and liquor stores. For factories and warehouses, there are four visits per year, and for liquor stores, there are three visits. In these visits, the authorities make sure for sanitary conditions of the places and take random samples of a suspect batch of liquor, where these samples cannot be less than 1 liter or a bottle generally 750 cc. The first step in this process is to check the security seals and the label, the second step is to look for suspended solids inside the bottle, and the last step is to carry the sample to the corresponding authority lab and make physiochemical tests [5].

In this work, the use of a sensor based on a quartz crystal resonator in order to identify adulterated liquor in a simple and low cost-way is proposed. This sensor is capable of operating as a microviscometer knowing the density of the liquid to be analyzed [11–15]. Samples at different "levels of adulteration" have been used to get results that confirm the difference between an original alcoholic drink and adulterated one with methanol in different concentrations.

2. Theory

The piezoelectric sensors are considered highly efficient sensing systems not only because of their low manufacturing cost but also because of their high accuracy, biofunctionalization capacity, sensitivity, and reliability in measuring the deposited sample, either gaseous or liquid media [11].

In particular, the quartz crystal resonator (QCR) is a piezoelectric transducer sensitive to the deposition of rigid mass or liquids in contact with its surface (active electrode). The QCR is composed of two electrodes that are stimulated with AC voltage, and in consequence, the crystal vibrates. The crystal has a natural frequency of resonance that depends on its geometry and thickness; as a result, when the crystal is stimulated with a signal whose frequency is close to its natural resonance, the vibration increases. This phenomenon is visible by analyzing the behavior of the admittance or impedance of the crystal in the frequency domain [11, 16, 17].

When the mass or liquid sample is deposited on the crystal surface, its resonance frequency changes in accordance with two equations: Sauerbrey for mass deposition (1) and Kanazawa for fluid deposition (2).

$$\Delta f = -\frac{2{f_0}^2}{A\sqrt{\rho_q G_q}}\Delta m, \tag{1}$$

$$\Delta f = -\sqrt{n}{f_0}^{3/2}\sqrt{\frac{\rho_L \eta_L}{\pi \rho_q G_q}}, \tag{2}$$

where ρ_q and G_q are the specific density and the shear modulus of quartz, respectively; f_0 is the fundamental resonance frequency of the quartz, related to its thickness, and Δm is the thin film of mass deposited; A is the piezoelectrically active crystal area; ρ_L and η_L are the density and viscosity of the fluid, respectively; and n is the overtone number [11].

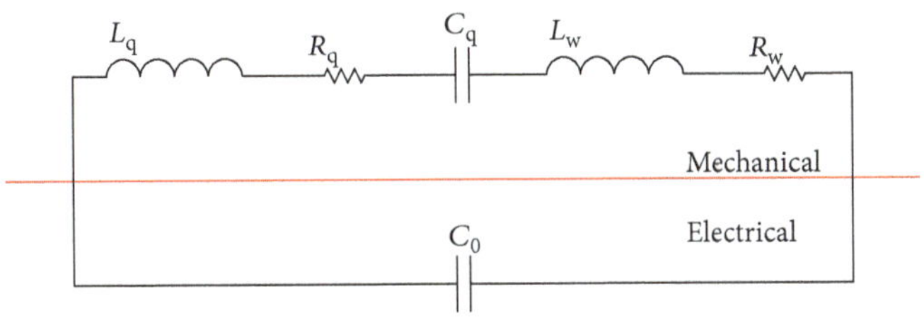

FIGURE 1: Basic electrical model (BVD) for QCM with liquid sample deposited.

In addition to Sauerbrey and Kanazawa equations, the quartz crystal behavior can be modeled as an electrical resonator circuit [11, 16, 17].

The electric model (BVD), shown in Figure 1, explains this phenomenon as an addition of impedance depending on the type of sample. For liquids in contact with the crystal, the equivalent impedance is an inductance and a resistance which are in series.

The value of this impedance is related to the characteristics of the samples according to the following equations:

$$\begin{aligned} C_q &= \frac{8 e_{26}{}^2 A}{(N\pi)^2 \overline{C_{66}} h_q}, \\ L_q &= \frac{1}{\omega_s{}^2 C_q}, \\ R_q &= \frac{\eta_q}{\overline{C_{66}} C_q}, \\ L_w &= \frac{\omega_s L_q}{N\pi}\left(\frac{2\rho_L \eta_L}{\omega_s \overline{C_{66}} \rho_q}\right), \\ R_w &= \frac{\omega_s L_q}{N\pi}\left(\frac{2\omega_s \rho_L \eta_L}{\overline{C_{66}}\, \rho_q}\right), \\ C_0 &= \frac{\varepsilon_{22} A_e}{h_q}, \end{aligned} \tag{3}$$

where $\overline{C_{66}}$ is the piezoelectrically stiffened elastic constant for loss-less quartz ($2947 \times 1010\,\mathrm{N/m^2}$), e_{26} is the piezoelectric stress constant for quartz ($953 \times 103\,\mathrm{A{\cdot}s/m^2}$), ε_{22} is the quartz permittivity ($3.982 \times 10^{-11}\,\mathrm{A^2{\cdot}s^4/kg{\cdot}m^{-3}}$), A is the active electrode area ($\mathrm{m^2}$), h_q is the quartz crystal thickness (m), η_q is the effective viscosity of quartz ($3.5 \times 10^{-4}\,\mathrm{kg/m{\cdot}s}$), and ω_s is the series resonance frequency ($2\pi F_s$). For the liquid, ρ_L and η_L are the density of the liquid sample ($\mathrm{kg/m^3}$) and the viscosity of the liquid sample (kg/m·s), respectively [17].

Figure 2 shows the conductance behavior (Geq) for the unperturbed crystal (kept in air) and the crystal in contact with liquid. The maximum value of conductance takes place in the frequency of resonance.

According to theoretical models, when the crystal electrode is in contact with liquid samples, the conductance curve changes, moving to the left in the axis of frequency, increasing its bandwidth and decreasing its conductance magnitude (Figure 2).

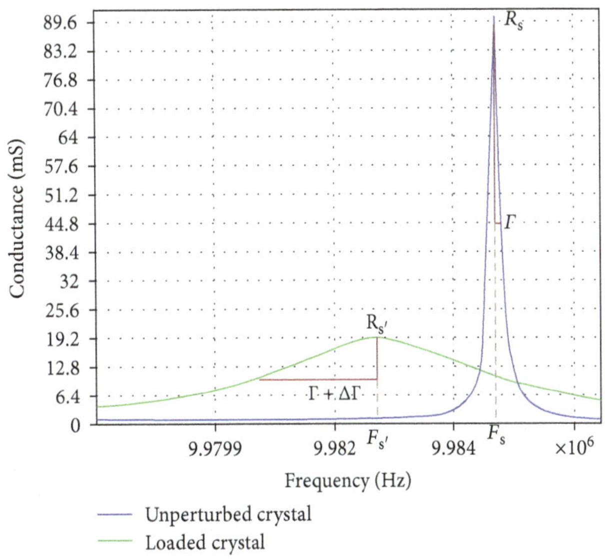

FIGURE 2: Conductance behavior for a QCM with and without the sample [12].

The values of ΔF and $\Delta\Gamma$ represent the changes in the behavior of the crystal due to the properties of the sample, and these are obtained using

$$\begin{aligned} \Delta F &= F_{s'} - F_s, \\ \Delta\Gamma &= \Gamma' - \Gamma, \\ \Delta R_s &= \frac{R_{s'}}{R_s}. \end{aligned} \tag{4}$$

3. Materials and Methods

3.1. Samples. The experiments were done with seven adulterated samples and three references; the commercial liquor (Ron) was purchased in a local market. It has 35% *v/v* of alcohol concentration and was used as one of the three references for the quartz crystal resonator (QCR) system. The other two references were ethanol (99.8% *w/w*, Merck) and methanol (99.8% *w/w*, Merck). The commercial liquor was adulterated in the Central University laboratory, but its alcohol concentration was kept, by addition of deionized water and methanol in levels of 10%, 25%, 35%, 50%, 55%, 65%, and 75% *v/v* to the pure commercial liquor (Figure 3).

Only 50 μL of volume of the sample was deposited in a uniform layer on the surface of the electrode of quartz crystal

FIGURE 3: Samples at different concentrations elaborated in the Central University chemical lab.

to be analyzed. The deposition has been achieved using a microfluidic system.

3.2. Crystals and the Holder Cell. The 10 MHz quartz crystals were purchased from International Crystal Manufacturing. According to the manufacturer, the crystal diameter is 13.66 mm, the electrode diameter is 5.1 mm, and the crystal thickness is 167 μm.

The holder cell has been designed and built in the mechanical laboratories at Central University in Bogotá, Colombia.

Figures 4(a)–4(c) show the crystal plate and the fluidic cells designed for depositing the fluid sample over the crystal electrode using a pressure pump. This technique permits the obtainment of measures related to the behavior of the crystal and the deposition of the samples with a constant flux without the need to manipulate the crystal. In this way, the reference measure is conserved. The sealing system is based on magnets, and the magnets permit a very reliable sealing and an easy way to open and close the cell.

3.3. QCR System. The development of the QCR system has been based on Nakamoto and Kobayashi's proposal [18]. In particular, the proposed circuit gets the voltage and current signal in the crystal for several frequencies in a configurable bandwidth close to the crystal resonance frequency (f_s) indicated by the manufacturer (10 MHz in the current study). The current system is connected to a laptop using a software interface designed in Labview in order to control the sensor and to acquire and to visualize the signals.

All systems have been developed by the coworker team, and all technical details for the measurement system and methods to obtain the changes in the admittance curve (f_s and Γ) are described in [11, 12].

Figure 5(a) shows the current QCR prototype system. In future work, the prototype will be miniaturized with the aim of increasing the portability of the system, so the user can analyze the sample in any place.

Figures 5(a) and 5(b) show the microfluidic system that was used on the experiments. The system consists of two syringe pumps: the first pump injects the buffer liquid on the fluidic system and the second pump injects the sample on the fluidic system. The sample has to pass through a loop where it is stored and waits for the valve activation; when that happens, the sample can pass through the crystal.

3.4. Viscosity of the Reference. For this work, the reference viscosity values have been obtained using the viscometer Fungilab Smart (Figure 6) and the spindle LCP because the samples have low viscosity range.

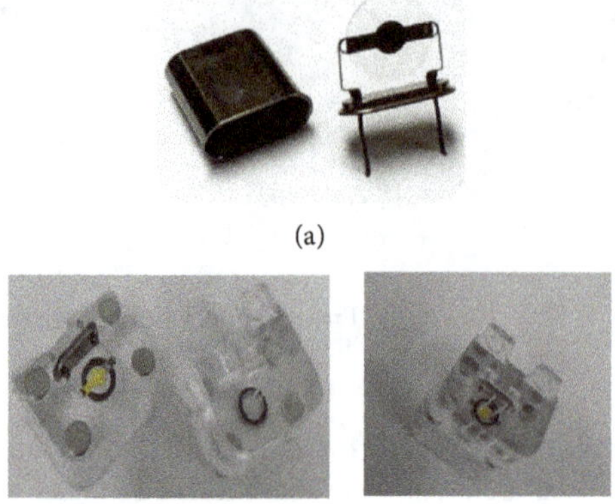

(a)

(b) (c)

FIGURE 4: (a) Crystals used in experimental tests, (b) holder developed (fluidic cells in parts), and (c) holder developed (fluidic cells assembled).

3.5. Measurement Protocol. The measurement protocol is as follows: to configure the frequency sweep using the system interface (frequency range, sweep step, and number of cycles) and to calculate the series resonance frequency (F_0) and the half-band half-width (Γ_0) using the developed code in Labview for each cycle. These measures are taken ten times for the bare crystal. The final values are the averages of the ten measures. To put the sample for analysis on the electrode surface of the quartz crystal (50 μL of volume), the microfluidic system has been used. The frequency range in the user interface needs to be adjusted in order to obtain the admittance values of the quartz crystal in contact with the liquid sample. Due to the sample effect, the curves are shifted in the frequency axis. For the results obtained in this work, the measures with the sample (F_s, $F_{r'}$, Γ_s, and $R_{s'}$) are taken 100 times (ca. 3 minutes) and the final values are the averages of the measures. To obtain $\Delta F_s = F_s - F_0$ and $\Delta\Gamma = \Gamma_s - \Gamma_0$. The density values were obtained using an analytical balance. The mass of each solution was found from 1 mL of its volume.

3.6. Cleaning Protocol. Before each experiment, the sensor crystal was cleaned in 96% acetone and 99.5% isopropyl alcohol for 5 min each; then, it is rinsed with Milli-Q water and dried in air.

4. Results and Discussion

According to the measurement process described in Materials and Methods, the results obtained are the following:

Figure 7 shows the behavior of ΔF_s and $\Delta\Gamma$ for each sample of adulterated alcohol. According to these results, the best variable to differentiate between grades of adulteration in the alcohol sample (percentage of methanol) is $\Delta\Gamma$ due to its more sensitivity (the curve slope is higher).

Table 1 shows the results for different tests done in an independent way. One hundred measures for each test have been done. The values have been averaged. In all cases, the value ($\Delta\Gamma$) decreases when the methanol quantity increases. In other words, the slopes for each test are similar to the

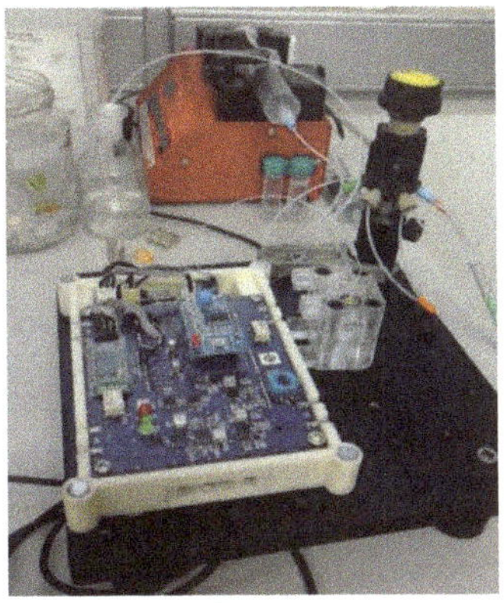

(a)

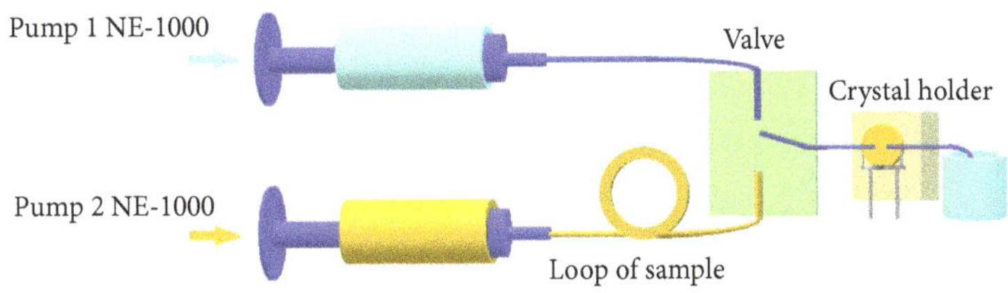

(b)

FIGURE 5: (a) QCR measurement system with the microfluidic system and holder cell and (b) schematic of the microfluidic system.

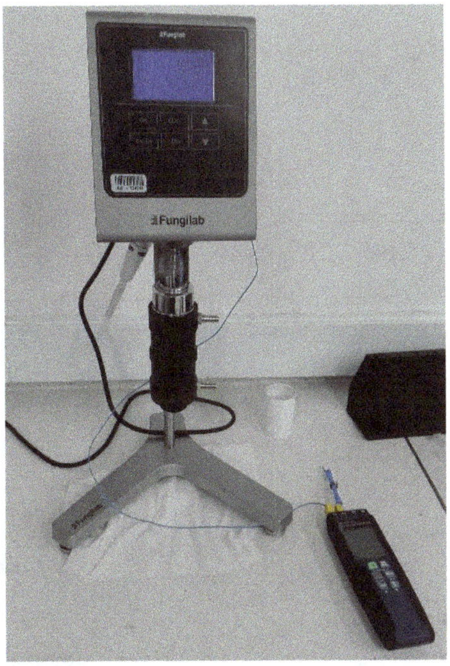

FIGURE 6: Viscometer for the reference (Fungilab Smart).

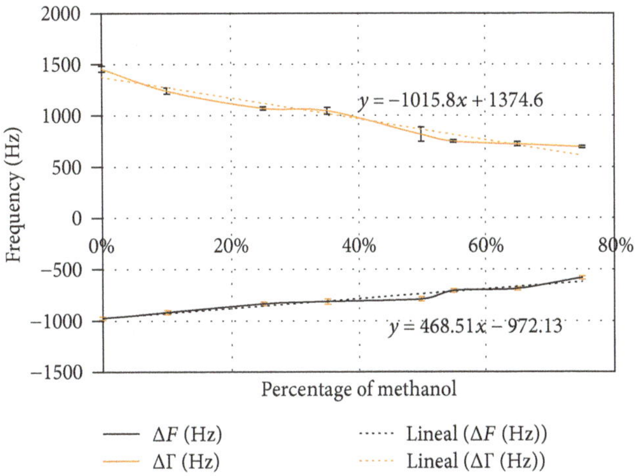

FIGURE 7: Changes in frequency for F_s and Γ variables for each sample.

trend (negative slope), which indicates that the results are coherent and reproducible and they have a good level of precision. According to these results, SD is less than 10% in all cases.

Figure 8 shows the resultant curve based on the average of the tests. Error bars are calculated using the standard deviation of data tests for each mixture. According to this figure, the prototype proposed can be used to determinate the degree of adulteration of each sample in a wide range. The reason for this capability is that the piezoelectric sensors (QCR) have high sensibility for viscosity changes. In particular, the adulteration process adds methanol to the liquor mixture generating a decrease in the viscosity of the liquor in relation to a legal liquor (negative slope in Figure 9). This decrease in the viscosity value (Figure 9, gray line) is related to the resonance of the quartz crystal; in this way, reading the frequency of resonance of the quartz crystal is possible, associating the change of this frequency with the change of the viscosity of the mixture.

According to Figure 9, the values obtained for all samples using the QCR sensor have been compared with those using the commercial equipment "viscometer Fungilab Smart" available in the material laboratory at Central University. Figure 9 permits the observation of the similar behavior between the results obtained with a commercial viscometer and the values obtained in the tests using the QCR sensor. The viscosity values have been found using (2) with the data shown in Figure 9. The density values were obtained using an analytical balance. The mass of each solution was determined from 1 mL of its volume.

In Figure 10, the red circle zone indicates that the QCR response is better than that of the viscometer because the Fungilab measures are very close in this range (25%–50% of adulterated mixture).

In accordance with the previous results, the prototype generates a very close response in comparison with the commercial equipment. In consequence, the data obtained are reliable. As an additional attribute, the quantity of the sample

TABLE 1: Results of ΔΓ for four independent tests (average ± SD). Each test contains one hundred measures.

Methanol (%)	Test 1 ΔΓ (Hz)		Test 2 ΔΓ (Hz)		Test 3 ΔΓ (Hz)		Test 4 ΔΓ (Hz)	
	Average	SD	Average	SD	Average	SD	Average	SD
0%	1374.6	27.87	1462.3	31.71	1364.6	29.51	1260.9	26.56
10%	1273.0	30.60	1368.7	33.95	1284.8	20.82	1187.3	18.06
25%	1120.7	15.72	1228.3	32.55	1165.1	104.91	1076.9	15.96
35%	1019.1	33.00	1134.8	24.98	1085.3	54.89	1003.3	34.06
50%	866.7	69.61	994.4	40.35	965.7	19.91	892.9	21.23
55%	815.9	12.94	947.6	21.22	925.8	15.86	856.1	19.26
65%	714.3	20.50	854.0	18.65	846.0	46.25	782.4	16.47
75%	612.8	10.10	760.4	14.45	766.2	21.06	708.8	41.42

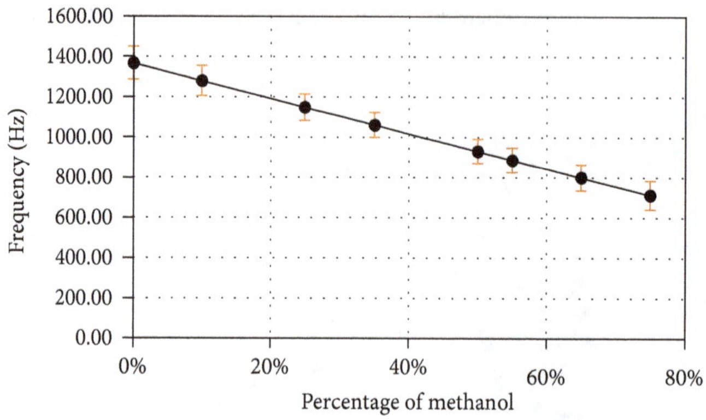

FIGURE 8: Average values of bandwidth shift for different independent tests.

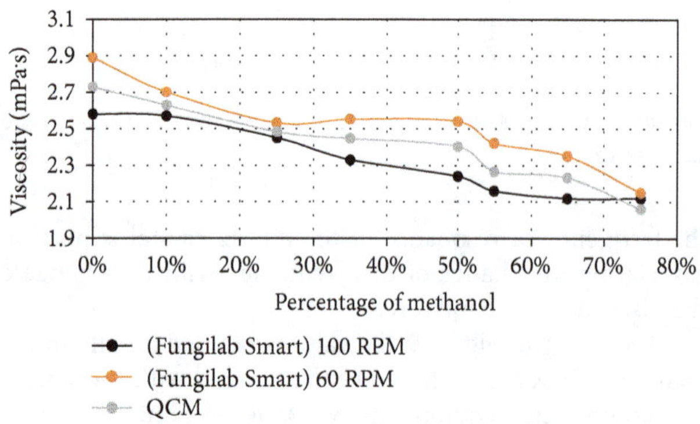

FIGURE 9: Comparison between the sensor QCR and Fungilab Smart.

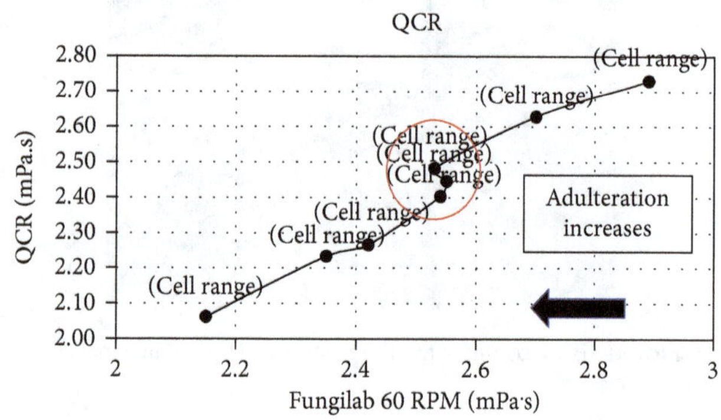

FIGURE 10: Comparison between the sensor QCR and Fungilab Smart.

used in the comparison between the prototype QCR and the commercial equipment is tens of microliters and tens of milliliters, respectively. In consequence, the prototype proposed is more efficient about the quantities of the samples used.

5. Conclusions

In the present work, the effectiveness of a low-cost piezoelectric sensor to detect fake liquor has been demonstrated through simple experiments. The capability of the quartz resonator to identify changes in the viscosity of liquid samples through an electronic transduction and characterization of its admittance response allows us to carry out the present study to detect in a clear way the changes in concentration of methanol in a common alcoholic drink like Ron. In consequence, this device could allow the identification of adulterated alcohol with a small amount of sample (tens of microliters).

In a future phase, the device will be optimized in size, its portability will be improved, and it will be connected to a smartphone. In addition, more tests with real adulterated liquor should be conducted with the aim of verifying the presence of minimum concentrations of methanol. The real adulterated liquor samples could be obtained through an agreement with the Colombian police.

Conflicts of Interest

The authors declare that there is no conflict of interest regarding the publication of this paper.

Acknowledgments

The current investigation work has been supported by COLCIENCIAS through Grant no. 0375–2013 with project number 6570577636375 and by Universidad Central (Bogotá, Colombia) through the Research Cluster on Converging Sciences and Technologies (NBIC).

References

[1] T. Robert, *Combatting Fake, Counterfeit, and Contraband Alcohol Challenges in the United Kingdom through the Alcohol Wholesaler Registration Scheme (AWRS)*, Federal Regulatory and Compliance Consultant Falls Church, Virginia, 2017.

[2] A. Tsiboukli, "Greek crisis: impact on drug treatment services, research and population," *Nordic Studies on Alcohol and Drugs*, vol. 32, no. 3, pp. 333–337, 2015.

[3] L. Riley, M. Marshall, World Health Organization, and Dept. of Substance Abuse, *Alcohol and Public Health in 8 Developing Countries*, World Health Organization, Geneva, Switzerland, 1999.

[4] D. W. Lachenmeier, *Advances in the Detection of the Adulteration of Alcoholic Beverages Including Unrecorded Alcohol*, Elsevier Ltd, 2016.

[5] J. Zapata, A. Sabogal, A. C. Montes, G. Rodriguez, and J. Castillo, *Una Estimación de la Adulteración y la Falsificación de Bebidas Alcohólicas en Colombia*, Fedesarrollo, 2012.

[6] J. L. Castellanos, A. B. Limón, A. R. Hernández, and D. E. F. Rojas, "Intoxicación por metanol, reporte de un Caso," *Archivos de Medicina de Urgencia de México*, vol. 1, pp. 67–73, 2009.

[7] Departamento Nacional de Planeación - DNP, "Estrategia para la Implementación del Sistema Unico Nacional de Información y Rastreo -SUNIR (Documento Conpes)," *Consejo Nacional de Política Económica y Social República de Colombia*, p. 33, 2009.

[8] Methanol Institute, "Methanol FACTS: adulterated alcohol poisoning: issue summary," 2016, http://www.methanol.org/wp-content/uploads/2016/06/Adulterated-Alcohol-Issue-Summary-20160322-EN-vfinal.pdf.

[9] D. Niño and J. Velandia, *Estudio Descriptivo De Las Intoxicaciones Por Metanol Reportadas En Sivigila 2010-2011 En Colombia*, Universidad de Ciencias Aplicadas y Ambientales (UDCA), 2014.

[10] M. J. C. Pontes, S. R. B. Santos, M. C. U. Araújo et al., "Classification of distilled alcoholic beverages and verification of adulteration by near infrared spectrometry," *Food Research International*, vol. 39, no. 2, pp. 182–189, 2006.

[11] L. A. Carvajal Ahumada, N. Peña Pérez, O. L. Herrera Sandoval, F. del Pozo Guerrero, and J. J. Serrano Olmedo, "A new way to find dielectric properties of liquid sample using the quartz crystal resonator (QCR)," *Sensors and Actuators A: Physical*, vol. 239, pp. 153–160, 2016.

[12] L. Ahumada, M. González, O. Sandoval, and J. Olmedo, "Evaluation of hyaluronic acid dilutions at different concentrations using a quartz crystal resonator (QCR) for the potential diagnosis of arthritic diseases," *Sensors*, vol. 16, no. 11, p. 1959, 2016.

[13] L. A. Carvajal Ahumada, J. E. Pazos Alonso, J. J. Serrano Olmedo, and O. L. Herrera Sandoval, "Diseño y Evaluación de un micro viscosímetro de bajo costo utilizando un resonador de cristal de cuarzo y Arduino," in *Congreso Internacional de Ciencias Básicas e Ingeniería de Universidad de los Llanos*, p. 11, Asociación Mexicana de Medicina de Urgencia, 2016.

[14] E. J. Calvo, R. Etchenique, P. N. Bartlett, K. Singhal, and C. Santamaria, "Quartz crystal impedance studies at 10 MHz of viscoelastic liquids and films," *Faraday Discussions*, vol. 107, pp. 141–157, 1997.

[15] M. Pax, J. Rieger, R. H. Eibl, C. Thielemann, and D. Johannsmann, "Measurements of fast fluctuations of viscoelastic properties with the quartz crystal microbalance," *Analyst*, vol. 130, no. 11, pp. 1474–1477, 2005.

[16] M. Cassiède, J. H. Paillol, J. Pauly, and J.-L. Daridon, "Electrical behaviour of AT-cut quartz crystal resonators as a function of overtone number," *Sensors and Actuators A: Physical*, vol. 159, no. 2, pp. 174–183, 2010.

[17] V. E. Granstaff and S. J. Martin, "Characterization of a thickness-shear mode quartz resonator with multiple nonpiezoelectric layers," *Journal of Applied Physics*, vol. 75, no. 3, pp. 1319–1329, 1994.

[18] T. Nakamoto and T. Kobayashi, "Development of circuit for measuring both Q variation and resonant frequency shift of quartz crystal microbalance," *IEEE Transactions on Ultrasonics, Ferroelectrics and Frequency Control*, vol. 41, no. 6, pp. 806–811, 1994.

The State-of-the-Art of Knowledge-Intensive Agriculture

Barun Basnet and Junho Bang

Department of IT Applied System Engineering, Chonbuk National University, Jeollabuk-do, Republic of Korea

Correspondence should be addressed to Junho Bang; jhbang@jbnu.ac.kr

Academic Editor: Wesley Beccaro

The application of sensors and information and communication technology (ICT) in agriculture has played a vital role in improving agricultural production and the value chain. Recently, the use of data analytics has shifted agriculture from input-intensive to knowledge-intensive as a large amount of agricultural data can be stored, shared, and analyzed to create information. In this paper, we have reviewed existing sensors and data analytics techniques used in different areas of agriculture. We have classified agriculture into five categories and reviewed the state-of-the-art technology in practice and ongoing research in each of these areas. Also, we have presented a case study of Korean scenario compared with other developed nations and addressed some of the issues associated with it. Finally, we have discussed current and future challenges and provided our views on how such issues can be addressed.

1. Introduction

The Food and Agriculture Organization of the UN (FAO) predicts that the global population will reach 9.2 billion by 2050, and food production must increase by 70 percent to keep the pace [1]. The income distribution in the world is uneven and hugely divided. In one part of the world, prosperity exists, and there is always demand for high-quality food. While in another part of the world, hunger and war exist, and there is always demand for a large quantity of foods. With limited farming land and freshwater resources, this quality and quantity crisis in food can only be addressed by the application of ICT in agriculture. Both small- and large-scale farming can benefit from introducing ICT into the agriculture value chain, having their productivity increased, quality improved, services extended, and costs reduced. Furthermore, ICT facilitates information- and knowledge-based approach rather than only focusing on input-intensive agriculture. As a result, agriculture becomes more networked, and decision making and resource utilization could significantly be leveraged.

ICT in agriculture is interchangeably used as e-agriculture, smart agriculture, precision agriculture (PA), or IoT (internet of things) in agriculture depending upon the context. Modern agriculture is hugely automated, controlled, and constantly monitored. Sensors are the heart of ICT, and various sensing devices used for this purpose generate a large volume of data continuously. The application of data analytics helps in solidifying the research in agriculture. It provides insights into various issues in the agriculture like weather prediction, crop and livestock disease, irrigation management, and supply and demand of agriculture inputs and outputs and helps in solving those problems. It can also provide valuable information for optimum resource utilization and production boosting. Our work reviews research articles focused on agricultural data and provides insights on several agricultural issues.

A wide variety of review literature is available, covering the topic: sensors and ICT in agriculture. Ojha et al. [2] reviewed the use and the state-of-the-art of wireless sensor networks (WSNs) in agriculture. Their work covers applications, design, standards, and technologies of WSNs used in agriculture. Also, another article by same authors [3] reviewed and proposed a sensor-cloud framework for the efficient addressing of various agricultural problems and applications. Another review article included key vision control techniques and their potential applications in fruit or vegetable harvesting robots [4]. In particular, it looked at various vision schemes and recognition approaches for harvesting robots. Similarly, Zion [5] reviewed on the use of computer vision technologies in aquaculture. The review highlighted on the measurement, stock identification, and monitoring of different gender and species of aquatic animals. Other reviews included the keywords "ICT" and "agriculture" but were more focused on models and architectures in agriculture absorbing ICT [6, 7]. A recent article by Wolfert et al. [8] reviewed the state-of-the-art of big data applications in smart farming and identifying socio-economic challenges associated with it. The article slightly touched the technological part but largely focused on socio-economic and governance issues for the design of suitable business models. Another recent article by Lan et al. [9] reviewed the state-of-the-art in precision agricultural aviation technology highlighting remote sensing, aerial spraying, and ground verification technologies. Likewise, a large number of research articles do exist, combining the use of artificial intelligence (AI), database, and advanced statistical tools in the agriculture [10–12]. However, a review article focused on sensors and data analytic techniques in the area of agriculture is still scarce in the literature.

This article aims to review the use of ICT especially sensors and data analytic techniques in the area of agriculture. Agriculture in this paper is used in a broader sense and covers research in crop cultivation, horticulture, animal husbandry, apiculture, and aquaculture. Our classification closely resembles the scientific classification of agriculture. By breaking agriculture into different subfields, and reviewing applied sensors and data analytics, we intend to complement existing reviews. The objective of this paper is to review research and development in the area of agriculture from the technological perspective highlighting its various subfields. Also, we intend to facilitate readers in comparing one subfield with the other subfields easily.

1.1. Associated Technical Terminologies. We briefly explain certain terms which frequently appears when we talk about ICT, sensors, and data analytics.

Sensors are electronic devices that measure a physical change in its environment and convert it into a suitable electrical form. Sensors like environmental sensors, airflow sensors, location sensors, electrochemical sensors, mechanical sensors, and optical sensors are used for acquiring various kinds of agricultural data. *Smart sensors* are capable of not only acquiring data but also store, process, and integrate such data. *Sensor fusion* is a combination of two or more sensor data to get additional insights or overcome the weakness of a single sensor. *Wireless sensor networks* (WSNs) are networks of such sensors connected wirelessly.

An *embedded system* is a microprocessor/controller embedded into an electro-mechanical system for performing a particular task. It is programmable and has limited memory and processing power. Most of the embedded systems are based on *sensing systems* consisting of sensors and actuators.

Internet of things (IoT) is a complex interconnected network of things that continuously exchange data. Here, "things" refer to any physical devices like sensors, cameras, wearables, vehicles, cell phones, and houses that are connected to the internet.

Cloud computing is an internet-based computing service where one can store, manipulate and retrieve data, and utilize resources from anywhere without actually owning the required hardware or software.

Data analytics techniques include analyzing and processing of acquired large datasets from the field and providing meaningful information so that any interested party may utilize them for their future work. Those large data sets are called *big data*, and the analytic technique is called *data mining*. A separate field of study called *data science* has emerged recently which combines computer algorithms and statistical methods for data analytics [13].

2. Research Methodology and Motivation

The primary purpose of this paper is to review the use of sensors and data analytics in different branches of agriculture. We frequently use ICT and sensors interchangeably in this paper as sensors are a key to the-state-of-the-art ICT, and sensor data are raw materials for any sort of data analytics. The terminology *sensor* used in this survey actually refers to *sensing systems* rather than the sensor devices itself. Likewise, agriculture is an enormous and vast area, and we do not attempt to review the use of ICT tools and techniques in every process, steps, fields, and subfields of agriculture. To narrow down, we have excluded *prefield* (genetics, seed/egg development) and postfield (distribution, processing, and consumer) and only reviewed the use of ICT tools and techniques for *In-field* applications, that is, planting/raising and harvesting. Also, the classification of the areas of agriculture in Section 3 is a general classification and might not exactly match with other sorts of classifications elsewhere. Figure 1 shows the evolution of different technologies used in the areas of the agriculture. However, we have excluded other (e.g., chemical, genetics, bioengineering) types of advancements made in the field of agriculture.

At large, only peer-reviewed research articles are provided as a reference for preparing this paper. However, reports and publications of various government and nongovernment organizations were also referred for understanding the perspectives outside academia in regard to challenges and future work direction. Also, our review does not contain research in agricultural machinery and robots but might cover certain underlying technologies if only they are relevant to sensors and data analytics tools.

3. Sensors and Data Analytics in Agriculture

3.1. Agronomy/Crop Farming. The general use of wireless sensor networks in crop monitoring and data acquisition can be found in various research and review articles [14–16]. Modern agronomical research and practice are becoming more and more data intensive. Data are continually collected, analyzed, and simulated to understand and predict crop growth and behavior under various circumstances.

Driemeier et al. [10] proposed a computational environment to support research in sugarcane precision agriculture.

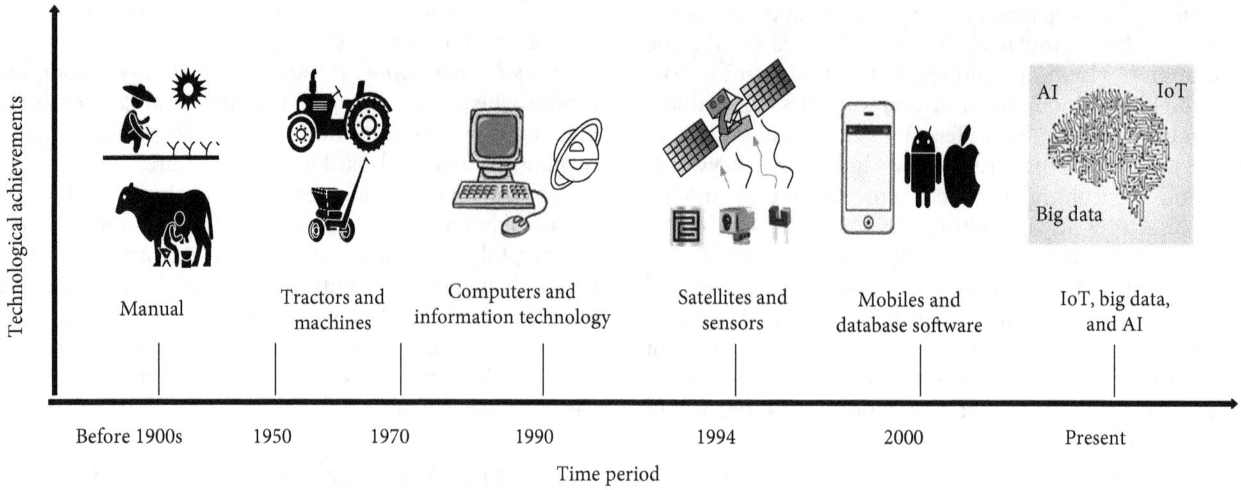

FIGURE 1: Evolution of different technologies used in the areas of the agriculture.

The work presented a data analysis workflow model for data acquisition, formatting, and verification. The model was employed to analyze three joint experiments comprised of soil attributes, sugarcane quality, and sugarcane yield. Yang et al. [11] reported the use of airborne multispectral and hyperspectral imagery and high-resolution satellite imagery for monitoring growth and estimating crop yield. They presented several application examples to demonstrate the advantages and limitations of different remote sensing and imagery analysis techniques. Similarly, Dong et al. [17] studied the feasibility of deriving spatially variable crop maximum light use efficiency (LUE_{max}) from satellite remote sensing data to improve crop biomass estimation. This study offered a new way to derive LUE_{max} for specific production efficiency models (PEM) and to improve the accuracy of biomass estimation using remote sensing. Equations (1), (2), and (3) represent an accuracy assessment model based on three statistical criteria.

$$\mathrm{RMSE} = \sqrt{\left(\frac{1}{n}\right) \times \sum_{i=1}^{n} (E_i - M_i)^2}, \qquad (1)$$

$$\mathrm{nRMSE} = \frac{\mathrm{RMSE}}{\bar{M}} \times 100, \qquad (2)$$

$$\mathrm{dindex} = 1 - \frac{\sum_{i=1}^{n} (E_i - M_i)^2}{\sum_{i=1}^{n} \left(\left|E_i - \bar{M}\right| + \left|M_i \bar{M}\right|\right)^2}. \qquad (3)$$

Here, RMSE, nRMSE, and d-index represent the root-mean-square error, the normalized RMSE, and the index of agreement, respectively. A smaller value for both RMSE and nRMSE gives the higher estimation accuracy. The parameters inside the equations n, M_i, E_i, and $\bar{M}$ represent the number of observations, the measured value, the estimated value, and the mean of all measured values.

Similarly, Ge et al. [18] reported a study to characterize the temporal dynamics of maize plants' growth and water use through RGB (red, green, blue) images and automated pot weights. These methods proved to help in quantifying plant leaf water content. Figure 2 represents the hyperspectral image analysis to extract leaf pixels and the average leaf reflectance to predict plant leaf water content. Kristen et al. [19] reported sensor-based approaches to facilitate cost-effective and site-specific management for soil health. The authors applied a sensor fusion approach (partial least square analysis) to estimate soil health indicators and soil management assessment framework (SMAF) scores using visible and near-infrared (VNIR) spectra in conjunction with electrical conductivity (EC_a) sensor data. The fusion of EC_a and CI data with VNIR improved estimation of the physical category and subsequently the overall SMAF soil health. However, chemical and fertility-related soil properties were not well estimated by this sensor fusion combination.

3.2. Horticulture/Plant Farming. Horticulture can sometimes be considered as a branch of agronomy concerned with the cultivation of plants, fruits, and vegetables rather than crops. The sensing technologies used in both areas are of similar nature. Various papers report design of greenhouse horticulture monitoring and control systems with WSN and commercially available embedded systems or IoT prototyping platforms [20–24].

Kim and Glenn [25] reported the development of a multimodal sensing system to identify the onset and severity of plant stress in young apple trees under different water treatments in a greenhouse. The data analysis result determined the spectral signature, and canopy temperature was highly correlated to plant water stress. Figure 3 represents thermal images of five different apple trees in the temperature range of 22.3°C to 40.7°C. The rectangle in each image indicates the region of interest (ROI) for the calculation of canopy temperature. In a similar manner, Tian et al. [26] reported the design of a growth cabinet using an LED light source for hydroponics cultivation of rape plants. The work's focus was to design a light source made up of blue and red LEDs to predict and provide enough energy required by plants for photosynthesis at different growth stages. The designed system could also control microclimatic parameters like temperature, humidity, light intensity, and

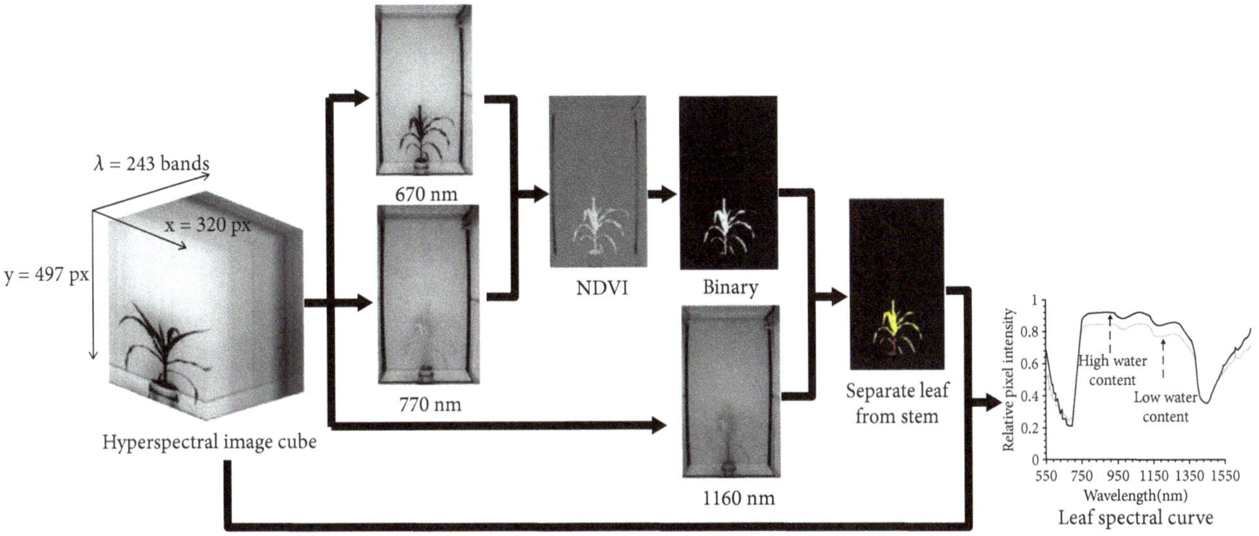

FIGURE 2: Hyperspectral image analysis to extract leaf pixels and the average leaf reflectance (pixel intensity) to predict plant leaf water content (Ge et al. [18]).

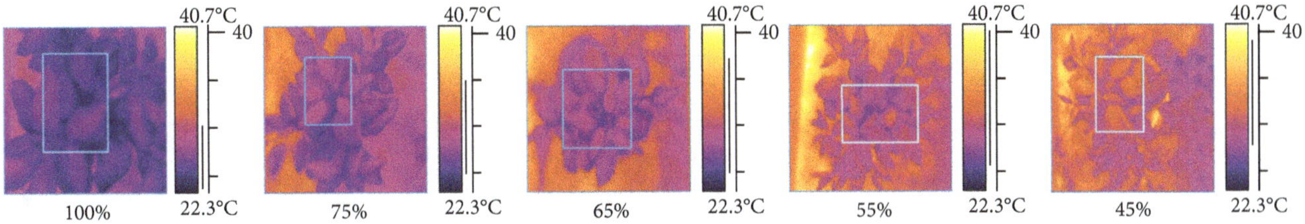

FIGURE 3: Thermal images of young apple trees in five different water treatments (Zhao et al. [4]).

CO_2 concentration level inside the growth cabinet. Their experimental results demonstrated that the plants grew very well and even superior to the plants that grew in the natural environments in terms of height, growth time, and taste. Basnet et al. adopted [27] a moving average algorithm for smoothing out variations in sensed data for a greenhouse automation system. The algorithm could greatly help in stabilizing fluctuations caused by rapid change in the environment or imprecise sensors and thus bringing stable output.

$$\mathrm{Avg}_{t1} = \frac{1}{n-1}\left[x_0^{t1} + x_1^{t1} + x_2^{t1} + \cdots + x_{n-1}^{t1}\right],$$
$$\mathrm{Avg}_{t2} = \frac{1}{n}\left[x_1^{t1} + x_2^{t1} + x_3^{t1} + \cdots + x_{n-1}^{t2}\right]. \tag{4}$$

Equation (4) represents the mathematical expression of the moving average algorithm presented in the work.

Rose et al. [28] researched the collection, classification, and quantization of phenotypic data of multiple vine rows using commercial multi-view-stereo software. Using a moving sensor platform (a track-driven vehicle, camera, GPS, and data acquisition hardware), morphological data of multiple vine rows were acquired. The authors claimed to complement existing 2D research with 3D solution-based image processing and demonstrated different data processing stages for predicting yields. Similarly, Zhao et al. [4] reviewed key techniques in vision-based control for harvesting fruits and vegetables. Their work presented an overview of various vision schemes (binocular, spectral, thermal, laser, etc.) and image processing algorithms (AdaBoost, Bayesian, fuzzy neural, etc.) for fruit recognition system.

Jesus et al. [29] proposed a field programmable gate array- (FPGA-) based wireless smart sensor for real-time photosynthesis monitoring system. A case study to monitor the photosynthetic response of chili pepper.

Capsicum annuum L. is made where the smart sensor acquires and fuses the primary sensor signals to measure temperature, relative humidity, solar radiation, CO2, air pressure, and air flow. The measurements are used to calculate net photosynthesis in real time and transmit the data via wireless communication to a sink node. In addition, the proposed smart sensor was equipped with signal processing ability, such as average decimation and *Kalman filters*, to the primary sensor readings so as to decrease the amount of noise in them as shown in Figure 4.

3.3. Animal Husbandry/Livestock Farming. Traditional animal husbandry requires the raising of a large flock of animals or cattle by experienced herdsmen. Modern animal husbandry called precision livestock farming (PLF) is about monitoring, tracking, and acquiring all sorts of data of livestock in real-time with the application of sensing systems and ICTs. These data are further processed and analyzed to

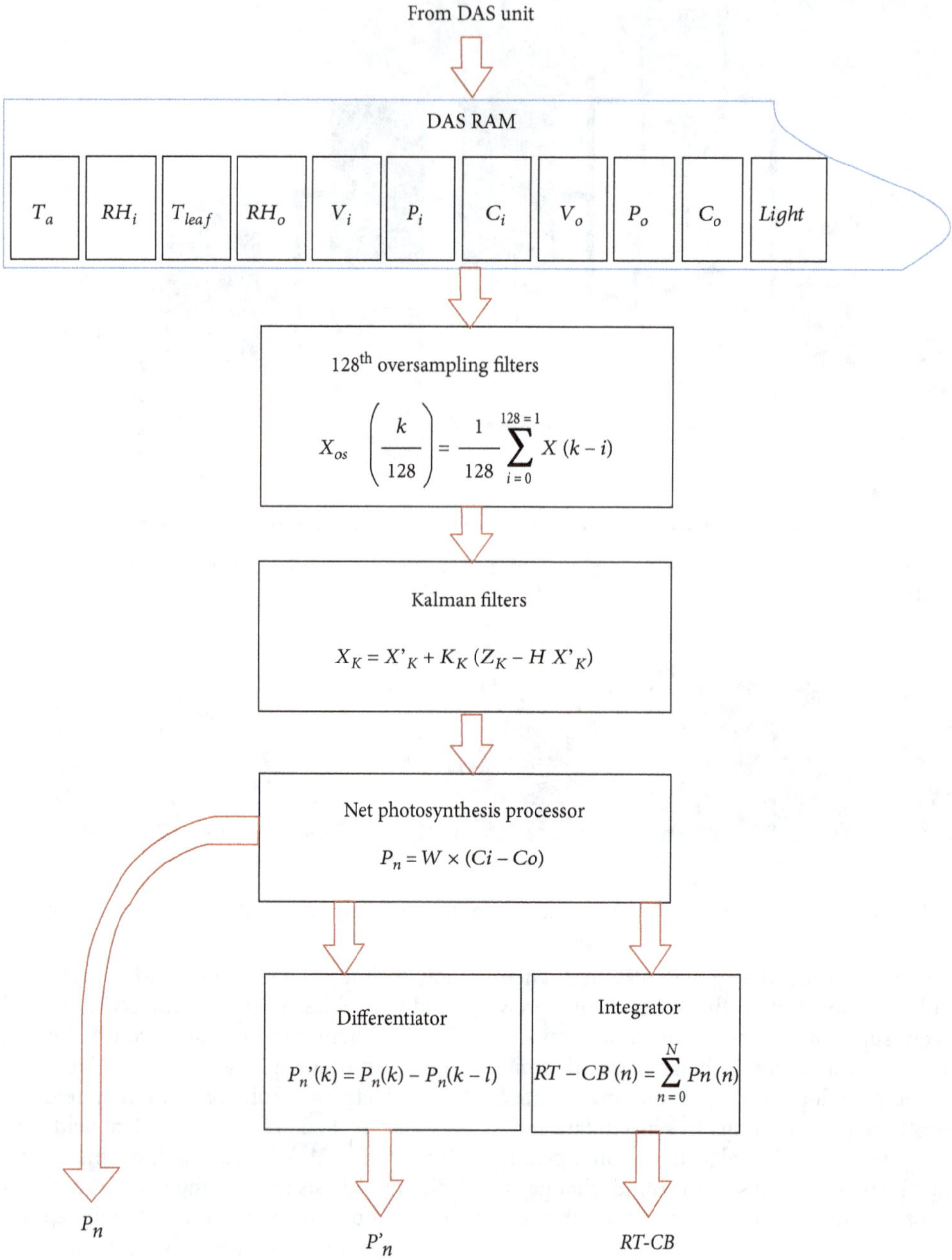

FIGURE 4: FPGA signal processing and smart processor unit (Jesus et al. [29]).

help in the decision-making process and provide a solution to various problems.

Aydin [30] reported a study to automatically assess the lameness of broiler by observing locomotion behaviors with the use of 3D vision camera and image process algorithm. The work is presented as a first attempt at assessing the lameness of broiler chickens with the use of 3D cameras having depth sensors as shown in Figure 5. Experiments were conducted to determine the *number of lying events* and *latency to lie down* of broiler chickens. The *gait scores* from 0 to 4 were chosen to rank lameness in the chicken. The proposed system had a high correlation between the output parameters against the manual labeling. Authors, therefore, asserted their work would be useful for developing an automatic animal monitoring and behavior analysis system to assess the health and welfare of broilers. Hongqian et al. [31] proposed a cloud-based data management system (CDMS) for automatic data collection for laying-hen farms. The CDMS facilitated asynchronous data transmission (*Kafka*-based), file distribution (*Hadoo*p-based), information collection and management (*MySQL*-based) for the farms located in different areas. The system was set with 8 networking nodes and experimented at a commercial egg farm. The work expected to enhance the modern poultry in the efficient management of big data and record-keeping in real-time. Nkwari et al. [32] presented an article where a cow behavior is modeled using data collected from GPS sensors to get the expected position of it. The authors used the *continuous-time*

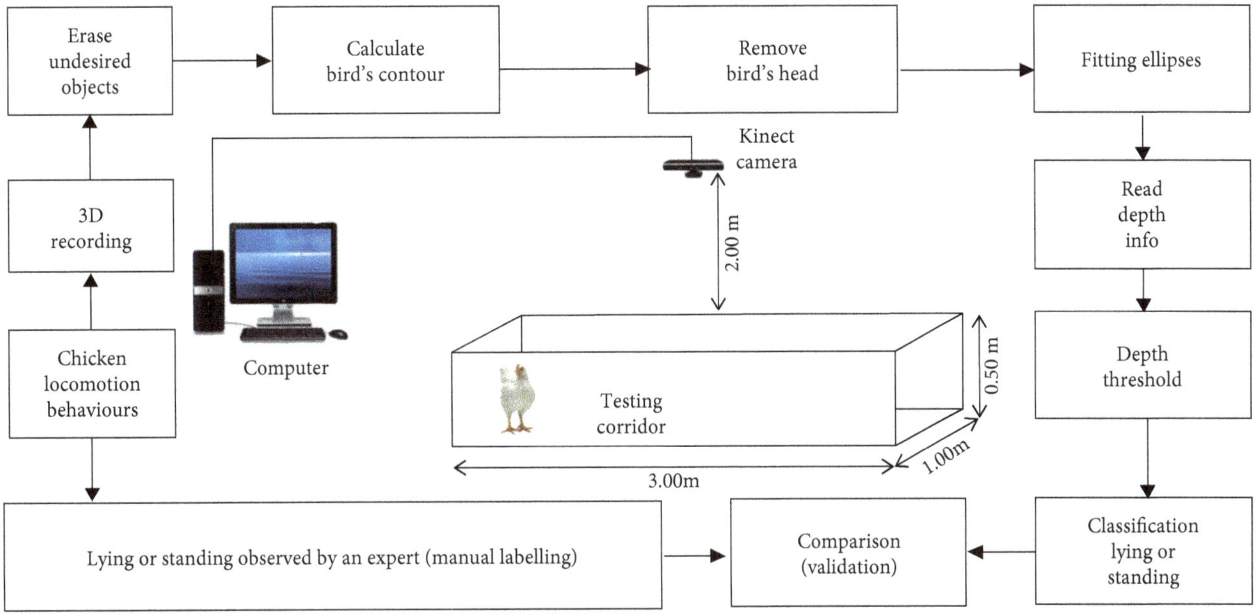

Figure 5: Flowchart of the image analysis and classification procedure using 3D vision camera (Aydin A 2017).

Markov process (CTMP) in order to model the random movement pattern (stochastic process) as shown in

$$P_{ij}^{(n)} = P_r(X_n = j|X_0 = i),$$
$$p_b = p_1 + p_2 + p_3 + p_4, \quad (5)$$
$$p_1 = P[a_1 \le r \le b_1 = \sum_k \frac{1}{n}.$$

Here, P_{ij} is the total probability that the cow moves from the location i to j. $P_{ij}^{(n)}$ represents the probability P_{ij} after the cow has taken n steps. p_1, p_2, p_3, and p_4 represent the probability the cow is in four different boundaries. a_1 and b_1 represent point limit of each boundary. By calculating which cow has a greater probability to get stolen, the work expects to help in preventing cattle rustling in farms.

Manteuffel et al. [33] presented a study which can help in preventing fatal piglet crushing events by mother sow. The work focuses on the extensive study of distress-specific vocalization to detect crushing events and thereby triggering for a mechanism to induce posture changes. Another study conducted by Feiyang et al. [34], a wireless network with the RFID tag collectors and weight sensors was employed to monitor chickens on a farm. The system detected sick chickens in the farm and classified them by studying their behaviors and parameters like the ability to snatch food, resting time, moving speed, and weight. The classification thus made and the extracted information is expected to facilitate precise husbandry and epidemic warning.

$$\text{Stay}(i) = \text{Stay}_{\text{last}}(i) - \text{Stay}_{\text{first}}(i),$$
$$D(i,j) = \sqrt{\sum_{p=1}^{n} (x_{ip} - x_{jp})^2}, \quad i = (x_{i1}, x_{i2}, \ldots, x_{ip}), j = (x_{j1}, x_{j2}, \ldots, x_{jp}). \quad (6)$$

Equation (6) represents the formula for chickens' resting time and the *Euclidean distance* formula applied in the *K-means clustering* method for the recognition of chickens' disease and quality, respectively. Here, Stay(i) represents the total resting time of chicken i from the beginning, that is, Stay$_{\text{first}}(i)$ to last period of stay, that is, Stay$_{\text{last}}$. In the same manner, $i = (x_{i1}, x_{i2}, \ldots, x_{ip})$ and $j = (x_{j1}, x_{j2}, \ldots, x_{jp})$ are two p-dimension objects with $D(i, j)$ Euclidean distance between them. The parameters $x_{i1}, x_{i2}, \ldots, x_{ip}$ are the chicken's ability to snatch food, weight, speed, and resting time of the chicken i. The same concept goes for another chicken j. Smaller Euclidean distance represents chickens of similar characteristics.

3.4. Apiculture/Beekeeping. Apiculture or beekeeping is a branch of agriculture where bee colonies are maintained in hives for harvesting honey. Various ICT technologies especially WSN were used previously to monitor the beehives and get various environmental data through sensors [33, 35–37]. Due to various environmental changes, their population is reported to be rapidly decreasing, and various interdisciplinary researches are going on to understand this phenomenon and provide possible solutions as well.

Murphy et al. [12] proposed a threshold-based algorithm and decision tree algorithms based on a biological study of bees to detect important hive changes and alert the beekeeper in a WSN for monitoring bee health. It classified hives as being in one of the ten possible states ranging from "normal" to "dead" which might or might not require an immediate response from the beekeeper. With the knowledge acquired, the beekeeper can automatically apply established beekeeping knowledge to the collected data, allowing early identification of poor health for improved colony health as well as analysis of the behavior. Figure 6 represents a *decision tree algorithm* to classify hive states of the bee colonies. Here,

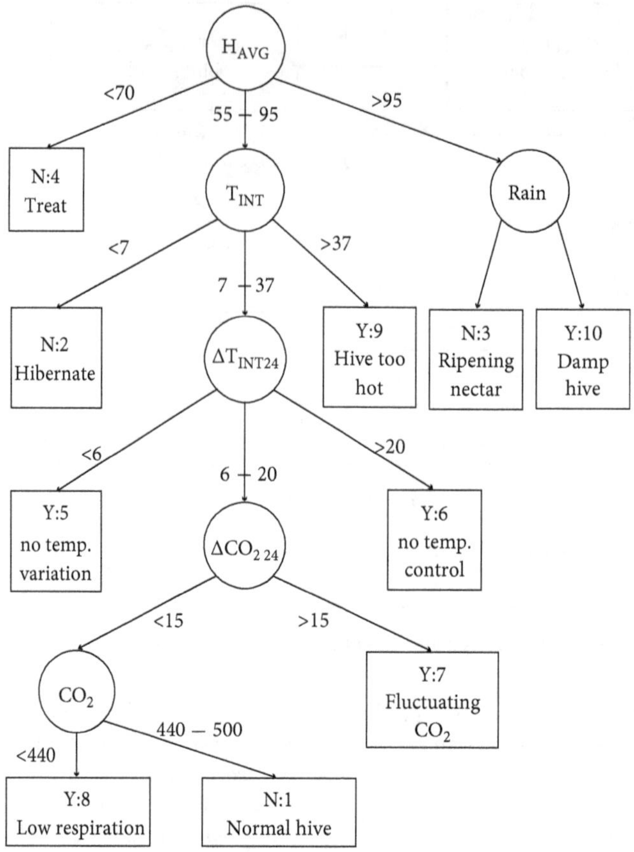

FIGURE 6: Decision tree induced from the training set for the hive classification (Murphy et al. [12]).

H_{AVG}, T_{INT}, ΔT, and ΔCO_2 represent average humidity, internal temperature, change in temperature, and change in CO_2 concentration level, respectively.

Similarly, another work presented by Kviesis et al. [38], a hierarchical three-level model consisting of a wireless node, a local data server, and a cloud data server called WBee is designed for monitoring honey bee colonies. The main distinguishing work presented in the paper is the design of the system acquiring synchronized samples from all the hives and being able to save data at each level in case there is a communication failure.

3.5. Aquaculture/Fishkeeping. Aquaculture outputs have high demands in the global food market, and the application of ICT has helped to increase its quality and the production in recent years. Water quality is the most important parameter in aquaculture since health, appetite, growth, and other activities of aquatic animals depend on it. Various subparameters like dissolved oxygen (DO), pH level, temperature, salinity, turbidity, and ammonia nitrogen content affect the quality of water. Monitoring of such parameters and recirculating water periodically is essential in maintaining the quality of water.

Lebrero et al. [39] reported the design of multiparameters monitoring system where DO sensor, pH electrode, pt1000 temperature sensor, and NH3-N sensor were used to monitor aquaculture water quality. The system would trigger an aerator ON or OFF if the sensor would read below or above the threshold (4 to 5.5 mg/L) of DO. Similar work by Hongpin et al. [40] reported on aquaculture monitoring system and control based on virtual instruments, additional features of power management, and networking solutions. The work has implemented sensor network nodes (dissolved oxygen sensor, temperature sensor, water level sensor, and pH sensor) in fish ponds for maximizing monitoring, control, and recording of the aquaculture system. With such benefits, the work reported on effectively reducing the probability of high risk of fish mortality, increase on economic benefit and consumer confidence, safety, and low energy consumption. The working concept of the designed system is presented in a flowchart as given in Figure 7.

Simbeye et al. [42] presented a multiple fish tracking systems in 3D space with structured-light (SL) sensor for acquisition of detailed information required for behavioral studies. Similarly, Saberioon and Cisar [41] used near-infrared imaging technique to observe feeding process and behavior of fish. Their work [41, 42] can help to quantify such behavior of fish and can help in developing an automatic feeding system in the future. Figure 8 represents the work of infrared imaging technique to observe the feeding process of fish. Zion reviewed the use of computer vision technologies in aquaculture and reported the satisfactory level of works done in the area of edible or ornamental fish farms [5]. However, in the case of sea cage farms, the author points out the challenges in the application of the technology because of various parameters like deep water level, high

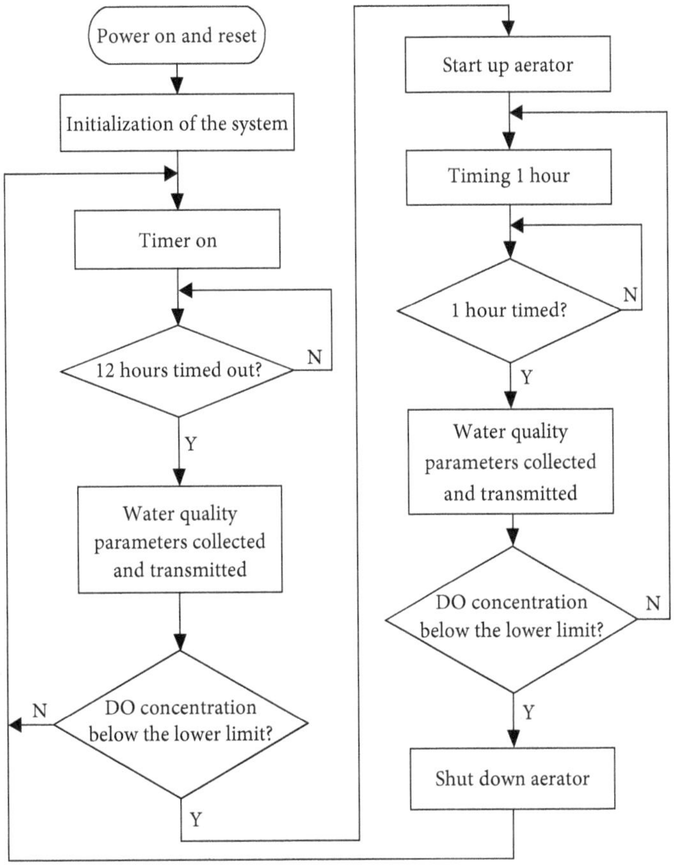

FIGURE 7: Flowchart of the remote monitoring system of aquaculture (Hongpin et al. [40]).

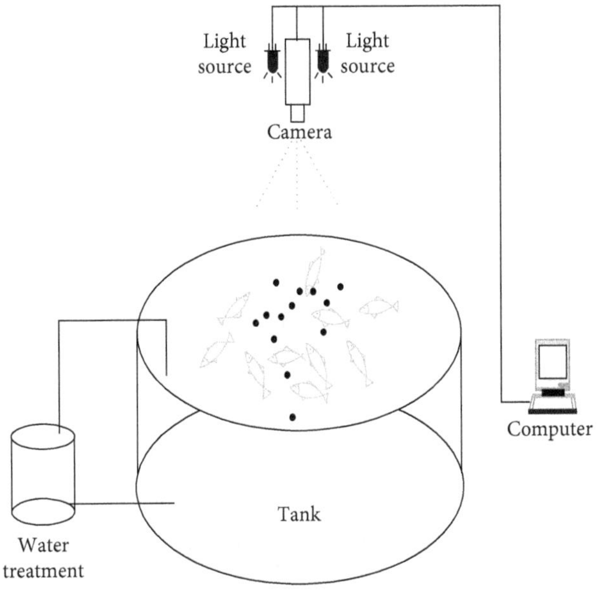

FIGURE 8: Infrared imaging technique to observe the feeding process (Saberioon and Cisar [41]).

turbidity, the presence of other aquatic animals, adjustment difficulty of feeding location, and lightening.

3.6. Chapter Conclusion. In this chapter, we reviewed various sensor-based technologies and data analytics techniques used in the field of agriculture. As mentioned in the previous chapters, agriculture is a very vast field and we do not attempt to review the application and development of different technologies in every process, steps, fields, and subfields of agriculture. We have only covered *In-field* applications.

There are other types of technological advancements made which has revolutionized the field of agriculture. Some of the noticeable things are advances in chemical fertilizers and pesticides, genetic engineering, soil and irrigation technology, agricultural tools and machinery, and production and distribution technology. However, review of such technologies is beyond the scope of this article. Table 1 shows the state-of-the-art use of different sensors and data analytics techniques presented in this work ([11–42]). Tables 2 and 3 show the state-of-the-art technology of different sensing system platforms and big data applications in smart farming and key issues, respectively.

4. Korean Scenario

South Korea is a highly industrialized country and hosts some of the world's huge leading tech-giants like Samsung, LG, and Hyundai. The Korean government has initiated steps in transferring the technology used in the industrial sector for the development of the agricultural sector as well. A report published by the Korean Rural Economic Institute emphasized the role of Korean agriculture in the nation's

Table 1: Use of sensors and data analytic techniques in different areas of agriculture.

Analytics and ICT	Sensors used	Purpose	Area	References
Image processing	Satellite data Sensors	Monitoring/estimating yield, water content check, feeding process observation	Agronomy, aquaculture	[4, 11, 17, 41]
Computer vision	3D camera, depth sensors	Lameness assessment, harvesting fruit	Animal husbandry, horticulture	[27, 28, 34]
Pattern recognition	GPS/sensors, RFID/ weight sensors, structured light sensor	Preventing cattle rustling classifying/epidemic warning, behavioral studies	Animal husbandry, apiculture	[31, 32, 42]
WSN	Environmental sensors	Detect hive changes, monitoring	Apiculture, aquaculture	[35, 38, 40]
Cloud computing	Smart devices, wireless nodes	Remote access, monitoring bee colonies	Animal husbandry, apiculture	[30, 39]
Others	Satellite LED and sensors	Crop biomass estimation hydroponics cultivation	Agronomy, horticulture	[18, 26]

Table 2: The state-of-the-art of the sensing platforms in agriculture (Ojha et al. [2]).

Feature	MICAZ	TelosB	IRIS	LOTUS	Imote2	SunSPOT
Processor	ATmega128	IMSP430	ATmega128	Cortex M3 LPC 17xx	Marvell/ XScalePXA271	ARM 920T
Clock speed (MHz)	7.373	6.717	7.373	10–100	13–416	180
Bus width (bits)	8	16	8	32	32	32
System memory (kB)	4	10	4	64	256	512
Operating frequency band (MHz)	2400	2400	2400	2400	2400	2400
Transceiver chip	CC2420	CC2420	Atmel RF230	Atmel RF231	CC2420	802.15.4
Number of channels	Programmable	Programmable	Programmable	—	In steps of 5 MHz	—
Data rate (kbps)	250	250	250	250	250	250
I/O connectivity	UART, I2C, SPI, DIO	UART, I2C, SPI, DIO	UART, I2C, SPI, DIO	3xUART, SPI, I2C, I2S, GPIO, ADC	UART3x, SPI2x, I2C, I2S, GPIO, DIO, JTAG	DIO, I2C, GPIO

Table 3: The state-of-the-art of big data applications in smart farming and key issues (Wolfert et al. [8]).

Stages of the data chain	State of the art	Key issues
Data capture	Sensors, open data, unmanned aerial vehicles (UAV), biometric sensing, genotype information, reciprocal data	Availability, quality, formats
Data storage	Cloud-based platform, Hadoop Distributed File System (HDFS), hybrid storage systems, cloud-based data warehouse	Quick and safe access to data
Data transfer	Wireless, cloud-based platform, Linked Open Data	Safety, agreements on responsibilities and liabilities
Data transformation	Machine learning algorithms, normalization, visualization and anonymization	Heterogeneity of data sources, automation of data cleansing and preparation
Data analytics	Yield models, planting instructions, benchmarking, decision ontologies, cognitive computing	Semantic heterogeneity, real-time analytics, scalability
Data marketing	Data visualization	Ownership, privacy, new business models

economic growth for the past 70 years [43]. It also reported the country's policy in investing in new technologies including ICT to cope with global warming, lack of resources, and changes in human consumption patterns. Similarly, the Electronics and Telecommunications Research Institute (ETRI) reported emerging ICT technologies to combine agricultural products at each stage in smart farms [44].

Koo et al. [45] reviewed Korean and international research trend related to ICT-based horticultural facilities. With keywords precision agriculture, smart farm, ICT, and IoT, the paper has thoroughly investigated technologies used in Korean agricultural scenario as shown in Table 4. It has also provided various case studies on failures and given directions and solutions from several perspectives. It emphasizes

TABLE 4: Core technologies for the realization of smart farm based on ICT (Koo et al. [45]).

Research objectives	Research content
Standardization and open common platform	Standardization of sensor technology and device. Building an open service platform. User-oriented interface development.
Sensor and instrumentation technology	Development of crop monitoring technology through image processing. Development of measuring device for soil and crop biomaterial information. Developed a high-efficiency sensor and measuring device. Real-time monitoring of electricity usage in the greenhouse complex. Monitoring and responding to load on the surrounding ecosystem and environment. Establishment of a sensor network system and improvement of battery or signal amplification technology of wireless measurement equipment.
Control and automation	Development of greenhouse complex environment control system algorithm. Developed smart grid integrated greenhouse control algorithm linked with energy saving technology (renewable energy). Development of differentiated ventilation fan and circulation fan control system. Development of additional forced ventilation operation algorithm through ventilation analysis.
Big data utilization and performance evaluation	ICT-based big data standardization, DB construction method development database analysis and application software development. Smart farm model performance analysis program development.
Web and mobile services	Provides real-time management and production information through the web system. Development of pest detection system and alarm system using image processing.
Renewable energy and energy load reduction technology	Developing appropriate capacity of the unit farmhouse considering the economy. Energy saving technology development for greenhouse structure and operation. Development of energy efficient operation technology using renewable energy.

the development of an intelligent service system for management and control of all the process of agricultural production. Yeo [46] reported the development of a moving monitor and control system for crops in the greenhouse. The movable sensing units gather continuous data and the connection is made through the Wi-Fi where the data are saved and processed on the server as shown in Figure 9. The movable sensing units consist of high-resolution IP camera, environmental sensors, and Wi-Fi repeater and controlling units contain embedded PC, programmable logic controller (PLC), and BLDC motors. This work expects to provide a better solution in monitoring scheme and management of plants and crops in a greenhouse.

Another recent article published by Kim [47] researched on the estimation of the factors influencing the future prices of corn and wheat through *Bayesian model* averaging as shown in (7). With the application of probabilistic factors, the results of the study facilitate the improvement in the ability to forecast grain future prices.

$$\Pr(A|B) = \frac{\Pr(B|A)\Pr(A)}{\Pr(B)}. \tag{7}$$

Here, $\Pr(A)$ and $\Pr(B)$ represents the probabilities of the occurrence of the events A and B, respectively. Whereas, $\Pr(A \mid B)$ and $\Pr(B \mid A)$ represents a conditional probability of the likelihood of the occurrence of the event A given B is true and vice-versa.

5. Challenges and Future Work Direction

The use of sensors and data analytics gives a better hope for the agricultural sector, but challenges still remain. The factors associated with them which still need further attention are listed as Sections 5.1–5.6.

5.1. Sustainable Agriculture. While technological efforts are made in increasing the productivity of the agriculture, environmental and social factors are often not taken into account. In particular, developing countries prioritize on adopting technologies for fast development rather than a sustainable one [48]. Initiatives from the different sectors of society should be taken with a view to developing the sustainability of farming systems along with the adoption of the various technologies including sensors technology and ICT. Also, technology interventions should be designed to select appropriate and environment-friendly technologies for collecting, storing, recycling, and treating and disposing of e-waste [1].

5.2. Technological Constraint. Technology at hand determines the progress in each area of its application. The improvement in technology will add more precision, accuracy, speed, and reliability whereas reducing cost in the future. Standardization in technology is vital for the improvement in communication between farm equipment. Also, research and open source projects should be encouraged more to improve the overall quality of technological solutions [49].

5.3. End-User Oriented. Technological solutions should always be user-friendly. It should suit local contexts and needs. In the case of agricultural application, solution providers should take special care in making their products and solutions as easy, readable, and understandable as possible.

5.4. Big Data. Intelligent processing and analysis of a large amount of unstructured and heterogeneous data are required. Research show *dimensional reduction* in big data is necessary for core value extraction, and also the work is also equally challenging [50]. Open access to research and publication of these data are very important. But there exists ownership ambiguity of whom they belong to. Farmers?

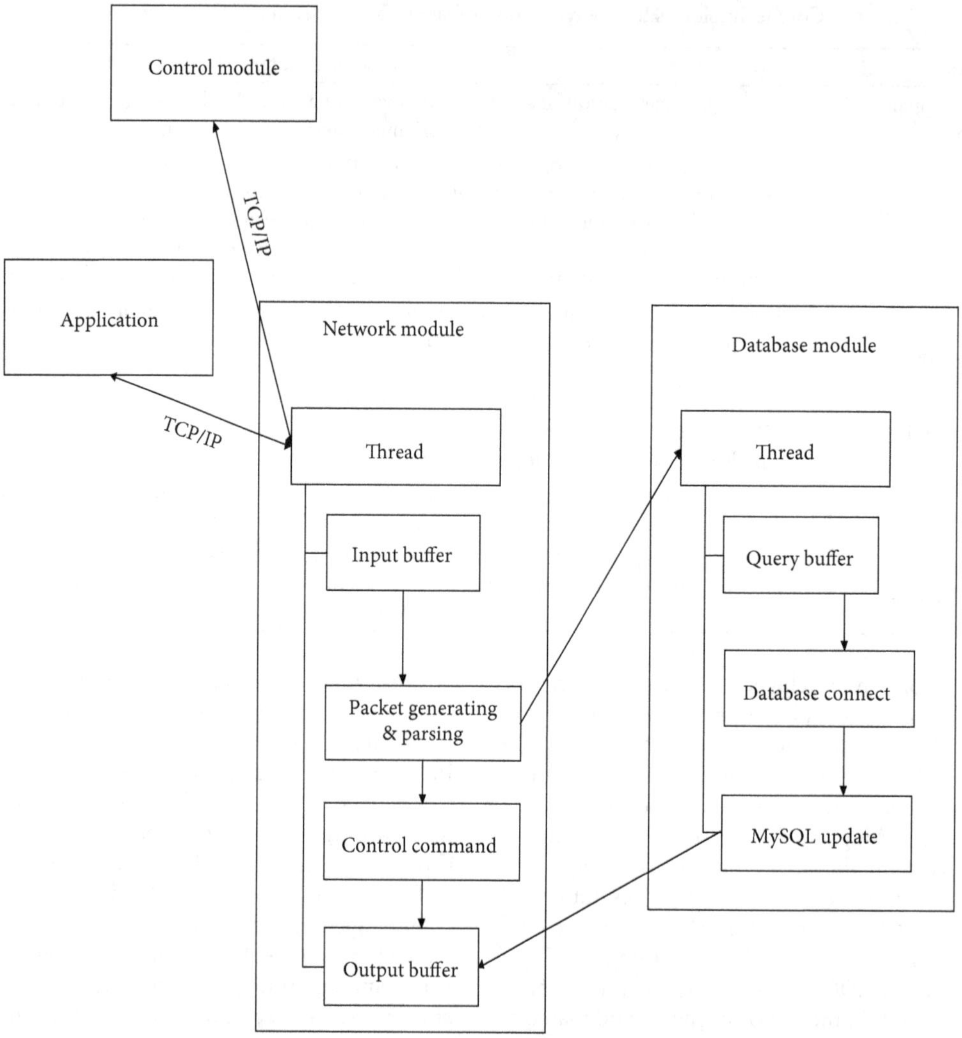

FIGURE 9: Server architecture proposed by Yeo et al. [46].

Companies? Or government? The problem should be addressed, but again if it is addressed too strictly, it can slow down innovations. Also, it is important to improve the understanding of big data usage. It is necessary to systematically promote the concept, its practical use, necessity, and value of use by expanding the education and sensitization of big data utilization [8].

5.5. Cost and Investment. Investing in the technology should not only make things easier to do but also help in increasing the return. Reduced cost increases the willingness to embrace the technology. Therefore, the cost of sensing systems and ICTs needs to be reduced, and their use needs to be financially sustainable. Farmers should be briefed about the economic consequences before and after the use of the technology [51].

5.6. Multidiscipline Collaboration. Not only through sensors technology and ICT but also many issues in the agriculture can be approached from various other disciplines. They can offer better solutions, enhance productivity, and provide other insights that agriculture scientists and other concerned parties might have overlooked. Collaboration and cooperation among experts from different fields will help in the betterment of the agricultural industry.

6. Discussions and Conclusions

In this paper, we presented a broad overview of the use of various sensors and data analytic techniques used in the area of agriculture. We divided the agriculture into five different subfields and reviewed on the state-of-the-art technology and research trends associated with each subfields. Sensor technologies especially IoTs and smart sensors have led to the exponential growth in sensor-based applications in agriculture. Our survey found the application of AI-based data analytics to be very active as many research was focused on these areas. As agriculture is becoming more data-intensive, these improvements in the technology have helped in the advancement of the area of agriculture. Many challenges still remain as we discussed in the "Challenge and Future work direction" section. However, at large, positive trends were identified, and we conclude that as sensors and data analytic techniques improve, it will bring more insights in solving a wide variety of agricultural issues.

Conflicts of Interest

The authors declare that there are no conflicts of interest regarding the publication of this paper.

Acknowledgments

This work is a part of a research project (1801000664) supported by the National Research Foundation of Korea. The authors also would like to thank Dr. R. P. Tasho and the reviewers for their suggestions on improving the quality of this paper.

References

[1] N. Alexandratos and J. Bruinsma, "World agriculture towards 2010/2050," in *The 2012 Revision*, , pp. 1–3, Food and Agriculture Organization of the United Nations, 2012, Chapter 1.

[2] T. Ojha, S. Misra, and N. S. Raghuwanshi, "Wireless sensor networks for agriculture: the state-of-the-art in practice and future challenges," *Computers and Electronics in Agriculture*, vol. 118, pp. 66–84, 2015.

[3] T. Ojha, S. Misra, and N. S. Raghuwanshi, "Sensing-cloud: leveraging the benefits for agricultural applications," *Computers and Electronics in Agriculture*, vol. 135, pp. 96–107, 2017.

[4] Y. Zhao, L. Gong, Y. Huang, and C. Liu, "A review of key techniques of vision-based control for harvesting robot," *Computers and Electronics in Agriculture*, vol. 127, pp. 311–323, 2016.

[5] B. Zion, "The use of computer vision technologies in aquaculture – a review," *Computers and Electronics in Agriculture*, vol. 88, pp. 125–132, 2012.

[6] S. J. C. Janssen, C. H. Porter, A. D. Moore et al., "Towards a new generation of agricultural system data, models and knowledge products: Information and communication technology," *Agricultural Systems*, vol. 155, pp. 200–212, 2017.

[7] J. W. Kruize, J. Wolfert, H. Scholten, C. N. Verdouw, A. Kassahun, and A. J. M. Beulens, "A reference architecture for farm software ecosystems," *Computers and Electronics in Agriculture*, vol. 125, pp. 12–28, 2016.

[8] S. Wolfert, L. Ge, C. Verdouw, and M. J. Bogaardt, "Big data in smart farming- a review," *Agricultural Systems*, vol. 153, pp. 69–80, 2017.

[9] Y. Lan, S. Chen, and K. F. Bradley, "Current status and future trends of precision agricultural aviation technologies," *International Journal of Agricultural and Biological Engineering*, vol. 10, no. 3, pp. 1–17, 2017.

[10] C. Driemeier, L. Y. Ling, G. M. Sanches, A. O. Pontes, P. S. G. Magalhaes, and J. E. Ferreira, "A computational environment to support research in sugarcane agriculture," *Computers and Electronics in Agriculture*, vol. 130, pp. 13–19, 2016.

[11] C. Yang, J. H. Everitt, Q. du, B. Luo, and J. Chanussot, "Using high-resolution airborne and satellite imagery to assess crop growth and yield variability for precision agriculture," *Proceedings of the IEEE*, vol. 101, no. 3, pp. 582–592, 2013.

[12] F. Edwards-Murphy, M. Magno, P. M. Whelan, J. O'Halloran, and E. M. Popovici, "B+WSN: smart beehive with preliminary decision tree analysis for agriculture and honey bee health monitoring," *Computers and Electronics in Agriculture*, vol. 124, pp. 211–219, 2016.

[13] H. Wimmer and L. M. Powell, "A comparison of open source tools for data science," in *Proceedings of the Conference on Information Systems Applied Research*, vol. 8, pp. 4–12, Wilmington, NC, USA, 2015.

[14] Z. Liqiang, Y. Shouyi, L. Leibo, Z. Zhen, and W. Shaojun, "A crop monitoring system based on wireless sensor network," *Procedia Environmental Sciences*, vol. 11, pp. 558–565, 2011.

[15] A. Sivasankari and S. Gandhimathi, "Wireless sensor based crop monitoring system for agriculture using Wi-fi network dissertation," *International Journal of Computer Science and Information Technology Research*, vol. 2, pp. 293–303, 2014.

[16] K. Y. Bendigeri and J. D. Mallapur, "Advanced remote monitoring of a crop in agriculture using wireless sensor network topologies," *International Journal of Electronics and Communication Engineering & Technology*, vol. 6, pp. 30–38, 2015.

[17] T. Dong, J. Liu, B. Qian et al., "Deriving maximum light use efficiency from crop growth model and satellite data to improve crop biomass estimation," *IEEE Journal of Selected topics in Applied Earth observations and Remote sensing*, vol. 10, no. 1, pp. 104–117, 2017.

[18] Y. Ge, G. Bai, V. Stoerger, and J. C. Schnable, "Temporal dynamics of maize plant growth, water use, and leaf water content using automated high throughput RGB and hyperspectral imaging," *Computers and Electronics in Agriculture*, vol. 127, pp. 625–632, 2016.

[19] K. S. Veum, K. A. Sudduth, R. J. Kremer, and N. R. Kitchen, "Sensor data fusion for soil health assessment," *Geoderma*, vol. 305, pp. 53–61, 2017.

[20] J. A. Enokela and T. O. Othoigbe, "An automated greenhouse control system using Arduino prototyping platform," *Australian Journal of Engineering Research*, vol. 1, no. 1, pp. 64–73, 2015.

[21] V. S. Jahnavi and S. F. Ahamed, "Smart wireless sensor network for automated greenhouse," *IETE Journal of Research*, vol. 61, no. 2, pp. 180–185, 2015.

[22] A. Salleh, M. K. Ismail, N. R. Mohamad, M. A. Aziz, M. A. Othman, and M. H. Misran, "Development of greenhouse monitoring uisng wireless sensor network through Zigbee technology," *International Journal of Engineering Science Invention*, vol. 2, pp. 2319–6734, 2013.

[23] R. Pahuja, H. K. Verma, and M. Uddin, "A wireless sensor network for greenhouse climate control," *IEEE Pervasive Computing*, vol. 12, no. 2, pp. 49–58, 2013.

[24] K. A. Eldhose, R. Antony, P. K. Mini, M. N. Krishnapriya, and M. S. Neenu, "Automated greenhouse monitoring system," *International Journal of Engineering and Innovative Technology*, vol. 3, pp. 164–166, 2014.

[25] J. Y. Kim and D. M. Glenn, "Measurement of photosynthetic response to plant water stress using a multi-modal sensing system," *Transactions of the ASABE*, vol. 58, no. 2, pp. 233–240, 2015.

[26] L. Tian, Q. Meng, L. Wang, and J. Dong, "A study on crop growth environment control system," *International Journal of Control and Automation*, vol. 7, no. 9, pp. 357–374, 2014.

[27] B. Basnet, I. Lee, M. Noh, H. Chun, A. Jaffari, and J. Bang, "An smart greenhouse automation system applying moving average algorithm," *The Transactions of The Korean Institute of Electrical Engineers*, vol. 65, no. 10, pp. 1755–1760, 2016.

[28] J. Rose, A. Kicherer, M. Wieland, L. Klingbeil, R. Töpfer, and H. Kuhlmann, "Towards automated large-scale 3D phenotyping of vineyards under field conditions," *Sensors*, vol. 16, no. 12, pp. 2–15, 2016.

[29] J. R. Millan-Almaraz, I. Torres-Pacheco, C. Duarte-Galvan et al., "FPGA-based wireless smart sensor for real-time photosynthesis monitoring," *Computers and Electronics in Agriculture*, vol. 95, pp. 58–69, 2013.

[30] A. Aydin, "Using 3D vision camera system to automatically assess the level of inactivity in broiler chickens," *Computers and Electronics in Agriculture*, vol. 135, pp. 4–10, 2017.

[31] C. Hongqian, H. Xin, T. Guanghui et al., "Cloud-based data management system for automatic real-time data acquisition from large-scale laying-hen farms," *International Journal of Agricultural & Biological Engineering*, vol. 9, no. 4, pp. 106–115, 2016.

[32] P. K. M. Nkwari, S. Rimer, and B. S. Paul, "Cattle monitoring system using wireless sensor network in order to prevent cattle rustling," in *IST-Africa 2014 Conference Proceedings*, Le Meridien Ile Maurice, Mauritius, May 2014.

[33] C. Manteuffel, E. Hartung, M. Schmidt, G. Hoffmann, and P. C. Schon, "Online detection and localisation of piglet crushing using vocalisation analysis and context data," *Computers and Electronics in Agriculture*, vol. 135, pp. 108–114, 2017.

[34] Z. Feiyang, H. Yueming, C. Liancheng, G. Lihong, D. Wenjie, and W. Lu, "Monitoring behavior of poultry based on RFID radio frequency network," *International Journal of Agricultural and Biological Engineering*, vol. 9, no. 6, pp. 139–147, 2016.

[35] J. F. Odoux, P. Aupinel, S. Gateff, F. Requier, M. Henry, and V. Bretagnolle, "ECOBEE: a tool for long-term honey bee colony monitoring at the landscape scale in West European intensive agroecosystems," *Journal of Apicultural Research*, vol. 53, no. 1, pp. 57–66, 2015.

[36] F. E. Murphy, E. Popovici, P. Whelan, and M. Mango, "Development of an heterogeneous wireless sensor network for instrumentation and analysis of beehives," in *2015 IEEE International Instrumentation and Measurement Technology Conference (I2MTC) Proceedings*, Pisa, Italy, May 2015.

[37] D. W. Fitzgerald, F. E. Murphy, W. M. D. Wright, P. M. Whelan, and E. M. Popovici, "Design and development of a smart weighing scale for beehive monitoring," in *2015 26th Irish Signals and Systems Conference (ISSC)*, Carlow, Ireland, June 2015.

[38] A. Kviesis, A. Zacepins, M. Durgun, and S. Tekin, "Application of wireless sensor networks in precision apiculture," *Engineering for Rural Development*, vol. 20, pp. 440–445, 2015.

[39] S. Gil-Lebrero, F. J. Quiles-Latorre, M. Ortiz-López, V. Sánchez-Ruiz, V. Gámiz-López, and J. J. Luna-Rodríguez, "Honey bee colonies remote monitoring system," *Sensors*, vol. 17, p. 55, 2017.

[40] L. Hongpin, L. Guanglin, P. Weifeng, S. Jie, and B. Qiuwei, "Real-time remote monitoring system for aquaculture water quality," *International Journal of Agricultural and Biological Engineering*, vol. 8, no. 6, pp. 136–143, 2015.

[41] M. M. Saberioon and P. Cisar, "Automated multiple fish tracking in three-dimension using a structured light sensor," *Computers and Electronics in Agriculture*, vol. 121, pp. 215–221, 2016.

[42] D. S. Simbeye, J. Zhao, and S. Yang, "Design and deployment of wireless sensor networks for aquaculture monitoring and control based on virtual instruments," *Computers and Electronics in Agriculture*, vol. 102, pp. 31–42, 2014.

[43] C. Zhou, B. Zhang, K. Lin et al., "Near-infrared imaging to quantify the feeding behavior of fish in aquaculture," *Computers and Electronics in Agriculture*, vol. 135, pp. 233–241, 2017.

[44] Korea Rural Economic Institute, *Agriculture in Korea*, 2015.

[45] H. S. Koo, J. H. Min, and J. Y. Park, "2015 electronics and telecommunications trends," in *Survey of ICT-Agriculture Convergence*, ETRI, 2014.

[46] U.-h. Yeo, I.-b. Lee, K.-s. Kwon et al., "Analysis of research trend and core technologies based on ICT to materialize smart-farm," *Protected Horticulture and Plant Factory*, vol. 25, no. 1, pp. 30–41, 2016.

[47] H. J. Kim, "Development of a moving monitor system for growing crops and environmental information in green house," *The Journal of Korea Institute of Information, Electronics, and Communication Technology*, vol. 9, no. 3, pp. 285–290, 2016.

[48] B. R. Prasad and S. Agarwal, "Comparative study of big data computing and storage tools: a review," *International Journal of Database Theory and Application*, vol. 9, no. 1, pp. 45–66, 2016.

[49] S. G. Dutia, *Agtech: Challenges and Opportunities for Sustainable Growth*, Ewing Marion Kauffman Foundation, 2014.

[50] T. H. M. d. Oliveira, M. Painho, V. Santos, O. Sian, and A. Barriguinha, "Development of an agricultural management information system based on open-source solutions," *Procedia Technology*, vol. 16, pp. 342–354, 2014.

[51] K. Sabarina and N. Priya, "Lowering data dimensionality in big data for the benefit of precision agriculture," in *International Conference on Intelligent Computing, Communication & Convergence*, Odisha, India, 2015.

13

Fault Detection of Wind Turbine Sensors using Artificial Neural Networks

Ayse Gokcen Kavaz and Burak Barutcu

Energy Institute, Istanbul Technical University, 34469 Istanbul, Turkey

Correspondence should be addressed to Ayse Gokcen Kavaz; agkavaz@itu.edu.tr

Academic Editor: Jesus Corres

This paper proposes a method for sensor validation and fault detection in wind turbines. Ensuring validity of sensor measurements is a significant part in overall condition monitoring as sensor faults lead to incorrect results in monitoring a system's state of health. Although identifying abrupt failures in sensors is relatively straightforward, calibration drifts are more difficult to detect. Therefore, a detection and isolation technique for sensor calibration drifts on the purpose of measurement validation was developed. Temperature sensor measurements from the Supervisory Control and Data Acquisition system of a wind turbine were used for this aim. Low output rate of the measurements and nonlinear characteristics of the system drive the necessity to design an advanced fault detection algorithm. Artificial neural networks were chosen for this purpose considering their high performance in nonlinear environments. The results demonstrate that the proposed method can effectively detect existence of calibration drift and isolate the exact sensor with faulty behaviour.

1. Introduction

In the last decades, both the size and capacity of wind turbines have increased by virtue of technological developments in wind energy field. This situation resulted in an increasing focus on topics such as wind turbine fault detection [1–6]. Condition monitoring and fault detection algorithms of wind turbines are important systems that contribute to reducing maintenance costs and downtime of wind power plants. Furthermore, wind power plants are generally located in distant sites which makes reliability even more important. Wind turbine faults lead to the need for repair and/or replacement actions and result in loss of energy production. Moreover, in some cases, failure in a component also affects other components and even the entire wind turbine. Considering all these facts, it is significant to detect and isolate wind turbine faults as early as possible to take required actions for preventing such undesired results. Besides, with the decrease in maintenance costs, wind energy will become commercially more competitive compared to other energy sources.

Fault detection methods can be classified into two different categories, namely, model-based and data-driven approaches. In model-based methods, firstly a mathematical model expressing wind turbine dynamics is generated. The output of that model is compared to real measurements from wind turbine sensors, and a fault alarm is given by analyzing residuals between outputs of the real system and model. Some examples of model-based fault detection studies in wind turbines can be found in [7, 8]. Model-based methods have the advantage of not requiring high sampling rate measurements [3]. However, the success of these methods depends on the consistency of the mathematical model and the real behaviour of system, and this approach meets challenges because in addition to systematic complexities, wind turbines have complex control systems which increase the difficulty of obtaining a reasonable mathematical model of the whole system [9].

Data-driven methods do not require explicit mathematical models identifying the physical system. Instead, they are based on the analysis of measurements gathered from sensors mounted on different parts of wind power plants. Different alternatives exist in terms of choosing sensor data to be processed. The first alternative is to use sensor outputs from the Supervisory Control and Data Acquisition (SCADA) system

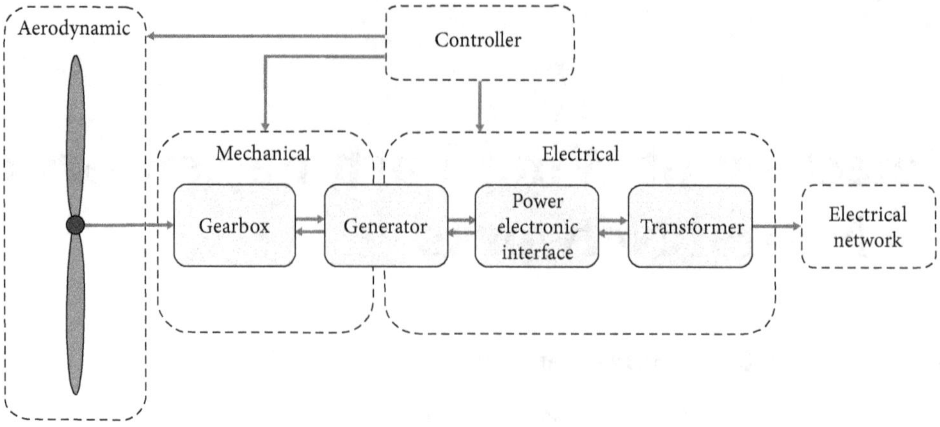

FIGURE 1: Subsystems of the wind turbine.

which is a built-in part of most modern wind turbines with no additional costs. Many attempts have been made to benefit from SCADA data for fault detection aims [9–15]. The other alternative is to use purpose-designed measurement systems having sensors specifically mounted for condition monitoring purposes. Common sensor types employed for this aim are vibration, acoustic emission, strain, torque, and bending moment sensors. Some past researches on wind turbine fault detection using specific sensors can be found in [16–24]. Both two data gathering methods have certain advantages and disadvantages. Sensor data from the SCADA system are of low frequency and generally have 10 min sampling intervals. Such a low frequency brings difficulties due to the loss of noise characteristics which carry important information on fault formation. Besides, generally there are many missing values and imperfections in data collected by the SCADA system. But as an advantage, the SCADA system is a built-in part of most of the large-scale wind turbines; therefore, no additional cost is required to reach these data. On the other hand, data output frequency for purpose-built wind turbine condition monitoring sensors can be selected high enough not to lose information in signal. However, this approach causes extra cost for maintenance and, as a result, becomes a disadvantage in terms of cost competitiveness against other energy conversion methods.

In this paper, data gathered from the temperature sensors of a wind turbine SCADA system were used to detect and isolate sensor faults. To build an effective fault detection system, it is crucial to distinguish between sensor errors from faults originated from other components of the turbine, because faulty behaviour in sensors can lead to errors in data evaluation and cause false alarms and performance degradation of fault detection system.

To compensate for the disadvantages of low-frequency sensor measurements from the SCADA system, a sophisticated signal processing method is required. Artificial neural networks (ANN) were selected for this purpose considering their high performance in modelling nonlinear systems. Different ANN architectures were built and compared to obtain the best option among different alternatives.

The layout of this paper is as follows. In Section 2, sensor selection and data collection parts of the research are presented. In Section 3, ANN architectures designed in this research are explained, and in Sections 4 and 5, results and discussions are presented, respectively.

2. Data Collection

Wind turbines are comprised of different interconnected subsystems. The primary subsystems can be listed as aerodynamic, mechanical, electronic, and control systems (Figure 1). A sensor fault in any of these components results in a degradation in a turbine's performance. Detecting and isolating sensor faults are an important part of the overall fault detection system as distinguishing the cause of abnormality is required for repair or replacement actions.

Various types of sensor faults can be encountered in subsystems of wind turbines. Typical sensor faults can be listed as multiplicative, additive, offset faults, and faults resulting in changing dynamics in the system [25]. In this paper, multiplicative fault in temperature sensor measurements is investigated. Multiplicative faults may arise from calibration drifts and act like a scaling factor on sensor measurements. This type of sensor fault was selected to be investigated because unless scaling factor of a multiplicative fault is too big, it does not cause an easily recognizable change comparing to normal response of sensors which makes them harder to detect. SCADA measurements used in this work supply some statuses of main faults such as mains failures, feeding faults, and pitch control errors; however, the SCADA system does not provide information on sensor faults. But ensuring measurement validation is a compulsory step in the design of an overall fault detection system. Therefore, the multiplicative sensor fault to be detected was artificially created in this research. Measurement differences in a form of a multiplicative effect can be originated either from sensor calibration drifts or real temperature deviations, so it is required to distinguish these two situations. For this reason, the method designed was tested in different cases representing both these situations.

Temperature measurements from the SCADA system of a 900 kW wind turbine were collected to train ANNs designed in this work. The data have 10 min sampling intervals and were gathered between 01 November 2015 and 30 November 2015. The SCADA system includes 10

Table 1: Temperature sensors from the SCADA system.

Description
Generator stator temperature
Generator rotor temperature
Nacelle ambient temperature
Nacelle temperature
Front hub bearing temperature
Rear hub bearing temperature*
Nacelle control cabinet temperature
Control cabinet temperature*
Tower temperature*
Transformer temperature*

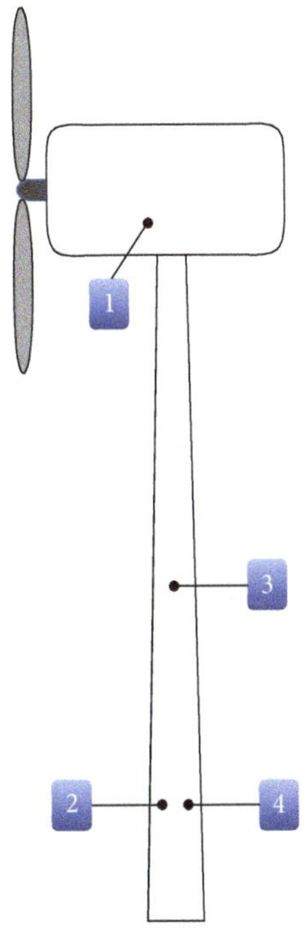

Figure 2: Locations of the sensors used in this research.

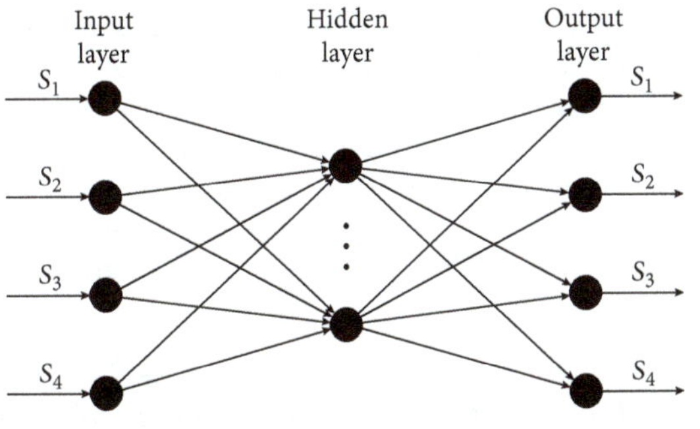

Figure 3: Autoassociative input-output structure.

temperature sensors mounted on different components of the turbine which are presented in Table 1.

Sensors with the * sign in Table 1 were selected for the sensor validation purpose in this work. They measure rear

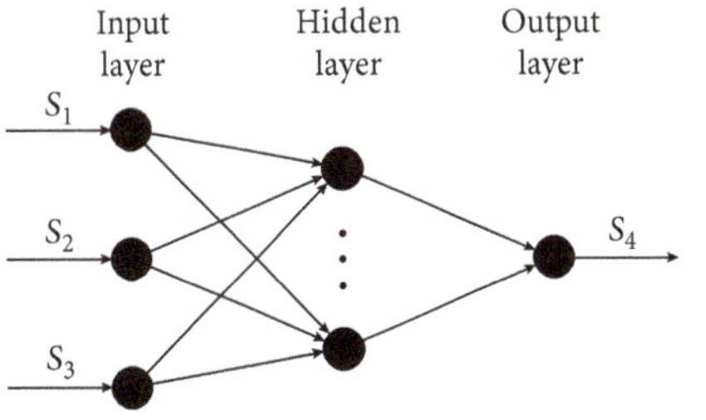

Figure 4: MISO input-output structure.

hub bearing temperature (S_1), control cabinet temperature (S_2), tower temperature (S_3), and transformer temperature (S_4). The algorithm proposed detects calibration drifts in one of these sensors based solely on the measurements of them without using any operational or environmental data. Therefore, this subset of sensors was selected considering the similar temperature characteristics of the areas they were installed in. Each of them is monitored by operators to find out if any abnormalities appear in the components around them. Locations of the sensors used in this research are presented in Figure 2.

Using measurements from these sensors, firstly ANNs were trained. For training purpose, 75% of the data was used. The remaining 25% of the data was used to form test data set to evaluate and compare the performance of the networks. ANN models were designed and trained in MATLAB environment.

3. Artificial Neural Network Architectures

ANNs have proven to be powerful algorithms for various aims including classification, estimation, and fault detection. Due to their capability of dealing successfully with nonlinearities and their advantages in real-time applications, it is convenient to design ANN models for detecting wind turbine faults. Several ANN models were trained and analyzed to obtain a successful network architecture that satisfies the requirements of this research.

In terms of input-output relations, 2 approaches have been used, namely, autoassociative and MISO (multiple-input single-output) structures. The autoassociative case, where the input vector is to be associated with itself is presented in Figure 3. As can be seen in the figure, all the sensors were set both as inputs and outputs of the autoassociative network. In the MISO case, 3 out of 4 selected sensors were set as inputs and the remaining sensor is set to be the output. The structure is repeated 4 times with each one of the sensors used as output separately. Figure 4 shows the MISO structure with S_4 as the output of the system. Similar networks with other sensors in the output layer were also created.

The input-output relations presented in Figures 3 and 4 were used in different neural network types, namely, back-propagation neural network (BPNN), radial basis function neural network (RBFNN), and general regression neural network (GRNN). Table 2 presents the computational principles and input-output relations of the networks designed.

TABLE 2: Neural network architectures.

Network	Computational principle	Input-output relation
Network 1	BPNN	Autoassociative
Network 2	BPNN	MISO
Network 3	RBFNN	Autoassociative
Network 4	RBFNN	MISO
Network 5	GRNN	Autoassociative
Network 6	GRNN	MISO

3.1. Backpropagation Neural Network. BPNN is a type of feedforward neural network where information flows from inputs to outputs without any feedback connections, whereas error propagates from outputs to inputs. Detailed mathematical foundations of BPNNs can be found in [26, 27]. This kind of neural network has been commonly used for fault detection purposes.

A typical BPNN consists of an input layer, one or multiple hidden layers, and an output layer. By taking requirements of problem into consideration, different activation functions can be used. With the help of nonlinear activation functions existing in hidden or output layers, neural networks become capable of learning nonlinear relationships between inputs and outputs. In this research, several BPNN architectures with one hidden layer were designed. Firstly, the type of activation functions for hidden and output layers was determined. For this reason, networks with different activation function pairs were designed and evaluated using a smaller sample data set, such as "logarithmic sigmoid-linear," "logarithmic sigmoid-logarithmic sigmoid," "tangent sigmoid-linear," and "tangent sigmoid-tangent sigmoid." The former function represents the hidden layer's activation function, and the latter represents the output layer's activation function. The best results were obtained by "logarithmic sigmoid-linear" activation function pair. Therefore, the detailed network developments in the following parts of the research were continued with these activation functions.

As presented in Table 2, autoassociative and MISO structures were used in the design of BPNN models. To obtain an autoassociative BPNN with high performance, different network architectures with 2 and 3 hidden neurons were designed and 10 trials were made with each architecture. The reason of multiple trials is that the performance of the network alters depending on the initial values of weights connecting different layers and to ensure the reach of a minimum point in the cost function space.

For MISO BPNNs, the number of hidden neurons was increased from 2 to 15 and again 10 trials with different random initial conditions for each architecture were held. The selection of the best network to use in the next parts of the research was made based on the comparison of R^2 goodness of fit values. Amongst the autoassociative BPNNs, the R^2 values of the best network were 0.999, 0.999, 0.996, and 0.993 for sensors 1 through 4, respectively. For MISO BPNNs, the highest scores were obtained as 0.836, 0.979, 0.984, and 0.985. Figure 5 shows the regression graphs of the test set for autoassociative and MISO BPNN with the best performances.

3.2. Radial Basis Function Neural Network. RBFNN is a type of feedforward neural network typically having a single layer of hidden units that are connected to linear output units. They are trained in a hybrid manner. Computational parts in the hidden layer use unsupervised learning, and each of them is described by a radial basis function. The size of the units in the hidden layer is the same as the size of the training vector. The output layer is trained in a supervised manner. Mathematical details of RBFNNs can be found in [28].

The output of a neuron of RBFNN is given in the following:

$$y = \sum_{k=11}^{N} w_k \varphi_k(\|x - c_k\|), \tag{1}$$

where N is the number of neurons in the hidden layer, φ_k is the kernel of radial basis function for each unit, c_k is the center of radial basis function vector for neuron k, and w_k is the weight of neuron k in the output layer. The most common choice for RBFNN kernel is Gaussian activation function which was also used in this research. It is presented in the following:

$$\varphi_k(\|x - c_k\|) = \exp\left(-\frac{\|x - c_k\|^2}{2\sigma_k^2}\right), \tag{2}$$

where $\|x - c_k\|^2$ is the squared Euclidian distance between the associated center vector and the input vector and σ_k is the width factor of the kth hidden unit in the hidden layer which controls the smoothness properties of interpolating function.

Several autoassociative and MISO networks with RBFNN approach were developed in this research. Width factors of the networks were selected amongst varying values between 0.1 and 100. The performances of the networks were evaluated based on the R^2 values. The best R^2 scores amongst the autoassociative RBFNNs were bigger than 0.999 for all sensors. The highest scores for MISO RBFNNs were 0.824, 0.979, 0.982, and 0.984. In the detailed analysis for the next parts of the research, the networks with the best performances were used. Regression plots for the test set of RBFNN networks with the highest performance results are presented in Figure 6.

3.3. General Regression Neural Network. GRNN computes the most probable value of an output y given only training vectors x. In this approach, a specific functional form to describe the relation between inputs and outputs is not required, instead the appropriate form is expressed as a probability density function which is determined from observed data. Since the parameters are directly calculated using examples, an iterative computation is not required [29].

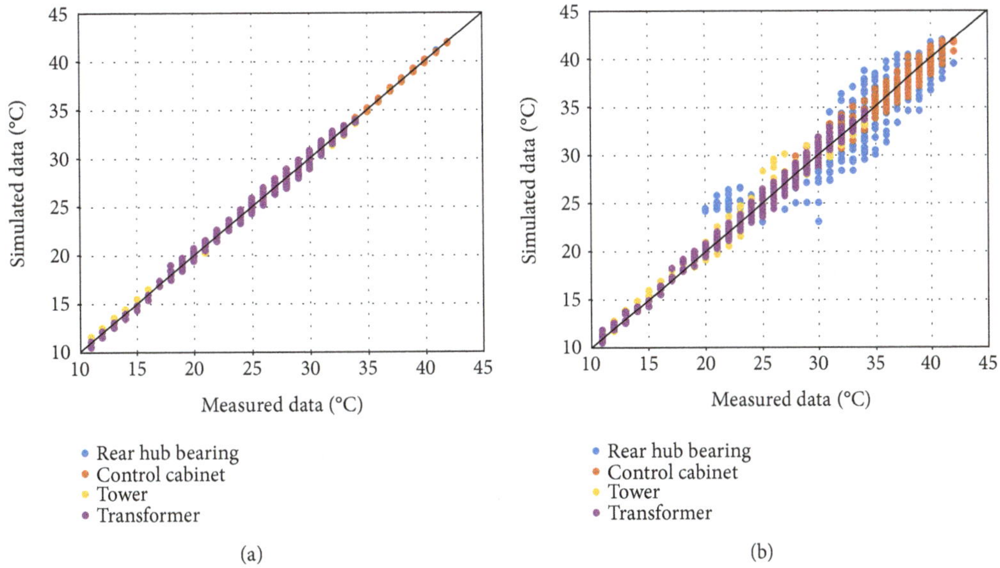

FIGURE 5: Regression plots for (a) autoassociative BPNN and (b) MISO BPNN.

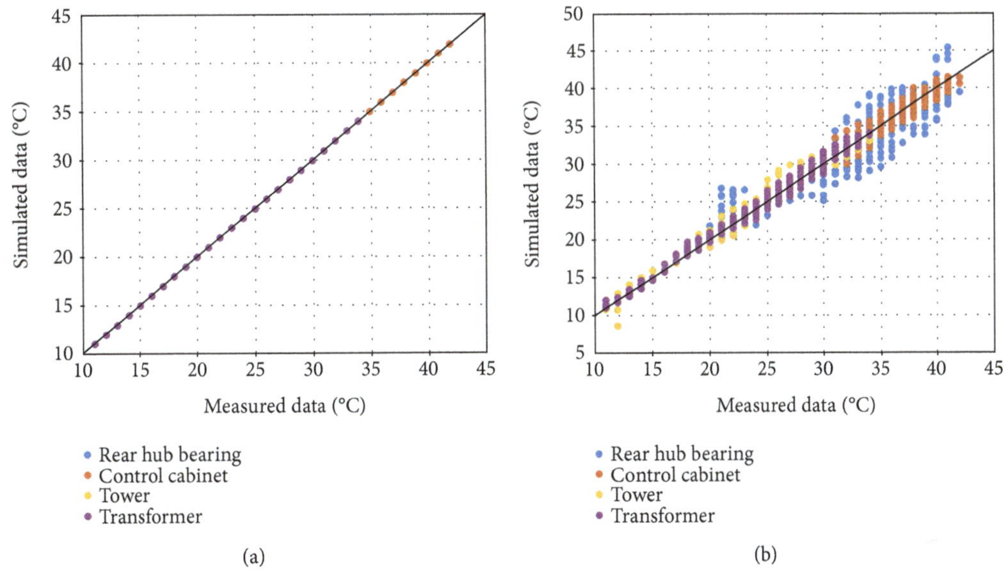

FIGURE 6: Regression plots for (a) autoassociative RBFNN and (b) MISO RBFNN.

The squared distance D_i^2 which is given in (3) is used to calculate the probable values.

$$D_i^2 = \left(x - x^i\right)^T \left(x - x^i\right). \quad (3)$$

The output is calculated by

$$\hat{y} = \frac{\sum_{i=1}^{n} y^i \exp\left(-\left(D_i^2/2\sigma^2\right)\right)}{\sum_{i=1}^{n}\left(-\left(D_i^2/2\sigma^2\right)\right)}, \quad (4)$$

where $\hat{y}$ is the estimate of the output which is a weighted average of all the observed samples y^i, x is the input vector, n is the number of sample observations, and σ is the width factor. Like the approach used in RBFNNs, GRNN models were also designed with varying width factors from 0.1 to 100. The highest R^2 values for autoassociative GRNN networks were 0.992, 0.990, 0.990, and 0.994 for sensors 1 through 4, respectively, and the best scores for MISO GRNNs were 0.808, 0.976, 0.973, and 0.981. Figure 7 shows the regression plots for autoassociative and MISO GRNN models with the highest goodness of fit results.

4. Results and Discussion

As stated in Section 2, SCADA data do not supply information on incipient temperature sensor faults. Therefore, a multiplicative fault representing a calibration drift was artificially created in this work. The models do not evaluate single-sensor measurements independently but instead evaluate

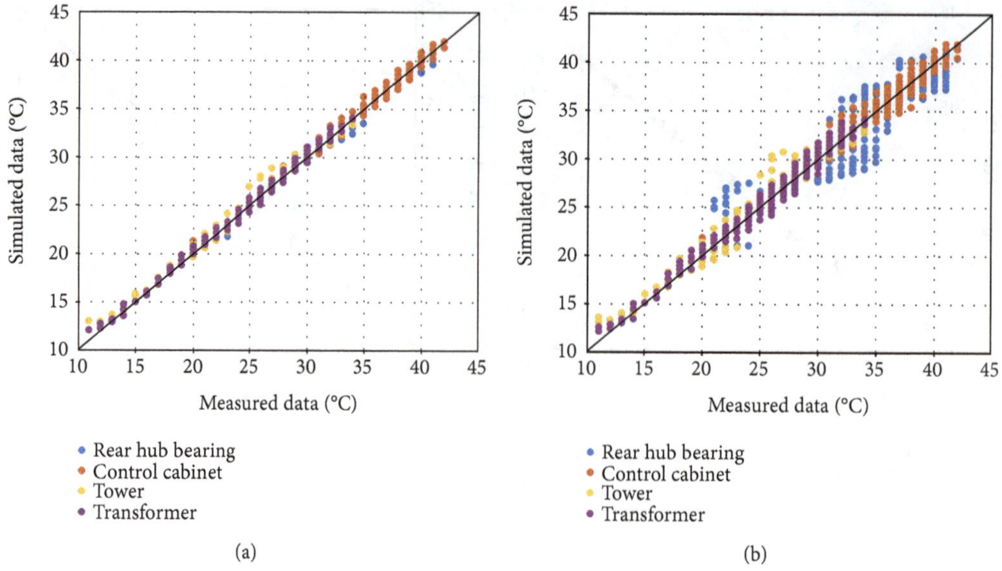

FIGURE 7: Regression plots for (a) autoassociative GRNN and (b) MISO GRNN.

TABLE 3: Root mean square error values (°C).

		Autoassoc. BPNN			Autoassoc. RBFNN			Autoassoc. GRNN	
Sensor	Case 1	Case 2	Case 3	Case 1	Case 2	Case 3	Case 1	Case 2	Case 3
S_1	0.1	0.4	0.5	$0.2e-4$	$0.3e-3$	$0.8e-3$	0.5	1.1	3.9
S_2	0.1	1.6	0.4	$0.1e-4$	1.3e-3	$0.5e-3$	0.4	4.2	2.3
S_3	0.3	0.3	0.7	$0.2e-4$	$0.5e-3$	$0.4e-3$	0.4	1.8	1.3
S_4	0.4	1.3	0.7	$0.1e-4$	$0.4e-3$	$0.5e-3$	0.4	1.8	1.3
		MISO BPNN			MISO RBFNN			MISO GRNN	
Sensor	Case 1	Case 2	Case 3	Case 1	Case 2	Case 3	Case 1	Case 2	Case 3
S_1	1.8	2.7	3.4	1.8	20.8	8.9	1.9	2.5	4.7
S_2	0.6	7.2	1.5	0.6	7.2	4.8	0.7	7.2	2.8
S_3	0.6	2.6	2.2	0.6	20.8	44.4	0.7	3.3	1.6
S_4	0.6	3.2	2.1	0.6	4.9	5.9	0.7	2.9	1.7

them by observing their relationship with other sensors in the group which were selected from physically relevant places, so it becomes possible to distinguish variations caused by calibration drifts from variations that appeared due to the real temperature changes. A temperature rise originated from the environmental effects would be seen in more than one sensor, whereas a calibration drift would result in a change only in the faulty sensor. To ensure that the proposed algorithm distinguishes the root reason for the temperature change, networks were tested in 3 different cases. Performances of the ANN models for the test set without any temperature drifts were observed in Case 1. In Case 2, a calibration fault was simulated by multiplying only one of the sensor's outputs with a constant factor. An overall shift in all temperature measurements was simulated in Case 3 which represents a change originated not from a fault but a real environmental temperature change.

The details and different expectations from ANN outputs for all cases are summarized as follows.

Case 1. No fault case.

In this case, temperature measurements gathered from the sensors were directly used as network inputs. 25% of the data that had been separated as the test set was used for this aim. The data presents the normal behaviour of the turbine without any faults, therefore the expectation from the networks is to produce as close output values as possible to the real sensor measurements.

Case 2. Multiplicative fault in one of the sensors.

Case 2 was designed to introduce a fault to be detected to the measurement system. A multiplicative fault was artificially created in one of the sensors. The test set for the measurements of the control cabinet temperature sensor (S_2) was multiplied by the constant term 1.2. The performances of the networks for this case were evaluated by analyzing the residuals between network outputs and sensor measurements. The expectation is to obtain greater residuals between

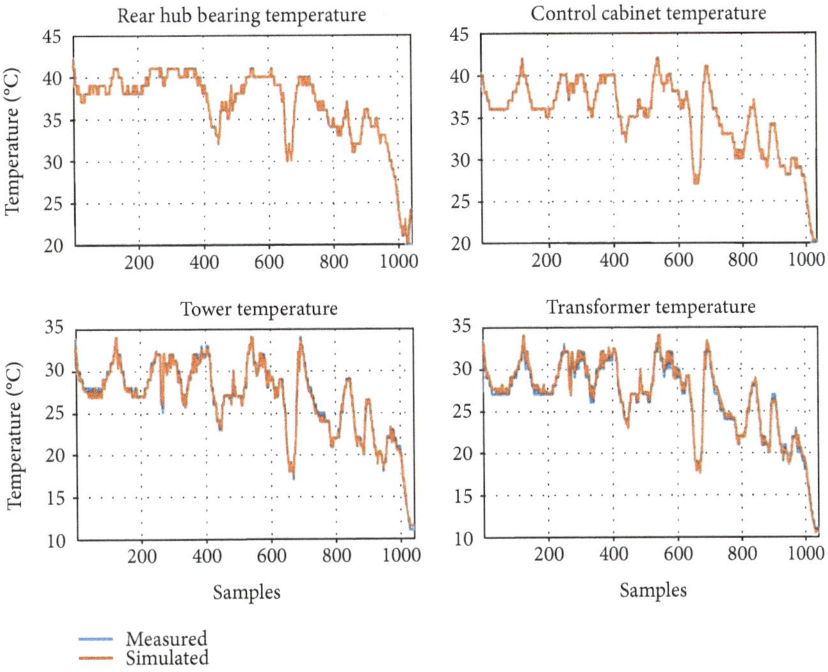

Figure 8: Measured and network values for Case 1 with autoassociative BPNN.

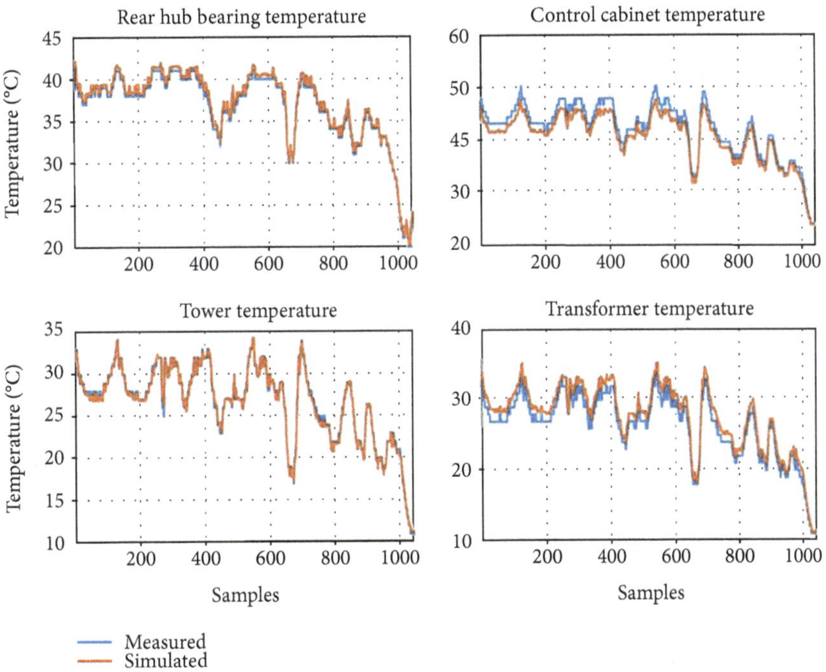

Figure 9: Measured and network output values for Case 2 with autoassociative BPNN.

measurement and estimation of the faulty sensor comparing to other sensors.

Case 3. No fault case. Overall shift in all measurements due to environmental temperature rise.

In this case, the measurements from all 4 sensors were shifted by multiplying by the constant term 1.2. The aim for this overall shift is to ensure that networks used for this fault detection algorithm do not produce a false fault alarm when none of the sensors are faulty but instead a real temperature rise is recorded. The expectation from networks for this case is again to produce as close estimation values as possible to actual temperature values.

Input data for all 3 cases were implemented to the resulting 6 models with the best R^2 values. Outputs of the networks were analyzed based on different requirements for each case.

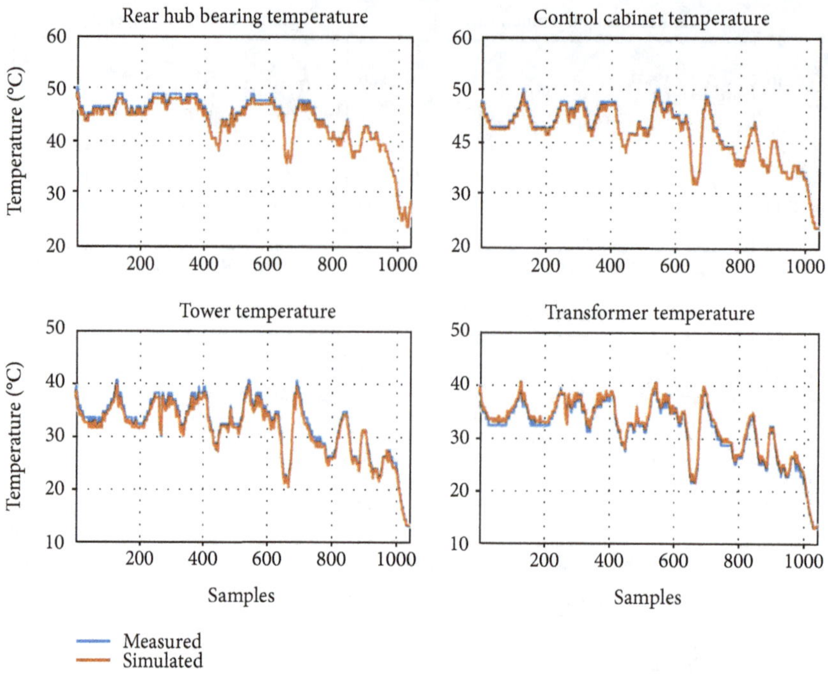

FIGURE 10: Measured and network output values for Case 3 with autoassociative BPNN.

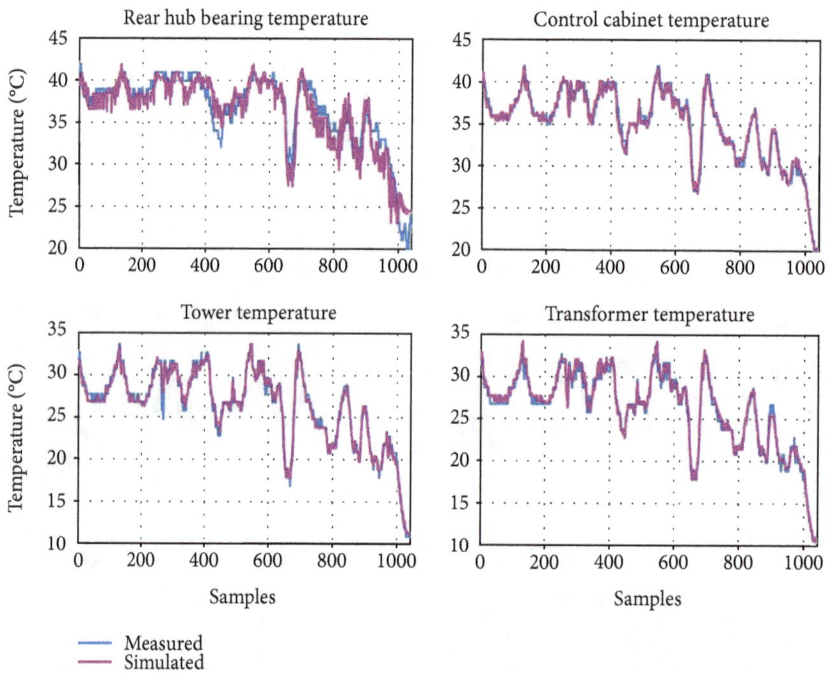

FIGURE 11: Measured and network values for Case 1 with MISO BPNN.

Table 3 presents the root mean square error (RMSE) values of ANNs for each case in °C. The results indicate that the most convenient network to distinguish the faulty case from nonfaulty ones is autoassociative BPNN. In Cases 1 and 3, the RMSE values between the real values and ANN outputs are less than 0.8°C for all sensors, whereas in Case 2, the RMSE values are 1.6°C and 1.3°C for sensors 2 and 4, respectively. Although autoassociative RBFNN outperforms the former network in Case 1 and Case 3, it was unable to produce the required residuals in the faulty case (Case 2). Figures 8–10 presents the autoassociative BPNN results for each case. As can be seen from Figures 8 and 10, real temperature values and ANN outputs are very close in Case 1 and Case 3. Figure 9 shows that the residuals are visible in Case 2, which is a sign of fault existence. However, the magnitude of residuals for control cabinet and transformer is very close to each other. This situation prevents finding the exact location of the fault. Therefore, to isolate the fault location, the

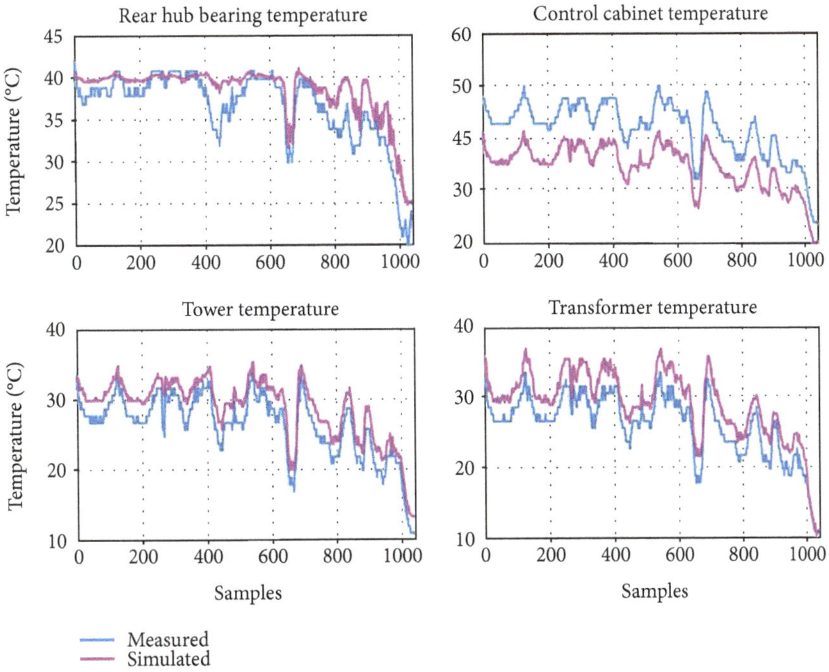

FIGURE 12: Measured and network values for Case 2 with MISO BPNN.

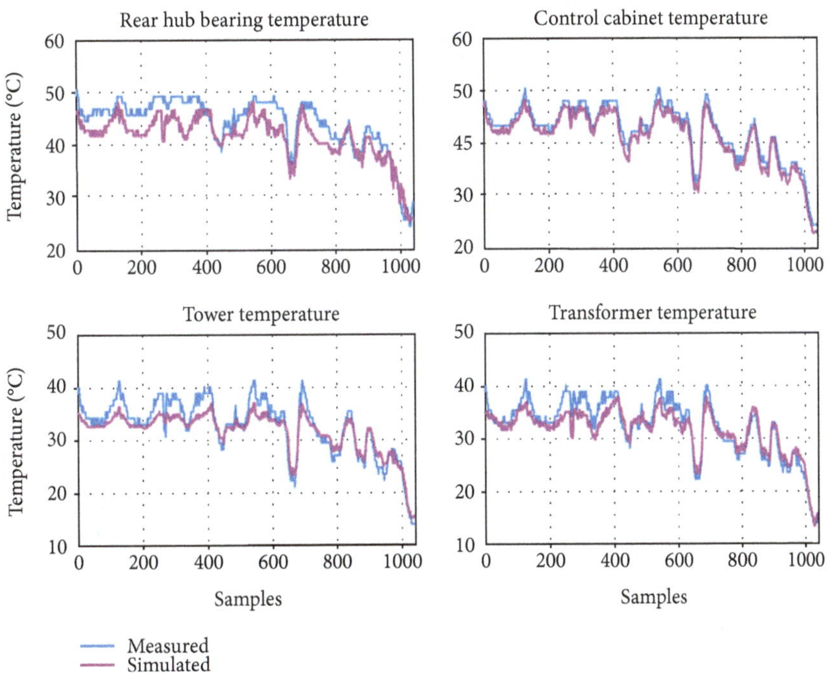

FIGURE 13: Measured and network values for Case 3 with MISO BPNN.

necessity of using different network estimations in a combined way has emerged.

The network providing the best results for fault isolation was found to be MISO BPNN, where the RMSE value of the faulty sensor (S_2) is significantly larger than other sensors in Case 2, and in other cases, the RMSE values are comparatively smaller. Figures 11–13 present the results obtained by the specified network. The corresponding figures show a combination of 4 neural networks with a different output sensor in each subplot due to the single-output architecture of MISO networks.

As shown in Figure 11, the real data and the network outputs are consistent in the normal operation case. Figure 12 shows that in Case 2, unlike the results from the autoassociative network, this time the residuals for the control cabinet temperature are significantly bigger than the residuals for other sensors with an RMSE value of 7.2°C. Therefore, this network can be used for the isolation of fault location. For

Case 3 (as presented in Figure 13), the networks produce residuals which is undesired for this scenario; however, the decision of fault existence would be made with the support of autoassociative BPNN to prevent false alarms. By this combined decision-making algorithm using different networks, the fault detection system would be able to distinguish faulty and nonfaulty situations in a more sensitive way and give information on the exact location of faults.

5. Conclusions

A data-driven method using temperature sensors of a wind turbine SCADA system to detect and isolate sensor faults is proposed in this paper. A significant advantage of this method is the use of sensor data from the SCADA system which is a built-in part in most large-scale modern wind turbines; therefore, the data collection system does not bring any additional hardware requirements and extra costs. However, since SCADA measurements are of low frequency for fault detection, an intelligent model was needed that was provided by artificial neural networks in this research. Several artificial neural network architectures were trained for this aim. Based on the evaluation of simulation results, it is proposed to benefit from the strengths of different architectures. Autoassociative BPNN has proven to be a successful network to distinguish fault situation from nonfaulty ones, and MISO BPNN gives satisfactory results in terms of finding the exact sensor with faulty behaviour. The results demonstrate that the proposed method is feasible and effective for sensor fault detection and isolation. As a future work, the research can be extended to cover other measurements gathered from wind turbines to build an overall fault detection system.

Conflicts of Interest

The authors declare that there is no conflict of interest regarding the publication of this paper.

Acknowledgments

This work was partly supported by the Scientific and Technological Research Council of Turkey (TUBITAK).

References

[1] P. Tchakoua, R. Wamkeue, M. Ouhrouche, F. Slaoui-Hasnaoui, T. Tameghe, and G. Ekemb, "Wind turbine condition monitoring: state-of-the-art review, new trends, and future challenges," *Energies*, vol. 7, no. 4, pp. 2595–2630, 2014.

[2] W. Yang, P. J. Tavner, C. J. Crabtree, Y. Feng, and Y. Qiu, "Wind turbine condition monitoring: technical and commercial challenges," *Wind Energy*, vol. 17, no. 5, pp. 673–693, 2014.

[3] M. L. Wymore, J. E. Van Dam, H. Ceylan, and D. Qiao, "A survey of health monitoring systems for wind turbines," *Renewable and Sustainable Energy Reviews*, vol. 52, no. 1069283, pp. 976–990, 2015.

[4] W. Qiao and D. Lu, "A survey on wind turbine condition monitoring and fault diagnosis–part I: components and subsystems," *IEEE Transactions on Industrial Electronics*, vol. 62, no. 10, pp. 6536–6545, 2015.

[5] W. Qiao and D. Lu, "A survey on wind turbine condition monitoring and fault diagnosis-part II: signals and signal processing methods," *IEEE Transactions on Industrial Electronics*, vol. 62, no. 10, pp. 6546–6557, 2015.

[6] G. d. N. P. Leite, A. M. Araújo, and P. A. C. Rosas, "Prognostic techniques applied to maintenance of wind turbines: a concise and specific review," *Renewable and Sustainable Energy Reviews*, vol. 81, pp. 1917–1925, 2018.

[7] P. F. Odgaard, J. Stoustrup, R. Nielsen, and C. Damgaard, "Observer based detection of sensor faults in wind turbines," in *Proceedings European Wind Energy Conference 2009*, pp. 1–10, Marsielle, France, 2009.

[8] X. Wei and M. Verhaegen, "Sensor and actuator fault diagnosis for wind turbine systems by using robust observer and filter," *Wind Energy*, vol. 14, no. 4, pp. 491–516, 2011.

[9] W. Yang, R. Court, and J. Jiang, "Wind turbine condition monitoring by the approach of SCADA data analysis," *Renewable Energy*, vol. 53, pp. 365–376, 2013.

[10] M. Schlechtingen, I. F. Santos, and S. Achiche, "Wind turbine condition monitoring based on SCADA data using normal behavior models. Part 1: system description," *Applied Soft Computing*, vol. 13, no. 1, pp. 259–270, 2013.

[11] K. Leahy, R. Lily Hu, I. C. Konstantakopoulos, C. J. Spanos, and A. M. Agogino, "Diagnosing wind turbine faults using machine learning techniques applied to operational data," in *2016 IEEE International Conference on Prognostics and Health Management (ICPHM)*, pp. 1–8, Ottawa, ON, Canada, 2016.

[12] Y. Zhao, D. Li, A. Dong, J. Lin, D. Kang, and L. Shang, "Fault prognosis of wind turbine generator using SCADA data," in *NAPS 2016-48th North Am. Power Symposium*, pp. 1–5, Denver, CO, USA, 2016.

[13] P. Sun, J. Li, C. Wang, and X. Lei, "A generalized model for wind turbine anomaly identification based on SCADA data," *Applied Energy*, vol. 168, pp. 550–567, 2016.

[14] P. B. Dao, W. J. Staszewski, T. Barszcz, and T. Uhl, "Condition monitoring and fault detection in wind turbines based on cointegration analysis of SCADA data," *Renewable Energy*, vol. 116, Part B, pp. 107–122, 2018.

[15] L. Wang, Z. Zhang, H. Long, J. Xu, and R. Liu, "Wind turbine gearbox failure identification with deep neural networks," *IEEE Transactions on Industrial Informatics*, vol. 13, no. 3, pp. 1360–1368, 2017.

[16] P. Caselitz and J. Giebhardt, "Rotor condition monitoring for improved operational safety of offshore wind energy converters," *Journal of Solar Energy Engineering*, vol. 127, no. 2, p. 253, 2005.

[17] M. J. Schulz and M. J. Sundaresan, *Smart Sensor System for Structural Condition Monitoring of Wind Turbines: 30 May 2002–30 April 2006*, Tech. Rep. National Renewable Energy Lab (NREL), Golden, CO, USA, 2006.

[18] D. J. Comboni and B. Fazenda, "Acoustic condition monitoring of wind turbine blades: tip faults," *The Journal of the Acoustical Society of America*, vol. 132, p. 1956, 2012.

[19] Y. Qu, E. Bechhoefer, D. He, and J. Zhu, "A new acoustic emission sensor based gear fault detection approach," *International Journal of Prognostics and Health Management*, vol. 4, pp. 1–14, 2013.

[20] J. Ruan, S. C. M. Ho, D. Patil, and G. Song, "Structural health monitoring of wind turbine blade using piezoceremic based active sensing and impedance sensing," in *Proceedings of the 11th IEEE International Conference on Networking, Sensing and Control*, pp. 661–666, Miami, FL, USA, 2014.

[21] O. M. Bouzid, G. Y. Tian, K. Cumanan, and D. Moore, "Structural health monitoring of wind turbine blades: acoustic source localization using wireless sensor networks," *Journal of Sensors*, vol. 2015, Article ID 139695, 11 pages, 2015.

[22] Y. Qian and R. Yan, "Gearbox fault diagnosis in a wind turbine using single sensor based blind source separation," *Journal of Sensors*, vol. 2016, Article ID 6971952, 14 pages, 2016.

[23] Y. Guo, X. Chen, S. Wang, R. Sun, and Z. Zhao, "Wind turbine diagnosis under variable speed conditions using a single sensor based on the synchrosqueezing transform method," *Sensors*, vol. 17, no. 5, p. 1149, 2017.

[24] A. Coscetta, A. Minardo, L. Olivares et al., "Wind turbine blade monitoring with Brillouin-based fiber-optic sensors," *Journal of Sensors*, vol. 2017, Article ID 9175342, 5 pages, 2017.

[25] P. F. Odgaard, J. Stoustrup, and M. Kinnaert, "Fault-tolerant control of wind turbines: a benchmark model," *IEEE Transactions on Control Systems Technology*, vol. 21, no. 4, pp. 1168–1182, 2013.

[26] B. Krose and P. Van Der Smagt, *An Introduction to Neural Networks*, University of Amsterdam, 1996.

[27] K. Patan, *Artificial Neural Networks for the Modelling and Fault Diagnosis of Technical Processes*, vol. 377, Springer, 2008.

[28] S. Haykin, *Neural Networks and Learning Machines*, Prentice Hall, 3 edition, 2009.

[29] D. F. Specht, "A general regression neural network," *IEEE Transactions on Neural Networks*, vol. 2, no. 6, pp. 568–576, 1991.

14

Position-Based Feature Selection for Body Sensors regarding Daily Living Activity Recognition

Nhan Duc Nguyen,[1] **Duong Trong Bui**,[1] **Phuc Huu Truong**,[2] **and Gu-Min Jeong**[1]

[1]*Department of Electrical Engineering, Kookmin University, Seoul, Republic of Korea*
[2]*Korea Institute of Industrial Technology, Ansan, Republic of Korea*

Correspondence should be addressed to Gu-Min Jeong; gm1004@kookmin.ac.kr

Academic Editor: Eduard Llobet

This paper proposes a novel approach to recognize activities based on sensor-placement feature selection. The method is designed to address a problem of multisensor fusion information of wearable sensors which are located in different positions of a human body. Precisely, the approach can extract the best feature set that characterizes each activity regarding a body-sensor location to recognize daily living activities. We firstly preprocess the raw data by utilizing a low-pass filter. After extracting various features, feature selection algorithms are applied separately on feature sets of each sensor to obtain the best feature set for each body position. Then, we investigate the correlation of the features in each set to optimize the feature set. Finally, a classifier is applied to an optimized feature set, which contains features from four body positions to classify thirteen activities. In experimental results, we obtain an overall accuracy of 99.13% by applying the proposed method to the benchmark dataset. The results show that we can reduce the computation time for the feature selection step and achieve a high accuracy rate by performing feature selection for the placement of each sensor. In addition, our proposed method can be used for a multiple-sensor configuration to classify activities of daily living. The method is also expected to deploy to an activity classification system-based big data platform since each sensor node only sends essential information characterizing itself to a cloud server.

1. Introduction

Daily living activity recognition is one of the most important research topics that can provide valuable information for various applications such as health monitoring, security, intelligent assistance in daily life, fall detection, and surveillance [1, 2]. It is obvious that the quantity and placement of wearable sensors on the human body are factors that affect the accuracy of activity classification [3–5]. Each kind of activity has different characteristics which are specific to the particular location on the human body.

The development of embedded sensor technology in wearable devices, that is, smartphone, smart band, and smart shoe, has featured a variety of daily activity recognition-related researches [6–8]. A major issue of these studies is that there is a difficulty in acquiring experimental results that carry all real-world applications while activities are conducted in a controlled environment [9]. Some recent publications [6, 7, 10], which utilized a single sensor module attached to different body positions, can achieve a high classification accuracy. However, a single sensor module-based platform cannot be sufficient to recognize complex daily living activities. In particular, characterizations of the activities may not be true using one sensor even though a high classification accuracy is achieved [11]. To overcome that problem, researchers have tended to attached many sensors to different parts of a human body [4, 11–14]. These approaches can recognize both dynamic and transitional activities. Furthermore, a multiple-sensor system, which contains up to three sensors, can provide a good recognition accuracy [11]. The fact remains that researchers have made efforts to deal with a fusion of multiple-sensor data. In addition, challenges of a variety of monitoring activities and precision of classification are directly impacted by the number and the placement of inertial sensors attached to the human body [15]. On-body device position was investigated to determine its influence on the performance of activity recognition [16, 17]. Figueira et al. [18] explored the sensor-position awareness of monitoring activities. Atallah et al. [19] discussed the sensor positions and significant features for each activity group according to the compendium of physical activities. However, a recognition system with many sensors can contain numerous redundant features which degrade the classification performance and may even confuse the classifier. Recent studies [19–22] have approved feature selection techniques as a solution to deal with the redundant features. There have been various activity recognition topologies designed with feature selection algorithms. Appropriately, a novel scheme based on feature selection is introduced to achieve a high activity classification performance.

In this paper, we propose a new scheme which is evaluated on the DaLiAc dataset for human daily living activity recognition based on sensor-placement feature selection. We first apply a low-pass filter to the raw data. The filtered data are segmented by a 10-second sliding window with no overlap to obtain a sample. Then, we extract valuable features from the filtered data. Feature selection techniques are applied separately on each sensor location to obtain the best feature set for each body position. We optimize the feature set by investigating the correlation of the features. Finally, a k-nearest neighbor algorithm is applied to an optimized feature set containing features from four body positions to classify thirteen activities. The key of our framework lies in the feature selection layer. In this work, we implement feature selection on an individual set of sensor features. This approach optimizes the features for specific sensors, and therefore, facilitates correct selection of the features specific to each activity. The contributions of the paper are as follows:

(i) It proposes a new method of daily living activity recognition that achieves a higher accuracy rate in a benchmark dataset in comparison with recent publications

(ii) It discovers sensor positions that most affects the recognition rate in a benchmark dataset

(iii) It explores meaningful features by utilizing three feature selection methods, that is, Laplacian score, SFFS, and Relief-F, as well as correlation-based feature optimization for four sensor locations in the ankle, chest, hip, and wrist

The rest of the paper is organized as follows. In Section 2, we present some related works on the recognition of daily living activities with one or more sensors. The proposed method is described in Section 3. Furthermore, we go into the details of data preprocessing, feature extraction, future selection, feature optimization, choice of classifier, and evaluation criteria. All comparisons of the selected set of features and the performance of the proposed method will be shown in Section 4. Finally, Section 5 presents the conclusion of this paper and future works.

2. Related Works

Many authors have reported activity recognition issues when using one sensor to monitor activities of daily life [6, 7, 10]. In our previous work [6], a multiclass support vector machine (SVM) algorithm demonstrated 97.3% accuracy when we used one module smart band with a triaxial acceleration stuck on the wrist to collect data during the performance of five wrist activities: texting, calling, hand in pocket, hand swinging, and suitcase carrying. Nguyen et al. [7] proposed three novel plantar-pressure sensor-based features to classify five ambulatory activities using smart insoles. By implementing a k-nearest neighbor classifier, these walking activities can be recognized with an accuracy rate of 97.84% at the sixth step. Zheng [10] presented a hierarchical framework based on the least squares support vector machine and naive Bayes algorithm to classify ten activities using an inertial module attached on the hip. Their implemented method, which recognized activities regarding four groups (bidimensional walking, tridimensional walking, plane motion, and static activities), obtained a high accuracy rate of 95%. These studies can achieve high activity classification rates using a small size of feature vectors and simple classifiers. However, this approach, which used data collected from one sensor location, only attained the best recognition performance to predict activities that relate directly to the sensor's location. This limits the number of activities for consideration.

To address the drawback of a single-sensor configuration, various studies have utilized multiple inertial sensors attached to different body locations to classify human activities [4, 9, 11–14]. Sztyler et al. [9] improved an activity recognition performance by being aware of the device position. Their method was implemented on a dataset of eight activities collected from seven on-body locations. In the experimental results, they obtained an F-measure of 89%. In addition, an awakening of sensor locations can enhance the performance of activity recognition compared with subject-specific and cross-subject approaches. Khan et al. [11] presented a cloud-based system to recognize thirteen activities using the WARD dataset [23]. Effects of on-body sensor position were explored by implementing three classifiers for each activity. Their analysis showed that activity recognition is significantly dependent on the positions of on-body devices. Furthermore, utilizing up to three sensor modules is needed to recognize more dynamic and transient activities. Schuldhaus et al. [12] proposed a decision level fusion-based method to classify seven daily living activities. Four sensor nodes, which were located on the chest, hip, wrist, and ankle, separately performed activity classification. Then, an activity was predicted by majority voting of each classifier decision of the sensor nodes. The designed system achieved an overall recognition error rate of 6.1%. Baños et al. [13] presented a metaclassifier framework which included an insertion and rejection algorithm of weights for base (classifier) and source (activity) levels, respectively. A voting algorithm is designed to fuse the decisions of individual classifiers to predict activities. To compare with other fusion models, this scheme can deal with misclassification caused by a failure of one or more sensors. The method which was evaluated on the REALDISP benchmark dataset [24] can improve the recognition performance compared with a traditional feature fusion, hierarchical-weighted classifier [25] approach to classify thirty-three activities. Gao et al. [14] developed a framework consisting of sensor selection and hierarchical classifier module to address the dynamic multisensor fusion problem. They investigated the recognition performance and energy consumption of the method on a dataset of eight activities. The experimental results demonstrated that a system can save 37.5% of the energy consumption and reduce the classification rate by 2.8% compared with an original four-sensor classifier approach. Bao and Intille [4] used features extracted from data collected using five biaxial accelerometers attached to the left elbow, right wrist, hip, left thigh, and right ankle of twenty different subjects to recognize twenty different activities. The overall accuracy using a decision tree was 84% without the utilization of a feature selection technique. The main advantage of multiple sensors is that they can recognize many daily living activities with high precision. Moreover, it is more suitable to recognize complex activities, that is, dynamic and transitional activities [26]. However, the main issue of this configuration is an optimal sensor combination to achieve a high classification performance and robustness [25]. In addition, the multiple-sensor systems can become uncomfortable for users who wear them for a long time [13, 25].

The wearable sensor community recently investigated feature selection techniques which demonstrated an efficient solution to deal with the problem of redundant features and computational expense [27] in recognizing human activities [19–22]. Feature selection algorithms were implemented in many ways, that is, performing on a total set of features extracted from various activities, applying to each activity group, and undertaking each activity. Atallah et al. [19] employed six wearable accelerometers positioned at six different body locations in the ear, chest, wrist, ankle, knee, arm, and waist. They used a filter method-based feature selection technique, that is, Relief [28], Simba [29], and minimum redundancy maximum relevance (MRMR) [30] to evaluate and rank the features. The K-nearest neighbor with $k = 5$ and $k = 7$ and the Bayesian classifier were used to classify fifteen daily activities which were combined into five groups of activities labeled as very low level, low level, medium level, high level, and transitional activities. The performance of both algorithms was similar at approximately 82%. Zhang and Sawchuck [20] acquired data using a 6-DOF inertial measurement unit (IMU), which integrated a triaxial accelerometer, gyroscope, and magnetometer. This study utilized three selection techniques, that is, Relief [28], a wrapper method based on a single feature (SFC) [31], and a wrapper method based on sequential for selection (SFS) [32] to select the most important features. The SVM algorithm was then deployed to recognize twelve activities including walking, forward, walking left, walking right, going upstairs, going downstairs, jumping up, running, standing, and sitting. The accuracy of the single-layer classifier was up to 90%. Pirttikangas et al. [21] configured four triaxial accelerometers attached to the right wrist, left wrist, right thigh, and neck and one heart rate sensor attached to the right wrist of thirteen different users. The collected data from 17 different activities were characterized into five features, that is, acceleration mean, standard deviation, correlation, mean crossing, and heart rate mean. The forward-backward search was utilized to select a subset of the best features. The author used the k-nearest neighbor as a classifier and the algorithm achieved 90.61% activity recognition accuracy. Huynh and Schiele [22] showed that the performance can be improved by implementing feature selection techniques on individual features for each activity such as walking, standing, jogging, skipping, hopping, and riding a bus. Data was collected using an integrated sensor board containing sensors for 3D acceleration, audio, temperature, IR/visible/high-frequency light, humidity, barometric pressure, and a digital compass. The board was placed on the shoulder strap of a backpack and the subjects were required to wear the backpack while performing the measured activities. The author applied cluster precision to evaluate the features and the best accuracy of recognition was about 90% when the best features and window length of jogging and walking were considered. For the other activities, the recognition accuracy was about 85%. The authors applied feature selection to individual activities, but the results did not show improved efficiency when the best set of features for all activities was combined into one global set as the input for the classifier.

Some state-of-art methods evaluated on the DaLiAc dataset [33], which obtained acceptable classification accuracy rates to recognize thirteen activities, can be found in [33, 34]. Leutheuser et al. [33] achieved an average classification rate of 85.8% utilizing a hierarchical multisensor-based activity recognition system. Furthermore, their proposed method can distinguish between complex daily living activities, that is, treadmill running and rope jumping. Zdravevski et al. [34] proposed a systematic feature engineering method that uses a two-phase feature selection layer. The method achieved an accuracy rate of 91.2% with 60 selected features to classify thirteen activities.

3. Activity Recognition Algorithm for Daily Living Activities

In this section, we first present a comparison between the proposed and conventional algorithms. Then, data preprocessing, feature extraction, selection, optimization, choice of classifier, and evaluation criteria are discussed.

3.1. Feature Selection Based on Body-Sensor Placement. Figures 1(b) and 1(a) respectively illustrate the proposed and standard approaches to recognize human activity using wearable devices. In the proposed architecture, the feature selection layer is different from that of the standard approach. In this layer, the proposed method separately performs feature selection on each sensor position, whereas the standard method considers all extracted features at once. The main advantage of the proposed method is that each sensor has a different set of features which is better for depicting the activities with regard to the position of the sensors. Moreover, the computation requirement of feature selection is significantly decreased compared with the standard approach.

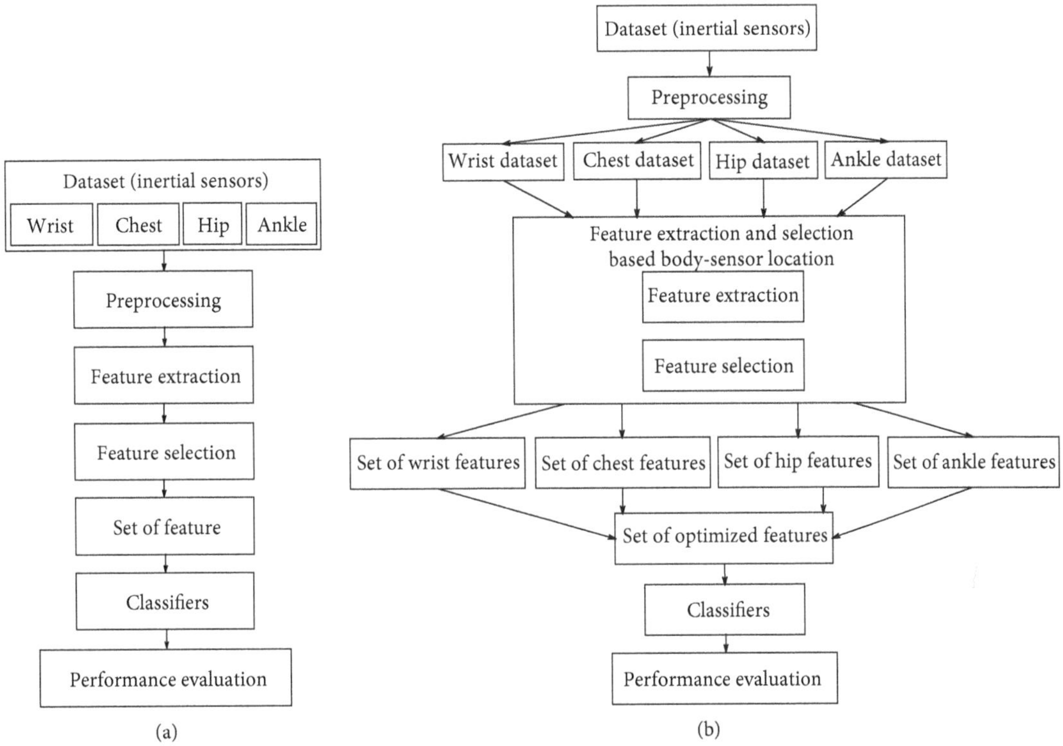

FIGURE 1: Human activity recognition approaches: (a) standard approach and (b) proposed approach.

In addition, the standard approach deployed in most studies [19, 21, 33] achieves medium classification accuracy (below 95%). The disadvantage of the standard approach is that it becomes computationally expensive as the number of features increases. We also perform optimization based on the Pearson correlation coefficient to take high-correlation features into account in the proposed method.

3.2. Data Preprocessing. In this study, inertial data were uniformly resampled with a sampling frequency of 200 Hz. We cut out a segment of the inertial signal using a 10 s sliding window with no overlap to obtain the activity samples. A low-pass filter was then applied to these samples to remove noise. Table 1 shows the distribution of the human activity samples. Examples of the acceleration and gyroscope signals for sitting are shown in Figure 2.

3.3. Feature Extraction. Feature extraction transforms high-dimensional raw data into a meaningful representation data that can depict the characteristics of different activities. This technique enables machine learning algorithms to deal with high-dimensional data. We obtain a large set of features calculated using the filtered data for each activity. To perform feature extraction, we implement eight features on smaller windows of 2 s data, that is, signal magnitude area, intensity of movement, mean trend, windowed mean difference, variance trend, windowed variance difference, standard deviation trend, and windowed standard deviation difference. In addition, the 10 s long inertial signals are divided into 5 windows with no overlap. The remaining features were extracted directly without breaking the samples into smaller windows.

TABLE 1: Statistics of preprocessed dataset.

Number	Classes	Subjects	Sample size
1	Sitting	19	237
2	Lying	19	189
3	Standing	19	188
4	Washing dishes	19	891
5	Vacuuming	19	211
6	Sweeping	19	515
7	Walking	19	791
8	Ascending stairs	19	187
9	Descending stairs	19	113
10	Treadmill running	19	438
11	Bicycling on an ergometer (50 W)	19	470
12	Bicycling on an ergometer (100 W)	19	656
13	Rope jumping	19	160

The details of the feature are shown in Table 2.

The standard deviation trend and windowed standard deviation have different features, which are computed similar to the mean trend and windowed mean difference, and are given below:

(i) Standard deviation trend

$$\sigma T = \sum_{i=2}^{5} (|\sigma_i - \sigma_{i-1}|). \tag{1}$$

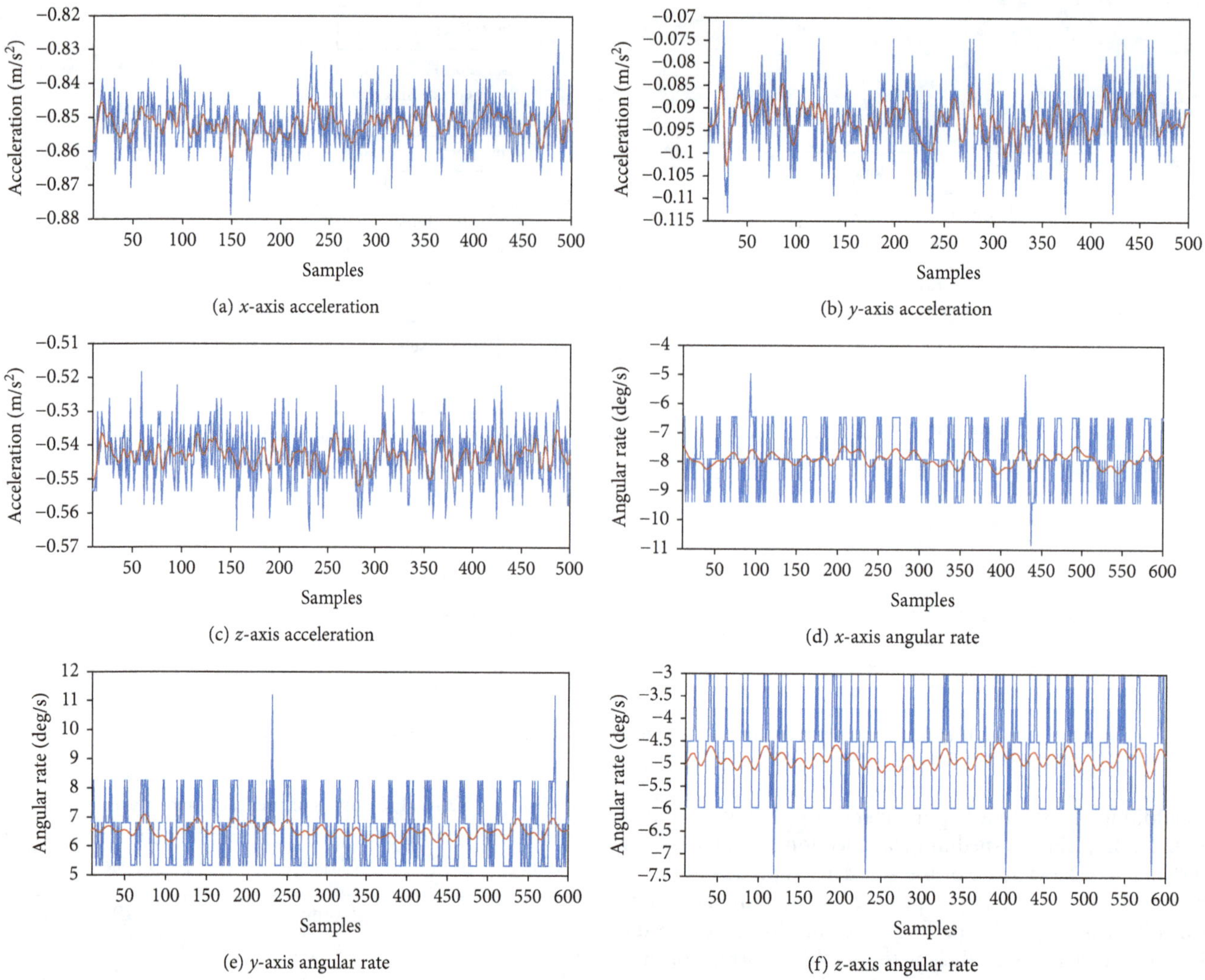

FIGURE 2: Sample signals of sitting activity of ankle position (blue: raw data; red: filtered data).

(ii) Windowed standard deviation difference

$$\sigma D = \sum_{i=1}^{5} \left(\left| \sigma - \sigma_i \right| \right). \tag{2}$$

The extracted features are not distributed normally and are sparse which can affect the performance of estimators. Therefore, feature normalization is a common requirement for many machine learning algorithms. In our study, we normalized all the features to zero mean and unit variance.

$$x_{\text{normalized}} = \frac{x_{\text{raw}} - \mu}{\sigma}, \tag{3}$$

where μ and σ are the mean and standard deviation of a particular feature, respectively.

3.4. Feature Selection. We obtain a total of 536 features including 192 statistical features, 120 magnitude-based features, 24 correlation-based features, 56 frequency-domain features, and 144 trend-variance features after performing feature extraction. In order to improve the classification accuracy and reduce the expense calculation time, the redundant (features are closely related) and irrelevant features should be removed [27, 35]. This task can be performed with feature selection techniques. There are three common groups of feature selections: filter method, wrapper method, and embedded method [36]. Filter methods which are computationally efficient select features that are independent of any machine learning algorithms. On the contrary, wrapper methods evaluate the quality of the selected features based on the performance of a predefined learning algorithm. Embedded methods embed the feature selection with model learning. In this paper, we use 2 filter methods and 1 wrapper method, that is, Relief-F, Laplacian Score, and Sequential Forward Floating Selection (SFFS) to rank the features for each sensor location. These methods are briefly described below.

Relief-F [37] is a popular filter method and is extended from the original feature selection method Relief [28]. This

TABLE 2: Derived features with symbols and implemented data.

Feature	Notation	Applied data
Mean	μ	
Standard deviation	σ	
Variance	σ^2	
Mean absolute deviation	MAD	
Range	r	
Skewness	γ_1	
Kurtosis	γ_2	
4th moment	μ_4	
5th moment	μ_5	
Root mean square	RMS	$a_x^w, a_y^w, a_z^w, w_x^w, w_y^w, w_z^w,$
Sum of absolute values	SA	$a_x^c, a_y^c, a_z^c, w_x^c, w_y^c, w_z^c,$
Signal magnitude area	SMA	$a_x^h, a_y^h, a_z^h, w_x^h, w_y^h, w_z^h,$
Intensity of movement	IM	$a_x^a, a_y^a, a_z^a, w_x^a, w_y^a, w_z^a$
Dominant frequency	f_d	
Amplitude	A_{f_d}	
Mean trend	μT	
Windowed mean difference	μD	
Variance trend	σT^2	
Windowed variance difference	σD^2	
Standard deviation trend	σT	
Windowed standard deviation difference	σD	
Correlation	ρ	$(a_x^w, a_y^w), (a_x^w, a_z^w), (a_y^w, a_z^w),$ $(w_x^w, w_y^w), (w_x^w, w_z^w), (w_y^w, w_z^w),$ $(a_x^c, a_y^c), (a_x^c, a_z^c), (a_y^c, a_z^c),$ $(w_x^c, w_y^c), (w_x^c, w_z^c), (w_y^c, w_z^c),$ $(a_x^h, a_y^h), (a_x^h, a_z^h), (a_y^h, a_z^h),$ $(w_x^h, w_y^h), (w_x^h, w_z^h), (w_y^h, w_z^h),$ $(a_x^a, a_y^a), (a_x^a, a_z^a), (a_y^a, a_z^a),$ $(w_x^a, w_y^a), (w_x^a, w_z^a), (w_y^a, w_z^a)$
Average energy	AE	$(a_x^w, a_y^w, a_z^w), (w_x^w, w_y^w, w_z^w),$ $(a_x^c, a_y^c, a_z^c), (w_x^c, w_y^c, w_z^c),$ $(a_x^h, a_y^h, a_z^h), (w_x^h, w_y^h, w_z^h),$ $(a_x^a, a_y^a, a_z^a), (w_x^a, w_y^a, w_z^a)$

method finds the k nearest instances from different classes (k nearest hits/misses) using the Manhattan norm instead of only one nearest hit and miss when using the Euclidean norm and then updates the weight of each feature. Relief-F can also handle incomplete data and the multiclass problem, which are problems in the Relief algorithm. If feature A is considered, then the weight $W(A)$ will be updated following [37]:

$$W(A) := W(A) - \sum_{i=1}^{k} \frac{\mathrm{diff}(A, R, H_i)}{n \times k} + \sum_{C \neq \mathrm{class}(R)} \frac{P(C)}{1 - P(\mathrm{class}(R))} \sum_{i=1}^{k} \frac{\mathrm{diff}(A, R, M_i(C))}{n \times \mathrm{k}}, \quad (4)$$

where $\mathrm{diff}(A, R, H_i)$ calculates the difference in feature value A between two instances R and H_i in the near hit class (same class as R), and $\mathrm{diff}(A, R, H_i)$ calculates the difference in features A between two instances R and M_i (C) in the near miss class (different class from R).

Laplacian Score algorithm is another feature selection technique based on the filter method [38]. Suppose we have m data points, $x_1, x_2, \ldots, x_m$, and the ith node corresponds to x_i. In order to select the significance features, first, the Laplacian score algorithm constructs a k-nearest neighbor graph G with m nodes and defines an edge in the graph for the node if another node is one of its k-nearest neighbors. Second, it calculates the weight matrix with the element $S_{ij} = e^{-\|x_i - x_j\|^2/t}$ (t is constant) if any two nodes x_i and x_j are connected. In [38] the Laplacian score of the rth feature was defined as

$$L_r = \frac{\tilde{\mathbf{f}}_r^T L \tilde{\mathbf{f}}_r}{\tilde{\mathbf{f}}_r^T D \tilde{\mathbf{f}}_r}, \quad \text{where } \tilde{\mathbf{f}}_r = \mathbf{f}_r - \frac{\mathbf{f}_r^T D 1}{1^T D 1} 1, \quad (5)$$

where $\mathbf{f}_r = (f_{r1}, f_{r2}, \ldots, f_{rm})^T$ denotes the vector of m samples for the rth feature. L is a Laplacian matrix defined as $L = D - S$ [39] and D is a diagonal matrix defined as $D_{ii} = \sum_j S_{ij}$.

The third feature selection technique used in the paper is Sequential Forward Floating Selection [40]. This technique is extended from the Sequential Forward Selection (SFS) method. Let us assume that we have a set of n dimensional features $Y = \{y_1, \ldots, y_n\}$. The SFS will output a subset of m dimensional features $X = \{x_1, \ldots, x_m\}$ with $m < n$ and assessed by a criterion function $J(x_i)$ with $x_i \in X$. First, SFS takes all m dimensional features from the original feature set as its input and will start adding the first feature x_i which has the highest significance subject to

$$x_1 = \arg \max_{y_j \in Y} J(y_j). \quad (6)$$

From the initial subset X_1, SFS extends more features according to

$$x_{i+1} = \arg \max_{y_j \in Y - X_i} J(y_j, X_i). \quad (7)$$

```
for i = 1 : 3 % We run three feature selection algorithms
        CVO = LeaveOneOutCV ( ); %Leave-one-out cross-validation
        for j = 1:CVO. NumofTestSets
                % Get training and test sets
                X_train = Features {i} (training Index,:);
                Y_train = Activities (training Index);
                X_test = Features {i} (testIndex,:);
                Y_test = Activities (testIndex);
                k_idex = 0;
                for k = 1:CVO. MaxNumOfFeature (i)
                        k_idx = k_idex + 1;
                        % Get features according to rank of feature
                        fs_index = FeatureRanking {i} (1 : k);
                        % Evaluate the feature importance by K-NN algorithm.
                        Prediction = knnModel (X_train (:, fs_index), ....
                                        Y_train, X_test (:, fs_index), 1);
                        cvAccuracy (j, k_idx) = sum (double (Prediction == Y_test)) ....
                                                                        /TestSize (j);
                end
        end
        Accuracy {1,i} (1: size (cvAccuracy, 2)) = mean (cvAccuracy);
end
```

Algorithm 1

Then, the new subset of features becomes

$$X_{i+1} = X_i + x_{i+1}, \quad i = i + 1. \tag{8}$$

The disadvantage of SFS is that once a feature is selected, it cannot be removed from the subset. SFFS can solve this problem by checking and removing the least significant feature x_w in the subset for each iteration, that is

$$\begin{aligned} x_w &= \arg \min_{x_i \in X_{n+1}} J(x_i, X_{i+1}), \\ \text{then}\, X_{i-1} &= X_i - x_w, \quad i = i - 1 \text{ if } x_w \neq x_{i+1}, \\ X_{i+1} &= X_{i+1}, \quad i = i + 1 \text{ if } x_w = x_{i+1}. \end{aligned} \tag{9}$$

This procedure is repeated until the number of selected features equals the desired number of features or $i = n$.

After ranking the features for each sensor location, we use k-NN to consider various features from high to low rank based on the model accuracy of each feature selection technique. Algorithm 1 presents the model accuracy procedure used to select the important features. A leave-one-out cross-validation and a k-NN classifier are utilized in the algorithm.

3.5. Feature Set Optimization. Relief-F and Laplacian Score select features by evaluating the capabilities of the features in preserving sample similarity. Therefore, they are unable to handle feature redundancy, which may have a high correlation with each other [41]. In addition, SFFS which cannot eliminate redundant features generated in the searching process, does not consider the feature correlation [42]. To address the redundancy caused by high-correlation features, we use the Pearson correlation coefficient [4, 43] with a threshold equal to 1 in our study. Let us assume that we have two data segments $X = x_1, x_2, \ldots, x_n$ and $Y = y_1, y_2, \ldots, y_n$. Then, the Pearson correlation is defined as

$$\rho_{XY} = \frac{n\left(\sum_{i=1}^{n} x_i y_i\right) - \sum_{i=1}^{n} x_i \sum_{i=1}^{n} y_i}{\left(n\sum_{i=1}^{n} x_i^2 - \left(\sum_{i=1}^{n} x_i\right)^2\right)\left(n\sum_{i=1}^{n} y_i^2 - \left(\sum_{i=1}^{n} y_i\right)^2\right)}. \tag{10}$$

3.6. Classifier. K-nearest neighbor (k-NN) and support vector machine (SVM) are popular supervised classification algorithms in machine learning problems. SVM is robust and delivers a unique solution because of convex optimization. However, this technique is more computationally expensive when dealing with high-dimensional data [44, 45]. In contrast, k-NN, which is a simple and robust algorithm, does not require a learning process [15, 46]. It uses the principle of similarity (Euclidean distance) between the training set and the new observation to be classified. Therefore, it is easy to implement and less computation intensive in comparison to SVM. The problem with k-NN is that it is very sensitive to redundant features [45]. In this study, we use the k-NN classifier to choose the number of features and the classification.

3.7. Evaluation Criteria. We evaluate the performance of our proposed method by utilizing four metrics, that is, accuracy, precision, recall, and F1-score. The accuracy can be expressed for binary classification as follows:

$$\text{Accuracy} = \frac{TP + TN}{TP + TN + FP + FN}, \tag{11}$$

where true positive (TP) means that an instance labeled as positive is predicted as positive. A true negative (TN) is defined when the algorithm predicts a negative result for a label that is indeed negative. When the label is negative but is predicted as positive, the prediction is defined as a false positive (FP). The reverse situation for a false positive is called a false negative (FN).

$$\begin{aligned} \text{Precision} &= \frac{TP}{TP + FP}, \\ \text{Recall} &= \frac{TP}{TP + FN}. \end{aligned} \tag{12}$$

F1-score which is the combination of two metrics, that is, the precision and recall, is defined as follows:

$$\text{F1} - \text{score} = 2 \times \frac{\text{Precision} \times \text{Recall}}{\text{Precision} + \text{Recall}}. \tag{13}$$

4. Experimental Results

In this section, a brief description of the used dataset is given. Then, we present the results of the feature selection and feature optimization procedure. The performance of our proposed method and a comparison with the reference papers are also presented. In addition, three metrics, that is, precision, recall, and F1-score are utilized to evaluate the proposed method's performance. The comparison highlights the performance of the different methods in terms of the F1-score.

4.1. DaLiAc Dataset. In this study, we use the DaLiAc (Daily Living Activities) dataset, which is a benchmark dataset for the classification of daily life activities based on inertial data as described in [33]. Nineteen healthy subjects (8 women and 11 men, age 26 ± 8 years, height 177 ± 11 cm, weight 75.2 ± 14.2 kg) performed thirteen activities, that is, sitting (labeled as SI), lying down (LY), standing (ST), washing dishes (WD), vacuuming (VC), sweeping (SW), walking (WK), ascending stairs (AS), descending stairs (DS), treadmill running (RU), bicycling on an ergometer (50 W) (BC50), bicycling on an ergometer (100 W) (BC100), and rope jumping (RJ). Inertial data were collected from four Shimmer sensor nodes [26] placed on the right hip, chest, right wrist, and left ankle. Figure 3 illustrates the positions and coordinates of each sensor.

Each sensor node consisted of a 3D accelerometer and a 3D gyroscope. The range of the accelerometer was ±6g. The range of the gyroscopes was ±500 deg/s for the sensor nodes at the wrist, chest, and hip and ±2000 deg/s for the sensor node on the ankle. The sampling rate was set to 204.8 Hz.

4.2. Results of Feature Selection and Feature Optimization. The experimental results of feature selection are given in Figure 4 and Table 3. Figure 4 illustrates the classification accuracy according to the rank of the selected features for the four sensor locations. In detail, the obtained results are summarized in Table 3.

In Table 4, it is clear that the Laplacian Score outperforms SFFS and Relief-F in terms of the number of the selected

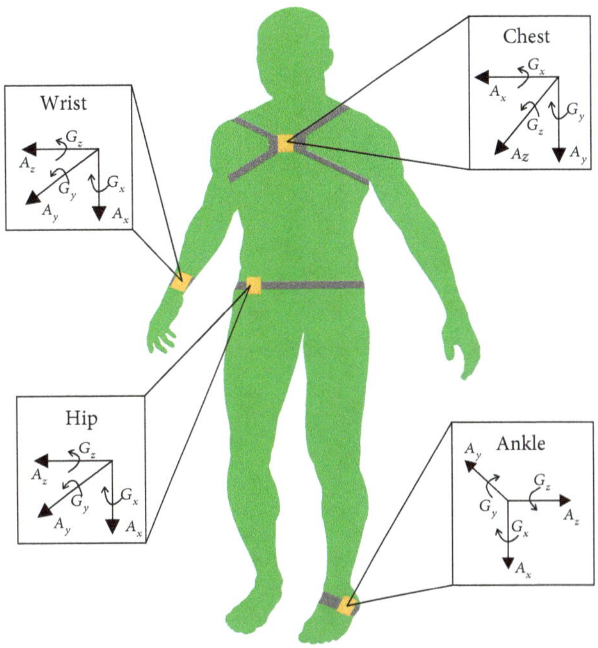

FIGURE 3: Body-sensor placement with each sensor-node coordinate (A_x, A_y, and A_z: x-, y-, and z-axes of an accelerometer; G_x, G_y, and G_z: x-, y-, and z-axes of a gyroscope).

features and accuracy. Among the sensor positions, the ankle and hip, which have the same number of selected features (29 features), give a high overall accuracy, that is, 96.85% and 96.91%, respectively. The accuracy rates of the three feature selection methods are very close, but the number of features selected by Laplacian Score is the lowest number, that is, 36 features in comparison with 87 features (SFFS) and 77 features (Relief-F). Considering the wrist sensor, Laplacian Score achieves an accuracy rate of 95.24% with 44 features, whereas SFFS and Relief-F attain classification rates of 91.60% (124 features) and 92.75% (73 features), respectively.

Figure 5 presents in detail the distribution of selected features for each sensor position after optimization. Wrist location contributes most of features (32%) for activity classification which is shown in Figure 5(a). According to Figure 5(b), the mean absolute deviation (MAD), range (r), correlation (ρ), and standard deviation (σ) features for one or more axes overwhelm four sensor positions, that is, the hip (71.43% in total), ankle (66.66%), chest (58.82%), and wrist (52.38%).

4.3. Performance of the Proposed Method and a Comparison with Reference Methods. Table 5 presents the confusion matrix for the recognition of thirteen daily living activities.

Table 3 shows the performance evaluation of our proposed method in detail. The proposed method recognizes washing dishes, descending stairs, treadmill running, and rope running with the highest accuracy rate of 100%. The method also achieves high accuracy for walking (99.62%), lying (99.47%), and ascending stairs (99.47%). Differences in intensity cause confusion for the algorithm when distinguishing bicycling on an ergometer (100 W) (98.93%) and

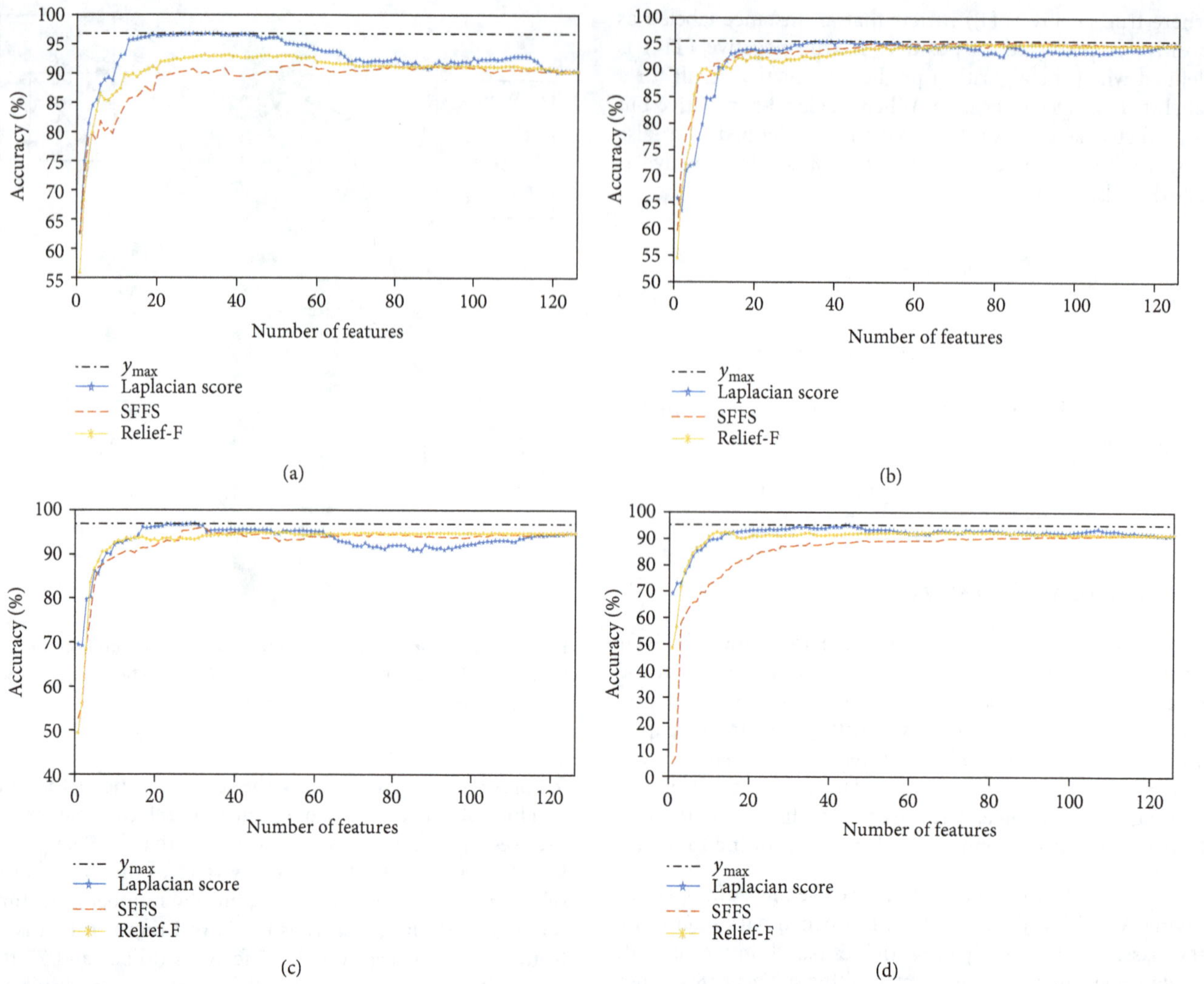

FIGURE 4: Classification accuracy rates depending on the number of features selected by the three feature selection methods for: (a) ankle, (b) chest, (c) hip, and (d) wrist.

TABLE 3: Performance evaluation.

Activity	Precision	Recall	F1-score
Sitting	96.62%	97.03%	96.83%
Lying	100%	99.47%	99.74%
Standing	97.87%	96.84%	97.35%
Washing dishes	99.66%	100%	99.83%
Vacuuming	95.73%	98.06%	96.88%
Sweeping	99.22%	98.27%	98.74%
Walking	100%	99.62%	99.81%
Ascending stairs	100%	99.47%	99.73%
Descending stairs	97.35%	100%	98.65%
Treadmill running (8.3 km/h)	100%	100%	100%
Bicycling on an ergometer (50 W)	98.51%	98.72%	98.62%
Bicycling on an ergometer (100 W)	99.09%	98.93%	99.01%
Rope jumping	100%	100%	100%

TABLE 4: Number of selected features and accuracy rates depending on the sensor positions and feature selection methods.

Number	Sensor position	Feature selection method	Number of selected features	Accuracy
		SFFS	56	91.46%
1	Ankle	Relief-F	32	93.16%
		Laplacian Score	29	96.85%
		SFFS	87	95.62%
2	Chest	Relief-F	77	95.11%
		Laplacian Score	36	95.62%
		SFFS	32	96.08%
3	Hip	Relief-F	52	95.01%
		Laplacian Score	29	96.91%
		SFFS	124	91.60%
4	Wrist	Relief-F	73	92.75%
		Laplacian Score	44	95.24%

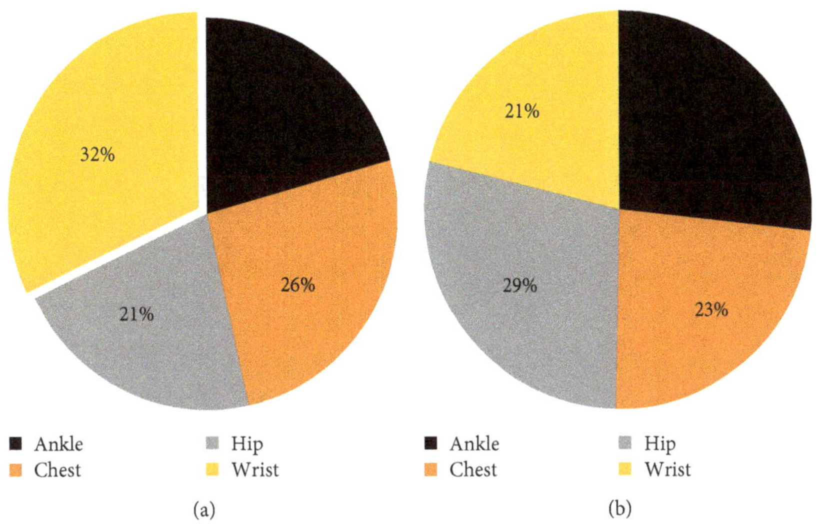

Figure 5: Statistics of selected features according to sensor positions: (a) distribution of number of features and (b) distribution of overwhelming features.

Table 5: Confusion matrix of our proposed method.

	SI	LY	ST	WD	VC	SW	WK	AS	DS	RU	BC50	BC100	RJ
SI	**229**	0	4	3	0	0	0	0	0	0	0	0	0
LY	1	**189**	0	0	0	0	0	0	0	0	0	0	0
ST	6	0	**184**	0	0	0	0	0	0	0	0	0	0
WD	0	0	0	**888**	0	0	0	0	0	0	0	0	0
VC	0	0	0	0	**202**	4	0	0	0	0	0	0	0
SW	1	0	0	0	8	**511**	0	0	0	0	0	0	0
WK	0	0	0	0	1	0	**791**	0	2	0	0	0	0
AS	0	0	0	0	0	0	0	**187**	1	0	0	0	0
DS	0	0	0	0	0	0	0	0	**110**	0	0	0	0
RU	0	0	0	0	0	0	0	0	0	**438**	0	0	0
BC50	0	0	0	0	0	0	0	0	0	0	**463**	6	0
BC100	0	0	0	0	0	0	0	0	0	0	7	**650**	0
RJ	0	0	0	0	0	0	0	0	0	0	0	0	**160**

bicycling on an ergometer (50 W) (98.72%). The confusion matrix in Table 5 clearly shows that these activities are misclassified as each other. The accuracy for sweeping (98.27%) and vacuuming (98.06%) are quite similar. Sitting and standing, which have the same perspective of orientation in gravity, are misclassified by 2.97% and 3.16%, respectively.

Table 6 gives a comparison of the proposed method with two reference papers, that is, Leutheuser et al. [33] and Zdravevski et al. [34]. As shown in Table 6, our proposed method compares well against the reference papers in terms of recognition of sitting, standing, washing dishes, walking, ascending stairs, descending stairs, treadmill running, and rope jumping. Overall, our proposed method achieved an accuracy rate of 99.13% in classifying 13 daily living activities, whereas the two reference methods of Leutheuser et al. and Zdravevski et al. achieved accuracy rates of 89.00% and 93.00%, respectively.

5. Conclusion

We proposed a novel approach that applies a feature selection algorithm on different sensor locations to recognize thirteen daily living activities using the DaLiAc benchmark dataset. To perform the recognition, we extracted features from a preprocessed dataset. Then, the features were ranked according to feature importance using a feature selection technique. In addition, we performed this process separately for each sensor-location dataset. Before feeding the features to a classifier, we optimized the selected features by using the Pearson correlation coefficient with a threshold equal to 1. We evaluated the performance of our proposed method with three metrics, that is, precision, recall, and F1-score. The results showed that the proposed method achieved a high precision rate of 100% for five activities, that is, lying, walking, ascending stairs, treadmill running, and rope

Table 6: Recognition performance for thirteen activities in terms of F1-score metric.

Activity	Leutheuser et al.	Zdravevski et al.	Proposed method
Sitting	94.10%	93.07%	96.83%
Lying	100%	100%	99.74%
Standing	90.93%	74.40%	97.35%
Washing dishes	96.50%	79.21%	99.83%
Vacuuming	85.13%	100%	96.88%
Sweeping	88.57%	100%	98.74%
Walking	98.89%	94.48%	99.81%
Ascending stairs	95.55%	98.50%	99.73%
Descending stairs	94.80%	91.50%	98.65%
Treadmill running (8.3 km/h)	100%	94.80%	100%
Bicycling on an ergometer (50 W)	63.50%	99.60%	98.62%
Bicycling on an ergometer (100 W)	58.75%	99.55%	99.01%
Rope jumping	100%	98.43%	100%
Overall accuracy	89.00%	93.00%	99.13%

jumping. The lowest precision rate was 95.73% in recognizing vacuuming activity. Furthermore, we compared the F1-score metric with the two reference papers. Overall, our proposed method outperformed the two reference methods with an accuracy rate of 99.13% in classifying thirteen activities. In contrast, Leutheuser et al. and Zdravevski et al. only achieved accuracy rates of 89.00% and 93.00%, respectively. The results are promising and show that the proposed method can be useful for a multiple-sensor configuration in human activity recognition.

In the future, we plan to improve the accuracy rate and reduce the feature dimension for classifying activities of daily living. In addition, we will extend our work by taking into consideration the feature correlation and the feature rank among the sensor positions.

Conflicts of Interest

The authors declare no conflict of interests.

Acknowledgments

This research was supported by the Basic Science Research Program through the National Research Foundation of Korea (NRF) funded by the Ministry of Education (NRF-2015R1D1A1A01060917). This research was also supported by the National Research Foundation of Korea (NRF) funded by the Korean Government (MSIP) (NRF-2016R1A5A1012966).

References

[1] K. Rajesh Kanna, V. Sugumaran, T. Vijayaram, and C. Karthikeyan, "Activities of daily life (ADL) recognition using wrist-worn accelerometer," *International Journal of Engineering and Technology*, vol. 8, no. 3, pp. 1406–1413, 2016.

[2] G. Koshmak, A. Loutfi, and M. Linden, "Challenges and issues in multisensor fusion approach for fall detection: review paper," *Journal of Sensors*, vol. 2016, Article ID 6931789, 12 pages, 2016.

[3] A. Özdemir, "An analysis on sensor locations of the human body for wearable fall detection devices: principles and practice," *Sensors*, vol. 16, no. 8, 2016.

[4] L. Bao and S. S. Intille, "Activity recognition from user-annotated acceleration data," *Lecture Notes in Computer Science*, vol. 3001, pp. 1–17, 2004.

[5] A. M. Khan, Y.-K. Lee, S. Y. Lee, and T.-S. Kim, "A triaxial accelerometer-based physical-activity recognition via augmented-signal features and a hierarchical recognizer," *IEEE Transactions on Information Technology in Biomedicine*, vol. 14, no. 5, pp. 1166–1172, 2010.

[6] N. D. Nguyen, P. H. Truong, and G.-M. Jeong, "Daily wrist activity classification using a smart band," *Physiological Measurement*, vol. 38, no. 9, pp. L10–L16, 2017.

[7] N. D. Nguyen, D. T. Bui, P. H. Truong, and G. M. Jeong, "Classification of five ambulatory activities regarding stair and incline walking using smart shoes," *IEEE Sensors Journal*, vol. 18, no. 13, pp. 5422–5428, 2018.

[8] D. Trong Bui, N. Nguyen, and G.-M. Jeong, "A robust step detection algorithm and walking distance estimation based on daily wrist activity recognition using a smart band," *Sensors*, vol. 18, no. 7, 2018.

[9] T. Sztyler, H. Stuckenschmidt, and W. Petrich, "Position-aware activity recognition with wearable devices," *Pervasive and Mobile Computing*, vol. 38, pp. 281–295, 2017.

[10] Y. Zheng, "Human activity recognition based on the hierarchical feature selection and classification framework," *Journal of Electrical and Computer Engineering*, vol. 2015, Article ID 140820, 9 pages, 2015.

[11] M. U. S. Khan, A. Abbas, M. Ali et al., "On the correlation of sensor location and human activity recognition in body area networks (BANs)," *IEEE Systems Journal*, vol. 12, no. 1, pp. 82–91, 2018.

[12] D. Schuldhaus, H. Leutheuser, and B. M. Eskofier, "Classification of daily life activities by decision level fusion of inertial sensor data," in *Proceedings of the 8th International Conference on Body Area Networks*, pp. 77–82, Brussels, Belgium, 2013.

[13] O. Baños, M. Damas, H. Pomares, and I. Rojas, "Activity recognition based on a multi-sensor meta-classifier," in *Advances in Computational Intelligence*, I. Rojas, G. Joya, and J. Cabestany, Eds., pp. 208–215, Springer, Berlin, Heidelberg, 2013.

[14] L. Gao, A. K. Bourke, and J. Nelson, "Activity recognition using dynamic multiple sensor fusion in body sensor networks," in *2012 Annual International Conference of the IEEE Engineering in Medicine and Biology Society*, pp. 1077–1080, San Diego, CA, USA, September 2012.

[15] F. Attal, S. Mohammed, M. Dedabrishvili, F. Chamroukhi, L. Oukhellou, and Y. Amirat, "Physical human activity recognition using wearable sensors," *Sensors*, vol. 15, no. 12, pp. 31314–31338, 2015.

[16] K. Kunze, P. Lukowicz, H. Junker, and G. Tröster, "Where am I: recognizing on-body positions of wearable sensors," in *Location- and Context-Awareness*, T. Strang and C. Linnhoff-Popien, Eds., pp. 264–275, Springer, Berlin, Heidelberg, 2005.

[17] A. Vahdatpour, N. Amini, and M. Sarrafzadeh, "On-body device localization for health and medical monitoring applications," in *2011 IEEE International Conference on Pervasive Computing and Communications (PerCom)*, pp. 37–44, Seattle, WA, USA, March 2011.

[18] C. Figueira, R. Matias, and H. Gamboa, "Body location independent activity monitoring," in *Proceedings of the 9th International Joint Conference on Biomedical Engineering Systems and Technologies*, pp. 190–197, Rome, Italy, 2016.

[19] L. Atallah, B. Lo, R. King, and G.-Z. Yang, "Sensor positioning for activity recognition using wearable accelerometers," *IEEE Transactions on Biomedical Circuits and Systems*, vol. 5, no. 4, pp. 320–329, 2011.

[20] M. Zhang and A. A. Sawchuk, "A feature selection-based framework for human activity recognition using wearable multimodal sensors," in *Proceedings of the 6th International ICST Conference on Body Area Networks*, vol. 6, pp. 92–98, Beijing, China, June 2012.

[21] S. Pirttikangas, K. Fujinami, and T. Nakajima, "Feature selection and activity recognition from wearable sensors," in *Ubiquitous Computing Systems*, pp. 516–527, Springer, Berlin, Heidelberg, 2006.

[22] T. Huynh and B. Schiele, "Analyzing features for activity recognition," in *Proceedings of the 2005 Joint Conference on Smart Objects and Ambient Intelligence Innovative Context-Aware Services: Usages and Technologies—sOc-EUSAI '05*, pp. 159–163, Grenoble, France, October 2005.

[23] A. Y. Yang, R. Jafari, S. S. Sastry, and R. Bajcsy, "Distributed recognition of human actions using wearable motion sensor networks," *Journal of Ambient Intelligence and Smart Environments*, vol. 1, no. 2, pp. 103–115, 2009.

[24] O. Banos, M. Damas, H. Pomares, A. Prieto, and I. Rojas, "Daily living activity recognition based on statistical feature quality group selection," *Expert Systems with Applications*, vol. 39, no. 9, pp. 8013–8021, 2012.

[25] O. Banos, M. Damas, H. Pomares, F. Rojas, B. Delgado-Marquez, and O. Valenzuela, "Human activity recognition based on a sensor weighting hierarchical classifier," *Soft Computing*, vol. 17, no. 2, pp. 333–343, 2013.

[26] L. Gao, A. K. Bourke, and J. Nelson, "Evaluation of accelerometer based multi-sensor versus single-sensor activity recognition systems," *Medical Engineering & Physics*, vol. 36, no. 6, pp. 779–785, 2014.

[27] J. Li, K. Cheng, S. Wang et al., "Feature selection: a data perspective," *ACM Computing Surveys*, vol. 50, no. 6, pp. 1–45, 2018.

[28] K. Kira and L. A. Rendell, "A practical approach to feature selection," in *Machine Learning Proceedings 1992*, Morgan Kaufmann Publishers Inc., 1992.

[29] R. Gilad-Bachrach, A. Navot, and N. Tishby, "Margin based feature selection-theory and algorithms," in *ICML '04 Proceedings of the Twenty-First International Conference on Machine Learning*, vol. 21, pp. 43–50, Banff, Alberta, Canada, July 2004.

[30] H. Peng, F. Long, and C. Ding, "Feature selection based on mutual information: criteria of max-dependency, max-relevance, and min-redundancy," *IEEE Transactions on Pattern Analysis and Machine Intelligence*, vol. 27, no. 8, pp. 1226–1238, 2005.

[31] I. Guyon and A. Elisseeff, "An introduction to variable and feature selection," *Journal of Machine Learning Research*, vol. 3, pp. 1157–1182, 2003.

[32] A. W. Whitney, "A direct method of nonparametric measurement selection," *IEEE Transactions on Computers*, vol. C-20, no. 9, pp. 1100–1103, 1971.

[33] H. Leutheuser, D. Schuldhaus, and B. M. Eskofier, "Hierarchical, multi-sensor based classification of daily life activities: comparison with state-of-the-art algorithms using a benchmark dataset," *PLoS One*, vol. 8, no. 10, article e75196, 2013.

[34] E. Zdravevski, P. Lameski, V. Trajkovik et al., "Improving activity recognition accuracy in ambient-assisted living systems by automated feature engineering," *IEEE Access*, vol. 5, pp. 5262–5280, 2017.

[35] L. Yu and H. Liu, "Feature selection for high-dimensional data: a fast correlation-based filter solution," in *Proceedings of the 20th International Conference on Machine Learning (ICML-03)*, pp. 856–863, Washington, DC, USA, 2003.

[36] Y. Saeys, I. Inza, and P. Larranaga, "A review of feature selection techniques in bioinformatics," *Bioinformatics*, vol. 23, no. 19, pp. 2507–2517, 2007.

[37] I. Kononenko, E. Simec, and M. Robnik-Sikonja, "Overcoming the myopia of inductive learning algorithms with relief," *Applied Intelligence*, vol. 7, no. 1, pp. 39–55, 1997.

[38] X. He, D. Cai, and P. Niyogi, "Laplacian score for feature selection," *Advances in Neural Information Processing Systems*, vol. 18, 2006.

[39] F. R. K. Chung, "Spectral graph theory," in *Regional Conference Series in Mathematics*, vol. 92, American Mathematical Society, 1996.

[40] M. Kudo and J. Sklansky, "Comparison of algorithms that select features for pattern classifiers," *Pattern Recognition*, vol. 33, no. 1, pp. 25–41, 2000.

[41] Z. Zhao, L. Wang, H. Liu, and J. Ye, "On similarity preserving feature selection," *IEEE Transactions on Knowledge and Data Engineering*, vol. 25, no. 3, pp. 619–632, 2013.

[42] A. M. De Silva and P. H. W. Leong, *Grammar-Based Feature Generation for Time-Series Prediction*, Springer, 2015.

[43] J.-Y. Yang, J.-S. Wang, and Y.-P. Chen, "Using acceleration measurements for activity recognition: an effective learning algorithm for constructing neural classifiers," *Pattern Recognition Letters*, vol. 29, no. 16, pp. 2213–2220, 2008.

[44] C.-W. Hsu and C.-J. Lin, "A comparison of methods for multi-class support vector machines," *IEEE Transactions on Neural Networks*, vol. 13, no. 2, pp. 415–425, 2002.

[45] P. Cunningham and S. J. Delaney, "*K*-nearest neighbor classifiers," *Technical Report UCD-CSI-2007-4*, School of Computer Science and Informatics, Ireland, 2007.

[46] H. Bhaskar, D. C. Hoyle, and S. Singh, "Machine learning in bioinformatics: a brief survey and recommendations for practitioners," *Computers in Biology and Medicine*, vol. 36, no. 10, pp. 1104–1125, 2006.

15

Acoustic Emission Source Localization System using Fiber Bragg Grating Sensors and a Barycentric Coordinate-Based Algorithm

Dandan Pang,[1,2] Qingmei Sui,[3] Ming Wang,[1,2] Yaozhang Sai,[3] Rongyao Sun,[1] and Yikang Wang[1]

[1]*School of Information and Electrical Engineering, Shandong Jianzhu University, Jinan 250101, China*
[2]*Shandong Provincial Key Laboratory of Intelligent Buildings Technology, Jinan 250101, China*
[3]*School of Control Science and Engineering, Shandong University, Jinan 250061, China*

Correspondence should be addressed to Dandan Pang; pangdandan@sdjzu.edu.cn

Academic Editor: Stephen James

Structures and machines are very susceptible to some barely visible defects; acoustic emission (AE) is an effective technique of detecting defects and examining growth and location of these defects. In this paper, a new method was proposed to predict the position of the AE source. The dynamic strain signal from the AE source was acquired by fiber Bragg grating (FBG) sensors. Complex Morlet wavelet transform was used to extract narrowband signals from AE waves. It was proposed that the barycentric coordinate-based localization algorithm can be used to predict the coordinates of the AE source. The validation tests based on the designed AE detection system were performed on a 500 mm × 500 mm × 2 mm Al-alloy plate. The experimental results show that the proposed method is feasible.

1. Introduction

The degradation of structural properties is well considered as an enormous threat for security and performance of structures and machines. Structural health monitoring (SHM) utilizes several kinds of sensors attached to or embedded in the monitored structures to detect the appearance, location, and severity of damage. Thus, SHM is imperative for damage detection and structural failure. Various methods have been researched for structural damage detection. These methods can be divided into two categories: model-based and signal-based approaches. The former develops suitable model and analyzes changes that relate to damage in the model; the latter extracts features and establishes a relationship between these features and potential damage [1–3]. Acoustic emission (AE) is one of the effective signal-based means implemented for damage detection [4].

AE wave is generated by energy release due to a propagating crack or friction in structures. AE technique is used to detect, locate, and assess defects. To obtain AE wave, some AE sensors have been widely used, such as piezoelectric sensors (PZT) and capacitive AE sensors [5–8]. However, the traditional AE sensors have the disadvantages such as complex structure, weak antielectromagnetic interference, and unsuitability for distributed measurement. At present, fiber Bragg grating (FBG) sensors have been expected as an alternative because of their excellent performance. FBG sensors can be easily embedded in the monitored structure without destructive effect, due to their small size and light weight [9]. Furthermore, FBG sensors are immune to electromagnetic interference [10]. In addition, long-distance distributed measurement can be realized by using FBG sensors [11]. Therefore, some scholars have done several researches on AE source localization using FBG sensors [12, 13]. Yu et al. [14] investigated the effective length of the FBG sensor from the AE source on the composite plate. Yu et al. [15] proposed an AE source identification method based on the algebraic reconstruction algorithm and 3D imaging technique.

Recently, an increasing number of AE source localization methods was developed based on the intelligent algorithm [16, 17]. Cheng et al. [18] applied an optimized wavelet

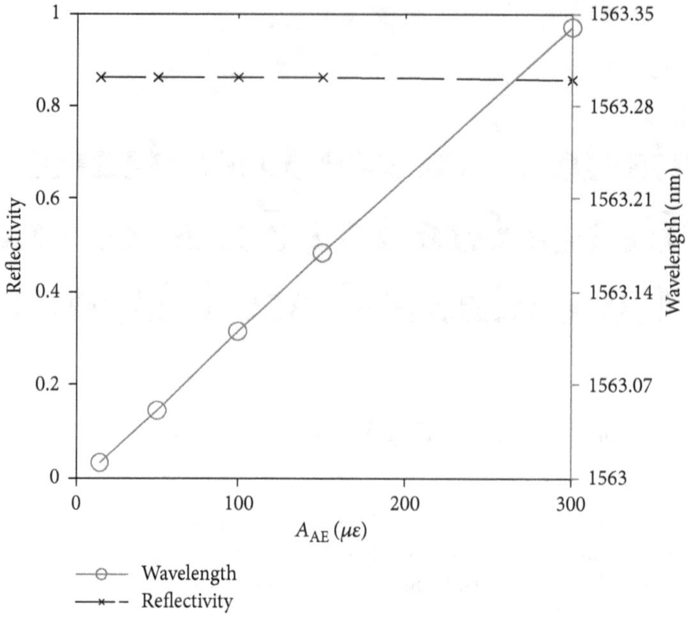

FIGURE 1: Relation between A_{AE} and reflectivity and wavelength of FBG.

neural network to realize AE source location in rotating machinery. Sadegh et al. [19] applied the genetic algorithm and artificial neural networks to extract features of AE signals for monitoring lubrication conditions of a journal bearing. However, in the intelligent algorithms used in those studies, lots of training samples were needed, which leaded to inefficiency and complexity of the localization process. Furthermore, AE source localization in plate-like structures usually requires lots of sensors. In particular, those approaches are usually accurate only within the convex area surrounded by sensors.

Based on the previous studies, this paper presented a novel AE detection system based on FBG sensors and a new AE source location method. The AE signal was collected by the FBG sensor which was pasted on the structural surface. Barycentric coordinate-based location method was used to predict positioning. Different from the approaches mentioned above, the proposed method was developed based on only range measurements between sensors, not training samples. The AE source is not required to lie inside the convex area formed by sensors. At last, AE source location experiments were carried out to verify this new designed system and localization algorithm. The results show that it is an efficient and feasible method for AE source localization.

2. Theoretical Principle

2.1. FBG AE Sensing Principle. The dynamic strain field along the FBG generated by AE wave can be simplified as a time-dependent cosine-Gauss function:

$$\varepsilon_{AE}(t) = A_{AE}e^{-((t-t_0)^2/a^2)} \cos\left(\frac{2\pi}{\lambda_{AE}}z - \omega_{AE}t\right), \quad (1)$$

where A_{AE} is the amplitude of AE wave propagating in a material to be monitored, λ_{AE} is the wavelength of AE wave, ω_{AE} is the angular frequency of AE wave, t_0 is the arrival time, and a is the duration of AE wave.

The alternation of wavelength of FBG is mainly attributed to two factors: one is the modulation of the grating period Λ (geometric effect), which causes the axial deformation of FBG, and the other is the change of the effective refractive index of FBG $\bar{n}_{eff}$ (photoelastic effect).

In consideration of the geometric effect and photoelastic effect, the expression of the effective refractive index of FBG under AE wave is obtained:

$$\begin{aligned} n'''_{eff}\left(z',t\right) &= n_{eff0} - \Delta n \sin^2\left(\frac{\pi}{\Lambda_0}z\right) \\ &\quad - \frac{n^3_{eff0}}{2}\cdot\left[P_{12} - \nu(P_{11}+P_{12})\right] \\ &\quad \cdot A_{AE}e^{\left(-(t-t_0)^2/a^2\right)} \cos\left(\frac{2\pi}{\lambda_{AE}}z' - \omega_{AE}t\right), \end{aligned} \quad (2)$$

where $z' = z + \int_0^z \varepsilon(\xi)d\xi$, P_{11} and P_{12} are the photoelastic coefficients of FBG, and ν is Poisson's ratio [20].

Based on the mentioned theory above, the spectrum characteristics of FBG under AE waves are simulated and analyzed. When the wavelength of AE wave λ_{AE} is 100 mm, the length of grating L is 10 mm and the amplitudes of AE waves A_{AE} are 15 $\mu\varepsilon$, 50 $\mu\varepsilon$, 100 $\mu\varepsilon$, 150 $\mu\varepsilon$, and 300 $\mu\varepsilon$, respectively. Relation between A_{AE} and reflectivity and wavelength of FBG is shown in Figure 1. The center of FBG shifts towards the longer wavelength as A_{AE} increases when λ_{AE} is greater than L. Results also show that there is a good linear relation between central wavelength and A_{AE}. The maximum reflectivity of the FBG reflection spectrum keeps invariant as the change of A_{AE}.

2.2. Complex Morlet Wavelet Transform. The collected AE waves often contain some mutation-shaped peaks or nonstationary components. Morlet wavelet transform is employed to extract the narrowband signal of AE waves.

The continuous wavelet transform of AE wave $\varepsilon_{\mathrm{AE}}(t)$ can be defined as

$$W_{\varepsilon}(\alpha,\tau)=\frac{1}{\sqrt{\alpha}}\int_{-\infty}^{+\infty}\varepsilon_{\mathrm{AE}}(t)\overline{\phi}\left(\frac{t-\tau}{\alpha}\right)dt, \tag{3}$$

where $1/\sqrt{\alpha}\overline{\phi}(t-\tau/\alpha)$ is the complex conjugate of the scaled wavelets $1/\sqrt{\alpha}\phi(t-\tau/\alpha)$ obtained by zooming and panning the mother wavelet, α is the zoom factor, and τ is the shift factor.

The complex Morlet wavelet can be expressed as [21, 22]

$$\phi(t)=\frac{1}{\sqrt{\pi f_{\mathrm{b}}}}e^{-\left(t^{2}/f_{\mathrm{b}}\right)}e^{i2\pi f_{c}t}. \tag{4}$$

The Fourier transform of the above function is defined by

$$\varphi(\omega)=e^{-\pi^{2}f_{\mathrm{b}}(\omega-f_{c})^{2}}, \tag{5}$$

where f_{b} and f_{c} are the bandwidth factor and center frequency of the complex Morlet wavelet, respectively. The wavelet is a complex exponential function with the shape of a Gaussian window. Hence, the narrowband signal with center frequency f_{c} can be extracted by Morlet wavelet transform. The bandwidth is limited in the range of $((\omega_{\mathrm{c}}-\omega_{\mathrm{b}})/2,(\omega_{\mathrm{c}}-\omega_{\mathrm{b}})/2)$. The complex Morlet wavelet filter can be constructed as

$$WT(f_{\mathrm{b}},f_{\mathrm{c}})=F^{-1}[E_{\mathrm{AE}}(\omega)\overline{\varphi}(\omega)], \tag{6}$$

where F^{-1} denotes the inverse Fourier transform, $E_{\mathrm{AE}}(\omega)$ and $\varphi(\omega)$ are the Fourier transform of $\varepsilon_{\mathrm{AE}}(t)$ and $\phi(t)$, respectively.

When considering the propagation distance z, the AE wave $\varepsilon_{\mathrm{AE}}(z,t)$ can be express as

$$\varepsilon_{\mathrm{AE}}(z,t)=e^{-j(n_{1}z-\omega_{1}t)}+e^{-j(n_{2}z-\omega_{2}t)}, \tag{7}$$

where n_1 and n_2 are the wave numbers. Denote

$$\begin{aligned}&\frac{n_1+n_2}{2}=n_0,\\&\frac{n_1-n_2}{2}=\Delta n,\\&\frac{\omega_1+\omega_2}{2}=\omega_0,\\&\frac{\omega_1-\omega_2}{2}=\Delta\omega,\\&\overline{\varphi}(\alpha\omega)=\overline{\varphi}(\alpha\omega_1)=\overline{\varphi}(\alpha\omega_2).\end{aligned} \tag{8}$$

The module value of AE wave is acquired by the complex Morlet wavelet transform

$$|WT(z,\alpha,\tau)|=\sqrt{2a}|\varphi(\alpha\omega_0)|\sqrt{1+\cos\left(2\Delta nz-2\Delta\omega\tau\right)}. \tag{9}$$

When $\tau=\Delta nz/\Delta\omega=z/V$, the module value is maximum, where V is the group velocity of AE wave. Therefore, the arrival time difference Δt can be obtained by finding the peak values of the module values:

$$\Delta t=\tau_1-\tau_2. \tag{10}$$

According to the arrival time difference and the distance difference, the wave velocity v_{AE} can be calculated on the monitored material:

$$v_{\mathrm{AE}}=\frac{\Delta z}{\Delta t}. \tag{11}$$

2.3. Barycentric Coordinate-Based Location Algorithm. In 1827, the barycentric coordinate was developed to characterize the relative position of a point with respect to other points [23–26]. In the plane, for a point k and other three points l, m, and n, the barycentric coordinate of point k with respect to points l, m, and n can be defined as $\{a_{kl}, a_{km}$, and $a_{kn}\}$ that satisfies

$$p_k=a_{kl}p_l+a_{km}p_m+a_{kn}p_n, \tag{12}$$

where p_k, p_l, p_m, and p_n are the Euclidean coordinates of points k, l, m, and n, respectively.

Consider an AE event k occurs on the plate with three sensors l, m, and n nearby, the AE signal is detected by each of the sensors. Then, the absolute values of the barycentric coordinates a_{kl}, a_{km}, and a_{kn} can be computed as

$$\begin{aligned}|a_{kl}|&=\frac{|S_{\Delta kmn}|}{|S_{\Delta lmn}|},\\|a_{km}|&=\frac{|S_{\Delta knl}|}{|S_{\Delta lmn}|},\\|a_{kn}|&=\frac{|S_{\Delta klm}|}{|S_{\Delta lmn}|},\end{aligned} \tag{13}$$

where $S_{\Delta kmn}$, $S_{\Delta knl}$, $S_{\Delta klm}$, and $S_{\Delta lmn}$ are the signed areas of the corresponding triangles Δkmn, Δknl, Δklm, and Δlmn. These areas can be solved according to

$$S_{\Delta kmn}^{2}=-\frac{1}{16}\begin{vmatrix}0&1&1&1\\1&0&d_{km}^{2}&d_{kn}^{2}\\1&d_{mk}^{2}&0&d_{mn}^{2}\\1&d_{nk}^{2}&d_{nm}^{2}&0\end{vmatrix}, \tag{14}$$

where d_{km}, d_{kn}, and d_{mn} are the distance measurements among the AE source k and sensors m and n, respectively.

In the plane, three adjacent sensors l, m, and n are not collinear to avoid the case that $S_{\Delta lmn} = 0$. Thus, it remains to determine the sign patterns of the barycentric coordinates for obtaining the barycentric coordinates. In addition, note that $a_{kl} + a_{km} + a_{kn} = 1$. Therefore, given $|a_{kl}|$, $|a_{km}|$, and $|a_{kn}|$, the problem of determining the sign of the barycentric coordinate is equated with solving the following equation:

$$\sigma_{kl}|a_{kl}| + \sigma_{km}|a_{km}| + \sigma_{kn}|a_{kn}| = 1, \tag{15}$$

where σ_{kl}, σ_{km}, and σ_{kn} take values of either 1 or −1. The available information to determine the sign patterns should be limited to the pairwise distance measurements between the AE source k and sensors l, m, and n.

As shown in Figure 2, there are only 7 possible sign patterns (σ_{kl}, σ_{km}, and σ_{kn}) of the barycentric coordinates (as the pattern (−1, −1, −1) is not a possible solution for (15)). Therefore, the plane can be divided into 7 zones: I (1, 1, 1), II (1, −1, −1), III (1, 1, −1), IV (−1, 1, −1), V (−1, 1, 1), VI (−1, −1, 1), and VII (1, −1, 1). It turns out that sometimes the sign pattern can be uniquely determined from (15), but sometimes this cannot be done. When the sign pattern cannot be uniquely solved from (15), we consider two cases in the following.

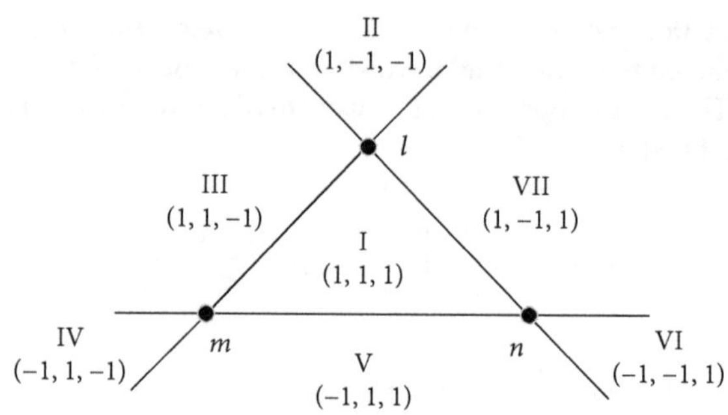

FIGURE 2: 7 sign patterns for (σ_{kl}, σ_{km}, and σ_{kn}).

In the first case, one of $|a_{kl}|$, $|a_{km}|$, and $|a_{kn}|$ equals to zero. That is, the AE source k lies on the line aligned with one of the three edges of the triangle formed by the three adjacent sensors, according to (13). Without loss of generality, say $a_{kl} = 0$. For this case, we can take $\sigma_{kl} = 1$ (without loss of generality). Then, the other two signs σ_{km} and σ_{kn} can be determined according to the following criterion.

$$(\sigma_{kl}, \sigma_{km}, \sigma_{kn}) = \begin{cases} (1, 1, 1) & \text{if } |a_{km}|, |a_{kn}| \le 1, \\ (1, 1, -1) & \text{if } |a_{km}| > 1, |a_{km}| > |a_{kn}|, \\ (1, -1, 1) & \text{if } |a_{kn}| > 1, |a_{kn}| > |a_{kl}|. \end{cases} \tag{16}$$

In the second case, one of $|a_{kl}|$, $|a_{km}|$, and $|a_{kn}|$, saying a_{kl}, satisfies $|a_{kl}| = 1$ and $|a_{km}| = |a_{kn}| \ne 0$. For this case, suppose $\angle lmn$ is an acute angle, the sign pattern (σ_{kl}, σ_{km}, and σ_{kn}) can be determined according to the following criterion.

$$(\sigma_{kl}, \sigma_{km}, \sigma_{kn}) = \begin{cases} (-1, 1, 1) & \text{if } d_{mk} = d_{ln}, d_{nk} = d_{lm}, d_{lk}^2 = 2d_{lm}^2 + 2d_{ln}^2 - d_{mn}^2, \\ (1, 1, -1) & \text{if } d_{mk}^2 < d_{lm}^2 + d_{lk}^2, \\ (1, -1, 1) & \text{if } d_{mk}^2 > d_{lm}^2 + d_{lk}^2. \end{cases} \tag{17}$$

3. AE Source Location Experiments

In this work, all the experiments are performed on a 6061 aluminum alloy (Al-alloy) plate. The dimension of the Al-alloy plate is 500 mm × 500 mm × 2 mm. The diagram of the experimental setup is shown in Figure 3. The data sampling frequency is set at 2 MHz. Three FBGs are glued on the preselected points of the Al-alloy plate, whose coordinates are A = 100 mm, 200 mm; B = 400 mm, 200 mm; and C = 250 mm, 350 mm. A normalized pencil lead break source (HSU) is used as an AE source.

According to Section 2.3, the location method based on barycentric coordinates needs the distance measurements among AE source S, point A, point B, and point C. The AE source location geometry is shown in Figure 4. As the distance measurements between three sensors (A, B, and C) are known before the experiment, there are three unknown distances, d_{SA}, d_{SB}, and d_{SC}.

Because of the isotropic characteristic of the Al-alloy plate, the AE wave travels at the same speed in all directions. As shown in Figure 4, AE wave velocity measurement is carried out at point P. Point P is laid on the line BC and the coordinate is (200, 200). When there is an AE event happens at point P, the AE signals obtained by FBGs (B and C) are shown in Figure 5. The frequency responses of FBG signals are shown in Figure 6. It is known that the AE signal is a broadband signal with a range from 50 kHz to 300 kHz. The Morlet wavelet transform is utilized to extract a narrowband signal with central frequency of 100 kHz. As shown in Figure 7, the module

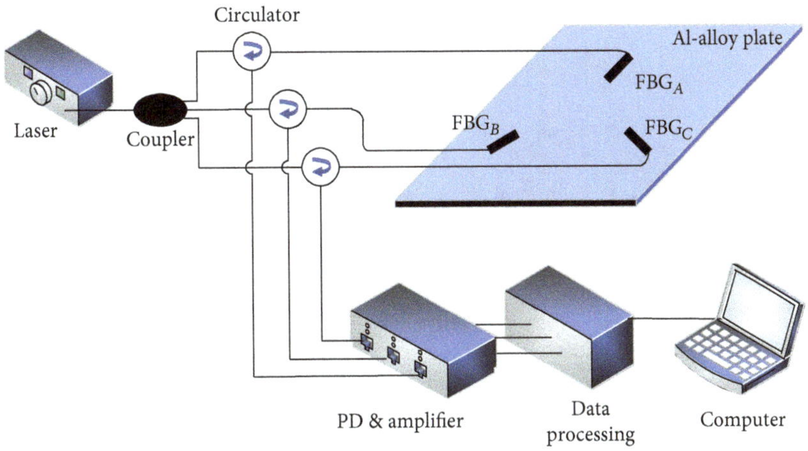

FIGURE 3: The schematic of the experimental setup.

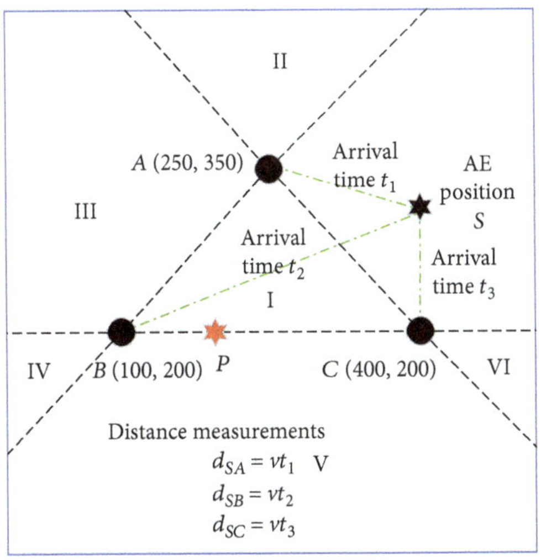

FIGURE 4: The AE source location geometry.

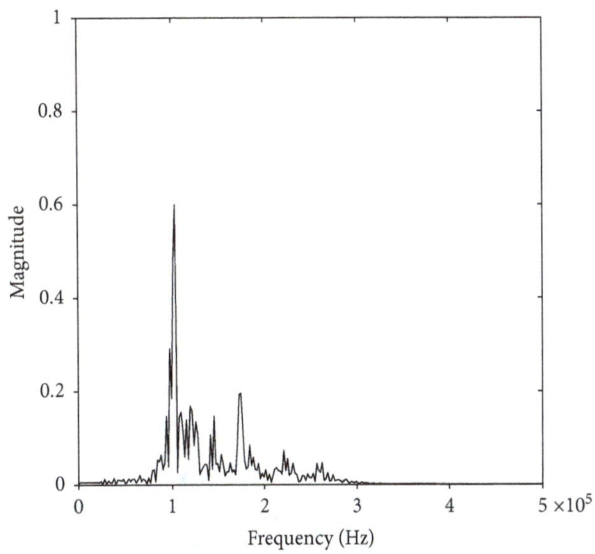

FIGURE 6: Frequency response of the detected signal.

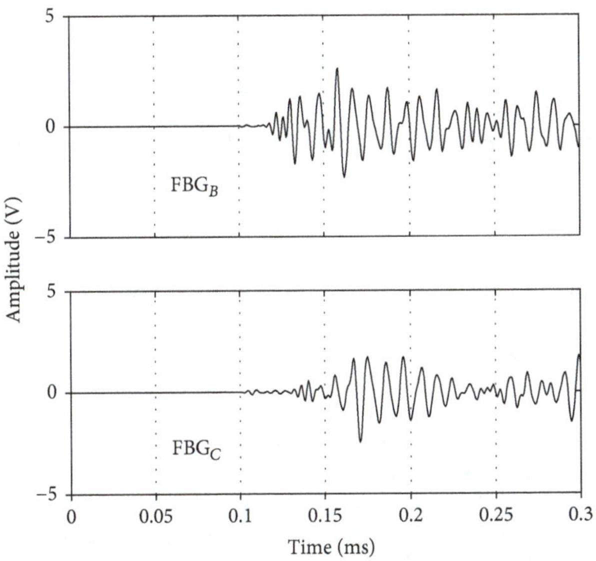

FIGURE 5: AE waves detected by FBG$_B$ and FBG$_C$.

values of the extracted narrowband signals are obtained and the first peaks of the module values are used to compute the time difference Δt between FBGs (B and C). Depending on the time difference and the distance difference measured between PB and PC, the AE wave velocity v_{AE} can be obtained. In order to reduce the measurement error, the same experiment is repeated for 10 times and the results are shown in Figure 8. Based on the above method, the average wave velocity is 2454.1 m/s and the estimation error is less than 0.25%.

The threshold value method is an effective and simple technique to obtain the arrival times. At the beginning, according to the detected velocity v_{AE} and the arrival time of the AE wave from point P to point C, t_{PC} is calculated as $t_{PC} = d_{PC}/v_{AE}$. In Figure 7, the absolute module value corresponding to t_{PC} is set as the threshold value. In the next experiment, when the module value of the FBG signal is greater than the preset threshold value, this moment is the arrival time of the AE wave. Based on this process,

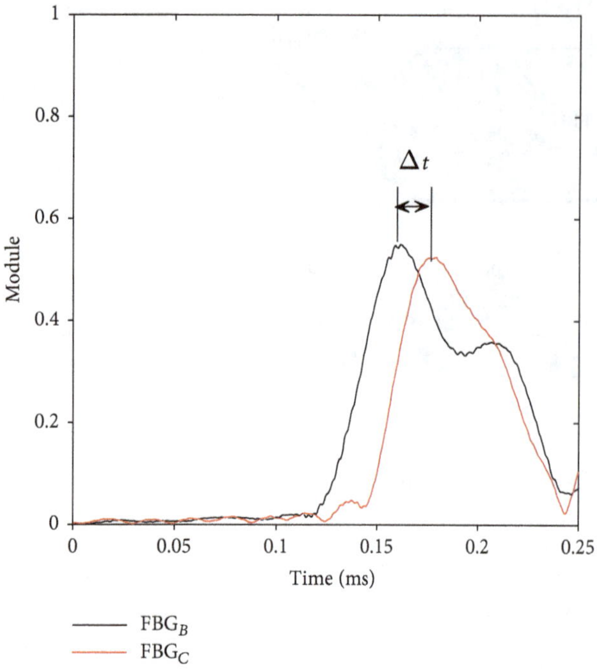

FIGURE 7: Module values of the extracted narrowband signals.

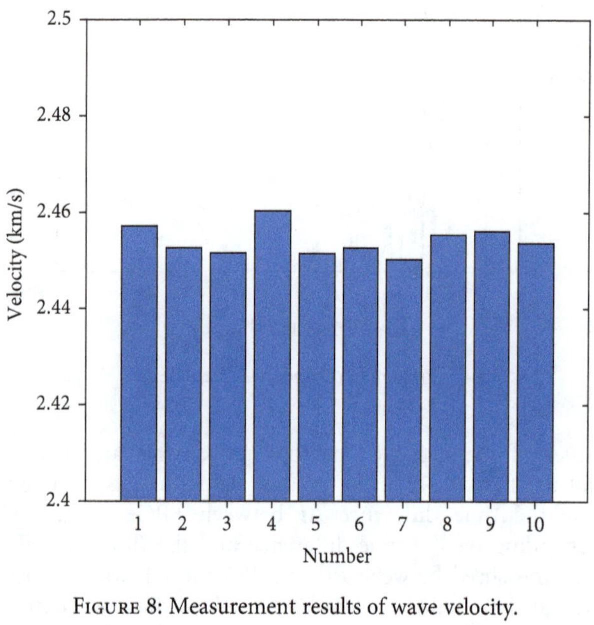

FIGURE 8: Measurement results of wave velocity.

arrival times (t_1, t_2, and t_3) of AE waves in Figure 4 can be computed.

4. Results

Impact is applied on the plate using a steel ball at randomly selected location of S (400 mm, 300 mm), as shown in Figure 4. The AE waves detected by FBGs are shown in Figure 9. After complex Morlet wavelet transform, the narrowband signals are illustrated in Figure 10. According to the narrowband signals, the arrival times of AE signals

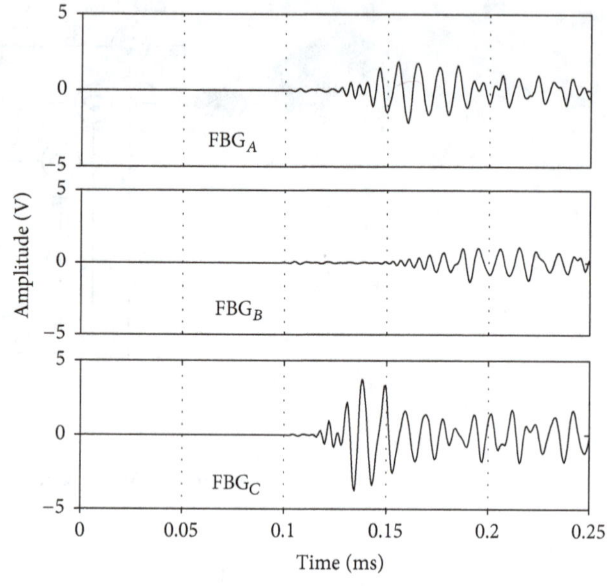

FIGURE 9: AE waves detected by FBGs.

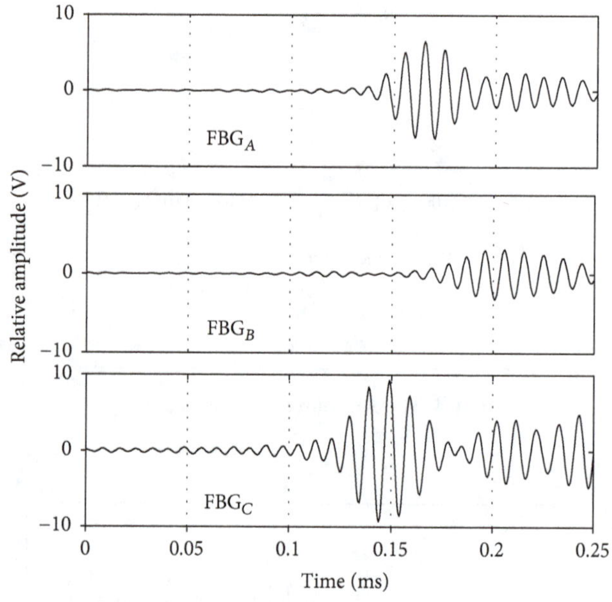

FIGURE 10: Narrowband signals extracted from the detected signals.

are computed. Based on the calculated velocity v_{AE}, the distance measurements among AE source S, point A, point B, and point C are $d_{AB} = 150\sqrt{2}$, $d_{AC} = 150\sqrt{2}$, $d_{BC} = 300$, $d_{AS} = 50\sqrt{10}$, $d_{BS} = 100\sqrt{10}$, and $d_{CS} = 100$, respectively. Then, absolute values of the barycentric coordinates $|a_{SA}|$, $|a_{SB}|$, and $|a_{SC}|$ are calculated. According to (15), (16), and (17), the sign patterns (σ_{SA}, σ_{SB}, and σ_{SC}) can be deduced as (1, −1, 1); therefore, the position of impact is correctly found at the zone VII. The Euclidean coordinate p_S of the AE source S is calculated.

In each zone, two positions are randomly selected to verify the results of the proposed location method. The location results are presented in Figure 11 and Table 1.

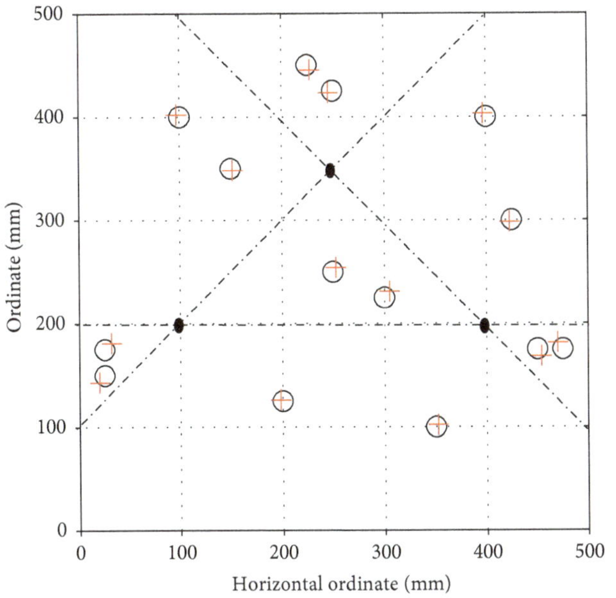

Figure 11: The localization results.

Table 1: Location results in 7 zones.

	Actual position (mm)	Predicted position (mm)
I	(250, 250)	(253, 254)
	(300, 225)	(305, 231)
II	(250, 425)	(246, 423)
	(225, 450)	(228, 445)
III	(150, 350)	(152, 348)
	(100, 400)	(97, 402)
IV	(25, 175)	(32, 181)
	(25, 150)	(20, 143)
V	(200, 125)	(198, 126)
	(350, 100)	(352, 102)
VI	(450, 175)	(454, 168)
	(475, 175)	(470, 181)
VII	(425, 300)	(423, 298)
	(400, 400)	(397, 403)

The results show that the average and maximum location errors that correspond to the horizontal direction are, respectively, 3.6 and 7 mm. The average and maximum location errors that correspond to the vertical direction are 3.9 and 7 mm, respectively. Also, mean square deviation is employed to calculate the AE source location error using the proposed localization method.

$$E = \sqrt{\left(x_p - x_a\right)^2 + \left(y_p - y_a\right)^2}, \tag{18}$$

where (x_p, y_p) is the Euclidean coordinate of the predicted AE source and (x_a, y_a) is the Euclidean coordinate of the actual AE source. The location errors based on the proposed location method are indicated in Figure 12. The results show that the average and maximum location errors are 5.4 and 9.2 mm, respectively.

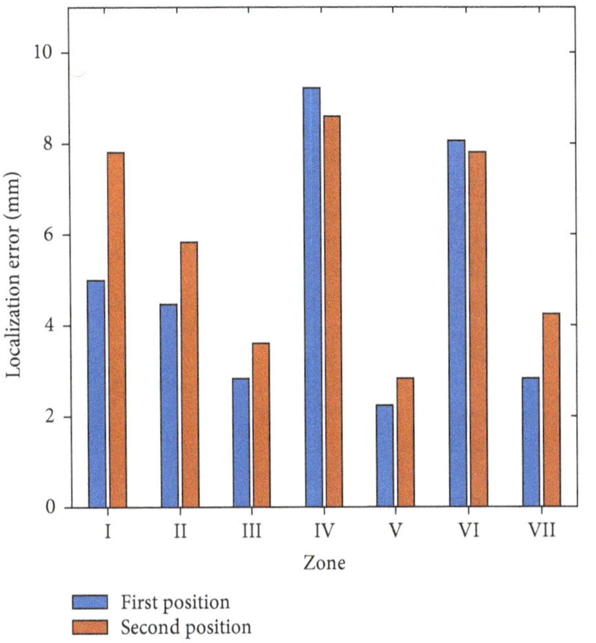

Figure 12: The AE source localization errors.

5. Conclusions

In this paper, we have introduced an easy realizable AE source localization method based on the AE waves obtained by FBG sensors. The edge filter method-based FBG interrogation system is used to satisfy the high-speed signal demodulation needs. To realize the barycentric coordinate-based location method, the Morlet wavelet transform is utilized for velocity and arrival time calculation. Then, the AE source location method based on barycentric coordinates is proposed. In the end, impact experiments with random selected positions are carried out for verification of the proposed location method. This research provides an easy and efficient method to safety monitoring of structures.

Conflicts of Interest

The authors declare that they have no conflicts of interest.

Acknowledgments

This research is supported by the National Natural Science Foundation of China (Grant nos. 61503218, 61773241, 61573226, and 41472260) and the Doctoral Research fund of Shandong Jianzhu University (no. XNBS1423).

References

[1] K. Dziedziech, L. Pieczonka, P. Kijanka, and W. J. Staszewski, "Enhanced nonlinear crack-wave interactions for structural damage detection based on guided ultrasonic waves," *Structural Control and Health Monitoring*, vol. 23, no. 8, pp. 1108–1120, 2016.

[2] T. Stepinski, T. Uhl, and W. Staszewski, *Advanced Structural Damage Detection: From Theory to Engineering Applications*, John Wiley & Sons, 2013.

[3] U. Dackermann, Y. Yu, E. Niederleithinger, J. Li, and H. Wiggenhauser, "Condition assessment of foundation piles and utility poles based on guided wave propagation using a network of tactile transducers and support vector machines," *Sensors*, vol. 17, no. 12, p. 2938, 2017.

[4] T. Druet, B. Chapuis, M. Jules, G. Laffont, and E. Moulin, "Passive SHM system for corrosion detection by guided wave tomography," in *Sensors, Algorithms and Applications for Structural Health Monitoring*, IIW Collection, B. Chapuis and E. Sjerve, Eds., pp. 21–29, Springer, Cham, 2018.

[5] D. Ozevin, D. W. Greve, I. J. Oppenheim, and S. P. Pessiki, "Resonant capacitive MEMS acoustic emission transducers," *Smart Materials and Structures*, vol. 15, no. 6, pp. 1863–1871, 2006.

[6] A. Perelli, L. De Marchi, A. Marzani, and N. Speciale, "Acoustic emission localization in plates with dispersion and reverberations using sparse PZT sensors in passive mode," *Smart Materials and Structures*, vol. 21, no. 2, article 025010, 2012.

[7] L. De Marchi, N. Testoni, and A. Marzani, "Spiral-shaped piezoelectric sensors for Lamb waves direction of arrival (DoA) estimation," *Smart Materials and Structures*, vol. 27, no. 4, article 045016, 2018.

[8] Q. Zhang, Y. Zhu, X. Luo, G. Liu, and M. Han, "Acoustic emission sensor system using a chirped fiber-Bragg-grating Fabry–Perot interferometer and smart feedback control," *Optics Letters*, vol. 42, no. 3, pp. 631–634, 2017.

[9] P. A. Fomitchov and S. Krishnaswamy, "Response of a fiber Bragg grating ultrasonic sensor," *Optical Engineering*, vol. 42, no. 4, p. 956, 2003.

[10] G. Zhao, S. Li, H. Hu, Y. Zhong, and K. Li, "Impact localization on composite laminates using fiber Bragg grating sensors and a novel technique based on strain amplitude," *Optical Fiber Technology*, vol. 40, pp. 172–179, 2018.

[11] M. Kurohiji, S. Ichiriyama, N. Yamasaku et al., "A robust fiber Bragg grating hydrogen gas sensor using platinum-supported silica catalyst film," *Journal of Sensors*, vol. 2018, Article ID 5810985, 8 pages, 2018.

[12] T. Fu, Z. Zhang, Y. Liu, and J. Leng, "Development of an artificial neural network for source localization using a fiber optic acoustic emission sensor array," *Structural Health Monitoring*, vol. 14, no. 2, pp. 168–177, 2015.

[13] M. Jiang, S. Lu, Y. Sai, Q. Sui, and L. Jia, "Acoustic emission source localization technique based on least squares support vector machine by using FBG sensors," *Journal of Modern Optics*, vol. 61, no. 20, pp. 1634–1640, 2014.

[14] F. Yu, Y. Okabe, Q. Wu, and N. Shigeta, "Fiber-optic sensor-based remote acoustic emission measurement of composites," *Smart Materials and Structures*, vol. 25, no. 10, article 105033, 2016.

[15] J. Yu, F. Xu, and X. Bingsheng, "Acoustic emission tomography based on simultaneous algebraic reconstruction technique to visualize the damage source location in Q235B steel plate," *Mechanical Systems and Signal Processing*, vol. 64-65, pp. 452–464, 2015.

[16] Y. Sai, X. Zhao, D. Hou, and M. Jiang, "Acoustic emission localization based on FBG sensing network and SVR algorithm," *Photonic Sensors*, vol. 7, no. 1, pp. 48–54, 2017.

[17] C. Li, Z. Ding, D. Zhao, J. Yi, and G. Zhang, "Building energy consumption prediction: an extreme deep learning approach," *Energies*, vol. 10, no. 10, p. 1525, 2017.

[18] X. Cheng, X. Zhang, L. Zhao et al., "The application of Shuffled Frog Leaping Algorithm to Wavelet Neural Networks for acoustic emission source location," *Comptes Rendus Mecanique*, vol. 342, no. 4, pp. 229–233, 2014.

[19] H. Sadegh, A. N. Mehdi, and A. Mehdi, "Classification of acoustic emission signals generated from journal bearing at different lubrication conditions based on wavelet analysis in combination with artificial neural network and genetic algorithm," *Tribology International*, vol. 95, pp. 426–434, 2016.

[20] X. Yu, Y. Yu, M. Zhang, Y. Liao, and S. Lai, "Study on the strain and temperature sensing characteristics of FBG packaged by the copper slice," *Acta Photonica Sinica*, vol. 35, no. 9, pp. 1325–1328, 2006.

[21] S. Tao, X. Zhou, and Z. Zhang, "On fault feature extraction of a gear by complex Morlet wavelet transform and coefficient correlation," *Mechanical Science and Technology for Aerospace Engineering*, vol. 5, p. 19, 2010.

[22] N. G. Nikolaou and I. A. Antoniadis, "Demodulation of vibration signals generated by defects in rolling element bearings using complex shifted Morlet wavelets," *Mechanical Systems and Signal Processing*, vol. 16, no. 4, pp. 677–694, 2002.

[23] H. Coxeter, *Introduction to Geometry*, Wiley, New York, NY, USA, 1969.

[24] Y. Diao, Z. Lin, and M. Fu, "A barycentric coordinate based distributed localization algorithm for sensor networks," *IEEE Transactions on Signal Processing*, vol. 62, no. 18, pp. 4760–4771, 2014.

[25] J. Warren, S. Schaefer, A. N. Hirani, and M. Desbrun, "Barycentric coordinates for convex sets," *Advances in Computational Mathematics*, vol. 27, no. 3, pp. 319–338, 2007.

[26] T. Han, Z. Lin, R. Zheng, and M. Fu, "A barycentric coordinate-based approach to formation control under directed and switching sensing graphs," *IEEE transactions on cybernetics*, vol. 48, no. 4, pp. 1202–1215, 2018.

16

A Review of Wrist-Worn Wearable: Sensors, Models, and Challenges

Rasha M. Al-Eidan, Hend Al-Khalifa, and Abdul Malik Al-Salman

College of Computer and Information Sciences, King Saud University, Riyadh, Saudi Arabia

Correspondence should be addressed to Rasha M. Al-Eidan; drasha3@gmail.com

Academic Editor: Evangelos Hristoforou

Wearable technology impacts the daily life of its users. Wearable devices are defined as devices embedded within clothes, watches, or accessories. Wrist-worn devices, as a type of wearable devices, have gained popularity among other wearable devices. They allow quick access to vital information, and they are suitable for many applications. This paper presents a comprehensive survey of wearable computing as a research field and provides a systematic review of recent work specifically on wrist-worn wearables. The focus of this research is on wrist-worn wearable studies because there is a lack of systematic literature reviews related to this area. This study reviewed journal and conference articles from 2015 and 2017 with some studies from 2014 and 2018, resulting in a selection of 54 studies that met the selection criteria. The literature showed that research in wrist-worn wearables spans three domains, namely, user interface and interaction studies, user studies, and activity/affect recognition studies. Our study then concludes with challenges and open research directions.

1. Introduction

Wearable technology impacts the daily life of its users. On a daily basis, humans perform many physical and cognitive activities, such as decision-making, eating, studying, walking, and communication with others. New technologies are involved in many aspects of our lives, such as communication (through social networks) or shopping (through e-commerce websites). In 1995, a new field of research called affective computing, which considers human affects [1], was introduced by Picard.

Wearable technology, as a type of affective computing, is mainly used for activity recognition [2, 3] and feeling or affect detection [4, 5]. Wristwear device technology has been studied more recently, for example, the WearWrite system for smartwatches [6]. Tomo is an example of an ad hoc wristwear system that uses hand gesture recognition [7]. Some other studies expand the interface of commercial devices, such as [8], where the interface of a wristwear device is extended to the user's skin. In this article, we aim to provide a review of the studies based on only wrist-worn devices (WWDs).

The field of wearable computing has spawned many conferences and research groups. The Conference on Human Factors in Computing Systems, the International Symposium on Wearable Computers, and the Enterprise Wearable Technology Summit are examples of high impact conferences. Also, popular research groups exist at Carnegie Mellon, Columbia University, Georgia Tech, MIT, Bremen University, Darmstadt University, ETH Zurich, Lancaster University, University of South Australia, and NARA in Japan.

The structure of this paper is as follows: Initially, in Section 2, we provide background information on wearable computing (definitions, fields, devices, etc.). Then, in Section 3, we explain the methodology of this literature review, which includes search strategy and inclusion criteria. Section 4 presents a general overview of recent review studies of wearable computing. Section 5 focuses on experimental papers that are based on wrist-worn wearables. Sections 6 and 7 present discussion, challenges, and open directions. At the end of this review, Section 8 presents our conclusion and future work.

2. Background

Wearable computers are any devices that can be worn on the

body. There is no specific definition of wearable computers, but they can be defined by their distinct characteristics [9]. Rhodes in 1997 and Hendrik Witt in 2008 defined wearable computers by describing many of their properties, such as portability, limited capability, context awareness, operational constancy, and hands-free or limited use of hands. In 2014, Genaro Motti et al. gave a simple definition of wearable computers as body-worn devices, such as clothing and accessories, that integrate computational capabilities to provide specific features to users [10]. The term *wearables,* as well as the terms *wearable technology* and *wearable devices,* is indicative of consumer electronics technology that is based on embedded computer hardware that is built into products that are worn on the outside of one's body [11].

In 1980, Professor Steve Man built a prototype of wearable personal computer-imaging system [12]. It consists of lens, mirror, a partly silvered mirror, reflections off eyeglasses, and two antennas for communication. Over 16 years, from 1980 until 1997, his system had passed through many developments until becoming a prototype consisting of eyeglasses, a handheld control, and a computer worn in the back under the shirt.

The first report on a wearable computer was authored by Thad Starner in 1995 and was called "The Cyborgs are Coming" [9]. His concern was with wearable computer interfaces, and he identifies two main characteristics: persistence and constancy [9]. Persistence describes the permanent availability of wearable computers and the ability to use them while simultaneously performing other tasks. Constancy describes how one wearable computer can be used in every situation.

In 1998, Professor Kevin Warwick implemented a sensor embedded in the median nerves of his left arm [9]. This work has been applied to controlling a wheelchair and an artificial hand by measuring transmitted signal and creating artificial sensors through electrodes on the arm.

Wearable computing is not an independent research area; research questions from different disciplines must be raised depending on the goal of a study. As shown in Figure 1, the three significant fields that contribute to wearable computing are computer science, electrical engineering, and psychology [9]. Artificial intelligence, human-computer interaction (HCI), and hardware design are branches of computer science. HCI is related strongly to psychology, while hardware design has its roots in electrical engineering. Wearable computer interfaces are essentially related to HCI and phycology rather than electrical engineering [9].

Wearable computers are related to the internet of things (IoT). The IoT is a concept that states that everything that can be connected will be connected [13]. This evolved from both ubiquitous computing and pervasive computing [13]. Wearable devices provide this critical functionality; they are IoT devices in the sense that they are always connected to the Internet, even if they are through a device such as a phone or a tablet. Many people already own or plan to purchase wearables for fitness or medical reasons, and eventually, wearables will become essential work tools.

Wearable hardware devices include smartwatches, smart glasses, textiles (also called smart fabrics), hats, caps, shoes, socks, contact lenses, earrings, headbands, hearing aids, and jewelry, such as rings, bracelets, and necklaces [11].

Over the course of our literature review, we observed that the most investigated wearable devices are as follows:

(i) Smartwatch

(ii) Smart eyewear (e.g., smart glasses and head-mounted displays) [14]

(iii) Egocentric vision devices [15]

(iv) Light-based devices (e.g., LED) [16–24]

(v) Fabrics, textiles, and skin-based devices [25–28]

(vi) Tactile gloves [29]

(vii) Hair and nail-based devices [30]

(viii) Magnetic inputs (e.g., Google cardboard) [31, 32]

These wearable devices' main challenges are networking, power and heat, display, and mobile input. All of them should be affordable for low-income earners and small in size and consume a small amount of battery power [33]. They can be used to communicate via wireless technologies such as Wi-Fi, Bluetooth, Zigbee, and NFC [33].

3. Methodology

In this research, we review studies on wearables and focus specifically on WWD studies. In order to find the most recent and representative papers, our search strategy and inclusion criteria are as follows.

3.1. Search Strategy. This section describes our method for obtaining literature on wearable computers from books, theses, and recent journal and conference articles with a particular focus on WDDs. As can be seen in Figure 2, our search method can be divided into two categories: organization/individual-based and keyword-based. Next, we will explain each in detail.

3.1.1. Organization/Individual-Based Search. This search included university websites, conferences, research groups, and staff homepages. For university websites, we obtained the top universities in computer science as ranked by the Top Universities website and the complete university guide. From within each university website, we searched for research groups using the terms HCI, ubiquitous computing, people and technology, wearable, IoT, and so on. Then, we searched group publications, staff pages, and any related links. For example, at Georgia Tech, there is the People and Technology research group. This group develops technologies related to health care, modern society, education, and community [34]. Also, it contains "interdisciplinary teams

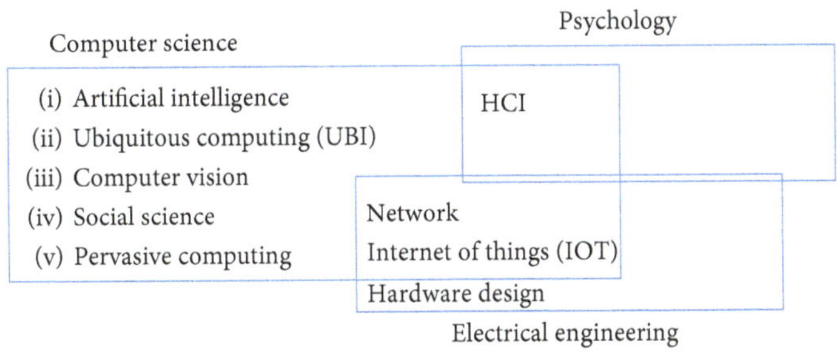

FIGURE 1: Fields of wearable computing.

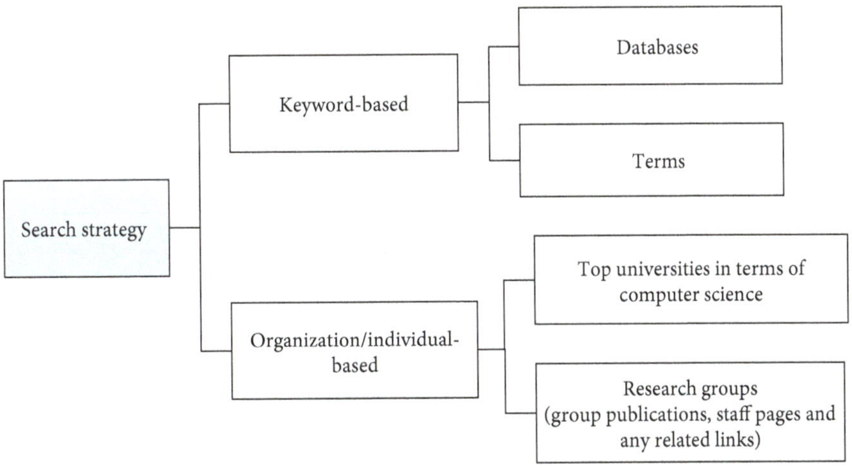

FIGURE 2: Our search strategy.

of computer scientists, system scientists, and engineers partnered with psychologists, sociologists, architects, designers, economists, medical professionals, government officials, and others to develop technologies that empower people in all walks of life" [34]. Similarly, within each conference website, we looked at all papers presented in these venues from the venues' websites or conference proceedings at ACM; for example, ISWC 2016 [35], ISWC 2015 [36], and UbiComp 2016 [37].

3.1.2. Keyword-Based Search. This search used certain search terms on popular online libraries and portals. We considered the following websites: ACM, IEEE, ScienceDirect, Springer, and Google Scholar. The following general terms were used for searching: wearable computer, wearable, smart textile, smart light, smart watch, sensor devices, mobile and wearable, child and wearable, elderly and wearable, activity recognition and wearable, and wrist-worn. Specific terms used for the health domain as an example were wearable and health, mental health, wearable and bipolar disorder, and wearable and dementia. Our systematic reviewing protocol is illustrated in Figure 2.

3.2. Inclusion and Exclusion Criteria. After the searching process, we followed certain criteria in order to select the appropriate articles for review. Figure 3 shows a flow diagram for the inclusion and exclusion criteria based on the PRISMA statement.

This study is aimed at reviewing recent work; therefore, we only searched for journal and conference articles published no later than 2017, although there are some small studies from 2014 and 2018 included. Once the duplicated records were removed, we screened each paper by looking at the abstract and quickly reading it. Then, we grouped the papers based on their type: review paper or experimental paper.

Review papers included all papers that review wearables under any domain. We discuss this classification in detail in the next section. After we found the relevant review papers, we observed that few of them were based on only wrist-worn wearables.

Therefore, for experimental papers, we excluded any papers that did not discuss wrist-worn wearables, such as ones on textiles or eyeglasses only. Finally, we read each article in full and classified them based on the classification schema that we introduce in the next section. Figure 4 shows how we clustered papers in our systematic review. In the first phase, we grouped papers in terms of their type. Then, we further classified each paper based on study domain or topic.

4. Review Papers

After reviewing the literature on wearable computing, we found that many studies are from different disciplines. For the most recent studies from 2015 to 2018, we summarized thirteen of them, as shown in Table 1, which gives a

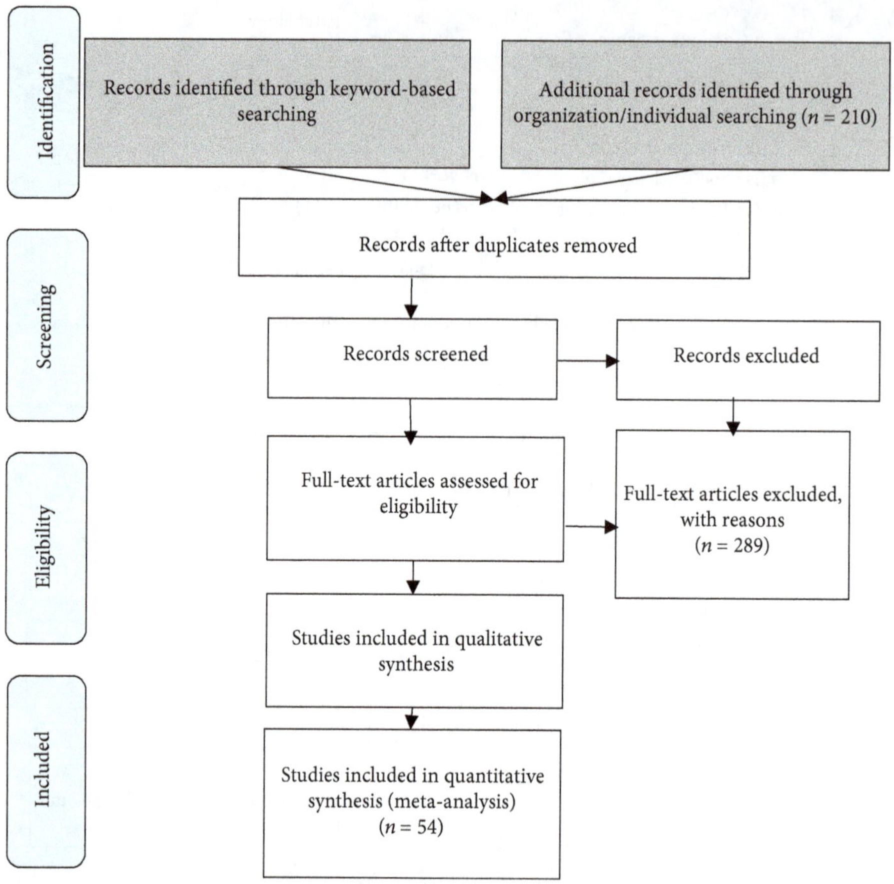

FIGURE 3: PRISMA flow diagram for our systematic review.

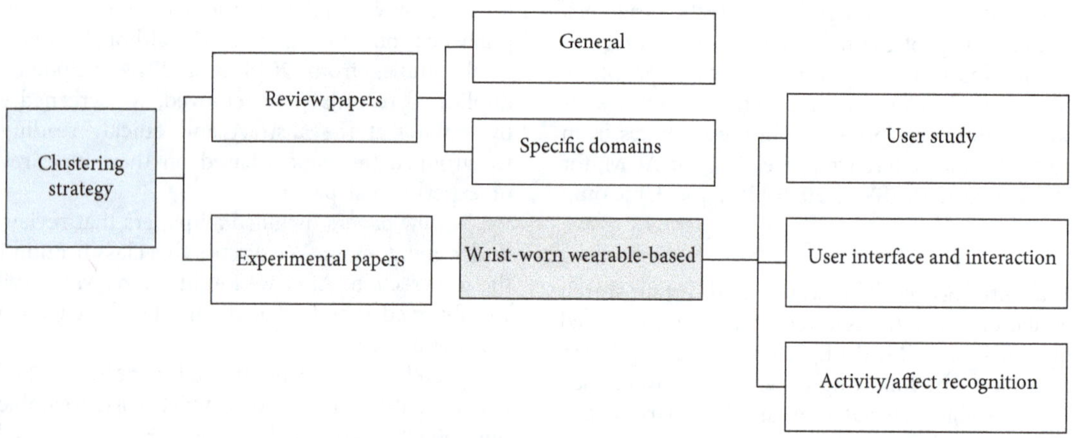

FIGURE 4: Our clustering strategy.

comparison of studies according to year, number of studies, domain, and outcomes.

We can see from Table 1 and Figure 5 that the two main domains of wearable review studies are health and activity recognition. Few studies review wearables for the education domain [46] or biometric recognition systems. Each study can be described as having one of two approaches: general or specific.

Regarding the health domain, a specific health approach is one that targets a specific chronic disease, specific people or medical specialty such as rehabilitation, impairment, Parkinson's disease [45], and Ambient Assistance Living systems (AALS) in elderly people [15]. In terms of activity recognition studies, they are either on wearable activity recognition in general [42] or on a specific kind of activity recognition, such as activity trackers [43].

In contrast, there is only one study that reviews the wearables in general [44] over all domains. This study reviews the current trend of wearable technology and found that wrist-worn devices have gathered much attention recently. To the

Table 1: Wearable computing review studies.

Study	Domain	Number of reviewed studies	Objectives and outcomes
2015 [38]	Health (rehabilitation and impairment)	Not available (N/A)	(i) The review found that wearable haptic devices are implemented for different clinical applications including rehabilitation, prosthetics, vestibular loss, osteoarthritis, vision loss, and hearing loss. (ii) Need development of haptic wearables based on clinical needs, multimodal haptic displays, low battery requirements, and long-term usage.
2015 [39]	General health	N/A	(i) Review of recent developments and applications of low-power technologies in wearable telecare and telehealth systems; divided different approaches into hardware-based approaches and firmware-based approaches. (ii) Need first to realize these systems in the wild, and then increase power efficiency. (iii) Low-power technologies will benefit people's daily lives.
2015 [40]	General health	N/A	(i) It discusses opportunities and challenges of wearable applications of health care and behavior changes particularly.
2015 [41]	General health	N/A	(i) Overview of current methods used within wearable applications to monitor and support positive health and wellbeing within an individual. (ii) Highlight issues and challenges outlined by previous studies and describe the future focuses of work.
2015 [42]	Activity	N/A	(i) Need to develop lightweight physiological sensors to have comfortable wearable devices that enable monitoring of different ranges of activities of inhabitants.
2015 [43]	Activity tracker	22	(i) A systematic review of 22 studies to evaluate validity and reliability of popular consumer wearable activity trackers (Fitbit and Jawbone). (ii) Determine trackers' ability to predict steps, distance, physical activity, energy expenditure, and sleep. (iii) Results: higher validity of steps, few studies on distance and physical activity, and lower validity for energy expenditure and sleep. High interdevice reliability for steps, distance, energy expenditure, and sleep for certain Fitbit models.
2016 [44]	General	793 historical and 103 current	(i) Two-phase survey of the application space of wearable technology by assessing its applications observed in research or industrial activities within two time periods: (a) Historical (up to 2014) (b) Current (2014-2015) (ii) Explore and assess product types, application categories, and the availability of wearable application at the body surface. (iii) Consider the differences in product price and gender. (iv) Discuss the effects of changes between the two time periods. (v) Wrist-worn devices that stand out from the current trends.
2016 [15]	Elderly people	133	(i) Explore frameworks and sensor systems of AALS relative to care and clinical systems. (ii) Most systems focused on activity monitoring for helping instance risks only. (iii) Lack of long-term care systems that must add the environmental factors for analytics and decision-making. (iv) Need to further explore the distributed storage and access to wearable devices and sensors. (v) Take account of social issues: acceptability and usability. (vi) Need to consider privacy and cybersecurity issues.
2016 [45]	Health (Parkinson's disease)	113	(i) Use accelerometer and gyroscope data. (ii) Improvement of battery life, movement sensors, and information technology to create a long-use clinical device.
2016 [46]	Education	N/A	(i) Require more research to know the needs of wearable technology for education.
2016 [47]	Biometric recognition	N/A	(i) Review and give a categorization of wearable sensors useful for capturing biometric signals. (ii) Computational cost of the different signal processing techniques. (iii) Review and classify the recent studies in the field of wearable biometrics.
2017 [48]	Wearable haptic	N/A	(i) Review the wearable haptic systems for only fingertip and hand. (ii) Summarize the main characteristics of these systems and discuss the main challenges in developments.

Table 1: Continued.

Study	Domain	Number of reviewed studies	Objectives and outcomes
2018 [49]	Elderly people	13	(i) Provide a framework for fall detection assessment system that focuses on three factors: sensor placement, task and feature category. (ii) Summarize the trends of wearable inertial sensors features and provide statistical analysis and meta-analysis for these features.

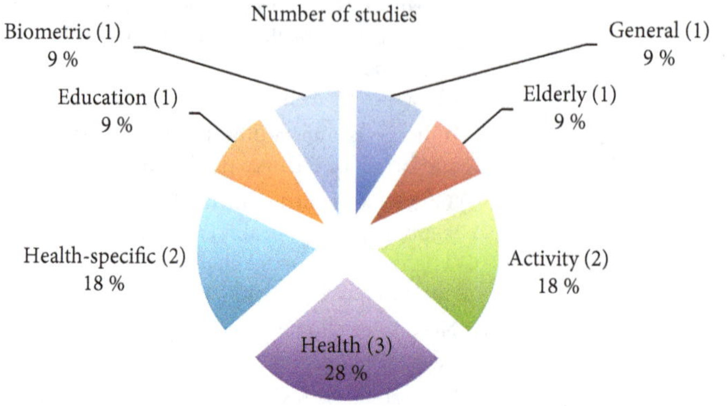

Figure 5: Percentages of review papers based on their domain.

best of our knowledge, there are no other studies that review wearable wrist-worn devices. Therefore, the next section focuses on wrist-worn wearable studies.

5. Wrist-Worn Wearable Studies

WWDs have gained more popularity than other wearables [44]. They allow quick access and are more suitable for many applications than other wearables. WWDs can be categorized as commercial or ad hoc devices. Commercial devices have three types: smart watch, fitness tracker, and armband [50]. Each of them has many applications from gesture recognition to authentication [51].

The literature on WWDs can be categorized into three types based on its goal: user interface and interaction studies, user studies, and activity/affect recognition studies. Next, we will explain each category in detail.

5.1. User Interface and Interaction Studies. Studies [52–55] in this category are aimed at supporting the interaction of users with WWDs. They depend on a wrist-worn interface (WWI), either for input or output techniques.

There are two main challenges ahead for WWIs: their physical limitations and the contexts of their use. In order to overcome these drawbacks, two novel wrist-worn interaction paradigms were proposed by Motti and Caine in 2015 [56]. The first paradigm is the microinteractions that enable users to complete tasks in less than four seconds and attempting to have many smaller tasks rather than one big task to minimize the cognition and attention required. Examples of these microinteractions are audio, gesture, graphics, tactile, and vibratory wrist-worn interfaces. The second paradigm is the multidimensional graphical user interface for both input and output. For example, make the virtual extension to a graphical user interface. The main user interface recommendations by this study are very brief text, display content or navigation, and only using short actions to complete tasks.

Table 2 gives a comparison of user interface–based studies regarding their goal, sensor used, devices, subjects, algorithms, interactions, and their outcome.

5.2. User Studies. These studies [51, 57–62] focus on understanding several issues related to WWDs from users' viewpoints and concerns. Therefore, they analyze users' reviews, answers, and usage in order to obtain different recommendations and limitations. As a result, the recent methods in WWD user studies can be categorized as one of the following:

(i) Review analysis [51, 58]

(ii) Online survey [57, 60]

(iii) Interview [59, 61]

(iv) Record of usage [59]

Table 3 shows a comparison of these studies.

5.3. Activity and Affect Recognition Studies. Studies of this type are aimed at providing extra abilities to WWDs so they can recognize different activities or affects by proposing different algorithms and using different sensors.

The aspects that should be considered to develop WWDs with activity or affect recognition are sensors and devices, modeling techniques, testing duration, sampling rate,

Table 2: Comparisons of user interface-based studies for WWDs.

Study	Application	Sensors	Devices	Subjects (M : F)	Method	Interactions	Outcome
2016 [52]	WatchOut New three gesture families	Inertial measurement unit (IMU) sensors with 200–250 Hz frequency	LG Urbane smartwatch and Sony SmartWatch 3	7 to build 2 classification models test: 12 subjects (8 : 4) and 6698 gesture samples collected.	SVM with 192 features	Side, bezel, and band of the watch.	88.7% to 99.4%
2016 [53]	Watch MI Augments normal touch input on a smartwatch	IMU with sensor fusion composed of an Invensense M651 6-axis accelerometer and gyroscope, and an Asahi Kasei AK8963 3-axis compass sensor with 100 Hz frequency	LG Urbane smartwatch	User study: 1152 valid trials (12 participants × 24 regions × 4 repetitions). Experiment: 12 volunteers (9 : 3) (3 left-handed) aged (20–36) (M: 25.8, SD: 5.2).	Analyzes in real-time	Omni-direction pressure touch, bi-direction twist, and omni-direction panning.	98.4%
2016 [54]	Circular selection proposes a list selection user interface designed for small round touchscreens.	Touchscreen	Android Wear platform displayed on a Motorola Moto 360 smartwatch	User study 1: 24 User study 2: 15	—	Circular selection list with 3 selection methods of fixed ring, movable ring, and jump back ring.	Outperforms traditional smartwatch list interfaces for user preference and task completion time (66% large lists and 45% small lists)
2016 [55]	Whoosh Uses nonvoice acoustic input for microinteractions on smartwatches	48 kHz frequency	LG G Watch (Android), single monophonic microphone, and Motorola Droid Turbo (Android) smartphone to explore multidevice interaction	8 participants in the laboratory and 4 participants in-the-wild	SVM: total of 52 features + 26 features based on the deltas of the MFCC coefficients	Blow, sip-and-puff, and directional air swipes "sip" on the watch and "puff" on the phone. (A) short blow, (B) double blow, (C) long blow, (D) swipe up, (E) swipe down, (F) clockwise blow, (G) shoosh, (H) open exhale, and (I)-(J) sip-and-puff.	Unmodified watch was 90.5% with a single classifier - Instrumental watch case with 14 additional interactions was 91.3% with per-user cross-validation

TABLE 3: Comparison of user studies of WWDs.

s	Goal	Devices	Dataset	Method	Outcome
[57] 2015	Study 1: examine usage behaviors Study 2: analyze battery usage data	Smartwatch	Study 1: 59 smartwatch users Study 2: 17 Android Wear smartwatch users	Online survey	(1) Many users are satisfied with current battery life. (2) The drain rate of the smartwatch battery is relatively low compared to that of the smartphone even with very frequent interactions. (3) Users usually recharge their smartwatch once a day.
[51] 2015	Understand the users' concerns regarding the interaction with WWDs	11 devices, including fitness trackers, armbands, and smartwatches:	1349 comments between May 2014 and November 2014 from 59 online sources (e-commerce websites and company forums)	Analysis of online reviews	297 key users' concerns emerged from a bottom-up content analysis involving both the input and the output of data. 10 design recommendations that can aid to improve the interaction design in novel WWDs.
[58] 2015	How the context of use impacts the user experience and interaction with WWDs	10 popular WWDs:	545 comments between 2010 and 2016 from Amazon	Analysis of online comments A qualitative analysis (coding) was combined with a quantitative approach (frequencies of occurrences)	Analysis of 31 interaction problems related to the users' contexts (platform, computational, and technological issues). Suggests that most problems are classified as significant or catastrophic, leading to both the users' frustration and task interruptions. Propose design implications: customization, adaptation, and personalization are essential in the UX design process for WWDs.
[59] 2016	How smartwatches are used, what for, and in what contexts.	Two wearable cameras and Apple watch	34 days 12 participants of ages 23–36	Recordings and interviews: capture and analysis using wearable cameras to record daily use of smartwatches using a small "sensor bag" to capture the users' body and wrist.	1009 watch uses. Most prominent uses for the watch in turn: timekeeping, notifications, activity tracking, and applications (third-party apps: Instagram, Twitter, and Nike Plus).
[60] 2016	Study of real-life experiences with three wearable activity trackers	Fitbit, Jawbone Up, and Nike+ FuelBand	133 responses by activity trackers in the US (median 30 years old, 35% female) on Amazon Mechanical Turk	Online survey	Revealed a user's experience derived from the needs of physical thriving or relatedness.
[61] 2015	Explore current smartwatch use	2 Samsung Galaxy Gears, 2 Pebbles, and a Moto 36	5 from the San Francisco - 2 females and 3 males (ages 18–51)	Semistructured interview	Should take as much care in designing a beautiful app that will be "worn" as device.
[62] 2015	Effectiveness of a smartphone application and wearable device for weight loss	Standard diet intervention, smartphone application, and wearable device	70 primary care patients aged 18 or older, Control group, $n = 35$ Experimental group, $n = 35$ Time: 12 months	Experimental group and control group	Change in body weight at 3, 6, and 12 months in the experimental group, compared with the control group. Effectiveness of most popular free apps and wearable devices for weight loss.

Table 4: Sensors used for activity recognition.

Sensor types	Activity studies
Accelerometer	[2, 3]
3-Axis accelerometer	[64–67]
Accelerometer and geomagnetic	[68]
Inertial measurement units: (i) Accelerometers (ii) Gyroscopes (iii) Magnetometers (sometimes)	[69]
Electric potential sensor	[70]
Multimodal sensors: (i) Activity (accelerometer and gyroscope) (ii) Ambient environment (temperature, atmospheric pressure, and humidity) (iii) Location context (Bluetooth message reception)	[63, 71]
Piezoresistive sensors and triaxial accelerometers	[72]
Ambient light	[16]
Heart rate (HR)	[67]

Table 5: Signals used for affective recognition.

Measurements	Affect studies
Body movement	[4, 5]
Prior physical activity (steps and workouts)	[63]
Electrodermal activity (EDA) = galvanic skin response (GSR) = skin conductivity (SC)	[5, 16, 73–75]
Electrocardiogram (ECG)	[76–78]
Body/skin temperature	[5, 16, 79]
Heart rate (HR)	[5, 16, 63, 77, 79–81]
Heart rate variability (HRV)	[5]
Pulse	[74]
Respiration (RSP)	[5, 75, 82]
Electromyogram (EMG)	[75, 78]
Electroencephalogram (EEG)	[74, 76, 81]
Electrooculogram (EOG), magnetoencephalography (MEG), and near-infrared facial videos	[78]
Photoplethysmogram (PPG)	[73]
Self-reports	[74, 78, 81, 83, 84]
Location	[63, 74, 83] [78]
Time	[83] [78]
Call	[78]
SMS	[78]
Semantic description	[78]
Ambient temperature, atmospheric pressure, humidity, light, posture, social interaction, and type of interaction	[5]

elicitation methods, extracted features, and experimental setup; next, we will discuss these aspects in more detail.

5.3.1. Sensors and Devices. The first step in recognizing activities or affects is to select the appropriate sensors and signals for measuring them. Sensing data can be classified as either direct or indirect.

(i) Direct sensing means, "tracking the parameters that are related to the human subject themselves" [15]. Examples of direct sensing are sound capture, video camera, motion sensors, and wearable body sensors

(ii) Indirect sensing "focuses on identifying environmental conditions and spatial features" [15]

For direct sensing, ambient intelligence techniques can be used to embed the sensing data into the environment. Such techniques can be classified as remote, mobile, or wearable sensing [63].

(i) *Remote sensing* is used for visual analysis such as a webcam for recognizing facial expression or blood flow under the skin. The main advantage of remote sensing is that it is easy to apply to the community without the need for any mobile or wearable devices [15]. However, there are disadvantages related to the inability to sense data away from a desk or in a remote area [63]

(ii) *Mobile sensing* collects data from mobile phones

(iii) *Wearable sensing* employs wearable devices to sense data from close to the body. Therefore, the availability of wearable sensing devices has increased more than mobile sensing devices. Any wearable sensor consists of three main components: sensor, processor, and display [42]. Wearable devices capture the sensor's data and send it to the processor. Then, any actions are output through the display unit. If the sensor uses wireless technology, the sensor data

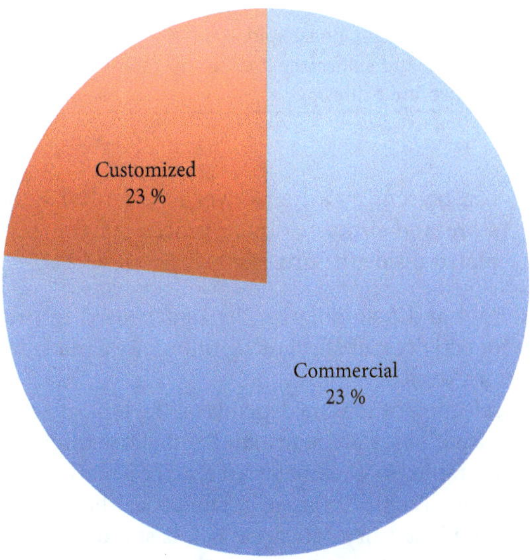

Figure 6: Percentage of using WWDs: customized (7 devices) or commercial (23 devices).

Table 6: Devices used for activity recognition.

Study	Activities	WWD	Phones
[64]	Falling	Shimmer platform with triaxis accelerometer MMA7461	—
[65]	Step	Customized WWD, for comparison: Samsung Gear S, Motorola Moto 360, SKT Smart Band, and Xiaomi MI Band	iPhone 6 for comparison
[69]	Step	SparkFun 9DoF Razor IMU	Smartphone for data registering
[71]	22 complex fine-grained activity contexts	Customized WWD	Samsung Galaxy S4 for data collection, Bluetooth beacon location tags in the physical environment
[16]	Two computing activities: (i) Keystrokes (ii) Web browsing	Moto 360 smartwatch Android 5.1	Nexus 5X smartphone Android 6.0, Dell XPS laptop, 15-inch screen, and Linux Debian 8
[68]	Holding mobile phone on left or right	Galaxy Gear Live smartwatch	Samsung Galaxy Note 4
[2]	Cardiopulmonary resuscitation (CPR), 30/2 (30 compressions 2 rescue breaths)	LG G Watch R Smart	—
[70]	Two habitual movements: (i) Hair touch (ii) Restless leg movement	Data-logging platform and the electric potential sensor extension board. Can be integrated to smartwatch	—
[3]	Two eating activity: (i) Hand-to-mouth gestures when eating (ii) Unique head motions during chewing, chewing range: 1-2.5 Hz	Pebble Watch, Google Glass, preinstalled app developed to collect accelerometer data	—
[66]	4 ambulation activities: walking, standing, sitting, and lying	Smartwatch, a fitness band or a clip on sensor	—
[6]	Useful computing tasks: writing a research paper	Smartwatch	Mobile and Google server
[72]	Mood recognition	PSYCHE wearable monitoring platform	—
[67]	Sitting, standing, household activities, and stationary cycling with two intensities	A PulseOn CLOUD PPG-based heart rate monitoring wristband (PulseOn, Espoo, Finland) with an embedded triaxial accelerometer	One Mobile phone

can be sent by a transceiver to a central station to store and process, or the processing can be completed inside the processor

Tables 4 and 5 summarize the signals used in previous studies for activity and affect recognition by wearable sensing (wrist-worn devices only).

The selection of sensors depends on the type of activity to be recognized. There are many studies that use only one sensor. For example, accelerometer, electric potential, and ambient light have each been used in studies as the only sensor type. The multisensory data approach is still used if the aim is to recognize multiple activities at the same time. Choosing the best wearable devices is important to capture the desired signals in an accurate, comfortable, and affordable way. WWDs can be classified as we mentioned before to commercial or customized. As shown in Figure 6, the percentage of using the commercial WWDs was 77% larger than using the customized devices (23%).

Tables 6 and 7 list the WWDs used in activity and affect recognition studies in addition to information about smartphones if they were used in the study. Furthermore, Table 8 lists the commercial WWD name, picture, price, their goal affect or activity, and sensors.

5.3.2. Modeling Techniques. The main step of a recognition system is classification. There are many classification methods used for activity recognition systems of WWDs; we can divide them into two approaches: machine learning-based or threshold-based. The most common and accurate machine learning algorithm is the support vector machine (SVM). Naïve Bayes and decision tree J48 are also popular to use. A deep neural network was used in only one WWD activity recognition system and gave promising results. While there are many machine learning–based studies, threshold-based studies have been conducted many times for classification. Table 9 lists the modeling techniques used in activity or affect recognition studies. Figure 7 illustrates

TABLE 7: Devices used for affective recognition.

Study	WWD	Phones
[4]	Customized on right wrist and ankle	—
[80]	Mio LINK wristband on left wrist	Motorola Moto G (2nd generation)
[77]	Shimmer 2R with flour electrocardiogram with sampling rate of 150 Hz	Smartphone to collect data
[83]	Pebble watch with sampling rate of 25 Hz	Smartphone
[73]	Customized	—
[76]	Customized, 3 electrocardiogram electrodes (right and left wrist and left leg), and electroencephalogram with 19 channels	—
[79]	Toshiba Silmee TM Bar Type (left wrist) and Toshiba Silmee TM W20/W21 wristbands (right wrist)	App on smartphone to collect data
[75]	Customized: SC with 2 electrodes (the index and ring fingers), RSP (placed tight enough in the abdominal area above the navel), and 2-channel EEG	
[78]	Customized	—
[81]	Apple Watch Sports Edition (heart rate) with sampling rate of 12 samples per minutes	iPad to play arcade and puzzle games.
[63]	Smartwatch	Smartphone
[74]	Wrist device, electrocardiogram device, GSR, and Pulse sensor	Smartphone
[82]	E4 wristband	—
[85]	Affectiva's Q sensor	—
[84]	Google Glass and Samsung Galaxy Gear smartwatch with constant sampling rate of 256 Hz.	Samsung Galaxy S4 smartphone
[5]	Samsung Galaxy Gear smartwatch, Affectiva Q smartwatch, narrative clip (torso), BioPatch (torso), and Google Glass	Samsung Galaxy S4 smartphone

clearly this comparison in line chart for both activity and affect recognition together. We can see that affect recognition studies preferred to use a different kind of machine learning algorithms, while most activity recognition studies preferred to use the threshold-based algorithms.

5.3.3. Testing Duration. Testing duration is essential in evaluating the performance of a system. Table 10 shows the time of test for each study. The time of testing the WWDs per user for each study ranges from two seconds up to 14 weeks. Usually, the testing time for activity recognition is in seconds or minutes for each user. In contrast, most health diagnosis systems for a specific disease, such as bipolar disorder [72], require more time to test to provide accurate results. In contrast, affect studies range from minutes to two months. Affect recognition systems take a longer time to test than activity recognition systems.

5.3.4. Extracted Features. WWD studies depend on sensory data. After the preprocessing phase, they extract the desired feature from the raw sensor data. Extracted features from sensor data have three main types: time domain, frequency domain, and district domain [86]. Time domain features have lower computational cost than those of the frequency domain [87]. The mean feature from the time domain is commonly used for WWD activity recognition systems. In addition, the statistical measures of standard deviation, minimum, and maximum have been commonly used.

5.3.5. Sampling Rate. Sampling rate is used to determine the range of activity frequencies. Therefore, it is important to set the correct sampling rate. Accuracy, power consumption, and other selected features are factors that are affected by the sampling rate [87]. The range of sampling rates is different based on the type of activity being detected. For example, step and fall detection usually uses 20 Hz as a sampling rate. This is due to the range of human movement being in the range 0-20 Hz [87]. In addition, using a small sampling rate reduces power consumption [69]. This review found different sampling rates being used for WWD activity recognition; they ranged from 1 to 1000 Hz. Different sampling rates used, along with their related studies, are shown in Table 11 and Figure 8.

5.3.6. Elicitation Method Used for WWD Affect Recognition. In order to collect the data, an emotion must be induced by a stimulus. Table 12 shows the different kinds of elicitation methods used in previous studies. Many recent studies are aimed at collecting data naturally without any elicitation by continuously collecting data over the participants' daily lives.

5.3.7. Experimental Setup. To evaluate the performance of activity recognition systems, we need to collect representative datasets via experiments. This dataset includes a number of participants wearing the WWDs and the target activities or feelings of the subjects. The participants might be hospital patients, elderly people, students, etc. The participant demographic depends on the study domain: activity tracker, home monitoring, health, security, education, etc. Most of the affect studies applied to participants at regular daily lives and work lives. There is a lack of affect recognition studies that deal with diseases.

There are different places that an experiment can be conducted. Few experiments were applied in real-world

TABLE 8: Commercial WWDs used a barometer (baro), thermometer (thermo), gyroscope (gyro), accelerometer (accel), magnetometer (magnet), pedometer (pedo), and altimeter (alti).

Name	Picture	Price $ (name of the resource website of price)	For affect/activity recognition	Sensors
Shimmer IMU http://www.shimmersensing.com/products/shimmer3-development-kit		444 shimmer	Activity	accel, gyro, magnet, and alti,
Samsung Gear S		219 Amazon	Activity	Accel, gyro, proximity, compass, HR, baro, UV light S3: Accel, gyro, HR, baro
Motorola Moto 360 2gen Motorola Moto 360 sport	 	99 125 193 Amazon	Activity	Accel, gyro, HR Sport: Accel, alti, gyro, HR, Ambient light sensor
SKT smartband	NA	50–53	Activity	Accel, LED matrix, vibration motor, cortex M3 (ARM CPU, activity tracking algorithm)
Xiaomi MI band		23–32 Amazon not ship to SA	Activity	HR, pedo, sleep management; SMS reply, anti-loss, rigorous test function.
SparkFun 9DoF Razor IMU		49.95	Activity	Accel, gyro and magnet
Samsung Gear Live Smartwatch https://www.samsung.com/us/support/gear/gear-live-support/ https://www.gsmarena.com/samsung_gear_live-7714.php		269.99 eBay	Activity + affect	Accel, gyro, HR, compass
LG G Watch R Smart		153.59 Amazon	Activity	Accel, gyro, proximity, HR, baro

Table 8: Continued.

Name	Picture	Price $ (name of the resource website of price)	For affect/activity recognition	Sensors
Pebble Watch Black Pebble 2 + Heart Rate Smart Watch -	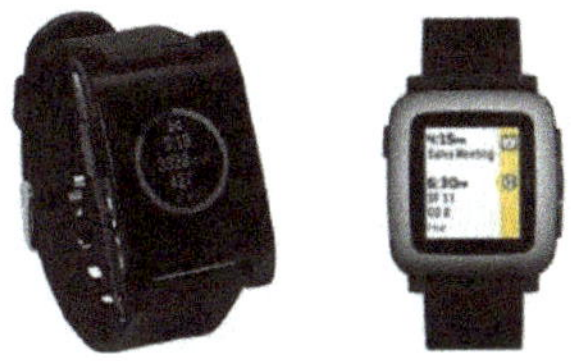	39 −89.95 Amazon	Activity + affect	Microphone Accel, Gyro, Magnet Ambient light sensor Compass

Table 9: Modeling techniques used for affective studies.

Modeling techniques	Activity studies	Affect studies
Threshold-based	[2, 16, 64, 65, 69, 70]	
SVM	[3, 66, 68]	[4, 5]
Decision tree	[66, 68]	[4, 79]
k-Nearest neighbors	[66]	[75, 79, 85]
Multilayer perceptron	[66]	[73, 75, 76]
Data mining		[63, 77, 81]
Random forest	[66]	[4]
Fuzzy logic		[73, 78]
Naïve Bayes	[66]	
C4.5	[3]	
Random tree		[4]
Regression		[83]
Linear discriminant analysis		[75]
Ontology-based		[78]
Deep learning neural network	[71]	

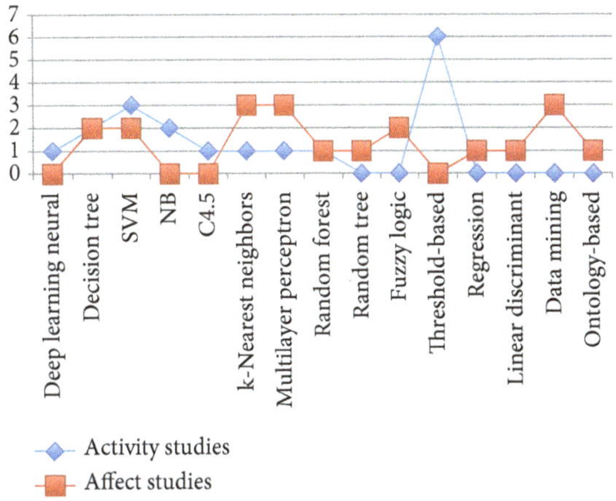

Figure 7: Comparison between modeling techniques used for either affect or activity recognition.

Table 10: Testing duration.

Testing duration	Activity study	Affect study
2 seconds	[64]	
5.5 seconds	[66]	
5 minutes		[4]
10 minutes	[68]	
45 minutes	[67, 71]	
3 days	[16]	
5 days		[5]
1 week	[6]	
Twice a week for 14 days	[72]	
30 days		[85]
75 days		[84]

conditions due to the difficulty to deal with changes and noise. Most experiments are implemented in a laboratory environment. For example, affect recognition studies have applied in different places: a quiet room, an office, a "natural" environment, and a real-life stress environment [5].

The number of participants for WWD activity recognition studies ranged from 1 to 41. In contrast, affect recognition studies had up to 123 participants.

Moreover, classes used to recognize the type and level of an affect vary between studies. There is no agreed-upon way to best categorize different kinds of emotion [1]. For affective computing researches, the best choice is whatever suits the application best [1]. More details about participant demographic are summarized in Table 13.

6. Discussion

In the previous sections, we reviewed 54 studies based on wearables with a particular focus on WWDs. The WWD studies take various directions, and thus, it is not possible to provide an absolute comparison of them in this discussion. Moreover, even studies that followed the same direction are not easily comparable. For example, within the activity recognition studies, we found that the activities being recognized varied, which hinders comparison. However, this discussion highlights and describes the significant challenges faced in studying wearables.

We found that wearable computing, in general, has been researched many times and in different domains, such as sport, health, education, and security. Most WWD studies

Table 11: Sampling rate for activity studies.

Sampling rate (Hz)	1 to 2.2	5	10	20	30	25	50	100	200	250	1000
Studies	[3]	[69]	[69]	[64, 65, 69]	[66]	[67, 69]	[69]	[69]	[71]	[68]	[70]

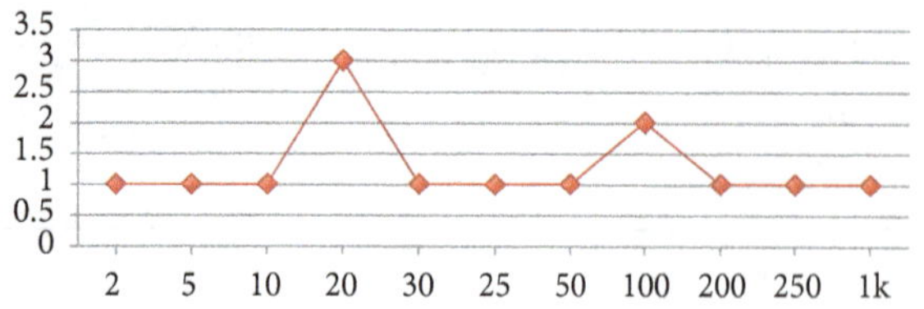

Figure 8: Sampling rate line chart (Y: number of studies, X: rates).

Table 12: Elicitation methods for affective studies.

Type	Studies
Image from the International Affective Picture System	[75, 76]
Video	[4, 73, 78, 80]
Sound	[77]
Games	[5]
None (continuous monitoring in a natural environment)	[5, 63, 74, 78, 79, 83]

were conducted in the domain of health [38–41, 45], activity tracking [43], and home monitoring [15, 49], and there is a lack of studies related to education [46], security [47], and child care.

Long-term usage is considered to be the main challenge of wearable technology [40]. This challenge can be overcome by considering the issues of battery life, user acceptance, safety, privacy, weight, and fault tolerance. Ambient intelligence is an important concept to implement in WWDs.

The recent WWD studies were divided, as shown in the previous section, based on three categories: user interface and interaction studies, user studies, and activity/affect recognition studies.

Regarding WWIs, we found many limitations that must be managed or resolved by researchers in order to improve user interface and interaction. Microinteractions and multidimensional interfaces are a new proposed solution for WWIs [56].

On the other hand, conducting a user study is very important to build the theoretical framework of WWDs as well as understanding user requirements. This would be beneficial in terms of providing sufficient ideas for the design of better interfaces and for implementing the appropriate applications.

We have noticed that many WWD researchers preferred to use one sensor for activity recognition in order to reduce the power consumption and increase WWD simplicity. In contrast, some studies used multisensory data to recognize more activities for intelligent life, such as elderly care within smart homes.

Features extracted from sensor data are varied. Although the sensitivity of features is important to recognize the target activity, many factors should be considered when selecting features themselves. These factors include battery consumption and availability of feature resources.

In the classification phase, most studies used a machine learning-based approach; however, researchers are still using threshold-based approaches in some studies [2, 16, 64, 65, 69, 70]. This is because a threshold-based approach must determine the critical points of classification, which are affected by changes in context and long usage needs. Consequently, we have seen that most studies for recognizing a single activity, such as falling [64], steps [65, 69], computing activity [16], CPR [2], and hair touch detection [70], used the threshold-based approach. On the other hand, to recognize the more complex activities that are sensitive to context, such as the activity of ambulation [66], eating [3], and mood [72], the machine learning approach was used. In addition, the machine learning approach used in the study on bipolar disorder diagnosis requires long-term usage. Machine learning techniques show promising results for user interface and activity recognition systems. Deep learning algorithms have been used recently and provided more efficient results [71].

Collection of a data is mandatory in order to train a classifier in the machine learning approach or for testing a system. The collection of data should be done in a real-world environment instead of a laboratory environment, which most studies used.

7. Challenges and Open Directions

After reviewing the literature of wearable computing, we found many critical challenges and issues. As we mentioned, the main technical challenge faced by wearable devices is long-term usage. This challenge is related to many aspects: battery life, user acceptance, safety, weight, fault tolerance, and privacy concerns. On the other hand, there are challenges related to creating a main standard of wearable systems. Furthermore, there are challenges in establishing a relationship with the commercial sector to ensure installation and process completion. In this section, we provide suggested solutions for overcoming each challenge in future work.

7.1. Weight. We could use energy harvesting technology to remove the need for a battery, which will decrease total weight.

7.2. Battery Life. This challenge could be overcome with respect to hardware, software, or interface. Hardware could also use energy harvesting technology, which extracts power from the surroundings via solar, kinetic, and electromagnetic emission energy [42]. Regarding software, we could reduce the power consumption of the system firmware, which can be categorized as event-driven, duty cycle, feature selection,

Table 13: Experimental setup for WWD studies.

Study	Goal	Dataset (M : F)	Domain
[64]	Falling	12 (3 groups of 4 people of different ages).	Elderly people living
[65]	Step	1 Commercial devices: walk on the treadmill at 4.5 km/h Prototype experiments: walk 120 seconds on the treadmill	Activity tracker (sport)
[69]	Step	1 Straight line walks of 30 steps at different paces	Activity tracker (sport)
[71]	22 Complex fine-grained activity contexts (i) Locomotive (walk indoors and run indoors); (ii) semantic (use refrigerator, clean utensil, cooking, sit and eat, use bathroom sink, standing, and talking); (iii) transitional (indoor to outdoor, outdoor to indoor, walk upstairs, and walk downstairs); and (iv) postural/relatively stationary (just stand, stand and lean on wall, lying on bed, sit on bed, sit on desk chair, lying on floor, sit on floor, lying on sofa, sit on sofa, and sit on commode).	2 two separate home environments. User 1: 22 activities, user 2: 19 activities. Every user's series of selected activities consisted of an average of 45 minutes of sensor data collection.	Home living
[16]	Keystrokes and web browsing	1 activity 1: 60 key presses per character, repeat in 3 days. Activity 2: capturing the lux readings for 10 popular websites, each one minute	Security
[68]	Holding mobile phone on left or right	24 14: control study, 10 minutes each 10: user study to receive feedback	Adaptive user interface: one-handed interaction
[2]	Cardiopulmonary resuscitation (CPR)	41 (24: 17) age (24-70, average: 37)	Health: training
[70]	Hair touch and restless leg movement	1 two different floorings: carpet and vinyl	Health: consumer care product research
[3]	Two eating activities	10 (7: 3) age (15-52, average: 28.7), four food groups: different chewing motions due to different food textures	Health: weight management
[66]	Walking, standing, sitting, and lying	3 triaxial acceleration data in the dataset of Activity Recognition Challenge [16], 5.5 seconds. 300,000 records or 2300 windows (window length is 128) for each wrist of a subject.	Home living
[6]	Writing a research paper	7 (2 users, 5 crowd workers), full weak	Education
[72]	Mood recognition	14 patients	Health: mood changes of bipolar disorder
[67]	Sitting, standing, household activities and stationary cycling with two intensities	25 healthy people	Home living
[4]	3 (neutral, happy, and angry)	123 (45: 78)	Daily living
[80]	7 (amusement, sadness, anger, fear, disgust, surprise, and neutral)	14 (5: 9) age (20-28)	Recommendation and sharing
[77]	Computed dental radiography (CDR)	40 (15: 25) average age (25: 29)	Health
[83]	8 (upset, stressed, tense, excited, happy, bored, tired, and relaxed)	18 (2: 16) (10 university students, 6 researchers/staff, 1 software engineer, 1 professor)	Daily living
[73]	16 (pride, elation, joy, satisfaction, relief, hope, interest, surprise, sadness, fear, shame, guilt, envy, disgust, contempt, and anger)	DEAP dataset	Social living
[76]	4 (calm, happy, fear, sad) Valence, Arousal, and Dominance (VAD)	12	Daily living
[79]	8 (excitement, happiness, calmness, tiredness, boredom, sadness, stress, and anger)	4	Office living

Table 13: Continued.

Study	Goal	Dataset (M : F)	Domain
[75]	VAD (2: A: +/−, 3: V: +/−/0, 5: VA: 0−/++/−+/+−/−−, 10: VAD)	20 (9: 11) Age (22–76) (average age 47.4, SD = 23.16). Right-handed, healthy, and had normal vision or corrected normal vision.	Work living
[78]	VAD A: 0 (very calm) to 4 (very aroused) V: −2 (unpleasant) to 2 (very pleasant). 7 (amusement, anger, disgust, excitement, fear, fun, and shock)	30: train DECAF dataset (VAD) 30: test 600 individual records.	Daily living
[81]	2 (frustration and satisfaction)	3 (friends and family), age (26, 24, and 44) Big data: 176 minutes of sampled data, 10,560 seconds of raw data, and electroencephalogram with 5 million lines	Daily living
[63]	N/A	N/A	Daily living
[74]	2 (mania and depression)	N/A	Mental health
[82]	Only propose	Only propose	Child care
[85]	Happy-sad	68	Daily living
[84]	VAD	15 (7 : 8)	Daily living
[5]	Low stress and high stress	15	Daily living

or sensor selection. For interface, the microinteractions used to complete the task should be done in less than 3 seconds.

7.3. Lack of Standards. This challenge makes wearable technology unstable and difficult to adapt and use. Al-Shaqi et al. said regarding this issue that "*Adaptability of different system components from sensors, communication protocol, decision support, and subject interaction method or language. Availability of standards will help system designers to integrate efforts and provide the market with the necessary devices and systems to meet the subject defined requirements.*" There is also a lack of commercial concern for establishing a relationship between academia and commercial sectors.

7.4. Safety. A novel nonlinear optimization framework has been presented to consider safety and sustainability requirements that depend on human physiology and derive system-level design parameters for wearable sensor applications. Reliability of data is another challenge. Therefore, we should use data collection that uses accurate thresholds with low or no fault tolerance. Fault tolerance of devices is also a concern in terms of resistance to impact, heat, cold, and water.

7.5. User Acceptance. This is the main property of wearable devices. User acceptance has many challenges with respect to personalization, interface, design, and data. Personalization is the ability to support a person's lifestyle, for example, allowing the user to create the form that they want. Also, the focus should be on interfaces that are accessible, simple, and easy to use. In addition, there is a need for customization and adaptivity to different settings based on user requirements, such as in ALS.

7.6. Design. Wearable devices must be designed in a special form with the following properties. Firstly, it is not very obvious and appears to others. Secondly, they should not disturb the users' daily activities, and frequent wearing must be comfortable. Finally, they must be close to the body in order to sense the required measurements accurately.

7.7. Data. It should be meaningful by displaying ambient feedback rather than showing the user lots of numerical values.

8. Conclusions

Many review studies on wearable technology have been conducted. Therefore, this paper presented a review of previous research on wearable computing studies. As a result, and to the best of our knowledge, there were no review studies on WWDs that consider many aspects. This paper discussed the different kinds of WWD studies, highlighted important issues, and suggested future works.

The next step is to attempt investigating more studies based on a specific domain or a specific topic that could be solved with wearable technology then explore and detect problems with the aid of a domain expert to determine requirements.

Conflicts of Interest

The authors declare that there is no conflict of interest regarding the publication of this paper.

Acknowledgments

The authors would like to thank the Deanship of Scientific Research for funding and supporting this research through the initiative of DSR Graduate Students Research Support (GSR). The authors also thank the Deanship of Scientific Research and RSSU at King Saud University for the technical support.

References

[1] R. W. Picard, *Affective Computing*, MIT Press, Cambridge, MA, USA, 1997.

[2] A. Gruenerbl, G. Pirkl, E. Monger, M. Gobbi, and P. Lukowicz, "Smart-watch life saver: smart-watch interactive-feedback system for improving bystander CPR," in *Proceedings of the 2015 ACM International Symposium on Wearable Computers*, pp. 19–26, Osaka, Japan, September 2015.

[3] X. Ye, G. Chen, and Y. Cao, "Automatic eating detection using head-mount and wrist-worn accelerometers," in *2015 17th International Conference on E-Health Networking, Application Services (HealthCom)*, pp. 578–581, Boston, MA, USA, October 2015.

[4] Z. Zhang, Y. Song, L. Cui, X. Liu, and T. Zhu, "Emotion recognition based on customized smart bracelet with built-in accelerometer," *PeerJ*, vol. 4, article e2258, 2016.

[5] J. Hernandez Rivera, *Towards Wearable Stress Measurement*, Massachusetts Institute of Technology, 2015.

[6] M. Nebeling, A. To, A. Guo et al., "WearWrite: crowd-assisted writing from smartwatches," in *Proceedings of the 2016 CHI Conference on Human Factors in Computing Systems*, pp. 3834–3846, San Jose, CA, USA, May 2016.

[7] Y. Zhang and C. Harrison, "Tomo: wearable, low-cost electrical impedance tomography for hand gesture recognition," in *Proceedings of the 28th Annual ACM Symposium on User Interface Software & Technology*, pp. 167–173, Charlotte, NC, USA, November 2015.

[8] Y. Zhang, J. Zhou, G. Laput, and C. Harrison, "SkinTrack: using the body as an electrical waveguide for continuous finger tracking on the skin," in *Proceedings of the 2016 CHI Conference on Human Factors in Computing Systems*, pp. 1491–1503, San Jose, California, USA, May 2016.

[9] H. Witt, *User Interfaces for Wearable Computers: Development and Evaluation*, Vieweg and Teubner, Germany, 2008.

[10] V. Genaro Motti, S. Kohn, and K. Caine, "Wearable computing: a human-centered view of key concepts, application domains, and quality factors," in *Proceedings of the 16th International Conference on Human-Computer Interaction with Mobile Devices & Services*, pp. 563-564, Toronto, ON, Canada, September 2014.

[11] W. Barfield and T. Caudell, *Fundamentals of Wearable Computers and Augumented Reality*, L. Erlbaum Associates Inc., Hillsdale, NJ, USA, 2000.

[12] S. Mann, "Wearable computing: a first step toward personal imaging," *Computer*, vol. 30, no. 2, pp. 25–32, 1997.

[13] D. C. Ruiz and A. Goransson, *Professional Android Wearables*, Wrox Press Ltd., Birmingham, UK, 1st edition, 2015.

[14] A. Nassani, H. Bai, G. Lee, and M. Billinghurst, "Tag it!: AR annotation using wearable sensors," in *SIGGRAPH Asia 2015 Mobile Graphics and Interactive Applications*, pp. 12:1–12:4, Kobe, Japan, November 2015.

[15] R. Al-Shaqi, M. Mourshed, and Y. Rezgui, "Progress in ambient assisted systems for independent living by the elderly," *Springerplus*, vol. 5, no. 1, p. 624, 2016.

[16] A. Holmes, S. Desai, and A. Nahapetian, "LuxLeak: capturing computing activity using smart device ambient light sensors," in *Proceedings of the 2nd Workshop on Experiences in the Design and Implementation of Smart Objects*, pp. 47–52, New York City, NY, USA, October 2016.

[17] N. T. Ly, R. Tscharn, J. Preßler et al., "Smart lighting in dementia care facility," in *Proceedings of the 2016 ACM International Joint Conference on Pervasive and Ubiquitous Computing: Adjunct*, pp. 1636–1639, Heidelberg, Germany, September 2016.

[18] R. Tscharn, N. Ly-Tung, D. Löffler, and J. Hurtienne, "Ambient light as spatial attention guidance in indoor environments," in *Proceedings of the 2016 ACM International Joint Conference on Pervasive and Ubiquitous Computing: Adjunct*, pp. 1627–1630, Heidelberg, Germany, September 2016.

[19] H. Yoon, S.-H. Park, and K.-T. Lee, "Lightful user interaction on smart wearables," *Personal and Ubiquitous Computing*, vol. 20, no. 6, pp. 973–984, 2016.

[20] N. Zhao and J. A. Paradiso, "HALO: wearable lighting," in *Adjunct Proceedings of the 2015 ACM International Joint Conference on Pervasive and Ubiquitous Computing and Proceedings of the 2015 ACM International Symposium on Wearable Computers*, pp. 601–606, Osaka, Japan, September 2015.

[21] D. Aliakseyeu and J. Mason, "Tap sensor: evaluating a new physical user interface for connected lighting," in *Proceedings of the 2016 ACM International Joint Conference on Pervasive and Ubiquitous Computing: Adjunct*, pp. 1614–1619, Heidelberg, Germany, September 2016.

[22] Y. Choi, K. Yoo, S. J. Kang, B. Seo, and S. K. Kim, "Development of a low-cost wearable sensing glove with multiple inertial sensors and a light and fast orientation estimation algorithm," *The Journal of Supercomputing*, vol. 74, no. 8, pp. 3639–3652, 2018.

[23] Z. Cochran, B. Tomlinson, D.-W. Chen, and K. Patel, "LightWeight: wearable resistance visualizer for rehabilitation," in *Proceedings of the Adjunct Publication of the 27th Annual ACM Symposium on User Interface Software and Technology*, pp. 101-102, Honolulu, Hawaii, USA, October 2014.

[24] M. Caon, L. Angelini, O. A. Khaled, D. Aliakseyeu, J. Mason, and E. Mugellini, "Tangible interaction with light in the IoT," in *Proceedings of the 2016 ACM International Joint Conference on Pervasive and Ubiquitous Computing: Adjunct*, pp. 1646–1651, Heidelberg, Germany, September 2016.

[25] J. Haladjian, K. Bredies, and B. Brügge, "Interactex: an integrated development environment for smart textiles," in *Proceedings of the 2016 ACM International Symposium on Wearable Computers*, pp. 8–15, New York, NY, USA, September 2016.

[26] G. Valenza, L. Citi, C. Gentili, A. Lanatá, E. P. Scilingo, and R. Barbieri, "Characterization of depressive states in bipolar patients using wearable textile technology and instantaneous heart rate variability assessment," *IEEE Journal of Biomedical and Health Informatics*, vol. 19, no. 1, pp. 263–274, 2015.

[27] K. Ueda, T. Terada, and M. Tsukamoto, "Input interface using wrinkles on clothes," in *Proceedings of the 2016 ACM International Symposium on Wearable Computers*, pp. 56-57, New York, NY, USA, September 2016.

[28] M. Chen, Y. Ma, J. Song, C.-F. Lai, and B. Hu, "Smart clothing: connecting human with clouds and big data for sustainable health monitoring," *Mobile Networks and Applications*, vol. 21, no. 5, pp. 825–845, 2016.

[29] A. Jylhä, Y.-T. Hsieh, V. Orso, S. Andolina, L. Gamberini, and G. Jacucci, "A wearable multimodal interface for exploring urban points of interest," in *Proceedings of the 2015 ACM on International Conference on Multimodal Interaction*, pp. 175–182, New York, NY, USA, November 2015.

[30] K. Vega and H. Fuks, *Beauty Technology: Designing Seamless Interfaces for Wearable Computing*, Springer International Publishing, 2016.

[31] C. Lee, *Simple Magnetic Trigger Hack for Google Cardboard*, 2016, https://www.youtube.com/watch?v=1UA7I8uuA9M.Y.

[32] K. Lyons, "2D input for virtual reality enclosures with magnetic field sensing," in *Proceedings of the 2016 ACM International Symposium on Wearable Computers*, pp. 176–183, New York, NY, USA, September 2016.

[33] A. Nag and S. C. Mukhopadhyay, "Wearable electronics sensors: current status and future opportunities," in *Wearable Electronics Sensors. Smart Sensors, Measurement and Instrumentation*, S. Mukhopadhyay, Ed., vol. 15, pp. 1–35, Springer, Cham, 2015.

[34] "Georgia Institute of Technology," 2016, http://ipat.gatech.edu/.

[35] *Proceedings of the 2016 ACM International Symposium on Wearable Computers, ISWC'16, ACM, 201*, New York, NY, USA, 2016, https://dl.acm.org/citation.cfm?id=2971763&picked=prox.

[36] *Proceedings of the 2015 ACM International Symposium on Wearable Computers, ISWC'15, ACM, 201*, New York, NY, USA, 2015, https://dl.acm.org/citation.cfm?id=2802083.

[37] *Proceedings of the 2016 ACM International Joint Conference on Pervasive and Ubiquitous Computing, UbiComp'16*, New York, NY, USA, 2016, https://dl.acm.org/citation.cfm?id=2971648.

[38] P. B. Shull and D. D. Damian, "Haptic wearables as sensory replacement, sensory augmentation and trainer - a review," *Journal of NeuroEngineering and Rehabilitation*, vol. 12, no. 1, p. 59, 2015.

[39] C. Wang, W. Lu, M. R. Narayanan, S. J. Redmond, and N. H. Lovell, "Low-power technologies for wearable telecare and telehealth systems: a review," *Biomedical Engineering Letters*, vol. 5, no. 1, pp. 1–9, 2015.

[40] V. Genaro Motti and K. Caine, "An overview of wearable applications for healthcare," in *Proceedings of the 2015 ACM International Joint Conference on Pervasive and Ubiquitous Computing and Proceedings of the 2015 ACM International Symposium on Wearable Computers - UbiComp '15*, pp. 635–641, New York, NY, USA, September 2015.

[41] K. Hänsel, N. Wilde, H. Haddadi, and A. Alomainy, "Challenges with current wearable technology in monitoring health data and providing positive behavioural support'," in *Proceedings of the 5th EAI International Conference on Wireless Mobile Communication and Healthcare, ICST*, pp. 158–161, Brussels, Belgium, December 2015.

[42] S. C. Mukhopadhyay, "Wearable sensors for human activity monitoring: a review," *IEEE Sensors Journal*, vol. 15, no. 3, pp. 1321–1330, 2015.

[43] K. R. Evenson, M. M. Goto, and R. D. Furberg, "Systematic review of the validity and reliability of consumer-wearable activity trackers," *International Journal of Behavioral Nutrition and Physical Activity*, vol. 12, no. 1, p. 159, 2015.

[44] M. E. Berglund, J. Duvall, and L. E. Dunne, "A survey of the historical scope and current trends of wearable technology applications," in *ISWC '16 Proceedings of the 2016 ACM International Symposium on Wearable Computers*, pp. 40–43, New York, NY, USA, September 2016.

[45] C. Ossig, A. Antonini, C. Buhmann et al., "Wearable sensor-based objective assessment of motor symptoms in Parkinson's disease," *Journal of Neural Transmission*, vol. 123, no. 1, pp. 57–64, 2016.

[46] B. Sandall, "Wearable technology and schools: where are we and where do we go from here?," *Journal of Curriculum, Teaching, Learning and Leadership in Education*, vol. 1, no. 1, 2016.

[47] J. Blasco, T. M. Chen, J. Tapiador, and P. Peris-Lopez, "A survey of wearable biometric recognition systems," *ACM Computing Surveys*, vol. 49, no. 3, pp. 1–35, 2016.

[48] C. Pacchierotti, S. Sinclair, M. Solazzi, A. Frisoli, V. Hayward, and D. Prattichizzo, "Wearable haptic systems for the fingertip and the hand: taxonomy, review, and perspectives," *IEEE Transactions on Haptics*, vol. 10, no. 4, pp. 580–600, 2017.

[49] L. Montesinos, R. Castaldo, and L. Pecchia, "Wearable inertial sensors for fall risk assessment and prediction in older adults: a systematic review and meta-analysis," *IEEE Transactions on Neural Systems and Rehabilitation Engineering*, vol. 26, no. 3, pp. 573–582, 2018.

[50] V. G. Motti and K. Caine, "Smart wearables or dumb wearables?: understanding how context impacts the UX in wrist worn interaction," in *SIGDOC '16 Proceedings of the 34th ACM International Conference on the Design of Communication*, pp. 10:1–10:10, New York, NY, USA, September 2016.

[51] B. Lowens, V. Motti, and K. Caine, "Design recommendations to improve the user interaction with wrist worn devices," in *2015 IEEE International Conference on Pervasive Computing and Communication Workshops (PerCom Workshops)*, pp. 562–567, St. Louis, MO, USA, March 2015.

[52] C. Zhang, J. Yang, C. Southern, T. E. Starner, and G. D. Abowd, "WatchOut: extending interactions on a smartwatch with inertial sensing," in *ISWC '16 Proceedings of the 2016 ACM International Symposium on Wearable Computers*, pp. 136–143, New York, NY, USA, September 2016.

[53] H.-S. Yeo, J. Lee, A. Bianchi, and A. Quigley, "WatchMI: applications of watch movement input on unmodified smartwatches," in *MobileHCI '16 Proceedings of the 18th International Conference on Human-Computer Interaction with Mobile Devices and Services Adjunct*, pp. 594–598, New York, NY, USA, September 2016.

[54] K. Plaumann, M. Müller, and E. Rukzio, "CircularSelection: optimizing list selection for smartwatches," in *Proceedings of the 2016 ACM International Symposium on Wearable Computers*, pp. 128–135, New York, NY, USA, September 2016.

[55] G. Reyes, W. Keith Edwards, D. Zhang et al., "Whoosh: non-voice acoustics for low-cost, hands-free, and rapid input on smartwatches," in *Proceedings of the 2016 ACM International Symposium on Wearable Computers*, pp. 120–127, New York, NY, USA, September 2016.

[56] V. G. Motti and K. Caine, "Micro interactions and multi dimensional graphical user interfaces in the design of wrist worn wearables," *Proceedings of the Human Factors and Ergonomics Society Annual Meeting*, vol. 59, no. 1, pp. 1712–1716, 2015.

[57] C. Min, S. Kang, C. Yoo et al., "Exploring current practices for battery use and management of smartwatches," in *ISWC '15 Proceedings of the 2015 ACM International Symposium on Wearable Computers*, pp. 11–18, New York, NY, USA, September 2015.

[58] K. Lyons, "What can a dumb watch teach a smartwatch?: informing the design of smartwatches," in *ISWC '15 Proceedings of the 2015 ACM International Symposium on*

Wearable Computers, pp. 3–10, New York, NY, USA, September 2015.

[59] S. Pizza, B. Brown, D. McMillan, and A. Lampinen, "Smartwatch in vivo," in *CHI '16 Proceedings of the 2016 CHI Conference on Human Factors in Computing Systems*, pp. 5456–5469, New York, NY, USA, May 2016.

[60] E. Karapanos, R. Gouveia, M. Hassenzahl, and J. Forlizzi, "Wellbeing in the making: peoples' experiences with wearable activity trackers," *Psychology of Well-Being*, vol. 6, no. 1, p. 4, 2016.

[61] S. Schirra and F. R. Bentley, "It's kind of like an extra screen for my phone," in *Proceedings of the 33rd Annual ACM Conference Extended Abstracts on Human Factors in Computing Systems - CHI EA '15*, pp. 2151–2156, New York, NY, USA, 2015.

[62] E. Granado-Font, OBSBIT Study Group, G. Flores-Mateo et al., "Effectiveness of a smartphone application and wearable device for weight loss in overweight or obese primary care patients: protocol for a randomised controlled trial," *BMC Public Health*, vol. 15, no. 1, p. 531, 2015.

[63] K. Hänsel, A. Alomainy, and H. Haddadi, "Large scale mood and stress self-assessments on a smartwatch," in *Proceedings of the 2016 ACM International Joint Conference on Pervasive and Ubiquitous Computing Adjunct - UbiComp '16*, pp. 1180–1184, New York, NY, USA, 2016.

[64] S. Zhou, J. Chen, X. Wang, L. Zhou, B. Zhen, and J. Cui, "Inclination gradient-based fall detection algorithm for wrist-worn device," in *2015 IEEE International Conference on Consumer Electronics*, pp. 148-149, Taiwan, June 2015.

[65] Y. Cho, H. Cho, and C. M. Kyung, "Design and implementation of practical step detection algorithm for wrist-worn devices," *IEEE Sensors Journal*, vol. 16, no. 21, pp. 1–7730, 2016.

[66] M. Nguyen, L. Fan, and C. Shahabi, "Activity Recognition Using Wrist-Worn Sensors for Human Performance Evaluation," in *2015 IEEE International Conference on Data Mining Workshop (ICDMW)*, pp. 164–169, November 2015.

[67] S. Mehrang, J. Pietilä, and I. Korhonen, "An activity recognition framework deploying the random forest classifier and a single optical heart rate monitoring and triaxial accelerometer wrist-band†," *Sensors*, vol. 18, no. 3, 2018.

[68] H. Lim, G. An, Y. Cho, K. Lee, and B. Suh, "WhichHand: automatic recognition of a smartphone's position in the hand using a smartwatch," in *Proceedings of the 18th International Conference on Human-Computer Interaction with Mobile Devices and Services Adjunct*, pp. 675–681, New York, NY, USA, 2016.

[69] L. E. Diez, A. Bahillo, A. D. Masegosa et al., "Signal processing requirements for step detection using wrist-worn IMU," in *2015 International Conference on Electromagnetics in Advanced Applications (ICEAA)*, pp. 1032–1035, September 2015.

[70] A. Yazdan, R. Prance, H. Prance, and D. Roggen, "Wearable electric potential sensing: a new modality sensing hair touch and restless leg movement," in *UbiComp'16 Proceedings of the 2016 ACM International Joint Conference on Pervasive and Ubiquitous Computing: Adjunct*, pp. 846–850, New York:, 2016.

[71] P. Vepakomma, D. De, S. K. Das, and S. Bhansali, "A-Wristocracy: deep learning on wrist-worn sensing for recognition of user complex activities," in *2015 IEEE 12th International Conference on Wearable and Implantable Body Sensor Networks (BSN)*, pp. 1–6, June 2015.

[72] G. Valenza, M. Nardelli, A. Lanata et al., "Predicting mood changes in bipolar disorder through heartbeat nonlinear dynamics," *IEEE Journal of Biomedical and Health Informatics*, vol. 20, no. 4, pp. 1034–1043, 2016.

[73] J. A. Rincon, Â. Costa, P. Novais, V. Julian, and C. Carrascosa, "Using non-invasive wearables for detecting emotions with intelligent agents," in *International Joint Conference SOCO'16-CISIS'16-ICEUTE'16*, pp. 73–84, Springer, Cham, 2016.

[74] F. Gravenhorst, A. Muaremi, J. Bardram et al., "Mobile phones as medical devices in mental disorder treatment: an overview," *Personal and Ubiquitous Computing*, vol. 19, no. 2, pp. 335–353, 2015.

[75] L. Zhang, S. Rukavina, S. Gruss, H. C. Traue, and D. Hazer, "Classification analysis for the emotion recognition from psychobiological data," in *Proceedings of International Symposium on Companion-Technology*, pp. 149–154, 2015.

[76] A. M. AlzeerAlhouseini, I. F. T. Alshaikhli, A. Rahman, A. Wahab, and M. A. Dzulkifli, "Emotion detection using physiological signals EEG & ECG," *International Journal of Advancements in Computing Technology*, vol. 8, no. 3, pp. 103–112, 2016.

[77] R. Gravina and G. Fortino, "Automatic methods for the detection of accelerative cardiac defense response," *IEEE Transactions on Affective Computing*, vol. 7, no. 3, pp. 286–298, 2016.

[78] A. Cinquepalmi and U. Straccia, "An ontology-based affective computing approach for passenger safety engagement on cruise ships".

[79] A. Zenonos, A. Khan, G. Kalogridis, S. Vatsikas, T. Lewis, and M. Sooriyabandara, "HealthyOffice: mood recognition at work using smartphones and wearable sensors," in *2016 IEEE International Conference on Pervasive Computing and Communication Workshops (PerCom Workshops)*, pp. 1–6, March 2016.

[80] J. Vermeulen, L. MacDonald, J. Schöning, R. Beale, and S. Carpendale, "Heartefacts: augmenting mobile video sharing using wrist-worn heart rate sensors," in *Proceedings of the 2016 ACM Conference on Designing Interactive Systems*, pp. 712–723, New York, NY, USA, 2016.

[81] N. Costadopoulos, "Emotional intelligence via wearables: a method for detecting frustration," *Information Technology in Industry*, vol. 4, no. 1, pp. 19–25, 2016.

[82] S. Ramesh, "Using wearable technology to gain insight into children's physical and social behaviors," *Massachusetts Institute of Technology*, 2016.

[83] Z. Zhu, H. F. Satizabal, U. Blanke, A. Perez-Uribe, and G. Troster, "Naturalistic recognition of activities and mood using wearable electronics," *IEEE Transactions on Affective Computing*, vol. 7, no. 3, pp. 272–285, 2016.

[84] J. Hernandez, D. McDuff, C. Infante, P. Maes, K. Quigley, and R. Picard, "Wearable ESM: differences in the experience sampling method across wearable devices," in *Proceedings of the 18th International Conference on Human-Computer Interaction with Mobile Devices and Services*, pp. 195–205, New York, NY, USA, 2016.

[85] A. Sano, A. Z. Yu, A. W. McHill et al., "Prediction of happy-sad mood from daily behaviors and previous sleep history," in *2015 37th Annual International Conference of the IEEE Engineering in Medicine and Biology Society (EMBC)*, pp. 6796–6799, August 2015.

[86] D. Figo, P. C. Diniz, D. R. Ferreira, and J. M. P. Cardoso, "Preprocessing techniques for context recognition from accelerometer data," *Personal and Ubiquitous Computing*, vol. 14, no. 7, pp. 645–662, 2010.

[87] M. Shoaib, S. Bosch, O. Incel, H. Scholten, and P. Havinga, "A survey of online activity recognition using mobile phones," *Sensors*, vol. 15, no. 1, pp. 2059–2085, 2015.

17

A Survey on Energy Efficient Wireless Sensor Networks for Bicycle Performance Monitoring Application

Sadik Kamel Gharghan, Rosdiadee Nordin, and Mahamod Ismail

Department of Electrical, Electronic and System Engineering, Faculty of Engineering and Built Environments, Universiti Kebangsaan Malaysia (UKM), 43600 Bangi, Selangor, Malaysia

Correspondence should be addressed to Sadik Kamel Gharghan; sadiq@siswa.ukm.edu.my

Academic Editor: Laurent Francis

Wireless sensor networks (WSNs) have greatly advanced in the past few decades and are now widely used, especially for remote monitoring; the list of potential uses seems endless. Three types of wireless sensor technologies (Bluetooth, ZigBee, and ANT) have been used to monitor the biomechanical and physiological activities of bicycles and cyclists, respectively. However, the wireless monitoring of these activities has faced some challenges. The aim of this paper is to highlight various methodologies for monitoring cycling to provide an effective and efficient way to overcome the various challenges and limitations of sports cycling using wireless sensor interfaces. Several design criteria were reviewed and compared with different solutions for the implementation of current WSN research, such as low power consumption, long distance communications, small size, and light weight. Conclusions were drawn after observing the example of an advanced and adaptive network technology (ANT) network highlighting reduced power consumption and prolonged battery life. The power saving achieved in the slave node was 88–95% compared to the similar ANT protocol used in the medical rehabilitation.

1. Introduction

Cycling requires a very high level of endurance supported by a high fitness level and a suitable training program and teamwork. Cycling is a highly competitive sport in which the margins between victory and defeat are typically measured in seconds or milliseconds. Statistical information can be used to further improve and refine team skills and cyclist training programs to increase competitiveness in international tournaments. Acquired data may be processed to produce athlete performance profiles. Comparisons of athlete profiles can be made occasionally to measure an athlete's progress and to monitor and supervise him or her so that a coach can correct any deficient parameters.

In this paper, different types of wireless sensor networks (WSNs), which are used to monitor the biomechanical and physiological parameters of the bicycle and cyclist, respectively, will be critically reviewed and compared. A comparison between these wireless sensor technologies will be carried out on the basis of several performance metrics related to physical activities involving both the bicycle and cyclist. Some biomechanical and physiological parameters relevant to the sensors are covered in terms of their specifications and applications to cycling monitoring, including heart rate, speed, cadence, and pedal power.

The purpose of this review is not to criticize but rather to formulate potential extensions to existing approaches and to understand future research challenges. In addition, a methodical review of the use of WSN in cycling applications is used to describe current WSN research implementations. The review compares wireless technologies and performance metrics from recent research. Also, an ANT network example was introduced to overcome the limitations of power consumption and battery life.

The contribution of this paper is to present a comprehensive overview of the many challenges that face the wireless sensor networks in the bicycle performance monitoring system. In addition, comparisons between different wireless technologies were done based on various performance metrics. We attempted to highlight several wireless sensor network technologies that are used in wireless sensor bicycle monitoring, thus providing a comprehensive, up-to-date survey of wireless sensor networks research that is used in bicycle performance monitoring system. A new wireless sensor technology (i.e., ANT) that is used in bicycle performance monitoring was introduced, which aims to reduce the power consumption and prolong battery life of the bicycle performance monitoring system based on message rate reduction, as will be presented in Section 6. From the literature review, we found that ANT wireless technology offers significant potential in the area of energy efficient WSN transmissions. Our focus in this paper is on several wireless protocols which are connected to several bicycle-related sensors. Furthermore, this survey provides a good number of latest references for those interested in the area of wireless sensor bicycle performance monitoring. Based on our limited knowledge, this review paper is the first paper that deals with the extensive survey of wireless sensors for bicycle performance monitoring, focusing on energy efficient WSN protocols.

The rest of the paper is structured as follows. Section 2 describes the problem statement of the cycling team. The comparisons of types of wireless technologies used to transfer data from the cyclists will be presented in Section 3. The common biomechanical and physiological parameters in the cycling sport will be introduced in Section 4. Section 5 categorizes and describes previous work according to the type of wireless technology, highlighting the research methodology, advantages, and disadvantages of each research. In addition, a table of comparison for the research will be summarized. Furthermore, the biomechanical and physiological signals and sensors related to both bicycle and cyclist will be described in the same section. Section 6 will explain a proposed example of bicycle wireless sensor monitoring. Future works and a maturity evaluation will be introduced in Section 7. Finally, conclusions will be presented in Section 8.

2. Problem Statement and Challenges

Before advances in sensor development technologies, a cycling training program once depended mainly on the conventional stopwatch [1] to measure the performance of athletes. This method is inaccurate and uninformative when used to measure the performance and capability of an athlete on the track. Furthermore, this method also lacks scientific support, as there is no evidence that training with a stopwatch can improve the competitive performance of an athlete. There are many challenges when attempting to implement wireless sensor devices and networks specifically in bicycle performance monitoring systems as follows.

(i) The monitored parameters of the bicycle must be real time and accurate, so that coaching staff can give instant feedback to improve the performance of cyclists.

(ii) The wireless sensor monitoring system must operate at low power consumption (i.e., the battery must last for at least five hours of training).

(iii) There must be a flexible transmission range between the cyclist and coaching teams as it depends on the type of discipline, such as, track, road, or cross-country.

(iv) Ideally, lightweight and small-size wireless sensors and display units (mounted on the bicycle) are desired, especially for a highly competitive professional tournament.

(v) The system should expand to include additional (new) measurement parameters, such as body temperature, left/right axis (gyro sensors), specific athlete power profiles, and stress level.

(vi) An efficient algorithm must be introduced to reduce the power consumption of the three main consumers of power in the WSN, namely the sensors, microcontroller, and transceiver.

3. Wireless Technologies

There are many wireless technology standards which have been adopted for bicycle monitoring systems. Two of the most commonly available wireless protocols are ZigBee [2] and Bluetooth [3]. There is also alternative wireless standard ANT, which is commonly used in sensor networks [4, 5]. As is customary, wireless technology requirements such as transmitted power (dBm), data rate (kbps), the range of communications (m), power consumption (W), and weight (kg) differ among these standards. Practically, the power consumption, weight, and size must be as small as possible and the coverage area of the network must be as large as possible to improve the performance of the cyclist. For example, one worst-case scenario occurs when cyclists and coach are positioned too far apart.

Three wireless standards occupying the 2.4 GHz ISM band used in cycling monitoring are Bluetooth, ZigBee [6], and ANT, where ANT is commonly used in commercial products. Bluetooth operates in the 2.4 GHz ISM band, with a broadband speed 1 to 24 Mbps depending on the Bluetooth version [7] and a communication range of approximately 1 m/10 m/100 m [8]. Moreover, Bluetooth is a "lighter" standard that is extremely ubiquitous (e.g., it is built into most mobile phones, PDAs, and laptops) and can easily be adapted to several other services or networks [9].

ZigBee is a wireless technology that operates in the 868 MHz, 915 MHz, and 2.4 GHz frequency bands [10]. ZigBee was designed mainly for low-power-consumption applications, with additional features such as low cost, low data rate (maximum of 250 Kbps), and long battery life. In many ZigBee applications, the wireless device participation or contribution is very limited; the device spends most of its time in sleep mode. As a result, ZigBee device batteries are capable of working for several years and do not need to be replaced [11]. ZigBee enables broad-based deployment of wireless networks with an approximate communication range of 100 m. This allows it to work for years on small coin batteries for a host of control and monitoring applications, such as smart grid [12], patient monitoring [13], automation systems [14], control systems [15], and several other applications.

TABLE 1: Comparison between wireless technologies used in bicycle monitoring.

Parameters	Wireless technology		
	Bluetooth (IEEE802.15.1)	ZigBee/IEEE802.15.4	ANT
Frequency band(s)	2.4 GHz	868/915 MHz, 2.4 GHz	2.4 GHz
Data rate	1 Mbps[1] [104], 1.2 Mbps[2] [105] 3 Mbps[3] [106], 24 Mbps[4] [105]	20 kbps[a], 40 kbps[b], 250 kbps[c] [107]	1 Mbps
Output power	0 dBm[5], 4 dBm[6], 20 dBm[7] [108]	(−25–0) dBm	0 dBm
Transmission range	1 m[5], 10 m[6], 100 m[7] [108]	(10–100) m	30 m
Power consumption	(40–100) mW	1 mW	0.183 mW
Battery life (coin cell)	(1–7) days	(4–6) months	3+ years
Latency	<10 sec	(20–30) ms	≈Zero
Spreading type	FHSS	DSSS	Adaptive isochronous
Number of devices in network	7 (active)/255 (total) [109]	65535	2^{32}
Number of RF channels	79	One[a], 10[b], 16[c] [110]	125
Complexity	High	Low	Low
Network topology	Ad-hoc, point to point, star	Ad-hoc peer-to-peer, mesh, and star	Broadcast ad hoc peer to peer, mesh, and star

[1]Bluetooth LE, [2]Bluetooth 1.2, [3]Bluetooth 2.0 + EDR, [4]Bluetooth 3.0, and Bluetooth 4.0, [5]Class 3, [6]Class 2, [7]Class 1, [a]for 868 MHz, [b]for 915 MHz, [c]for 2.4 GHz.

The third type is ANT, an ultralow-power proprietary wireless technology that operates in the 2.4 GHz spectrum [16]. It was set up in 2004 by sensor company Dynastream [17]. Typically, the ANT transceiver device is treated as a black box and should not require much design potential to implement; that is, it is considered an infrastructureless network. ANT has been designed to cater for sports [18], the medical field [19], and industry [20], especially for performance monitoring of cyclists' and athletes' fitness levels. The ANT transceivers have been built into equipment such as cycle power meters, cadence meters, speed monitors, distance meters, and heart-rate belts to form personal area networks (PANs) for performance monitoring of users. For example, these networks have measured the performance metrics of speed, heart rate, and weight. The three key areas of operation of ANT are as follows: (i) sports, (ii) wellness, and (iii) home health [18]. It may also be used for data transfer to a number of devices; in this paper, a sport cycling application will be considered.

The main differences between the three protocols used in bicycle monitoring are summarized in Table 1 [21–31]. ZigBee and Bluetooth protocols are based on an IEEE 802 standard while ANT is not based on this standard. Classic Bluetooth, ZigBee, and ANT use a low data rate. Broadly speaking, Bluetooth and ZigBee are designed for a wireless personal area network (WPAN) communication range of approximately 100 m, while ANT is specified for a WPAN range of approximately 30 m [32]. Compared to Bluetooth, the ZigBee protocol is typical for WSNs or control applications, which require low data rate, and low complexity [33]. In addition, Zigbee can support for a large number of nodes in the network (up to 65,535), easy removal or addition of nodes of the network, a powerful mesh network such that the failure of one node does not bring down the rest of the network, and low power requirements [34].

A direct comparison of ANT to ZigBee is appropriate. ZigBee's/IEEE 802.15.4 standard for the 2.4 GHz ISM frequency band has an over-the-air data rate of 250 kbps [35, 36] compared to ANT's 1 Mbps [37]. Therefore, ZigBee is required to transmit over-the-air for longer than ANT for a given volume of data. ANT has been designed to stay in an ultralow-power mode (i.e., a sleep mode) for long periods, rapid wake-up [38], rapidly transmit for 150 μs, and return to sleep mode. In terms of latency the Bluetooth has high latency less than 10 second, ZigBee 30 ms [27], and ANT approximately zero latency [17, 39]. Latency is a very critical issue for cycling application because the cyclist parameters must be monitored in real time.

ZigBee and Bluetooth utilize direct sequence spread spectrum (DSSS) [40] and frequency-hopping spread spectrum (FHSS) [41] schemes, respectively, to ensure the safety of the wireless link, whereas ANT uses an adaptive isochronous [42] network technology to assure coexistence.

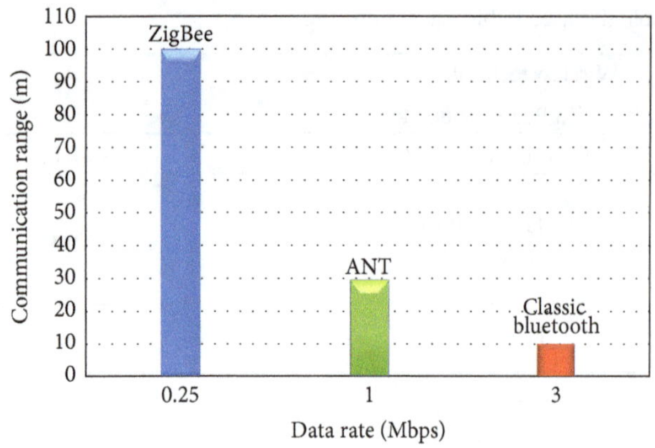

FIGURE 1: Data rates and communication ranges of several wireless sensor technologies.

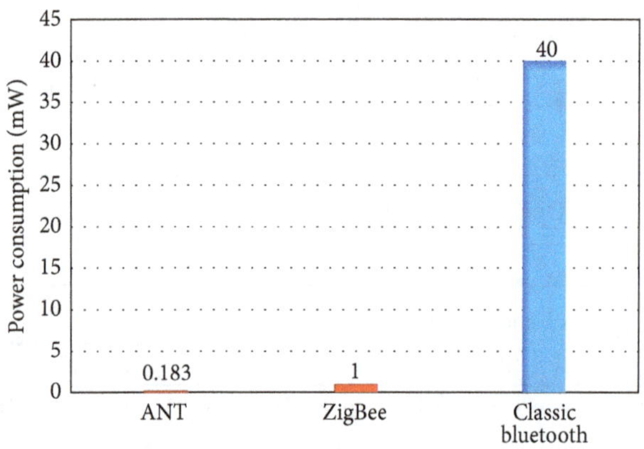

FIGURE 2: Power consumption of wireless sensor technologies.

Figure 1 shows a comparison of the data rates and communication ranges between wireless technologies used in wireless sensor bicycle monitoring, while Figure 2 shows the power consumption for these wireless technologies.

From the previous discussion, it is clear that the existing wireless standards, Bluetooth and ZigBee are unsuitable to address the wireless monitoring of sport cycling performance, on account of either power consumption or hardware complexity. In addition, latency is better for ANT compared to Bluetooth and ZigBee.

As a response to this topic, the sensor company Dynastream solved real problems in wireless sensor monitoring by introducing ANT wireless standard technology [43]. This wireless standard enjoys ultralow-power consumption [44] and small size and is therefore extremely suitable for WSN applications requiring long battery life, especially in sport and health monitoring.

4. Biomechanical and Physiological Parameters in Cycling

Most research works are concerned with four common bicycle parameters: (i) speed, (ii) heart rate, (iii) cadence, and (iv) power. These four parameters are key factors in improving the performance of cyclists, which are explained as follows.

4.1. Speed or Tire Rotation. The tire rotation or speed of the bicycle is the number of wheel rotations or revolutions per minute (RPM). Knowing the speed allows cyclists to avoid overtraining by regulating the training load [45] and it quantifies the external load [46]. The speed or tire rotation can be measured by using a magnetic sensor (reed switch) and a permanent magnet. The magnetic sensor is placed on the fork blade on the front wheel or on the chainstay of the rear wheel and the permanent magnet is fixed on one spoke on the front or rear wheel. The number of pulses generated by the sensor is directly proportional to the rotations of the rear or front wheel (RPM) and the ground speed of the bicycle v_b can be calculated in kilometers per hour (KPH) or miles per hour (MPH), based on the following equation:

$$v_b = \text{RPM} \cdot r \cdot \left(\frac{2\pi}{60}\right) = \omega \cdot r, \quad (1)$$

where r is the wheel radius in meter and RPM is the wheel angular velocity and $(2\pi/60)$ is the unit transformation of revolution per minute into radians per second.

4.2. Heart Rate. The heart rate (HR) is the number of heart beats per minute. The heart rate monitor device is usually mounted on the athlete's body in the form of a belt around the chest that contains a pulsed wireless sensor. The HR device is able to measure and evaluate the physical health condition of the athlete. Knowing the heart rate is important to allow each cyclist to achieve his or her training plan as much as possible; the difference between the training heart rate and the reference value of each cyclist must be as small as possible [47]. The heart rate is also used to assess the physical stress or internal load that results from an external load [46]. The heart rate in cycling is interchangeable with the workload (watts) made by the cyclist; for example, the heart rate is 91 beats/min when the workload is 25 W, while it becomes 137 beats/min when the workload is equal to 100 W [48].

4.3. Cadence. Cadence is the measure of the spinning speed of the pedals. The unit of measurement is revolutions per minute (RPM). The cycling power can be increased and the cyclist can be a productive part of the cycling team if the cyclist optimizes the cadence of the bicycle. Knowing the cadence allows cyclists to remain in a secure and an efficient range of revolutions per minute (RPM) to protect their knees [49], which is especially important over long distances. It is also possible to use the cadence to estimate the number of calories consumed during a bicycle trip. In addition, maintaining an ideal cadence (60–80) RPM helps a cyclist to consume calories efficiently. The cadence can be measured using a cadence sensor, which has two elements: a permanent magnet attached to the right crank arm of the bicycle and a magnetic sensor (reed switch) mounted on the chainstay. The pulses generated by the magnetic sensor are directly proportional to the crank rotations as measured in revolutions per minute (RPM).

4.4. Pedal Power. Pedal power is the conversion of energy from the human source into motion through the use of the foot pedal and crank system of the bicycle. The pedal power is mainly of interest in the high-power range (more than 75 W), which occurs during sports, rehabilitation activities, and exercise training. Knowing the cyclist power output allows the cyclist to optimize their performance, especially in competitions. Typically, the pedal power is used to quantify the external load. The pedal power can be measured by two methods. The first and most common method uses strain gauges attached to the right or left crank arm to measure the applied torque and at least one magnetic sensor that measures the RPM of the cadence. Therefore, the pedal power (P_c) can be calculated by using the following formula [50–52]:

$$P_c = T_c \cdot \text{CAD} \cdot \left(\frac{2\pi}{60}\right), \quad (2)$$

where T_c is the applied torque on the crank arm in N·m and CAD is the bicycle cadence or crank angular velocity in RPM.

The second method to measure pedal power is to calculate the sum of forces that face the cyclist during cycling. Many forces act on the cyclist and bicycle, but the major forces are air resistance force, rolling resistance force, gradient resistance force, and force due to the speed variables. Each force can be measured by using different sensors like temperature sensor, humidity sensor, pressure sensor, wind velocity sensor, and accelerometer sensor. The total pedal power P_{DT} can be calculated by multiplying the summation of these forces by the bicycle speed. Therefore, the total pedal power of the cyclist is modeled by [46, 53–55]

$$P_{\text{DT}} = \frac{\left[\left(F_W \cdot CF_{\text{draft}} + F_R + F_G + F_A\right) \cdot v_b\right]}{c_m}, \quad (3)$$

where F_W is the air resistance, CF_{draft} is the effect of drafting, F_R is the rolling resistance force, F_A is the force due to speed variation, F_G is the gradient resistance force, v_b is the bicycle speed in KPH, and c_m is the mechanical efficiency factor.

5. Previous Works

Over the past decades, the issues of mobility, size [56], capability of processing (i.e., microprocessors or microcontrollers), flexible memory, and robust modes of transmission have been solved in wireless sensors due to the advancement in wireless sensor technology. Many new wireless standards provide a reliable and secure platform for a sensor network. Such wireless sensors support wearable [57], unobtrusive [58] equipment and also benefit from remote monitoring. This development benefits many practical applications including those in the realm of sports. In sports applications, a WSN allows the gathering and use of data in real time and it is also possible to compare data among athletes.

The transfer of data from an athlete or patient to a remote data processor or storage may now be performed remotely. The most common way to transfer the data is simply point-to-point, which means that one transmitter on the side of the athlete and one receiver on the monitoring side communicate based on frequency-modulated transmission [6].

In this section, several types of bicycle wireless sensor monitoring (BWSM) are critically reviewed. This review will be categorized based on the type of wireless technology or platform.

5.1. Systems Based on ZigBee Platforms. Muscle work, sweat level, and heart rate were monitored by Armstrong [6]. This study represents a beneficial introduction to how the ZigBee wireless technologies can be employed to improve the measurement process.

Advantage. Power consumption is reduced.

Disadvantage. Small coverage area limits communication due to limitations of the ZigBee standard.

The pedal power, speed, cadence, and heart rate data were transmitted separately for each cyclist to his computer via a ZigBee network in the work of Kuhn et al. [45]. They compared these measured parameters while developing the ABT system with commercial Schoberer Rad Messtechnik (SRM) power meters for twelve to thirty hours of bicycle riding. They found this system to be quite convincing.

Advantages. The system has low cost and lightweight overhead for cyclists and the transceiver unit can be used with multiple radio channels. Thus, many networks may be active at the same place in the same time period.

Disadvantages. The transmission range is 50 m in outdoor conditions and the notebook size is large. Another limiting factor is battery capacity, especially when the power plug is not available.

Le et al. [47] built an assisted bicycle trainer (ABT) with a predictive controller to measure the parameters of the bicycle and rider: speed (RPM), position in the group, and heart rate (bpm) of the cyclists. The collected data were transferred via a ZigBee network to a notebook computer for monitoring and supervision.

Advantage. The predictive controls of the ABT system were an effective way of training the cyclist group.

Disadvantages. The system did not consider other parameters such as power, velocity, cadence of the bicycle, and wind velocity, which were left for future work.

Walker et al. [59] monitored pulse heart rate (bpm), SpO_2 (%), skin temperature (°C), ambient temperature (°C), bicycle speed or wheel speed (KPH or MPH), and bicycle location based on a remote mobile monitoring system (RMMS). In addition, this system may be used to monitor vital events for the elderly.

Advantages. Low-power wireless sensors (ZigBee module), light weight, and real-time application (the time to transmit data from the sensor location to the remote device is approximately 10 ms); monitoring can be performed by persons not admitted to the hospital or in a physician's office. This last advantage improves retrieval of the history of a person's status and allows a medical care center to service a huge number of people, reducing hospital visits and health care expenses, and enables the observer to recognize emergency situations quickly.

Disadvantage. The wireless sensor nodes (BioTE) have difficulty communicating when an obstacle blocks the line-of-sight between transmitter and receiver nodes.

Measurements of wheel spin rates (RPM), angular velocity (RPM) of the bicycle, steer angle, and the acceleration of three points on the frame were performed by Peterson et al. [60]. They were more ambitious than others in using a low-power-consumption Xbee-pro wireless module interfaced to a microcontroller based on the Arduino platform to transmit the measured parameters to a monitoring system, based on GUI toolkits.

Advantages. The system created an accurate assessment of the lean via numerical integration and precise filtering of gyroscope signals, and the optical encoders are frictionless and have high reliability at low speeds with a fast dynamic response.

Disadvantages. The roll of the bicycle frame is difficult to directly measure, while the high precision of the optical methods is countered by both the weight and expense of the head unit. Optical methods can also be affected by road surface bending; reflectance and a larger battery size were required to supply the system.

Location estimation and radio reachability of the bicycle were determined by Hayashi et al. [61] using the ZigBee WSN. The bicycle's location can be estimated based on base stations, internet protocol gateways, and network management servers.

Advantages. Low power consumption, wide-area sensor network, monitoring over a long distance of 3.5 km, remote reading, and remote control and bicycle management (including a temperature sensor, GPS sensor, luminance sensor, and acceleration sensor). The project found new services that are secure, eco-friendly, and safe.

Disadvantages. Large infrastructure, more components in the network, and the required line-of-sight between the transmission terminal and the base station.

Olieman et al. [62] measured vibrations of the bicycle produced from a combination of the wheels, speed, tire pressure, and road surface. Wireless inertial acceleration sensors (ProMove2) were used to measure the vibration of the bicycle in real time.

Advantages. Real-time measurements; the WSN nodes are connected to a computer by the fast gateway. GUI software is used to display the acquired data for further processing, while the receiver is designed to store the data every second, and automatic pausing is turned off.

Disadvantage. The experiment was partially unsuccessful because some of the frequencies were cancelled out more than others; for the best test and analysis parameters, more experiments are needed.

Marin-Perianu et al. [63] adopted a different approach from the others by focusing on determining the actual situation of the cyclist's lower limb segments in real-life and real-time conditions. A mobile WSN design was used to monitor lower-limb kinematics during cycling in real time. This design was based on ProMove (a low-power 2.4 GHz wireless radio connected to a central node). The results of this system were compared with those of a camera system.

Advantages. The system has high accuracy with a simple wireless strap-on and the miniaturized sensing system can help cyclists outdoors during racing or training. Feedback is instant, which provides major benefits directly to users by providing them with comprehensive insight into body kinematics in real-life situations. The proposed system can be used for any motion sport, including swimming, rowing, skiing, and running.

Disadvantages. The system has limited resources, sensing accuracy, computational power, battery capacity, and robustness of wireless communication; there was error accumulation in the inertial measurement.

5.2. Systems Based on Bluetooth Platforms. Yang et al. [64] monitored the movements and breathing of a cyclist in addition to the road condition. The system consisted of clip sensors, accelerometer, microcontroller, Bluetooth module, and PDA.

Advantages. The system was low cost, durable, washable, comfortable, and wearable; the cyclist's health, safety conditions, and work were monitored and analyzed over the long term.

Disadvantages. The belt is large and relatively heavy on the cyclist compared to other belt sensors.

Casamassima et al. [65] developed a new platform to improve the rhythm and amplitude of Parkinson's disease patients by utilizing gait rehabilitation based on a wearable system eligible to produce real-time audio feedback via vocal messages to the patient in a comfortable and efficient way. The system composed of three basic elements: Android mobile phone based on Bluetooth wireless technology, inertial sensor nodes, and a microcontroller based on the ARM Cortex-M3.

Advantages. The system is mobile and unobtrusive. It is comfortable to wear and able to be carried by the patient without restriction and real life monitoring in both outdoors and indoors conditions.

Disadvantages. Power consumed is relatively high and battery lifetime of 6 hours. This means the battery must be recharged or replaced when used more than six hours.

5.3. Systems Based on ANT Protocol. An ANT wireless sensor was used by Baca et al. [66] to monitor the biomechanical parameters of the bicycle: velocity (KPH), cadence (RPM), distance (Km), cyclist position, force (N), inclination, and motion amplitude and physiological parameters of the athletes: heart rate (bpm) and heart rate variability. The measured parameters are transmitted to a remote location (coach) via a framework system which consists of ANT protocol, mobile phone, mobile network GPRS or UMTS, and internet networks. The coach gives feedback messages as advices to the athletes in the form of an alert pop-up message or ringtone alarm in order to improve their performance.

Advantages. The system enjoys low power consumption and low cost because of using the ANT protocol and mobile phones; in addition, the proposed approach used the possibility of infrastructure of the mobile and internet networks.

Disadvantage. The system used the default value message rate (4 Hz) for ANT devices in the framework. It was possible to make it less than 4 Hz (e.g., 1 Hz), especially for physiological parameters where it does not change rapidly in order to reduce the power consumption and prolong battery lifetime of the sensor platforms.

The ANT transceiver was used by Johansson et al. [67] to analyze a Gait system for patients who need to rehabilitate gait or balance control after a stroke. They used ultralow-power wireless insole sensor biofeedback and the ANT transceiver interface with GSM/GPRS to send data to the home network via Bluetooth or to the Internet via GSM/GPRS. The gait system gives real-time biofeedback of squeeze allocation on the plantar surface during the stance phases of gait. The system helped to diagnose and treat these conditions and can also be used as a biofeedback system to monitor movement limits after joint surgery.

Advantages. The system showed good reliability and low power consumption and coexisted with both GSM/Wi-Fi and Bluetooth. The system has low latency and is based on an open-source mobile device. Therefore, the hardware is suitably available to build personal servers for a wireless body area network (WBAN) in a biofeedback gait-analyzing system.

Disadvantages. Higher power consumption at the high message rate (e.g., more than 4 Hz) and a small amount of time jitter in the data sampling, as it is based on the TDMA access method.

Moreover, there are some researches such as [46, 68] were adopted Wi-Fi technology to form an ad hoc network for bicycle monitoring, but these researches are out of this study, on account of either power consumption or size of the system.

This narrative description of previous research is summarized in Table 2, which is arranged consecutively according to the type of wireless technology, with the addition of some other information, including hardware configuration and power efficiency.

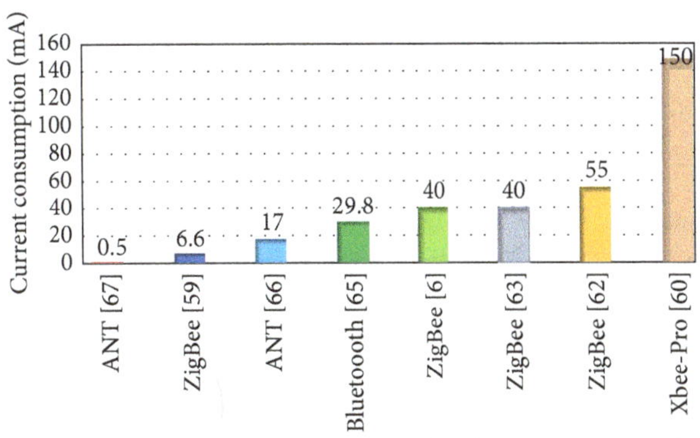

FIGURE 3: The current consumption of wireless sensor technology as listed in Table 2.

Through a review of previous researches which are summarized in Table 2, obviously, most of them supported the ZigBee wireless standard for bicycle monitoring. This protocol was most often selected as the wireless transmission technology because of its low cost [69], low power consumption [70, 71], higher network elasticity, a large number of nodes [72], greater coverage area, and secure networking [73]. Bluetooth is embedded in most portable mobile phones and computers [74] and has low cost, low EM interference [75], low-power applications [76], and low complexity [77], but it has appeared in few works considered in this study. ZigBee and Bluetooth are widely used in event monitoring because these two standards have low power consumption, which is a key feature of WSNs [78, 79].

ANT is the newest wireless standard technologies and is widely used in commercial cycling products such as SRM [80], PowerTap [81], Garmin Vector [82], and Pioneer [83]. However, there is still a lack of research literature related to bicycle monitoring that addresses this technology despite its use in health and rehabilitation monitoring as seen in Section 5.3. So, an example using this technology as an application for BWSM will be introduced in a later section.

Furthermore, a bar chart has been extracted from Table 2, as shown in Figure 3. This figure shows the current consumption comparison between wireless sensor technologies, which has been used in the previous research works. Figure 3 shows that the less power consumption was accomplished in research work in [67], which adopted ANT wireless standard technology. Some of research works aforementioned are related to the health (i.e., rehabilitation) applications, which based on Bluetooth and ANT wireless standard technologies are considered in this study. This is due to limited research works in sport cycling based on these technologies.

There are several performance metrics that have traditionally been chosen to quantitatively represent the achievements of the athlete and bicycle. In general, four of these performance metrics are most commonly used as demonstrated in the previous section. Measurement of the performance of these parameters depends on several sensing technologies, which are depicted in Table 3, and the relevant location of the main sensors (key factor parameters) on a bicycle is shown

TABLE 2: Comparison of previous works related to the bicycle wireless monitoring system.

Authors	Wireless Technology/ Platform	Hardware configuration	Biomechanical parameters	Physiological parameters	Power consumed
Armstrong [6]	ZigBee	ZigBee sensor network	N/A	(i) Muscle work (ii) Heart rate (bpm) (iii) Sweat rate	40 mA
Kuhn et al. [45]	ZigBee	A stationary forwarder node based on MicaZ motes with modified communication stack	(i) Pedal power (W) (ii) Cadence (RPM) (iii) Speed (RPM)	Heart rate (bpm)	N/A
Le et al. [47]	ZigBee	ZigBee sensor network configured as ad-hoc network with Notebook	(i) Power (W) (ii) Velocity (RPM) (iii) Cadence (RPM)	Heart rate (bpm)	N/A
Walker et al. [59]	ZigBee	BioTE Sensor with MSP430, CC2420, GPS RMMS, Cellular Network and MySQL database	(i) Bicycle speed or wheel speed (RPM or MPH) (ii) Bicycle location	(i) Heart rate (bpm) (ii) SpO_2 (%) (iii) Skin temp. (°C) (iv) Ambient temp. (°C)	6.6 mA
Peterson et al. [60]	Xbee	Xbee-PRO, I^2C, ADC, Arduino platform, and AVR	(i) Angular velocity (rad/s) (ii) Wheel spin rates (rad/s) (iii) Steer angle (degrees) (iv) Acceleration of three points fixed to the frame (m/s^2) (v) Roll angle (degree)	N/A	150 mA
Hayashi et al. [61]	ZigBee	ZigBee, sensor, base stations network, and server with internet protocol	(i) Location estimation (ii) Radio reachability	N/A	N/A
Olieman et al. [62]	ZigBee	Inertial sensors based on ProMove2 and GPS	Vibration of the bicycle	N/A	55 mA
Marin-Perianu et al. [63]	ZigBee	WSN platform based on ProMove	N/A	(i) Knee angle (degree) (ii) Ankle angle (degree)	40 mA
Yang et al. [64]	Bluetooth	Bluetooth, PDA, eight clip sensors, MSP430, and GPS	(i) Velocity (KPH) (ii) Step counts accumulated (iii) Distance (Km) (iv) Road conditions	(i) Amplitude of breathing (ii) Rate of breathing (iii) Posture determination (sitting, standing, walking, and cycling)	N/A
Casamassima et al. [65]	Bluetooth	Android mobile phone based on Bluetooth, microcontroller based on ARM Cortex-M3, and accelerometer, gyroscope and magnetometer sensors	N/A	(i) Steps per minute (ii) Step length (iii) Gait speed (iv) Gait asymmetry (v) Trunk flexion (vi) Clearance	29.8 mA
Baca et al. [66]	ANT	ANT (nRF24AP2) protocol, Mobile phone, mobile network GPRS/UMTS, and internet networks	(i) Velocity (KPH) (ii) Cadence (RPM) (iii) Distance (Km) (iv) Bicycle position (v) Inclination (vi) Force (N) (vii) Motion amplitude	(i) Heart rate (bpm) (ii) Heart rate variability	17 mA
Johansson et al. [67]	ANT	ANT network based on smart mobile phone, sole-integrated gait sensors, ATmega88, and ARM cortex	Pressure distribution on the patient's shoes	Balance control for post-stroke patients	500 μA

TABLE 3: Biomechanical and physiological sensors related to the athlete and bicycle.

Biomechanical and physiological signal type	Sensor type	Description of the measured data
Heart rate [111, 112]	Pulse belt sensor	Measurement of the heart rate in number of beats per minute.
Speed [66]	Magnetic	Calculation of the number of wheel rotations in revolutions per minute (RPM) or linear speed in KPH.
Cadence [66]	Magnetic	Calculation of the number of crank arm rotations in revolutions per minute (RPM).
Power [113–115]	Strain gauge	Measurement of the applied torque on the crank arm (N.m).
Bicycle location [61]	GPS	Determination of the bicycle's location on the road.
Skin temperature [116]	Integrated circuit LM35	Measurement of the athlete's temperature (°C).
Electrocardiogram (ECG) [117, 118]	Chest electrodes	Measurement of the electrical heart activity during cycling and stair-climbing.
Oxygen saturation (SPO_2) [119]	Pulse oximeter	Measurement the percentage (%) of the oxygen concentration in the cyclist's blood.
Rate of breathing [120]	Respiration sensor	Measurement of the rate of breathing (inspiration and expiration time) of the cyclist.
Ankle and knee joint angles [63]	Roll angle	Monitoring of the ankle and knee joints in order to control the power transmitted to the pedals.
Ambient temperature [121]	NFC thermistor	Measurement of the ambient air temperature (°C).
Bicycle position [122]	GPS or bicycle ID in wireless sensor network	Determination of the bicycle's position in the group.
Route of the road [123]	GPS	Monitoring of the road in order to avoid risks.
Road conditions [64]	Triaxial accelerometer	Measurement of triaxial acceleration to give the safety conditions.
Acceleration [123]	Tilt sensor	Measurement of X and Y accelerations of bicycle.
Distance [124]	Speed multiplied by time period	Measurement of the distance travelled by the bicycle.
Steer angle [60]	Potentiometers	Measurement of the angular displacement of the bicycle fork.
Vibration [62]	Inertial acceleration	Measure the vibration of bicycle in seat, wheel axle, and front and rear wheel.

in Figure 4. These sensors may be incorporated as part of a bicycle monitoring system, along with their corresponding measured biomechanical and physiological signals. The measurement of these signals and their subsequent processing for extracting some features is conducive to data collection of biomechanical and physiological parameters in real time, allowing an overall evaluation of the athlete and bicycle to be obtained at any time.

6. Example of the Proposed BWSM

In this section, a proposed example of BWSM based on ANT wireless standard technology will be highlighted. The ANT protocol operates in the 2.4 GHz ISM band with 125 channels of width equal to 1 MHz and a baud rate of up to 1 Mbps [22]. In order to establish a connection between ANT devices or modules, the channel configuration must be common for

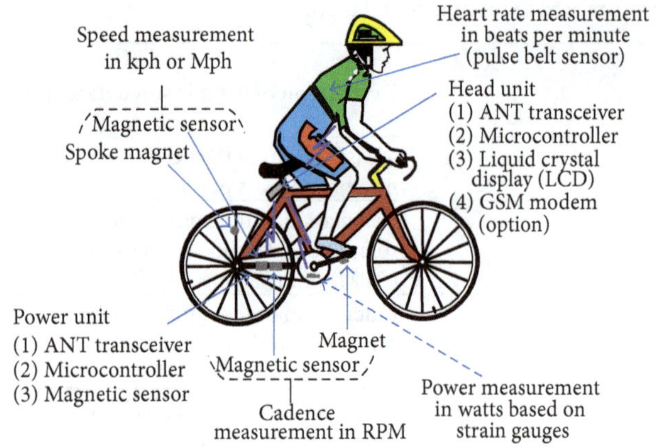

FIGURE 4: An overview of relevant wireless sensors and networks related to a bicycle and cyclist monitoring system.

these devices. There are several pieces of information related to the operating parameters of the channel that should be known for the ANT module in the network, which are [84] as follows.

(i) Network keys and network number ensure confidentiality and authorization access.

(ii) Radio frequency (RF): the radio frequency is 125 unique RF channels varying between "0 to 124," representing the offset from 2.4 GHz in 1 MHz increments and can be calculated by the following formula:

$$\text{RF freq. value} = \frac{\text{Desired RF freq. (MHz)} - 2{,}400\,\text{MHz}}{1\,\text{MHz}}. \quad (4)$$

The frequency value must be set the same for both the master and the slave nodes when the connection is created. For example, if the desired RF frequency is 2,466 MHz (default RF frequency), it represents the 66 RF frequency value.

(iii) Channel ID: in order for the ANT channel to be established, the host must specify its channel ID (whether it is master or slave).

(iv) Channel type: this specifies the type of communication (unidirectional adopted in this experiment) that will occur in the channel.

(v) Data types: there are four forms of data corroborated by ANT, namely, acknowledged, broadcast, burst, and advanced burst, each of which is transmitted in 8-byte packets over the desired RF channel.

(vi) Channel period: this represents the time interval between two successive data transmissions with broadcast data type. The channel message rate varies between 0.5 and 200 Hz and the default value is equal to 4 Hz. The channel period is a 16-bit field with its value determined by the following expression:

$$\text{Channel Period value} = \frac{32{,}768}{\text{Message Rate (Hz)}}, \quad (5)$$

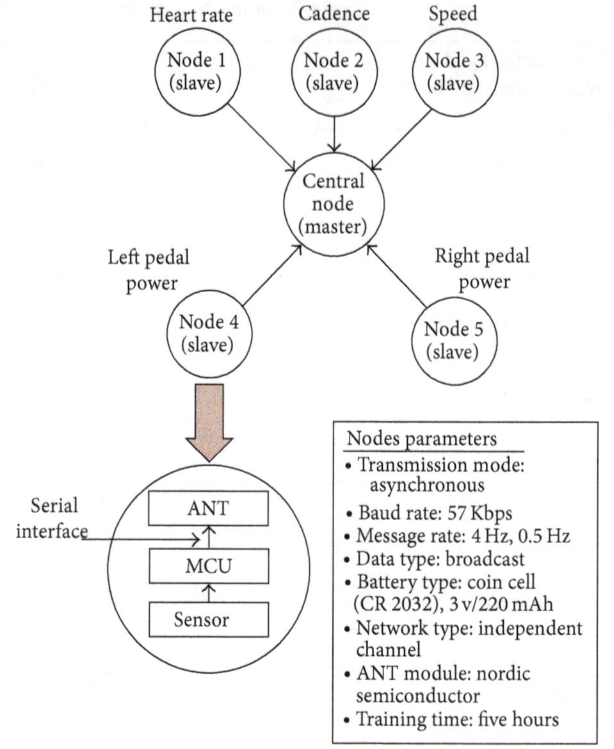

FIGURE 5: Example ANT network for bicycle wireless sensor.

where 32,768 represents an external clock source of the ANT module.

In the example of BWSM shown in Figure 5, six nodes form a simple independent channel network, five of them being slave nodes. Each node consists of a sensor, microcontroller, and ANT module. One node is the master node (central node) which consists of a microcontroller, ANT module and liquid crystal display (LCD). The slave nodes (1, 2, 3, 4, and 5) log data from relevant sensors of heart rate, cadence, speed, and left and right pedal power, respectively. The collected data of biomechanical and physiological parameters from multiple slave nodes will be broadcasted to the master node. The master node receives and analyses this data, which later on will be displayed on the LCD to be viewed by the cyclist. The slave nodes only establish one channel and can therefore be executed using a single channel ANT module, whereas a master node requires a five-received-channels ANT module.

The power consumption and battery life of master and slave nodes were estimated based on the Power Estimator Tool [85], and the Nordic Semiconductor nRF24AP2-8CH ANT module [86] was selected. The power consumption and battery life were estimated for two cases; in the first case, default values were used, while in the second case the message rate was changed.

First Case. By default, a broadcast (i.e., unidirectional) data packet will be sent by slave nodes and received by a master node in every time slot at the 4 Hz message rate; this message rate is chosen to provide more easily discoverable networks

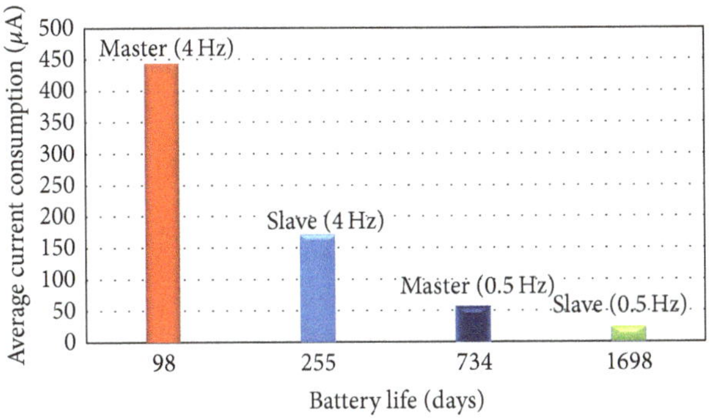

FIGURE 6: Current consumption and battery life for master and slave nodes.

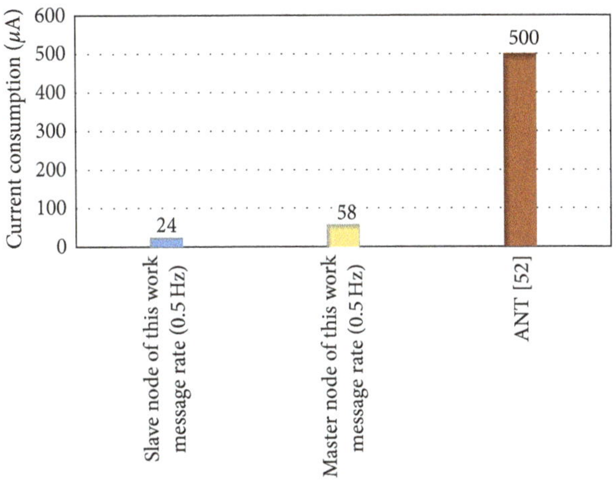

FIGURE 7: Current consumption comparison between proposed BWSM and research work [67] based on ANT.

with good latency and power consumption characteristics [84]. The baud rate at which the microcontroller can provide the data to ANT is set to 57 Kbps in order to achieve maximum throughput [87]; besides, a coin cell battery CR2032 (3 v/220 mAh) was chosen as the supply voltage to maintain the small size of the system.

The average current consumption for each slave node related to these parameters was approximately 170 μA and the battery life was 255 days for training five hours a day (daily training requirements for cyclists) [88], while the average current consumption for the master node was 443 μA and the battery life was 98 days.

Second Case. In typical practical use of a WSN application, the measured parameters do not change as rapidly as in cycling monitoring, and updating of the data every few seconds is acceptable. Therefore, a message rate of every 2 seconds (0.5 Hz) is adequate. In that case, the average current consumption of each slave node becomes 24 μA which could extend the battery life to 1,698 days (4.5 years) for five hour training per day, whereas the average current consumption for the master is was 58 μA, which could extend battery life to 734 days (2 years). It is clear that the second case is an appropriate choice in a BWSM network. Figure 6 shows a chart of two cases of average current consumption and battery life for master and slave nodes.

The energy efficient was achieved in proposed BWSM by 88% and 95% for master node and slave node, respectively, compared with similar wireless standard technology (i.e., ANT) adopted in research work in [67], as shown in Figure 7.

7. Future Work and Maturity Evaluation

To assess the current level of maturity for research into cycling monitoring systems, a set of important features was introduced which best describe the functionality and properties of bicycle monitoring systems. The feasible scope of research that cycling sport monitoring incorporates will greatly benefit from the ability to gather data unobtrusively, without impeding motion or performance. These data contain information about the biomechanical and physiological parameters of the bicycle and cyclist, respectively, in addition to other parameters such as position and altitude.

New wireless radio communication technologies could simplify this data collection by allowing data to be gathered synchronously from various locations on the bicycle and cyclist. This would enable data from several sensors to be streamlined using data fusibility to allow further analysis. These data can be displayed in a clear and flexible manner for both the cyclist and trainer. ZigBee and Bluetooth are both common wireless radio communication interfaces that allow the establishment of WSNs for the collection of data from the nodes. In practice, newer products use Bluetooth and ZigBee when their main application is sensor networks and they are far more common in the world of WSNs.

Bluetooth low energy (BLE) also called smart Bluetooth is extension of Bluetooth version 4.0 [89].

It was released in October 2006, uses the same frequency ISM band of classic Bluetooth (2.4 GHz) with 1 Mbps data rates in a 10 meter range [31, 38]. BLE is considered as an attractive technology for wireless sensor network applications demanding high data rates, but short range [44]. BLE enjoys several advantages over classic Bluetooth, which areas follows: (i) it is designed to transmit data at very low latency 15 times more efficiently than classic Bluetooth, (ii) only three channels are used for advertising, whereas 32 channels for classic Bluetooth, making BLE over 17 times more efficient than classic Bluetooth, (iii) saving a large amounts of energy because the slave devices do not stay in listening mode when it does not have any data to transmit [38, 90], and (iv) the power consumption of BLE is about one-tenth that of classic Bluetooth, which enables to use coin cell battery to power devices for more than a year [91].

BLE and ANT are both ultralow-power wireless sensor transmissions, both consume 15 mA in transmission mode, have similar sleep/active power, ability to operate for years on a single coin cell battery, data rate of 1 Mbps, wireless protocols operating in the range 2.4 GHz to 2.483 GHz, immunity to interference, and they are used in wireless personal area network (WPAN). On the contrary, they have differences in

TABLE 4: Dissimilarity in terms of technical specifications between BLE and ANT.

Specifications	BLE	ANT
Data channel	40	125
Transmission range	10 m	30 m
Latency	6 ms	≈Zero
Output power	(−20–+10) dBm	0/−6/−12/−18 dBm
Payload size byte	8–47	1–32
Number of nodes in the network	Limited by the application	2^{32}
Network standard	IEEE 802.15.1	Proprietary
Network topology	Simple star topology	Complex topologies

terms of specifications, as listed in Table 4 [44, 89–99]. On the other hand, the ANT is characterized by frequency agility, which is similar to frequency hopping schemes in Bluetooth technology. This property allows the ANT channel to change its operating frequency to improve coexistence with other wireless technology such as Bluetooth, Zigbee, and Wi-Fi. The ANT operating frequency automatically changes when significant degradation is observed in the channel [84, 100]. Furthermore, ANT requires less overhead to transmit the same amount of data, has more efficient scanning/initiating, and requires less transaction time per connections event. The BLE technology is still unproven, although many companies have declared their support of this technology [94], whereas the ANT is a proven technology with over 60 million devices in the market today [18].

This allows devices based on ANT are more efficient than BLE and they are preferred from different manufacturers to exchange data between ANT devices in several applications like health, sport, and fitness.

The ANT protocol has already been used in bicycle monitoring as in commercial products and has as-yet untapped potential in research. This type of wireless technology takes advantage of the cyclist's full capability and will move monitoring research into a new concept of WSNs as it has better features, such as ultra-low power consumption (a small coin-cell battery that lasts for 3+ years), low complexity, zero latency, can support large number of devices on the network, and compatible with different types of network topologies. Unfortunately, it currently has a short communication distance of approximately 30 m. Further hardware deployment using ANT for a bicycle monitoring system should be considered based on maturity evaluation.

8. Concluding Remarks

This study provided a survey of wireless sensor measurement systems for cycling based on three common wireless communication protocol standards (Bluetooth, ZigBee, and ANT) which are available commercially and academically. An assessment of the characteristics and behavior was introduced in terms of various performance metrics, including the data rate, output power, transmission range, power consumption, latency, complexity, network topology, and the number of devices in the network. A comparison of these wireless standard technologies was carried out to determine the most suitable for bicycle monitoring.

The results of the study point to ANT, which is a new wireless technology suitable for WSN in bicycle wireless monitoring. An example of this technology was adopted by the BWSM network; through this example, the power consumption was reduced compared to the similar ANT protocol in the medical rehabilitation application and the battery life was prolonged. Reducing power consumption and prolonging battery life were achieved by adjusting the message rate of ANT slave and master nodes and select the most suitable ANT chip that consumes less power (i.e., nRF24AP2-8CH). This led to an energy efficient in proposed BWSM by 88% and 95% for master and slave nodes respectively.

The major advantages of ANT are ultralow-power consumption, low latency, and high reliability. It may also be used in other applications that require longer battery life for continuous monitoring, such as a monitoring system of older adults [101], coal mine safety monitoring system [102], and plant disease remote monitoring [103] in medical, industrial, and agricultural applications respectively.

Conflict of Interests

The authors declare that there is no conflict of interests regarding the publication of this paper.

Acknowledgment

This work was supported by Malaysia's Ministry of Higher Education-Universiti Kebangsaan Malaysia (UKM) under Grant ref. no: PKT1/2013.

References

[1] C. Young, P. R. Fleming, S. Dixon, and M. J. Carré, "Shoe signature monitoring for advanced running technique," in *The Engineering of Sport 7*, M. Estivalet and P. Brisson, Eds., vol. 2, pp. 35–43, Springer, Paris, France, 2008.

[2] K. Malhi, S. C. Mukhopadhyay, J. Schnepper, M. Haefke, and H. Ewald, "A zigbee-based wearable physiological parameters monitoring system," *IEEE Sensors Journal*, vol. 12, no. 3, pp. 423–430, 2012.

[3] Y. Zhang and H. Xiao, "Bluetooth-based sensor networks for remotely monitoring the physiological signals of a patient,"

IEEE Transactions on Information Technology in Biomedicine, vol. 13, no. 6, pp. 1040–1048, 2009.

[4] E. Nemati, M. J. Deen, and T. Mondal, "A wireless wearable ECG sensor for long-term applications," *IEEE Communications Magazine*, vol. 50, no. 1, pp. 36–43, 2012.

[5] W. Zhang, P. Passow, E. Jovanov, R. Stoll, and K. Thurow, "A secure and scalable telemonitoring system using ultra-low-energy wireless sensor interface for long-term monitoring in life science applications," in *Proceedings of the IEEE International Conference on Automation Science and Engineering (CASE '13)*, pp. 617–622, Madison, Wis, USA, August 2013.

[6] S. Armstrong, "Wireless connectivity for health and sports monitoring: a review," *British Journal of Sports Medicine*, vol. 41, no. 5, pp. 285–289, 2007.

[7] A. H. Omre and S. Keeping, "Bluetooth low energy: wireless connectivity for medical monitoring," *Journal of Diabetes Science and Technology*, vol. 4, no. 2, pp. 457–463, 2010.

[8] R. Tesoriero, J. A. Gallud, M. D. Lozano, and V. M. R. Penichet, "Tracking autonomous entities using RFID technology," *IEEE Transactions on Consumer Electronics*, vol. 55, no. 2, pp. 650–655, 2009.

[9] A. Honkasuo, A. Hamalainen, P. Jappinen, J. Jaaskelainen, and J. Porras, "Mobile bicycle monitoring system," in *Proceeding of the 7th IEEE Workshop on Mobile Computing Systems and Applications (WMCSA '06)*, p. 44, Orcas, Wash, USA, August 2005.

[10] N. Baker, "ZigBee and Bluetooth strengths and weaknesses for industrial applications," *IEE Computing and Control Engineering*, vol. 16, no. 2, pp. 20–25, 2005.

[11] W. Guo, W. M. Healy, and M. Zhou, "Battery discharge characteristics of wireless sensors in building applications," in *Proceedings of the 9th IEEE International Conference on Networking, Sensing and Control (ICNSC '12)*, pp. 133–138, Beijing, China, April 2012.

[12] A. Kailas, V. Cecchi, and A. Mukherjee, "A survey of communications and networking technologies for energy management in buildings and home automation," *Journal of Computer Networks and Communications*, vol. 2012, Article ID 932181, 12 pages, 2012.

[13] S.-K. Chen, T. Kao, C.-T. Chan et al., "A reliable transmission protocol for zigbee-based wireless patient monitoring," *IEEE Transactions on Information Technology in Biomedicine*, vol. 16, no. 1, pp. 6–16, 2012.

[14] A. Betzler, C. Gomez, I. Demirkol, and J. Paradells, "A holistic approach to ZigBee performance enhancement for home automation networks," *Sensors*, vol. 14, pp. 14932–14970, 2014.

[15] K.-Y. Lian, S.-J. Hsiao, and W.-T. Sung, "Mobile monitoring and embedded control system for factory environment," *Sensors*, vol. 13, no. 12, pp. 17379–17413, 2013.

[16] P. Christ, B. Neuwinger, F. Werner, and U. Rückert, "Performance analysis of the nRF24L01 ultra-low-power transceiver in a multi-transmitter and multi-receiver scenario," in *Proceedings of the IEEE Sensors*, pp. 1205–1208, October 2011.

[17] P. Smith, "Comparisons between low power wireless technologies," US Patent CS-213199-AN, 2011.

[18] ANT Wireless, 2014, http://www.thisisant.com/.

[19] S. Sharma, A. L. Vyas, B. Thakker, D. Mulvaney, and S. Datta, "Wireless body area network for health monitoring," in *Proceedings of the 4th International Conference on Biomedical Engineering and Informatics (BMEI '11)*, pp. 2183–2186, IEEE, Shanghai, China, October 2011.

[20] G. Hong and S. Jianxiu, "Design of the temperature signal wireless receiver and display system on polishing interface in chemical mechanical polishing," *Procedia Engineering*, vol. 24, pp. 417–421, 2011.

[21] M. A. A. Hassan, "A review of wireless technology usage for mobile robot controller," in *Proceeding of the International Conference on System Engineering and Modeling (ICSEM '12)*, pp. 7–12, Singapore, June 2012.

[22] S. Khssibi, H. Idoudi, A. van den Bossche, T. Val, and L. A. Saidane, "Presentation and analysis of a new technology for low-power wireless sensor network," *International Journal of Digital Information and Wireless Communications*, vol. 3, no. 1, pp. 75–86, 2013.

[23] J.-S. Lee, Y.-W. Su, and C.-C. Shen, "A comparative study of wireless protocols: bluetooth, UWB, ZigBee, and Wi-Fi," in *Proceedings of the 33rd Annual Conference of the IEEE Industrial Electronics Society (IECON '07)*, pp. 46–51, Taipei, Taiwan, November 2007.

[24] M. Li and M. Zhuang, "An overview of Physical layers on wireless body area network," in *Proceedings of the International Conference on Anti-Counterfeiting, Security and Identification (ASID '12)*, pp. 1–5, Taipei, Taiwan, August 2012.

[25] M.-S. Pan and Y.-C. Tseng, *ZigBee Wireless Sensor Networks and Their Applications*, National Chiao Tung University, Hsin-Chu, Hsinchu, Taiwan, 2006.

[26] A. Pantelopoulos and N. G. Bourbakis, "A survey on wearable sensor-based systems for health monitoring and prognosis," *IEEE Transactions on Systems, Man and Cybernetics. Part C: Applications and Reviews*, vol. 40, no. 1, pp. 1–12, 2010.

[27] L. Ruiz-Garcia, L. Lunadei, P. Barreiro, and J. I. Robla, "A review of wireless sensor technologies and applications in agriculture and food industry: state of the art and current trends," *Sensors*, vol. 9, no. 6, pp. 4728–4750, 2009.

[28] "nRF24AP2 ultra-low power transceiver with ANT ultra-low power network protocol technical backgrounder," 2010, http://www.nordicsemi.com/.

[29] T. Vuorela, V.-P. Seppä, J. Vanhala, and J. Hyttinen, "Wireless measurement system for bioimpedance and ECG," in *Proceedings of the 13th International Conference on Electrical Bioimpedance and the 8th Conference on Electrical Impedance Tomography (ICEBI '07)*, vol. 17, pp. 248–251, Graz, Austria, September 2007.

[30] Z. Yiming, Y. Xianglong, G. Xishan, Z. Mingang, and W. Liren, "A design of greenhouse monitoring & control system based on ZigBee wireless sensor network," in *Proceedings of the International Conference on Wireless Communications, Networking and Mobile Computing (WiCOM '07)*, pp. 2563–2567, Shanghai, China, September 2007.

[31] T. Rault, A. Bouabdallah, and Y. Challal, "Energy efficiency in wireless sensor networks: a top-down survey," *Computer Networks*, vol. 67, pp. 104–122, 2014.

[32] F. Casamassima, E. Farella, and L. Benini, "Power saving policies for multipurpose WBAN," in *Proceedings of the 23rd International Workshop on Power and Timing Modeling, Optimization and Simulation (PATMOS '13)*, pp. 83–90, Karlsruhe, Germany, September 2013.

[33] D. M. Amzucu, H. Li, and E. Fledderus, "Indoor radio propagation and interference in 2.4 GHz wireless sensor networks: measurements and analysis," *Wireless Personal Communications*, vol. 76, no. 2, pp. 245–269, 2014.

[34] *ZigBee Wireless Interfaces and Modules*, 2013, http://www.kanda.com/zigbee-wireless.html.

[35] J. Wang, "Zigbee light link and its applicationss," *IEEE Wireless Communications*, vol. 20, no. 4, pp. 6–7, 2013.

[36] A. Z. Abbasi, N. Islam, and Z. A. Shaikh, "A review of wireless sensors and networks' applications in agriculture," *Computer Standards and Interfaces*, vol. 36, no. 2, pp. 263–270, 2014.

[37] M. Kohvakka, J. Suhonen, M. Kuorilehto, V. Kaseva, M. Hännikäinen, and T. D. Hämäläinen, "Energy-efficient neighbor discovery protocol for mobile wireless sensor networks," *Ad Hoc Networks*, vol. 7, no. 1, pp. 24–41, 2009.

[38] C. Buratti, A. Conti, D. Dardari, and R. Verdone, "An overview on wireless sensor networks technology and evolution," *Sensors*, vol. 9, no. 9, pp. 6869–6896, 2009.

[39] R. Tabish, A. Ben Mnaouer, F. Touati, and A. M. Ghaleb, "A comparative analysis of BLE and 6LoWPAN for U-HealthCare applications," in *Proceedings of the 7th IEEE GCC Conference and Exhibition (GCC '13)*, pp. 286–291, Doha, Qatar, November 2013.

[40] R. Nagarajan and R. Dhanasekaran, "Implementation of wireless data transmission in monitoring and control," in *Proceedings of the 2nd International Conference on Communication and Signal Processing (ICCSP '13)*, pp. 83–87, Melmaruvathur, India, April 2013.

[41] N. Langhammer and R. Kays, "Enhanced frequency hopping for reliable interconnection of low power smart home devices," in *Proceedings of the 8th IEEE International Wireless Communications and Mobile Computing Conference (IWCMC '12)*, pp. 305–310, Limassol, Cyprus, August 2012.

[42] S. Gasparrini, E. Gambi, and S. Spinsante, "Evaluation and possible improvements of the ANT protocol for home heart monitoring applications," in *Proceedings of the 2nd IEEE International Workshop on Measurements and Networking (M and N '13)*, pp. 214–219, Naples, Italy, October 2013.

[43] Digital sensing technology for wireless sport, fitness and health monitoring, 2013, http://www.dynastream.com/.

[44] P. Rawat, K. D. Singh, H. Chaouchi, and J. M. Bonnin, "Wireless sensor networks: a survey on recent developments and potential synergies," *The Journal of Supercomputing*, vol. 68, no. 1, pp. 1–48, 2014.

[45] T. Kuhn, T. Jaitner, and R. Gotzhein, "Online-monitoring of multiple track cyclists during training and competition," in *The Engineering of Sport 7*, M. Estivalet and P. Brisson, Eds., vol. 1, pp. 405–412, Springer, Paris, France, 2008.

[46] A. Le, L. Litz, and T. Jaitner, "A model predictive controller for sensor-based training optimization of a cyclist group," in *The Engineering of Sport 7*, M. Estivalet and P. Brisson, Eds., vol. 1, pp. 413–424, Springer, Paris, France, 2008.

[47] A. Le, T. Jaitner, and L. Litz, "Sensor-based training optimization of a cyclist group," in *Proceedings of the 7th International Conference on Hybrid Intelligent Systems (HIS '07)*, pp. 265–270, Kaiserlautern, Germany, September 2007.

[48] J. Chen, B. Kuo, and C. Chiang, "The development of wireless electrocardiography for bicycling and stairs-climbing," *Journal of Biomechanics*, vol. 40, p. S663, 2007.

[49] A. Zhan, M. Chang, Y. Chen, and A. Terzis, "Accurate caloric expenditure of bicyclists using cellphones," in *Proceedings of the 10th ACM Conference on Embedded Networked Sensor Systems (SenSys '12)*, pp. 71–84, Toronto, Canada, November 2012.

[50] R. B. Edmund, *High-Tech Cycling*, Human Kinetics, Champaign, Ill, USA, 2nd edition, 2003.

[51] A. S. Gardner, J. C. Martin, D. T. Martin, M. Barras, and D. G. Jenkins, "Maximal torque- and power-pedaling rate relationships for elite sprint cyclists in laboratory and field tests," *European Journal of Applied Physiology*, vol. 101, no. 3, pp. 287–292, 2007.

[52] W. Bertucci, R. Taiar, Y. Toshev, and T. Letellier, "Comparison of biomechanical criteria in cycling maximal effort test," *International Journal of Sports Science and Engineering*, vol. 2, no. 1, pp. 36–46, 2008.

[53] A. C. Lim, E. P. Homestead, A. G. Edwards, T. C. Carver, R. Kram, and W. C. Byrnes, "Measuring changes in aerodynamic/rolling resistances by cycle-mounted power meters," *Medicine and Science in Sports and Exercise*, vol. 43, no. 5, pp. 853–860, 2011.

[54] J. Lieh, "Determination of cycling speed using a closed-form solution from nonlinear dynamic equations," *Human Power eJournal*, vol. 10, no. 3, pp. 1–9, 2006.

[55] T. Dahmen and D. Saupe, "Calibration of a power-speed-model for road cycling using real power and height data," *International Journal of Computer Science in Sport*, vol. 10, pp. 18–36, 2011.

[56] A. B. Dahlin, "Size matters: problems and advantages associated with highly miniaturized sensors," *Sensors*, vol. 12, no. 3, pp. 3018–3036, 2012.

[57] S. Park and S. Jayaraman, "Smart textile-based wearable biomedical systems: a transition plan for research to reality," *IEEE Transactions on Information Technology in Biomedicine*, vol. 14, no. 1, pp. 86–92, 2010.

[58] D. Teichmann, A. Kuhn, S. Leonhardt, and M. Walter, "The main shirt: a textile-integrated magnetic induction sensor array," *Sensors*, vol. 14, no. 1, pp. 1039–1056, 2014.

[59] W. Walker, A. L. P. Aroul, and D. Bhatia, "Mobile health monitoring systems," in *Proceedings of the 31st Annual International Conference of the IEEE Engineering in Medicine and Biology Society (EMBC '09)*, pp. 5199–5202, September 2009.

[60] D. L. Peterson, J. K. Moore, D. Fintelman, and M. Hubbard, "Low-power, modular, wireless dynamic measurement of bicycle motion," *Procedia Engineering*, vol. 2, no. 2, pp. 2949–2954, 2010.

[61] H. Hayashi, O. Kagami, and M. Harada, "Results of field trials with wide-area ubiquitous network," in *Proceedings of the IEEE MTT-S International Microwave Symposium Digest (IMS '12)*, pp. 1–3, Montreal, Canada, June 2012.

[62] M. Olieman, R. Marin-Perianu, and M. Marin-Perianu, "Measurement of dynamic comfort in cycling using wireless acceleration sensors," *Procedia Engineering*, vol. 34, pp. 568–573, 2012.

[63] R. Marin-Perianu, M. Marin-Perianu, P. Havinga et al., "A performance analysis of a wireless body-area network monitoring system for professional cycling," *Personal and Ubiquitous Computing*, vol. 17, no. 1, pp. 197–209, 2013.

[64] C.-M. Yang, C.-C. Wu, C.-M. Chou, and C.-W. Yang, "Textile-based monitoring system for biker," in *Proceeding of the 9th IEEE International Conference on Information Technology and Applications in Biomedicine (ITAB '09)*, pp. 1–4, November 2009.

[65] F. Casamassima, A. Ferrari, B. Milosevic, P. Ginis, E. Farella, and L. Rocchi, "A wearable system for gait training in subjects with Parkinson's disease," *Sensors*, vol. 14, no. 4, pp. 6229–6246, 2014.

[66] A. Baca, P. Kornfeind, E. Preuschl, S. Bichler, M. Tampier, and H. Novatchkov, "A server-based mobile coaching system," *Sensors*, vol. 10, no. 12, pp. 10640–10662, 2010.

[67] A. Johansson, W. Shen, and Y. Xu, "An ANT based wireless body sensor biofeedback network for medical E-health care," in *Proceedings of the 7th International Conference on Wireless Communications, Networking and Mobile Computing (WiCOM '11)*, pp. 1–5, September 2011.

[68] K.-M. Yu, J. Zhou, C.-Y. Yu et al., "An event-based wireless navigation and healthcare system for group recreational cycling," in *Proceedings of the 5th International Conference on Mobile Ad-hoc and Sensor Networks (MSN '09)*, pp. 294–298, Fujian, China, December 2009.

[69] C. Raut and V. Giripunje, "The real-time monitoring system for in-patient based on biomedical data acquisition system," in *Proceedings of the International Conference on Information and Network Technology (ICINT '11)*, pp. 110–114, Singapore, 2011.

[70] D. K. Patil and R. Shastri, "Design of wireless electronic stethoscope based on zigbee," *International Journal of Distributed and Parallel Systems*, vol. 3, no. 1, pp. 351–359, 2012.

[71] B. Subramanian and A. Ramasamy, "Efficient zigbee based health care system for arrhythmia detection in electrocardiogram," *European Journal of Scientific Research*, vol. 69, no. 2, pp. 180–187, 2012.

[72] A. Sagahyroon, F. Aloul, A. R. Al-Ali, M. S. Bahrololoum, F. Makhsoos, and N. Hussein, "Monitoring patients' signs wirelessly," in *Proceedings of the 1st Middle East Conference on Biomedical Engineering (MECBME '11)*, pp. 283–286, Sharjah, UAE, February 2011.

[73] S. C. Mukhopadhyay, A. Gaddam, and G. S. Gupta, "Wireless sensors for home monitoring-a review," *Recent Patents on Electrical Engineering*, vol. 1, pp. 32–39, 2008.

[74] N. Belgacem and F. Bereksi-Reguig, "Bluetooth portable device for ECG and patient motion monitoring," *Nature and Technology*, vol. 4, pp. 19–23, 2011.

[75] N. Belgacem, S. Assous, and F. Bereksi-Reguig, "Bluetooth portable device and Matlab-based GUI for ECG signal acquisition and analisys," in *Proceedings of the 7th International Workshop on Systems, Signal Processing and their Applications (WOSSPA '11)*, pp. 87–90, Tipaza, Algeria, May 2011.

[76] A. Tura, M. Badanai, D. Longo, and L. Quareni, "A medical wearable device with wireless bluetooth-based data transmission," *Measurement Science Review*, vol. 3, pp. 1–4, 2003.

[77] V. T. Bai and S. Srivatsa, "Design and simulation of portable telemedicine system for high risk cardiac patients," *International Journal of Biomedical Science*, vol. 1, pp. 219–223, 2006.

[78] Y. V. Varshney and A. K. Sharma, "Design & simulation of Zigbee transceiver system using Matlab," *International Journal of Engineering Trends and Technology*, vol. 4, pp. 1316–1319, 2013.

[79] E. D. Zubiete, L. F. Luque, A. Rodriguez, and I. G. González, "Review of wireless sensors networks in health applications," in *Proceedings of the Annual International Conference of the IEEE Engineering in Medicine and Biology Society (EMBC '11)*, pp. 1789–1793, Boston, Mass, USA, August-September 2011.

[80] SRM, 2013, http://www.srm.de.

[81] Cycleops Power Meters, 2013, http://www.powertap.com/.

[82] Garmin Vector, 2013, http://sites.garmin.com/vector/#suite.

[83] Pioneer, 2013, http://www.pioneerelectronics.com/.

[84] ANT Message Protocol and Usage, 2014, http://www.thisisant.com/.

[85] "Power Estimator," 2013, http://www.thisisant.com/developer/components/power-estimator.

[86] Ultra low power wireless solutions, 2013, http://www.nordicsemi.com/eng/Products/ANT/nRF24AP2-8CH.

[87] ANT Burst Transfers, 2014, http://www.thisisant.com.

[88] J. Beer, "A special report from peak performance/training for cyclists/the road to success—training secrets of the pros," 2006, http://www.worldclassbodybuilding.com/.

[89] N. Langhammer and R. Kays, "Performance evaluation of wireless home automation networks in indoor scenarios," *IEEE Transactions on Smart Grid*, vol. 3, no. 4, pp. 2252–2261, 2012.

[90] C. Gomez, J. Oller, and J. Paradells, "Overview and evaluation of bluetooth low energy: an emerging low-power wireless technology," *Sensors*, vol. 12, no. 9, pp. 11734–11753, 2012.

[91] M. McGrath and C. Scanaill, "Wellness, fitness, and lifestyle sensing applications," in *Sensor Technologies*, pp. 217–248, Apress, 2013.

[92] I. Caballero, J. Vicente Sáez, and B. García Zapirain, "Review and new proposals for zigbee applications in healthcare and home automation," in *Ambient Assisted Living*, J. Bravo, R. Hervás, and V. Villarreal, Eds., vol. 6693, pp. 101–108, Springer, Berlin, Germany, 2011.

[93] H. Joh and I. Ryoo, "A hybrid Wi-Fi P2P with bluetooth low energy for optimizing smart device's communication property," *Peer-to-Peer Networking and Applications*, 2014.

[94] J. Wen and J. Nelson, "Modelling and performance analysis of an adaptive state-transition approach for power saving in Bluetooth," *Simulation Modelling Practice and Theory*, vol. 31, pp. 77–95, 2013.

[95] E. Mackensen, M. Lai, and T. M. Wendt, "Performance analysis of an Bluetooth Low Energy sensor system," in *Proceeding of the 1st IEEE International Symposium on Wireless Systems (IDAACS-SWS '12)*, pp. 62–66, Offenburg, Germany, September 2012.

[96] J. Haase, "Wireless network standards for building automation," in *Embedded Systems for Smart Appliances and Energy Management*, C. Grimm, P. Neumann, and S. Mahlknecht, Eds., vol. 3, pp. 53–65, Springer, New York, NY, USA, 2013.

[97] K. Mikhaylov, N. Plevritakis, and J. Tervonen, "Performance analysis and comparison of bluetooth low energy with IEEE 802.15. 4 and SimpliciTI," *Journal of Sensor and Actuator Networks*, vol. 2, pp. 589–613, 2013.

[98] M. Zareei, A. Zarei, R. Budiarto, and M. A. Omar, "A comparative study of short range wireless sensor network on high density networks," in *Proceedings of the 17th Asia Pacific Conference on Communications (APCC '11)*, pp. 247–252, Sabah, Malaysia, October 2011.

[99] S. Adibi, "Link technologies and BlackBerry mobile Health (mHealth) solutions: a review," *IEEE Transactions on Information Technology in Biomedicine*, vol. 16, no. 4, pp. 586–597, 2012.

[100] A. H. Omre, "Reducing healthcare costs with wireless technology," in *Proceeding of the 6th International Workshop on Wearable and Implantable Body Sensor Networks (BSN '09)*, pp. 65–70, Berkeley, Calif, USA, June 2009.

[101] R. Belchior, D. Júnior, and A. Monteiro, "ANT+ medical health kit for older adults," in *Wireless Mobile Communication and Healthcare*, B. Godara and K. Nikita, Eds., vol. 61 of *Lecture Notes of the Institute for Computer Sciences*, pp. 20–29, Springer, Berlin, Germany, 2013.

[102] J. Song, Y. Zhu, and F. Dong, "Automatic monitoring system for coal mine safety based on wireless sensor network," in *Proceedings of the Cross Strait Quad-Regional Radio Science and Wireless Technology Conference (CSQRWC '11)*, pp. 933–936, Harbin, China, July 2011.

[103] C. Zhang, X. Wang, and X. Li, "Design of monitoring and control plant disease system based on DSP&FPGA," in *Proceedings of the 2nd International Conference on Networks Security, Wireless Communications and Trusted Computing (NSWCTC '10)*, pp. 479–482, Wuhan, China, April 2010.

[104] A. C. Wong, M. Dawkins, G. Devita et al., "A 1 v 5 mA multi-mode IEEE 802.15.6/bluetooth low-energy WBAN transceiver for biotelemetry applications," *IEEE Journal of Solid-State Circuits*, vol. 48, no. 1, pp. 186–198, 2013.

[105] E. Georgakakis, S. Nikolidakis, D. Vergados, and C. Douligeris, "An analysis of bluetooth, Zigbee and bluetooth low energy and their use in WBANs," in *Wireless Mobile Communication and Healthcare*, J. Lin and K. Nikita, Eds., vol. 55, pp. 168–175, Springer, Berlin, Germany, 2011.

[106] J. Decuir, "Introducing bluetooth smart: part II: applications and updates," *IEEE Consumer Electronics Magazine*, vol. 3, pp. 25–29, 2014.

[107] R. Cavallari, F. Martelli, R. Rosini, C. Buratti, and R. Verdone, "A survey on wireless body area networks: technologies and design challenges," *IEEE Communications Surveys and Tutorials*, vol. 16, no. 3, pp. 1635–1657, 2014.

[108] J. Liu, G. Sun, D. Zhao, X. Yao, and Y. Zhang, "MCU-controlling based bluetooth data transferring system," *Procedia Engineering*, vol. 29, pp. 2109–2115, 2012.

[109] J. Padgette, K. Scarfone, and L. Chen, "Guide to bluetooth security," NIST Special Publication, vol. 800, p. 121, 2012.

[110] L. Mraz, V. Cervenka, D. Komosny, and M. Simek, "Comprehensive performance analysis of zigbee technology based on real measurements," *Wireless Personal Communications*, vol. 71, no. 4, pp. 2783–2803, 2013.

[111] V. Custodio, F. J. Herrera, G. López, and J. I. Moreno, "A review on architectures and communications technologies for wearable health-monitoring systems," *Sensors*, vol. 12, no. 10, pp. 13907–13946, 2012.

[112] G. C. May, A. R. Doherty, A. F. Smeaton, and D. G. Warrington, "Correlating multimodal physical sensor information with biological analysis in ultra endurance cycling," *Sensors*, vol. 10, no. 8, pp. 7216–7235, 2010.

[113] T. Reinhardt, W. Norbert, J. Thomas, and V. Roland, "Power measurement in cycling using inductile coupling of energy and data," in *The Engineering of Sport 7*, M. Estivalet and P. Brisson, Eds., vol. 1, pp. 397–403, Springer, Paris, France, 2008.

[114] P. Barratt, "SRM Torque Analysis of Standing Starts in Track Cycling," in *The Engineering of Sport 7*, M. Estivalet and P. Brisson, Eds., vol. 1, pp. 443–448, Springer, Paris, France, 2008.

[115] R. R. Bini, P. A. Hume, and A. Cerviri, "A comparison of cycling SRM crank and strain gauge instrumented pedal measures of peak torque, crank angle at peak torque and power output," *Procedia Engineering*, vol. 13, pp. 56–61, 2011.

[116] S. J. Selvarani, "Online health monitoring system using Zigbee," *International Journal on Computer Science and Engineering*, vol. 3, pp. 1578–1583, 2011.

[117] E. Jovanov, A. Milenkovic, C. Otto, and P. C. de Groen, "A wireless body area network of intelligent motion sensors for computer assisted physical rehabilitation," *Journal of Neuro-Engineering and Rehabilitation*, vol. 2, article 6, 2005.

[118] S. Hu, H. Wei, Y. Chen, and J. Tan, "A real-time cardiac arrhythmia classification system with wearable sensor networks," *Sensors*, vol. 12, no. 9, pp. 12844–12869, 2012.

[119] M. Tavakoli, L. Turicchia, and R. Sarpeshkar, "An ultra-low-power pulse oximeter implemented with an energy-efficient transimpedance amplifier," *IEEE Transactions on Biomedical Circuits and Systems*, vol. 4, no. 1, pp. 27–38, 2010.

[120] R. Trobec, A. Rashkovska, and V. Avbelj, "Two proximal skin electrodes—a respiration rate body sensor," *Sensors*, vol. 12, no. 10, pp. 13813–13828, 2012.

[121] C. A. Boario, M. Lasagni, and K. Römer, "Non-invasive measurement of core body temperature in Marathon runners," in *Proceedings of the IEEE International Conference on Body Sensor Networks (BSN '13)*, pp. 1–6, Cambridge, Mass, USA, May 2013.

[122] H.-Y. Shin, F.-L. Un, and K.-W. Huang, "A sensor-based tracking system for cyclist group," in *Proceedings of the 7th International Conference on Complex, Intelligent, and Software Intensive Systems (CISIS '13)*, pp. 617–622, Taichung, Taiwan, July 2013.

[123] S. B. Eisenman, E. Miluzzo, N. D. Lane, R. A. Peterson, G.-S. Ahn, and A. T. Campbell, "BikeNet: a mobile sensing system for cyclist experience mapping," *ACM Transactions on Sensor Networks*, vol. 6, pp. 87–101, 2009.

[124] J. Y. Chen, F. Wen, C. H. Lin, and D.-F. Lin, "A new automatic measurement system for a bicycle," in *Proceedings of the IEEE International Conference on Mechatronics (ICM '05)*, pp. 514–519, July 2005.

18

A State-of-the-Art Review of the Sensor Location, Flow Observability, Estimation, and Prediction Problems in Traffic Networks

Enrique Castillo,[1] Zacarías Grande,[1] Aida Calviño,[2] W. Y. Szeto,[3] and Hong K. Lo[4]

[1] *Department of Applied Mathematics and Computational Sciences, University of Cantabria, 39005 Santander, Spain*
[2] *Department of Computer Science and Mathematics, University of Rovira i Virgili, 43007 Tarragona, Spain*
[3] *Department of Civil Engineering, The University of Hong Kong, 999077 Pok Fu Lam, Hong Kong*
[4] *Department of Civil and Environmental Engineering, Hong Kong University of Science and Technology, 999077 Sai Kung District, Hong Kong*

Correspondence should be addressed to Zacarıas Grande; zgandrade@gmail.com

Academic Editor: Fanli Meng

A state-of-the-art review of flow observability, estimation, and prediction problems in traffic networks is performed. Since mathematical optimization provides a general framework for all of them, an integrated approach is used to perform the analysis of these problems and consider them as different optimization problems whose data, variables, constraints, and objective functions are the main elements that characterize the problems proposed by different authors. For example, counted, scanned or "a priori" data are the most common data sources; conservation laws, flow nonnegativity, link capacity, flow definition, observation, flow propagation, and specific model requirements form the most common constraints; and least squares, likelihood, possible relative error, mean absolute relative error, and so forth constitute the bases for the objective functions or metrics. The high number of possible combinations of these elements justifies the existence of a wide collection of methods for analyzing static and dynamic situations.

1. Introduction

The problem of knowing how traffic behaves in areas of different zones (cities, regions, etc.) has deserved the attention of researchers for a long period, not only because of its theoretical interest but because of its important practical implications. Observing, estimating, and predicting link, OD, and route flows are considered as very relevant problems in the traffic field, so that many papers have dealt with this problem. In this context, the role of sensors and their location and the flow observability, estimation, and prediction problems become important. Though sensors can be used for many different purposes in the traffic field (see, e.g., Gil Jiménez and Fernández-Getino García [1], Liu et al. [2], and Kong et al. [3]), in this paper, we use sensors to determine the traffic flow. In particular, it is important to distinguish between passive and active sensors. Following Gentili and Mirchandani [4], we call passive sensors to those sensors that simply count vehicles at points in a network and active sensors to those able to identify vehicles, so that routes can be determined. For a detailed description of the different types of sensors see Gentili and Mirchandani [4]. The difference between active and passive sensors is very relevant to the studied problem because the first provide much more information on flows and in particular on route flows, not provided by the last ones. Typical questions of location problems arising in a natural form are the following: determining how many, what type, and where they must be located to observe a given set of flows.

In this paper, we review the existing literature, aiming at trying to analyze the different approaches. However, before starting our study, we want to mention and recognize some previous works related to the reviews of existing works that can be useful to readers. For example, Branston [5] reviews the state of the art of link capacity functions, or volume-delay curves, for use in traffic assignment procedures; Abrahamsson [6] presents a survey of generic approaches and an annotated bibliography of some individual contributions on the OD matrix estimation problem based on traffic counts on links in the transport network and other available information; Montoya-Zamora et al. [7] compare different methods and algorithms; Li et al. [8] present an overview of dynamic traffic assignment models using the variational inequality method and describe the equivalent relations between the variational inequality model and the dynamic user optimal condition model; Peeta and Ziliaskopoulos [9] provide an important state-of-the-art review of dynamic traffic assignment models; Boyce et al. [10] review some analytical formulations of the dynamic traffic assignment problem under a variational inequality point of view; and Szeto and Lo [11] present a very complete and excellent analysis of the dynamic traffic assignment problem and suggest new lines of research.

In this work, we distinguish between observability, estimation, and prediction problems. In the first, we look for information derived exclusively from sensors and other general mathematical relations that need to be satisfied, such as conservation or balance laws, flow nonnegativeness relations, or different flow definition relations. In the second, we add to the previous sources of knowledge about the network other information such as prior flow information and/or statistical models that cover the lack of knowledge required for a proper estimation. Finally, in the third one, we use models to simulate travelers' route choice behavior and load traffic according to the network for prediction purposes.

(1) The Flow Observability Problem. Given a traffic network $(\mathcal{N}, \mathcal{L})$, where $\mathcal{N}$ and $\mathcal{L}$ are the sets of nodes and links, respectively, and assuming that a given subset $\mathcal{O}$ (observed links) of flows have been determined by installing a set of traffic measuring devices, a subset of unobserved flows $\mathcal{U}$ is said to be observable if its flows can be calculated in terms of the observed flows in $\mathcal{O}$. We note that in general the set $\mathcal{O}$ can include any type of flows, such as link, OD, and route flows.

In other words, given the target subset $\mathcal{U}$ of flows, the typical observability problem consists of determining which minimum set of flows $\mathcal{O}$ must be observed (measured) in order to be able to calculate the flows in $\mathcal{U}$.

The reason why the flows in $\mathcal{U}$ can be calculated based on the flows in $\mathcal{O}$ is that all these flows are not independent but related. Thus, the knowledge of all the relations among flows is crucial in the observability problem.

It is important to realize that in the observability problem we look for exact flows. This differentiates this problem from the other two, estimation and prediction of flows, where only some estimates (approximate but not exact flows) are the aim of the analyses.

The above observability problem, defined for the static case, can be extended to the dynamic case and then flows must be understood as flow-time functions. In other words, in the dynamic case we are interested in observing how flows change with time. For the dynamic case, the time delay in arrival from upstream links to downstream nodes needs to be incorporated for maintaining the conservation conditions. We must note, however, that almost all works related to the observability problem deal with the static case. Thus, some research for the dynamic case in this direction would be very useful.

For the sake of illustration, we include below a simple observability problem for the static case, consisting of observing a given subset of routes with a minimum number of cameras (active sensors) and determining where to locate them. The problem written as an optimization problem is (see Castillo et al. [12])

$$M_1 = \min_{\mathbf{y},\mathbf{z},n_c} n_c - \rho_1 \sum_{r\in\mathcal{R}} y_r \tag{1}$$

subject to

$$\sum_{a\in\mathcal{A}|\delta_a^r+\delta_a^{r_1}=1} z_a \geq y_r; \quad \forall r, r_1 \in R;\ r \neq r_1 \in \mathcal{R}, \tag{2}$$

$$\sum_{a\in\mathcal{A}} z_a\delta_a^r \geq y_r; \quad \forall r \in \mathcal{R}, \tag{3}$$

$$\sum_{a\in\mathcal{A}} z_a = n_c, \tag{4}$$

$$y_r = 1, \quad \forall r \in \mathcal{R}_{\text{obs}}, \tag{5}$$

$$n_c \leq n_c^{\max}, \tag{6}$$

$$\sum_{a\in\mathcal{A}} z_a\delta_a^r \leq n_{\max}; \quad \forall r \in \mathcal{R}, \tag{7}$$

where n_c is the minimum number of required cameras; z_a is a binary variable that equals one if link a contains a sensor and zero; otherwise y_r is a binary variable that equals one if route r is observed (can be distinguished from others) and zero; otherwise, $\mathbf{y}$ and $\mathbf{z}$ are the vectors of the corresponding variables; $\mathcal{R}$ is the set of routes; $\mathcal{A}$ is the set of links; δ_a^r is the element in ath row and rth column of the link-route incidence matrix; $\mathcal{R}_{\text{obs}}$ is the subset of routes that we want to observe; $n_c^{\max}$ is the maximum number of available cameras; $n_{\max}$ is the maximum number of scanned links per route; and ρ_1 is a nonnegative small number.

Objective function (1) minimizes the number of cameras and among all solutions chooses the one with the most observed routes. Constraint (2), if the binary variable y_r is equal to 1, guarantees that route r is able to be distinguished by the subset of scanned links from the other routes. Note that if $\delta_a^r + \delta_a^{r_1} = 1$, link a belongs only to one of the routes and that if $\sum z_a \geq y_r$ and $y_r = 1$, at least one scanned link has this property. On the other hand, if $y_r = 0$, the constraint always holds. Constraint (3) if $y_r = 1$ ensures that route r, which is able to be distinguished by the subset of scanned links from other routes, contains at least one scanned link. Constraint (4) guarantees that n_c cameras are used. Constraint (5) guarantees that each route in $\mathcal{R}_{\text{obs}}$ is identified by the subset of scanned links. Constraint (6) is a budget constraint, which limits the number of cameras to a maximum $n_c^{\max}$. Finally, constraint (7) limits the maximum number of scanned links in any route to $n_{\max}$.

The interested reader can see other simple observability problems for the static case in Yang and Zhou [13] or Gentili and Mirchandani [14].

(2) The Flow Estimation Problem. If flows are unobservable or there are variations in the observations, the observability methods are not sufficient and we must look for alternative options; then the estimation problem arises. The problem of estimating a set of flows consists of using methods to guess the unobservable flows in this set with maximum possible precision. To this end, observed flows and other sources of information, such as flow prior information or mathematical models and statistical methods, must be used. The estimation problem refers to past flow values, that is, flows that could have been measured but were not. Typical examples of traffic estimation problems are determining point estimates or confidence intervals of link, OD, or route flows.

A simple example of the flow estimation problem for the static case stated as an optimization problem is the following:

$$\min_{\mathbf{f},\mathbf{t},\mathbf{v}} Z = \rho_1 \sum_{r,e} \left(\frac{f_r - f_r^0}{f_r^0} \right) \omega_{r,e}^f \left(\frac{f_e - f_e^0}{f_e^0} \right) + \rho_2 \sum_{k,e} \left(\frac{\tau_k - \tau_k^0}{\tau_k^0} \right) \omega_{k,e}^\tau \left(\frac{\tau_e - \tau_e^0}{\tau_e^0} \right) + \sum_{a,e} \left(\frac{v_a - v_a^0}{v_a^0} \right) \omega_{a,e}^\ell \left(\frac{v_e - v_e^0}{v_e^0} \right) \quad (8)$$

subject to the constraints

$$\tau_k = \sum_{r \in \mathcal{R}_w} f_r; \quad k \in \mathcal{W}, \quad (9)$$

$$v_a = \sum_{r \in \mathcal{R}} \delta_r^a f_r; \quad a \in \mathcal{L}, \quad (10)$$

$$\widehat{v}_a = v_a; \quad a \in \mathcal{C}, \quad (11)$$

$$\widehat{w}_s = \sum_{r \in \mathcal{R}} f_r \eta_r^s; \quad s \in \mathcal{S}, \quad (12)$$

$$f_r \geq 0; \quad r \in \mathcal{R}, \quad (13)$$

where $\mathbf{f}$, $\mathbf{t}$, and $\mathbf{v}$ are the vectors of route, OD, and link flows, respectively, ρ_1 and ρ_2 are nonnegative numbers that measure the relative weight we give to the prior information of route and OD flows with respect to link flows, f_e, τ_e, and v_e and f_e^0, τ_e^0, and v_e^0 are the route, OD, and link flows and the corresponding previous values, respectively, $\omega_{r,e}^f$, $\omega_{k,e}^\tau$, and $\omega_{a,e}^\ell$ are the elements of the covariance matrices associated with route, OD, and link flows, respectively, v_a is the flow on link a, $\widehat{v}_a$ are the counted (observed) flows on link a, $\mathcal{C}$ is the subset of links where the flow is counted, $\mathcal{S}$ is the subset of all observed combinations of links associated with the different cameras, η_r^s is the element of route-scanned combination incidence matrix for route r, which equals one if route r contains the subset $s \subseteq \mathcal{S}$ of scanned links and no more scanned links and zero otherwise, and $\widehat{w}_s$ is the flow associated with the plate scanned link combination s.

Objective function (8) is a least squares function, constraints (9) and (10) are the relations between OD and link flows with route flows, respectively, constraints (11) and (12) force the corresponding flows to coincide with the observed flows, and (13) forces the nonnegativity of route flows.

Solving this problem, estimates of all flows f_r, v_a, and τ_k can be obtained. Note that since the observed flows are not sufficient to estimate all flows, we use extra information, normally previously observed flows, and a least squares metric to get the estimated flows. The interested reader can see other simple estimation problems in Cascetta [15] or Papola and Marzano [16].

Since observations play a key role in this type of problems, the optimization problem above can be complicated by including some location problems and then we obtain a special case of problem known as "sensor location flow estimation problem" (see Gentili and Mirchandani [14]). The problems analyzed in this paper are very general and include these as particular cases.

In addition, the flows can be replaced by flow-time functions and then we have dynamic estimation problems, which are stochastic estimation problems. Research in this direction is needed too.

(3) The Prediction Problems. The flow prediction problem consists in predicting a certain subset of flows that will take place under given circumstances. Prediction models are mainly based on statistical traffic data and on simulating the route choice behaviors of users or drivers, assuming either user equilibrium or system optimal or stochastic user equilibrium. In the first case, we have macro, meso, and micro models and also static and dynamic models, point queue, or spatial queue models, depending on the detail at which the traffic is modelled. Prediction models use mathematical equations to reproduce reality and previous data on which the predictions or the model parameters are based on. Typical examples of prediction problems involve questions such as the following: how do the traffic flows change if some changes in the traffic network are done?

Contrary to the estimation problem, the prediction problem refers to future flow values; that is, only flows that occurred at the prediction time can be measured.

In prediction problems the mathematical models reproducing the vehicle behavior play a relevant role and constitute its main components. However, in some cases, estimation models are complemented with these mathematical models and then they can be considered in the border between estimation and prediction models.

A very simple example of the flow prediction problem stated as an optimization problem, often referred to as the Beckmann transformation, is the following:

$$\min_{\mathbf{f},\mathbf{v}} Z = \sum_{a \in \mathcal{A}} \int_0^{v_a} c_a(v)\, dv = \sum_{a \in \mathcal{A}} C_a(v_a) \quad (14)$$

subject to the constraints

$$\sum_{r\in\mathscr{R}} \xi_{rw} f_r = \tau_w; \quad w \in \mathscr{OD},$$
$$\sum_{r\in\mathscr{R}} f_r \delta_a^r = v_a \quad a \in \mathscr{A}, \tag{15}$$
$$f_r \geq 0; \quad r \in \mathscr{R},$$

where **f** and **v** are the vectors of route and link flows, respectively, v is a dummy variable, $c_a(v_a)$ is the link a travel time function (e.g., the BPR function), $C_a(v_a)$ is its integral, ξ_{rw} is the element of the route-OD incidence matrix, which equals 1 if route r belongs to OD pair w and zero otherwise, and δ_a^r is one if link a belongs to route r and zero otherwise. Note that the travel time function $c_a(v_a)$ reproduces the congestion effect. Note also that the selection of the objective function in (14) implies a Wardrop solution; that is, vehicles with the same OD use only some routes that have the same associated minimum travel time.

This problem can be complicated by adding flow observation and sensor location constraints and variables and considering the dynamic case, that is, time-dependent flow travel time functions, rather than static travel time functions. Thus, in this paper we understand the prediction problem in a general sense.

2. An Integrated View of the Problems

With a great generality, we can state all the above as optimization problems. In fact, mathematical optimization provides a general framework to all of them. In order to understand an integrated view of all the previous problems, we dedicate this section to analyze the different components involved in traffic flow problems when considered as optimization problems.

As it becomes apparent from the previous three examples, an optimization problem has four main components as follows.

(1) Data. They mainly consist of the network topology and the observed flows but in some cases can include flow prior information and some flow parameters such as OD-link proportions, route-OD, or node-link probability choices.

(2) Variables. They are the unknown flows, the number of sensors, and their locations and can include some of the above parameters: OD-link proportions, route-OD or node-link probability choices, and so forth, when they are unknown.

The optimal variable values resulting from solving the optimization problem provide the solution to the problem being considered.

(3) Constraints. They are relations that must be satisfied by the unknowns for them to provide a credible and realistic solution.

There are several types of constraints:

(a) Conservation constraints: they express a flow balance that always must be satisfied. For example, if no flow is retained at the nodes, the total flow entering any node must be equal to the total flow exiting the node.

(b) Nonnegativity of flows: it is obvious that unidirectional flows cannot be negative.

(c) Bounded link capacities: these are a clear physical condition.

(d) Flow definition constraints: they are a consequence of the definition of certain flows such as link, OD, and route flows and their relations.

(e) Choice probability definition: this is a consequence of the definition of choice probabilities.

(f) Observation constraints: when flow data are observed, we can enforce the network flows to be consistent with the observations. These relations take a different form depending on the type of data (counts, plate scanned data, AVI, etc.).

(g) Flow propagation constraints: they depict how flow propagates within a link and reproduce how congestion increases the link travel time.

(h) Specific model constraints: if some models are used to improve the knowledge of the traffic behaviour, they provide relations that increase the number of constraints to be satisfied.

We note that the first six types of constraints are linear, but the rest can be nonlinear.

The model constraints play a very important role in observability, estimation, and prediction problems. In particular, if the system of constraints has a unique solution, the optimization problem degenerates and we can reduce our problem for solving the system of constraints to get the desired flow solution and/or sensor locations. This can occur, for example, when plate scanning data is sufficient to identify uniquely all flows involved in the problem.

However, the set of constraints normally has infinitely many solutions and we need to add more constraints if a unique solution is desired. The most common methods to add this complementary information are the least squares and the generalized least squares methods. For example, some authors minimize the sum of squares of the differences between the predicted and the prior matrices, correcting for the variances and covariances of each flow, as Cascetta [17], who uses the link traffic counts to estimate the OD matrix by means of an assignment model, proposes a generalized least squares estimator, and considers the measurement errors and the temporal variability in the observed flows (see also Cascetta and Nguyen [18], Doblas and Benítez [19], Wong et al. [20], Castillo et al. [21], and Mínguez et al. [22]).

It is also important to realize that the above constraints are not necessarily independent; that is, there can be some redundancies that must be taken into account. It is important to realize that redundancy can lead to incompatibility of the system of constraints in which case the resulting optimization problem has no solution. Incompatibility is real in practice when we have redundant constraints and/or observe redundant data because measurements have errors that can lead to inconsistency. In this case, the corresponding equality constraints must be replaced, for example, by least squares type constraints, which allow us to consider these errors and minimize them. For example, Li and Ouyang [23] propose a reliable facility location model to optimize traffic travel time estimation in addition to individual sensor flow coverage, while considering probabilistic sensor failures.

(4) Objective Function. This element is crucial in an optimization problem and expresses the criterion under which we select one optimal solution among all the consistent candidate solutions.

We note that when the system of constraints is consistent but undetermined, the optimality conditions (Karush, Kuhn, and Tucker (KKT) or other ones) complete and replace the required set of constraints to limit the set of feasible solutions and lead to a unique solution in many cases. Under this point of view the optimality conditions can be considered as an additional set of constraints that we finally impose to our problem.

In this paper we use as reference the above structure to discuss the different approaches used to solve the flow observability, estimation, and prediction problems.

Before analyzing the existing solutions given to the observability problem in the existing literature, it is convenient to discuss in detail the different elements: data, constraints, and objective functions that appear in traffic problems. This is the aim of the following sections.

2.1. Data Components. As previously indicated, data consists of several components. Probably, the most important data is the network topology that reveals and limits the possible vehicle displacements. If some models are used, the model parameters can also be a part of the data. Though difficult to obtain, sometimes this information is complemented with the allowable routes. In addition, the observed data obtained from active or passive sensors are basic to solve these problems. However, when observed data are not sufficient for observability, they can be complemented with prior information to allow for unique estimates. This supplementary information consists of prior or obsolete data f_r^0, τ_k^0, and v_a^0 of route, OD, and/or link flows, respectively.

In the case of underspecification, the only possibility for solving the problem and getting a unique solution consists of adding some extra information, which normally comes in the form of a combination of prior information and observed flows together with some optimization property.

2.2. Constraints. For the sake of generality and to facilitate the understanding, we first present the dynamic case and later we deal with the static case (a.k.a. steady state, fixed point, etc.). See Friesz et al. [24], Cantarella and Cascetta [25], and Lo and Szeto [26] for a particular example.

2.2.1. Traffic Flow Dynamics. We start by saying that obviously the traffic flow dynamics case is much more complicated than the static one (see Cascetta et al. [27]).

In dynamic traffic assignment we normally consider two components: travel choice and traffic flow. At first we analyze how the users select their routes. Second we analyze how traffic propagates in a network once the users have decided where to travel and which routes are used.

In this context, we have the following dynamic constraints.

(1) Dynamic Conservation Equations. They correspond to the flow conservation laws at the nodes. The conservation equations (total flow entering the node = total flow exiting the node) for the link flows are

$$\sum_{a\in\mathcal{I}_n} V_a^i(t) - \sum_{a\in\mathcal{O}_n} V_a^o(t) = I_n(t) - O_n(t); \quad n \in \mathcal{N};\ t \in \mathcal{T}, \tag{16}$$

where $V_a^i(t)$ and $V_a^o(t)$ are the cumulated flow that have entered and left link a at time t, respectively, $\mathcal{I}_n$ and $\mathcal{O}_n$ are the set of all links that supply or take flow from the node n, respectively, $I_n(t)$ and $O_n(t)$ are the cumulated external flows entering and exiting node n up to time t, respectively, and $\mathcal{T}$ is the set of selected times:

$$V_a^i(t) = V_a^o(t + d_a(t)) \tag{17}$$

and $d_a(t)$ is the travel time of a vehicle that enters link a at time t.

The set of (16) implies one conservation equation per node n (finite number) and time t in set $\mathcal{T}$ (finite or infinite). The main advantage of considering cumulated flows $V_a^i(t)$ instead of flow intensities is that we do not need to use differential equations but functional equations. In other words, this statement avoids derivatives. Replacing differential equations by functional equations is very relevant and should not be overlooked (see Wu et al. [28] and Castillo et al. [29]).

(2) Nonnegativity of Link Flows. Since flows cannot be negative, we must have an increasing cumulative flow function and

$$V_a^i(0) \geq 0; \quad a \in \mathcal{L}. \tag{18}$$

If this condition is not explicitly considered, we can have invalid (negative) flow solutions.

(3) Link Capacity Constraints. Since the link capacities are finite, we must have bounded link flows; that is,

$$V_a^i(t) - V_a^i(t-\varepsilon) \leq \varepsilon cp_a; \quad a \in \mathcal{L};\ t \in \mathcal{T}, \tag{19}$$

where cp_a is the capacity of link a and ε is an arbitrary small positive number.

This is another important condition that should not be forgotten. For example, Bierlaire [30] mentions that unbounded flows can arise in some models. This is the consequence of omitting this constraint.

(4) Flow Definition Constraints. The definitions of OD and route flows lead to the following set of constraints.

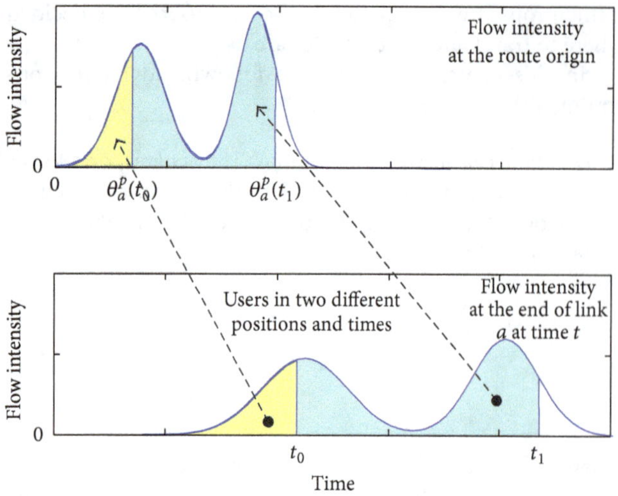

FIGURE 1: Illustration of how the traffic profiles associated with different points and times are related.

(i) OD-Link Flow Relations. The OD-link flow relations are given by

$$V_a^i(t) = \sum_{w\in\mathcal{OD}} \sum_{r\in\mathcal{R}_w} \delta_a^r p_{rw}(\theta_a^r(t))\, T_w(\theta_a^r(t)); \quad a \in \mathcal{L};\ t \in \mathcal{T}, \tag{20}$$

where $\mathcal{R}_w$ is the set of routes of OD pair w, δ_r^a is the element of the link-route incidence matrix and equals one if link a is on route r and equals zero otherwise; $\mathcal{W}$ is the set of all OD pairs and $p_{rw}(t)$ is the proportion of cumulated users choosing route r in OD pair w at time t and $T_w(t)$ is the cumulated flow that has entered a route in OD pair w at time t and $\theta_a^r(t)$ is the entry time to route r of a vehicle entering link a at time t (see Figure 1).

(ii) OD-Route Flow Relations. The OD-route flow relations are given by

$$T_w(t) = \sum_{r\in\mathcal{R}_w} F_r(t); \quad t \in \mathcal{T};\ w \in \mathcal{W}, \tag{21}$$

where $F_r(t)$ is the cumulated flow that has entered route r at time t.

These relations (21) can also be written as

$$F_r(t) = p_{rw}(t)\, T_w(t) \quad t \in \mathcal{T},\ r \in \mathcal{R}, \tag{22}$$

where $\mathcal{R}$ is the set of all routes.

(iii) Route-Link Flow Relations. The route-link flow relations are given by

$$V_a^i(t) = \sum_{r\in\mathcal{R}} \delta_r^a F_r(\theta_a^r(t)); \quad a \in \mathcal{L};\ t \in \mathcal{T}. \tag{23}$$

(5) Observation Constraints. When we observe data, the following relations must hold.

(i) Counting Data Information. The most common technique to count link users consists in using link counters that provide information on link flows.

Counting data can be taken into consideration by means of constraints such as

$$V_a^i(t) = \widehat{V}_a^i(t); \quad a \in \mathcal{C},\ t \in \mathcal{T}^*, \tag{24}$$

where $\widehat{V}_a^i(t)$ is the counted cumulated flow that has entered link a at time t, $\mathcal{C}$ is the subset of links where the flow is counted, and $\mathcal{T}^*$ is the set of times at which data are collected.

(ii) Plate Scanning Data Information. If a set of cameras are located to register car plate numbers of vehicles travelling on links, plate scanning data can also be considered by means of constraints:

$$\widehat{W}_s(t) = \sum_{s\in\mathcal{S}} \eta_r^s F_r(\theta_a^r(t)); \quad s \in \mathcal{S},\ t \in \mathcal{T}^*, \tag{25}$$

where $\mathcal{S}$ is the subset of all observed combinations of links associated with cameras, $\widehat{W}_s(t)$ is the total plate scanned observations in subset s at time t, and η_r^s equals one if route r contains the subset s of scanned links $s \subseteq \mathcal{S}$ and no more scanned links and zero otherwise.

(6) Flow Propagation Constraints. They describe how flow propagate within a link and how link travel time depends on congestion. There are several alternative approaches to formulate the constraints.

(i) Link-based travel time functions: they express the link travel time as a function of the link volume. Some examples are the temporal extensions of the BPR function:

$$\tau_a(t) = \tau_{0a}\left(1 + \kappa\left(\frac{x_a(t)}{x_a^{\max}}\right)^{\alpha}\right), \tag{26}$$

where $x_a(t)$ is the traffic volume on link a at time t, τ_{0a} is the free-flow travel time on link a, and $\tau_a(t)$ is the link a travel time of a vehicle that enters the link at time t.

Some authors as, for example, Boyce et al. [10], propose a mixed function of the form

$$\tau_a(t) = \tau_a(x_a(t), u_a(t), w_a(t)), \tag{27}$$

where $u_a(t)$ is the rate of flow entering link a at time t, $w_a(t)$ is the link exiting traffic density, and $x_a(t)$ is the link traffic volume.

Unfortunately, these models suffer from realism and consistency (see Merchant and Nemhauser [31] or Astarita [32]).

Some examples are the models used by Friesz et al. [33], Fernández and De Cea [34], and Wu et al. [35].

(ii) FIFO constraints: they guarantee the satisfaction of the FIFO rule (see Astarita [32] or Peeta and Ziliaskopoulos [9]):

$$w_a(t + \tau(t)) = \frac{u_a(t)}{1 + \tau_a'(t)}. \tag{28}$$

(iii) Exit flow functions: they were suggested by Merchant and Nemhauser [31] and express the flow density exiting link a at time t as a function of congestion; that is,

$$w_a(t) = g_a(x_a(t)). \tag{29}$$

(iv) Flow-density relations: Daganzo [36, 37] proposed the following piecewise linear relationship between traffic flow, y, and density, k:

$$y = \min\{Vk, Q, W(k_{\text{jam}} - k)\}, \tag{30}$$

where k_{jam}, Q, V, and W denote, respectively, jam density, inflow capacity (or maximum allowable inflow), free-flow speed, and the speed of the backward shock wave (or the backward propagation speed of disturbances in congested traffic).

(v) Different queue models: vehicles can be considered as simple points (point queue models) or as elements occupying a space in the links (physical queue models) (see Szeto and Lo [38], Lo [39], Rubio-Ardanaz et al. [40], Kuwahara and Akamatsu [41], Lo and Szeto [26, 42], Nogal [43], or Calviño [44]). The above constraints belong to point queue models. One possibility for physical queue models is to divide links into moving and queuing sections (see Castillo et al. [45]).

(7) Specific Model Constraints. If we use some model, such as UE or SUE models to reproduce traffic flows in the network, we must add the corresponding constraints. One example is the case of variational type models for route choices in which the constraints express a variational principle, such as

$$\begin{aligned} f_r(t)(\pi_r(t) - \pi_{w(r)}(t)) &= 0; \quad r \in \mathscr{R};\ t \in \mathscr{T}, \\ \pi_r(t) - \pi_{w(r)}(t) &\geq 0; \quad r \in \mathscr{R};\ t \in \mathscr{T}, \end{aligned} \tag{31}$$

where $\pi_r(t)$ and $\pi_{w(r)}(t)$ are the travel time for travelers entering route r at time t and the OD pair w containing route r, respectively.

As indicated, we must be careful in selecting the constraints of the model in order to avoid violations of the FIFO rule and causality that means that the speed and travel time of a vehicle must be affected only by the vehicles ahead. For example, Szeto and Lo [11] show that since

$$x_a(t + \tau_a(t)) = W_a(t + \tau_a(t)) - U_a(t), \tag{32}$$

the outflow $g_a(x_a(t + \tau_a(t)))$ depends on the vehicles that entered link a before $t + \tau_a(t)$, and then causality is violated.

Another important point to be considered is the possibility of redundancy or contradiction of constraints. This occurs, for example, in some mixed approaches that force exit and travel time function constraints that are contradictory.

2.2.2. Static or Steady State Case. The steady state case of the preceding case leads to the well-known system of equations for the static situation.

(1) Conservation Equations. They correspond to the flow conservation at the nodes. In this case the conservation equations (total flow entering the node = total flow exiting the node) for the link flows are

$$\mathbf{Bv} = \sum_{a \in \mathscr{I}_n} v_a - \sum_{a \in \mathscr{O}_n} v_a = I_n - O_n; \quad n \in \mathscr{N}, \tag{33}$$

where $\mathbf{B}$ is the node-link incidence matrix that contains zeroes, ones, and minus ones, v_a is the flow of link a, and I_n and O_n are the external flows entering and exiting node n, respectively.

The set of (33) implies one conservation equation per node. The rank of the node-link incidence matrix of a network is $n - 1$ (see Strang [46]).

(2) Nonnegativity of Link Flows. Since flows cannot be negative, we must have

$$v_a \geq 0; \quad a \in \mathscr{L}. \tag{34}$$

Unfortunately, sometimes this condition is not explicitly considered and this can lead to invalid (negative) flow solutions.

(3) Congestion and Link Capacity Constraints. Since the link capacities are finite, we must have bounded link flows; that is,

$$v_a \leq cp_a; \quad a \in \mathscr{L}, \tag{35}$$

where cp_a is the capacity of link a. This is another important condition that should not be forgotten.

(4) Flow Definition Constraints

(i) OD-Link Flow Relations. The OD-link flow relations are given by

$$v_a = \sum_{w \in \mathscr{W}} \sum_{r \in \mathscr{R}_w} \delta^r_a p_{rw} \tau_w; \quad a \in \mathscr{L}, \tag{36}$$

where τ_w is the flow of OD pair w, $\mathscr{W}$ is the set of ODs, and f_r is the flow on route r.

(ii) OD-Route Flow Relations. The OD-route flow relations are given by

$$\tau_w = \sum_{r \in \mathscr{R}_w} f_r; \quad w \in \mathscr{W}. \tag{37}$$

Since the OD flows τ_w; $w \in \mathscr{W}$ are nonnegative numbers, (37) shows that the OD flow vector $\mathbf{t}$ is a cone generated by the column vectors of the OD-route incidence matrix ξ_{rw}.

These relations can also be written as

$$f_r = p_{rw} \tau_w \quad r \in \mathscr{R}. \tag{38}$$

However, this is not convenient because p_{rw} values are difficult or impossible to obtain.

(iii) Route-Link Flow Relations. The route-link flow relations are given by

$$v_a = \sum_{r \in \mathscr{R}} \delta^a_r f_r; \quad a \in \mathscr{L}. \tag{39}$$

Since the route flows $f_r; r \in \mathcal{R}$ are nonnegative numbers, (39) shows that the link flow vector $\mathbf{v}$ is a cone generated by the column vectors of the link-route incidence matrix.

(5) Observation Constraints. The constraints associated with the observations in this case become as follows.

(i) Counting Data Information. Counting data can be taken into consideration by means of constraints such as

$$v_a = \hat{v}_a; \quad a \in \mathcal{C}, \tag{40}$$

where $\hat{v}_a$ is the counted (observed) flow on link a and $\mathcal{C}$ is the subset of links where the flow is counted.

(ii) Plate Scanning Data Information. Plate scanning data can also be considered by means of constraints:

$$\hat{w}_s = \sum_{r \in \mathcal{R}} f_r \eta_r^s; \quad s \in \mathcal{S}, \tag{41}$$

where $\mathcal{S}$ is the subset of all observed combinations of links associated with the different cameras and $\hat{w}_s$ is the number of observations in subset $s \in \mathcal{S}$.

Link flow estimates can be obtained as

$$\hat{v}_a = \sum_{s \in \mathcal{S}} \hat{w}_s \kappa_s^a = \sum_{s \in \mathcal{S}} \sum_{r \in \mathcal{R}} f_r \eta_r^s \kappa_s^a; \quad a \in \mathcal{L}, \tag{42}$$

where κ_s^a is one if link a appears in the link combination s and zero otherwise.

Since the route flows $f_r; r \in \mathcal{R}$ are nonnegative numbers, (41) shows that the set of plate scanned link flow vectors is a cone generated by the column vectors of the link-scanned combination incidence matrix.

(6) Flow Propagation Constraints. Since flows and travel times are dependent on congestion levels, cost or travel time functions exist that express the cost or travel time as a function of the congestion level. One example is the BPR function:

$$\tau_a = \tau_{0a} \left(1 + \kappa \left(\frac{v_a}{v_a^{\max}} \right)^{\alpha} \right), \tag{43}$$

where τ_a and τ_{0a} are the travel time and free travel time, respectively, of link a, κ and α are positive coefficients, and $v_a^{\max}$ is the practical capacity of link a.

(7) Specific Model Constraints. Some model constraints can be stated as optimization problems. Some examples are the UE and SUE models.

If the analytical solution of these models is known they can be incorporated directly. One example is the case of the logit model for which the probability of using route r in OD pair w becomes

$$p_{rw} = \frac{\exp(-\theta_w \psi_r)}{\sum_{s \in \mathcal{R}_w} \exp(-\theta_w \psi_s)}; \quad w \in \mathcal{W}, \ r \in \mathcal{R}, \tag{44}$$

where ψ_r is the travel time on route r and θ_w is a parameter that scales the perceived travel time by OD pair w.

Another example is the variational model for route choice:

$$\begin{aligned} f_r \left(\pi_r - \pi_{w(r)}\right) &= 0; \quad r \in \mathcal{R}, \\ \pi_r - \pi_{w(r)} &\geq 0; \quad r \in \mathcal{R}, \end{aligned} \tag{45}$$

where π_r and $\pi_{w(r)}$ are the travel time on route r and the minimum travel time in OD pair containing route r, respectively. The principle states that only routes with the minimum travel time are used for each OD pair.

(8) Budget Limit Constraints. This is a constraint that should not be forgotten when there are budget limits. Some examples can be found in Ehlert et al. [47] and Mínguez et al. [22] who present some optimization problems to locate link counters taking existing sensors into account, budget limitation, and prior OD data.

2.3. Objective Functions (Metrics). In the existing literature there exist statistical measures to quantify the quality of flow estimates. The most common examples are the root mean square error (RMSE) and the mean absolute error (MAE). For example, Sherali et al. [48] present two models, one based on a least squares estimation approach and the other based on a least absolute norm approach, for the estimation of split parameters that prescribe an OD matrix based on dynamic information regarding entering and exiting traffic volumes through an intersection or a small freeway segment.

Very common methods in the traffic area are the RMSE and MAE that measure the closeness between the observed (true) flows and the estimated flows, no matter if they are link, OD, route or, any other flow values.

In the following, some of the most important measures are described.

(1) Generalized Least Squares (GLS). The traffic flows can be obtained by solving optimization problems similar to the one in (8) (see Cascetta [15]) subject to some of the constraints, such as (33) to (41).

Note that we have used relative errors in the objective function, in order to get dimensionless ratios, and prior values have been selected for normalization in order to avoid the presence of zeros in the denominator.

If the system of constraints has a unique solution, then objective function (8) has no role and the prior data do not affect the solution of the optimization problem. Otherwise, the system of constraints has an infinite number of solutions and the optimal value of the problem depends on the prior values f_r^0, τ_k^0, and v_a^0. Thus, we can interpret that the prior f_r^0, τ_k^0, and v_a^0 complement the information lacking from the observations. Nevertheless, even in this case of an infinite number of solutions some particular flows can be unique and it is important to discover what the subset of these flows is.

Apart from those already mentioned, the least squares metric was used by other many authors, for example, Gentili and Mirchandani [4], Eisenman et al. [49], Castillo et al. [50], and Gentili and Mirchandani [14].

Expression (8) after being modified to take into account the sensor location problem becomes (see Mínguez et al. [22])

$$Z = \rho_1 \sum_{r\in\mathscr{R}} \varpi_r^f \left(\frac{f_r^0 - \sum_{r\in\mathscr{R}} f_r^0 y_r}{f_r^0} \right)^2 + \rho_2 \sum_{w\in\mathscr{W}} \varpi_k^t \left(\frac{\tau_w^0 - \sum_{r\in\mathscr{R}_w} f_r^0 y_r}{\tau_w^0} \right)^2 + \sum_{a\in\mathscr{A}} \varpi_a^\ell \left(\frac{v_a^0 - \sum_{r\in\mathscr{R}} \delta_a^r f_r^0 y_r}{v_a^0} \right)^2, \quad (46)$$

where τ_w^0, f_r^0, and v_a^0 are the prior OD, route, and link flows, respectively, R_w is the set of routes of OD pair w, and y_r is a binary variable equal to one if route r is identified (observed) uniquely through the scanned links and zero otherwise.

Note that because of y_r binary variable the covered flows are considered in (46).

(2) Mean Absolute Relative Error (MARE). This metric was proposed by Mínguez et al. [22]:

$$\mathrm{MA} = \frac{1}{|\mathscr{W}|} \sum_{w\in\mathscr{W}} \left| \frac{\tau_w^0 - \sum_{r\in\mathscr{R}_w} f_r^0 y_r}{\tau_w^0} \right| = \frac{1}{|\mathscr{W}|} \left(\sum_{w\in\mathscr{W}} \sum_{r\in\mathscr{R}_w} \frac{f_r^0 (1 - y_r)}{\tau_w^0} \right), \quad (47)$$

which implies that minimizing the MARE is equivalent to minimizing the sum of relative route flows of unobserved routes or equivalently maximizing the sum of relative route flows of observed routes.

(3) Possible Relative Error (PRE). Yang et al. [51] were concerned about the quality of the estimated OD matrix and proposed the concept of "Maximum Possible Relative Error" (MPRE) and showed that it can be formulated as a simple quadratic programming problem. This concept was used by Yang and Zhou [13] too.

The MPRE is the optimal value of the optimization problem:

$$\underset{w\in\mathscr{W}}{\text{Minimize}} \frac{1}{|\mathscr{W}|} \sum_{w\in\mathscr{W}} \left(\frac{\tau_w^* - \tau_w}{\tau_w} \right)^2 \quad (48)$$

subject to

$$\sum_{w\in\mathscr{W}} p_{aw} (\tau_w^* - \tau_w) = 0; \quad a \in \mathscr{A}, \quad (49)$$

where τ_w^* are the true or reference OD values.

Viti et al. [52] present a new method to optimize the position of traffic counts for reliable state estimation and prediction in complex networks by reformulating the MPRE to minimize the error in link traffic states. A simple solution algorithm to the problem was proposed that uses link flow and travel time correlations between links to select, in sequence, the most representative locations in the network.

(4) Total Demand Scale (TDS). It was proposed by Bierlaire [30] and used by Chen et al. [53] and Gan et al. [54].

The TDS measures the intrinsic undeterminate nature of the OD estimation problem and is independent of the OD estimation method. It is based on the route choice proportions, network topology, and traffic counts. It can be calculated by solving two linear programs:

$$\phi_{\max} = \underset{w\in\mathscr{W}}{\text{Max}} \sum_{w\in\mathscr{W}} \tau_w, \qquad \phi_{\min} = \underset{w\in\mathscr{W}}{\text{Min}} \sum_{w\in\mathscr{W}} \tau_w \quad (50)$$

subject to

$$\sum_{w\in\mathscr{W}} \sum_{r\in\mathscr{R}} \tau_w p_{rw} \xi_{rw} \delta_r^a = \hat{v}_a; \quad a \in \mathscr{O}, \qquad \tau_w \ge 0; \quad w \in \mathscr{W}, \quad (51)$$

where $\phi_{\max}$ and $\phi_{\min}$ are the maximum and minimum OD flows compatible with the constraints, respectively. Once these two problems have been solved, we can calculate the TDS measure as $\mathrm{TDS} = \phi_{\max} - \phi_{\min}$. We note that if TDS = 0 the set of observations has captured the total demand of the network; if TDS > 0 and finite, then the observations are not capable of capturing the total demand of the network; finally, if TDS is infinity, the flow of at least one OD pair cannot be captured by the observed data.

(5) Matrix Rank (MR). It is the rank matrix of the system of equations used to derive the traffic flows. This rank measures the number of linearly independent equations and thus the amount of information. This metric has been used by Gentili [55], Castillo et al. [50, 56], Ng [57, 58], or He [59].

(6) Flow Amount of Information (FAI). An interesting and not easy to solve problem consists of determining how many linearly independent equations are provided by each set of observations and in particular by a given subset $\mathscr{SC}$ of m counted or scanned links when the set of routes is fixed and known. This problem led to Castillo et al. [60] to introduce the concept of flow amount of information supplied by a set of scanned observations as the number of linear independent equations it provides.

While a set of m counted link data can provide at most m linearly independent equations with FAI = m, the rank of the scanning information matrix $\mathbf{H}_w$ associated with (41) is its number of rows. Thus, the "*flow amount of information*" of a subset of m scanned links is larger than or equal to m but normally much larger than m. The FAI can be easily calculated by the nontrivial formula (see Castillo et al. [12, 60]):

$$\mathrm{FAI} = \sum_{r|n_r^{\mathscr{SC}} \neq 0} \frac{1}{m - n_r^{\mathscr{SC}}}, \quad (52)$$

where $m = |\mathscr{R}|$ is the number of routes and

$$n_{r_1}^{\mathscr{SC}} = \sum_{r_2 \neq r_1} \min \left(\sum_{a\in\mathscr{SC}} (\delta_a^{r_1} + \delta_a^{r_2}) (1 - \delta_a^{r_1} \delta_a^{r_2}), 1 \right). \quad (53)$$

(7) Likelihood (LH). In statistical methods one of the most powerful methods is the maximum likelihood that generally provides good estimates. Thus, the likelihood is a good objective function to compare the different feasible solutions. Some examples of authors using this objective function are given next.

Spiess [61] proposes an independent Poisson model with unknown means and a maximum likelihood method for estimating an OD matrix from an observed sample matrix, when the volumes on a subset of the links of the network and/or the total productions and attractions of the zones are known.

Nihan and Davis [62] present a maximum likelihood-based method to estimate intersection turning and through movement probabilities from entering and exiting counts.

Vardi [63] deals with the problem of estimating the node-to-node traffic intensity from repeated measurements of traffic on the links of a network under Poisson assumptions and two types of traffic-routing regimens: deterministic and Markovian route selection models. They discuss maximum likelihood estimation and related approximations.

Parry and Hazelton [64] deal with the problems of estimation of OD matrices based on link count data and based on observed routes taken by each vehicle. They consider an intermediate problem in which link count data is supplemented by routing information for a fraction of vehicles on the network and develop a statistical model for these combined data sources and derive some tractable normal approximations thereof. They examine likelihood-based inference for these normal models under the assumption that the probability of vehicle tracking is known.

(8) Entropy (ET) Based Measures. Van Zuylen and Willumsen [65] propose two models, based on information optimization and entropy maximisation principles, to estimate OD matrices from traffic counts. The models assume that routes are known and use traffic counts to estimate the most likely OD matrix consistent with observed data and prior information about the trip matrix.

Since the entropy measure is proportional to the logarithm of the determinant of the covariance matrix, Zhou and List [66] suggest using this determinant, which is the product of the eigenvalues associated with the covariance matrix of estimates. It can be interpreted as a measure of the volume of a hyperellipsoid for unknown centered flow variables.

(9) Trace of Posterior Estimates (TPE). The trace of a covariance matrix is used to measure the amount of variations in random variables and corresponds to the circumference of the rectangular region that encloses the covariance matrix ellipsoid. It is the sum of the eigenvalues associated with the covariance matrix. Zhou and List [66] have suggested this measure and was used by Simonelli et al. [67].

(10) Flow Covered (FC). The flow covered principle seeks for the solution that maximizes the flow covered. Some authors using this measure are Yim and Lam [68], Chen et al. [53], and Fei and Mahmassani [69].

(11) Unobserved Variability (UV). It was proposed by Viti and Corman [70] and is defined as the sum of the maximum estimation errors on the unobservable flows. The interested reader can see details in Viti et al. [71].

(12) Correlation (CR) Based Measures. Castillo et al. [21] deal with the problem of estimating and updating the OD matrix and link flows from traffic counts and its optimal location. A combination (bilevel) of an OD pair matrix estimation model based on Bayesian networks and a Wardrop-minimum-variance model is used to estimate OD pair and unobserved link flows based on some observations of links and/or OD pair flows. The Bayesian network model is also used to select the optimal number and locations of the link counters based on minimum correlation among link counts.

(13) Number of Sensors (NS). It was used by Ng [57, 58] and Castillo et al. [72, 73].

(14) Synthetic Dispersion Measure (SDM). The synthetic dispersion measure (SDM) was proposed by Simonelli et al. [67] who consider the variability of the OD matrix estimate and is related to the trace of the covariance matrix of the posterior demand estimate conditional upon a set of sensor locations. In the case of a multivariate normal distribution for the prior demand estimate, the proposed SDM does not depend on the specific values of the counted flows, which are unknown in the planning stage, but just on the locations of such sensors.

The idea becomes clear now. We normally have a system of equations (constraints) that has infinitely many solutions and we want to add extra conditions to the system to get a unique solution if possible. The extra conditions are optimality conditions associated with an objective function or metric and set of constraints. Consequently, the objective function when optimized has the role of restricting the set of feasible solutions to optimal solutions. This raises the problem of how to select an adequate metric and points out the need of understanding what the different metrics mean and what are they suitable for.

Once we have defined what we mean by flow observability, estimation, and prediction problems, we have indicated that all of them can be formulated as optimization problems and we have described the types of data, variables, constraints, and objective functions that occur; we dedicate the following three sections to describe the most important proposed models to solve the three problems.

3. The Observability Problem

The observability problem has been intensively studied in the existing literature for the static case but needs research for the dynamic case. Because of their substantial differences, we deal with these two cases separately.

3.1. Some Approaches to the Observability Problem in the Static Case. The observability problem has been solved in the past by many different approaches, most of them assuming the static case, to which we refer in this section.

The observability problems can be classified (see Castillo et al. [12]) as follows: (a) link flow observability, (b) OD flow observability, (c) route flow observability, and (d) general case flow observability.

3.1.1. Link Flow Observability. The link flow observability problem (see Hu et al. [74]) consists of observing link flows based only on link flow observations.

However, depending on whether or not some route knowledge is assumed and the set of relations among link flows is used, there have been different approaches to this problem, such as (a) node based approaches, (b) OD based approaches, and (c) route based approaches. Next, some of them are discussed.

(1) Node Based Approaches. In this approach, only (33) is considered. The idea consists of determining a minimal subset of linearly independent link flows such that a given set of link flows can be calculated as linear combinations of them.

The set of solutions $\mathbf{v}$ of linear system (33) with (40) and/or (41) is an affine subspace; that is, it can be written as

$$\mathbf{v} = \mathbf{v}_0 + \mathbf{N}\boldsymbol{\rho}, \tag{54}$$

where $\mathbf{v}_0$ is a particular solution vector of (33) (any solution can be used), $\mathbf{N}$ is the null-space matrix associated with the system (33), and $\boldsymbol{\rho}$ is a column matrix of arbitrary real numbers (see Castillo et al. [75] or [76]).

The rows of matrix $\mathbf{N}$ play an important role in the observability problem because if row a is null (all its elements are zero), the corresponding link flow v_a is observable, and the number of nonnull elements in each row is the number of link flows that need to be observed to convert the corresponding link in observable. This permits sorting the links considering how far they are from observability.

Consequently, counting data is sufficient to identify link flows even though not all links are counted, because link flows are linearly dependent.

Ng [57] proposes for the first time a node based approach that has the advantage of avoiding route enumeration and provides an upper bound $\ell-\bar{n}$, where ℓ and $\bar{n}$ are the numbers of links and no-centroid nodes, respectively, for the number of links that need to be observed in order to get observability of all links. The networks considered by Ng [57] have only origin, destination, and intermediate nodes, but no node serves as both an origin and a destination at the same time. Ng [58] generalized the previous method to the case of partial observability. This is the best bound that can be obtained if only (33) and counted link flows are used. This problem is also discussed by Castillo et al. [77], who present alternative formulas for the link flow estimates and discuss in detail the partial observability problem. Since algebraic approaches are computationally intensive, a graphical method based on evaluating the minimum spanning tree was first suggested by Mori and Tsuzuki [78] who presented a new unified method for dealing with the problems of topological observability in power system state estimation.

Spanning trees were already used by He [59] to solve the link flow observability problem without considering routes. The idea consists of building a network with a virtual node and virtual links to reproduce node inflows and outflows and then working with spanning trees. However, a network with all centroid nodes cannot be analyzed using this method.

Castillo et al. [73] show that the minimum number of links to be counted for complete observability is $\ell-n+c$, where ℓ, n, and c are the number of links, nodes, and centroid nodes, respectively.

As indicated, node based approaches give the best upper bound for the number of links to be counted in order to have full link observability when route information is not available, that is, based on (33).

It is worth mentioning that system (33) with (40) and/or (41) could be incompatible. There are some methods to test the compatibility of a linear system without solving it (see Castillo et al. [75, 79]) and even ways of solving simultaneously all its subsystems (see Castillo et al. [80]). In particular, the latter allows us evaluating the contribution of each observation to the observability of the given subset of flows.

In addition, constraints (34) and (35) cannot be ignored. If constraints (34) are added to (33), the set of link flow vectors $\mathbf{v}$ consists of a vector (any particular positive solution of (33)) plus a cone.

If constraints (35) are added to (33) and (34), the set of link flow vectors $\mathbf{v}$ becomes a polytope (a linear convex combination of solution flow vectors).

(2) OD Based Approaches. In these approaches, only (36) together with (40) and/or (41) are considered. The idea consists of determining a minimal subset of linearly independent link flows such that all other link flows can be calculated as linear combinations of them and the OD flows τ_w are satisfied. Several cases can be considered here depending on the assumptions.

(1) The Proportions ρ_{aw} of OD Flow τ_w Using Link a Are Known. In this case, the resulting system of equations contains unknown links v_a and OD flows τ_w. We note that ρ_{aw} values cannot be given arbitrarily, because a set of routes consistent with these values must exist. In addition, the solution set is normally undetermined and then more link counters are needed in order to get a unique solution.

In this case we have

$$v_a = \sum_{w \in \mathscr{W}} \rho_{aw}\tau_w; \quad a \in \mathscr{L}. \tag{55}$$

(2) The Proportions ρ_{aw} of OD Flow τ_w Using Link a Are Unknown. In this case, the problem becomes nonlinear because of the product $\rho_{aw}\tau_w$ appearing in (36), unless the OD flows are known.

In this group, there are some important contributions. Some are given next.

Hodgson [81] deals with observability and flow estimation problems by finding the minimum number of sensors that guarantee full coverage and maximum captured flow with solutions of a minimal information redundancy (noncannibalizing solutions).

Morrison and Martonosi [82] provide a new necessary condition on the sensor location problem. This condition is not sufficient in general, but for a large class of problem instances the condition is sufficient.

Castillo et al. [50] present a detailed analysis of the plate scanning method to estimate OD flows considering the optimal location of cameras, the prior information, and error analysis.

Castillo et al. [83] deal with the problem of identifying which subset of OD pair and link flows that can be calculated is based on a subset of observed OD pair and link flows and related problems. Two algebraic methods for solving the observability problems are given: a global approach based on null spaces and a step-by-step procedure allowing updating the information once each piece of information (OD pair or link flow) becomes available. Castillo et al. [56] present a modified topological version of the previous existing algebraic method that is much faster, use much less memory, and present no rounding errors or zero test problems but identify fewer observable flows.

(3) Route Based Approaches. In this approach, (39) are used together with (40) and/or (42) and complete or partial route structure information. It is important to clarify that it is not necessary to include (33), (34), and (35) because they are automatically satisfied by nonnegative and finite route flows. Similarly, OD type equations such as (36) and (37) are automatically satisfied too.

Considering v_a and f_r as unknowns, the set of solutions to the system involving (39), (40), and (42) is an affine space and the associated null space provides information about the observability of any given sets of v_a or f_r, that is, whether or not they are observable and how far they are from observability. In this case, the required extra FAI for observability of any v_a or f_r coincides with the number of nonnull elements in the corresponding row of the null-space matrix.

Hu et al. [74] solve the whole link observability problem (observing all link flows) assuming route information and using the concept of "reduced row echelon form" (RREF), a technique based on the well-known Gaussian elimination method, which emphasizes its algebraic character. Castillo et al. [77, 84] provide a pivoting technique to solve the same problem for partial and whole observability and extend the method to plate scanned data; this problem has also been treated by Castillo et al. [72, 77, 84] and Ng [57] and consists of determining which subset of link flows can be calculated in terms of another subset of link flows but now using route information.

If route information is used, the bound given by node based approaches can be improved with savings reaching in some cases 16%, as demonstrated in Castillo et al. [72]. However, since enumerating all routes is a difficult task, the first method was improved by considering linearly independent route vectors by Castillo et al. [72]. More precisely, they show that only a subset of linearly independent routes is required and provide a method to select linearly independent route vectors. Finally, they present real examples in which more than 84% of the routes need not be enumerated.

Rinaldi et al. [85] observe that using k-shortest routes for each OD pair can result in an interesting geographical spread of the sensors and in favouring links near the centroids and at the major junctions. They also point out that using linearly independent routes may help in finding lower number of sensors guaranteeing full observability than the number indicated analytically by Ng [57].

Contrary to node information alone, route information remains fundamental, for example, for the following reasons. (a) It can be combined with flow estimation approaches (e.g., for optimizing location for OD estimation); (b) extra information, such as shortest routes, can be used to refine solutions; and (c) it can be extended to the powerful source of information provided by scanned links.

Some important conclusions resulting from the previous discussion are the following:

(1) Counting data is not sufficient to identify OD and route flows, because normally the number of these flow variables is huge compared with link flows and independent of them.

(2) Plate scanning permits following the car routes and identifying OD pairs and routes if they are conveniently located and in a sufficient number. They provide link, node, OD, and route flows.

3.1.2. OD, Route, and General Case of Flow Observability. One of the most common observability problems in the traffic literature is the OD pair observability problem, in which the OD pair flows are estimated in terms of other flows. Since in most cases the number of independent link flows is much smaller than the number of OD pair flows, even in the case in which all link flows are available, the OD observability problem based on counting links becomes underspecified; that is, the OD flow observability problem based on link flows has an infinite number of solutions. Thus, the OD flow observability problem based on counting links has no satisfactory solution in many cases of real practice, and when it has solution the situation is not realistic.

A more difficult observability problem is the route flow observability problem, in which we aim to observe all route flows. Since knowledge of the route flows leads to knowledge of the OD pairs and link flows, through the conservation equations, the observability problem of route flows is considered as the full observability problem. However, it is the most difficult observability problem, because its solution requires the maximum amount of information. Thus, the route flow observability problem presents the same underspecification difficulties as the OD observability problem, but even more exaggerated.

Therefore, similar methods, based on prior information, or more powerful techniques, such as the scanned link technology, can be used. The plate scanning technique opens the door to full OD and route flow observability, in the sense that installation of sufficient number of cameras permits calculating all OD and route flows.

However, a general case of observability problem can be stated, in which we aim to observe a given subset of flows $\mathcal{U}$ (including any type of flows) and we must determine the subset of observed flows $\mathcal{O}$ in order that this is possible.

Castillo et al. [77, 86] deal with the general observability problem in which link, OD, or route flows are involved. In these cases the use of active sensors is required; in particular, plate scanning is a very effective mean of obtaining flow data.

Table 1 gives the number of unknowns and equations or inequations of the systems of equations used in the different observability options and associated ranks. These ranks

TABLE 1: Number of unknowns and equations or inequations of different systems of equations and associated ranks.

Equation	System	# of unknowns v_a	τ_w	f_r	# of equations or inequations	Rank (r) Lower bound	Upper bound
					Equations		
(33)	Conservation equations	$\|\mathscr{L}\|$	—	—	$\|\mathscr{N}\|$	$\|\mathscr{L}\| - k_0 - h$	$\|\mathscr{L}\| - k_0 - h$
(40)	Counting data	$\|\mathscr{C}\|$	—	—	$\|\mathscr{C}\|$	1	$\|\mathscr{C}\|$
(41)	Plate scanned data	—	—	$\|\mathscr{R}\|$	$\|S\|$	$\|S\|$	$\|S\|$
(33), (40), (41)	Conservation, counts, and plate scanned equations	$\|\mathscr{L}\|$	—	$\|\mathscr{R}\|$	$\|\mathscr{N}\| + \|\mathscr{C}\| + \|\mathscr{S}\|$	$\|\mathscr{L}\| - k_0 - h + 1$	$\|\mathscr{L}\| - k_0 - h + \|\mathscr{C}\|$
(37)	OD-route flow relations	—	$\|\mathscr{W}\|$	$\|\mathscr{R}\|$	$\|\mathscr{W}\|$	$\|\mathscr{W}\|$	$\|\mathscr{W}\|$
(37), (40), (41)	Conservation, counts, and plate scanned equations	$\|\mathscr{L}\|$	$\|\mathscr{W}\|$	$\|\mathscr{R}\|$	$\|\mathscr{W}\| + \|\mathscr{C}\| + \|\mathscr{S}\|$	$\|\mathscr{W}\| + \|\mathscr{S}\| + 1$	$\|\mathscr{W}\| + \|\mathscr{S}\| + \|\mathscr{C}\|$
(39)	Route-link flow relations	$\|\mathscr{L}\|$	—	$\|\mathscr{R}\|$	$\|\mathscr{L}\|$	$\|\mathscr{L}\|$	$\|\mathscr{L}\|$
(39), (40), (41)	Conservation, counts, and plate scanned equations	$\|\mathscr{L}\|$	—	$\|\mathscr{R}\|$	$\|\mathscr{L}\| + \|\mathscr{C}\| + \|\mathscr{S}\|$	$\|\mathscr{L}\|$	$\|\mathscr{L}\| + \|\mathscr{C}\| + \|\mathscr{S}\|$
					Inequations		
(34)	Nonnegativity of link flows	$\|\mathscr{L}\|$	—	—	$\|\mathscr{L}\|$	—	—
(35)	Link capacity constraints	$\|\mathscr{L}\|$	—	—	$\|\mathscr{L}\|$	—	—

measure the amount of information associated with the corresponding systems of equations.

3.2. The Observability Problem in the Dynamic Case. The main aim of observability analysis in the dynamic case is to know the flows in a given network and how they change with time. The observability problem in this case is completely different from the observability problem in the static one, because the dynamic case implies observing flow-time functions instead of flow values. This means that at several locations we have different flow-time functions. Consequently, we need first to define the following: the flow (what), the location (where), and the time (when) to be observed.

While the observability problem in the static case has been widely discussed in the existing literature, as shown in the previous sections, it is very difficult to find specific research for the dynamic case. In this paper we want to point out the interest in working in this direction.

In the knowledge of how the flow evolves with time in a given network, the link exit time $w_a(t)$ and travel time functions $\tau_a(t)$ play a very relevant role, because knowledge of these functions together with some simplifying assumptions, such as FIFO, allows us to calculate $\theta_a^r(t)$ functions, that is, the times at which a vehicle entering link a at time t initiated its route r, and then the route travel time functions, that is, how the travel times of each route changes with departure times.

We must realize that we can have access to the large number of $\theta_a^r(t)$ functions using a reduced set of functions $\tau_a(t)$ that can be calculated from them, as follows:

$$\theta_a^r(t) = \tau_{a_1^r}^{-1}\left(\tau_{a_2^r}^{-1}\left(\cdots\tau_{a_{n-1}^r}^{-1}\left(\tau_{a_{n_r}^r}^{-1}(t)\right)\right)\right), \quad (56)$$

where $\tau_a^{-1}(t)$ is the link a entering time of a user who exits that link at time t, that is, the inverse of $\tau_a(t)$ function, and $a_1^r, a_2^r, \ldots, a_{n_r}^r$ and n_r are the links of route r and its number of links, respectively.

This suggests that we can state a wide collection of dynamic observability problems by solving the observability problem of functions $\tau_a(t)$ together with traffic volumes at some locations. This interesting problem is suggested for future research.

4. The Flow Estimation Problem

As indicated, when we are unable to observe the desired flows by direct observation and using the conservation and flow definition equations we must use additional information and/or models to estimate flows. The additional information tells us about previous experience so that we can use it for the actual situation assuming that the actual situation has not changed or suffered small changes.

4.1. Static Case. To measure the quality of the flow estimates we normally use a measure. Consequently, many of the existing methods can be classified by considering what this metric is. Some examples of the metrics used and a list of some works using them are given below.

Generalized Least Squares Measures. Least squares are very powerful and interesting methods to solve the estimation problems. Some examples are given below.

Lam and Lo [87] are concerned with the reliability of information and optimize it using generalized least squares; Bierlaire and Toint [88] propose an improvement of existing methods of OD matrix estimation by an explicit use of data, obtained from parking surveys, describing the structure of the matrix, and propose using the GLS metrics; Asakura et al. [89] deal with the problem of OD matrix estimation using the observed data with the AVI system and a least squares model. The results of license plate matching in a pair of AVI cameras are the input variables; Gan et al. [54] present some methods for OD matrix estimation and counter location with error bound analysis in an integrated form and use the MPRE, generalized least squares, and total demand scale (TDS) metrics; Marzano et al. [90] deal with the problem of correction of OD matrices using traffic counts and confirm that the main reason for the data insufficiency stands in the fact that the number of equations is lower than the number of unknowns. To solve the problem, they suggest moving to within-day dynamic contexts, where a much larger number of equations are generally available and using GLS methods; Mínguez et al. [22] apply least squares but suggest the use of dimensionless variables.

Likelihood Measures. The likelihood function, coming from the statistical field, has a lot of power for estimation purposes together with a powerful theory supporting it. Some authors using the likelihood as metrics in the classical sense are Spiess [61], Nihan and Davis [62], Vardi [63], and Parry and Hazelton [64]; Bell [91] describes a model that under certain circumstances yields the most likely OD matrix which is consistent with measurements of link traffic volumes; Watling and Maher [92] and Watling [93] suggest using active sensors together with maximum likelihood and best matching techniques to estimate OD matrices; Lo and Chan [94] propose a maximum likelihood procedure for the simultaneous estimation of an OD matrix and link choice proportions from OD survey data and traffic counts for congested networks recognizing that link choice proportions in a network change with traffic conditions; Papola and Marzano [16] using the maximum likelihood principle and the GLS methods show through laboratory experiments that the classical correction of the OD matrix from traffic counts is generally unable to provide effective correction of the OD matrix.

Maximum Coverage. Several authors realized the importance of covering with sensors a maximum of flows, so that the maximum coverage rule has been widely contemplated. For example, Church and ReVelle [95] deal with the maximal covering location problem; Fei et al. [96] identify a set of sensor locations that optimize the coverage of OD demand flows of the road network and maximize the information gains through observation data over the network, while minimizing the uncertainties of the estimated OD demand matrix; Larsson et al. [97] compare different methods for allocating link flow sensors with respect to the quality of the estimated OD matrix and compare three allocation methods. However, they conclude that maximizing the coverage of OD pairs seems to be unfavorable for the quality of the estimated OD matrix; Simonelli et al. [67] provide an innovative and theoretically sound methodology for solving the network sensor location problem, explicitly accounting for the variability of the OD matrix estimate.

Synthetic Dispersion Measure. Simonelli et al. [67] propose a specific measure, termed synthetic dispersion measure (SDM), related to the trace of the covariance matrix of the posterior demand estimate conditional upon a set of sensor locations and an algorithm to work with it.

Expected Information Gain. Zhou and List [66] discuss how to locate a limited set of traffic counting stations and automatic vehicle identification (AVI) readers in a network in order to maximize the expected information gain for the OD demand estimation problem subject to a budget constraint. The model takes into account several important error sources, such as the uncertainty in historical information, sensor measurement errors, and those associated with link proportions. Based on a mean square measure, the authors provide a stochastic optimization procedure and an algorithm to find suboptimal point and point-to-point sensor locations.

Rules for Reduction of Estimate Uncertainties. Since the very beginning, researchers were concerned with finding rules to obtain optimal locations for sensors. The most relevant were Yang and Zhou [13] who addressed the problem of how to determine the optimal number and the locations of sensors in a road network for a given prior OD distribution pattern and derived four rules: OD covering rule, maximal flow fraction rule, maximal flow-intercepting rule, and link independence rule. In addition, the authors presented integer linear programming models and heuristic algorithms to determine the counting links satisfying these rules; Mínguez et al. [22] provided techniques for obtaining the optimal number and location of plate scanning devices for a given prior OD distribution pattern under maximum route identifiability or budget constraints, developed two rules analogous to the counting location problem, and proposed several integer linear programming models fulfilling these rules.

Since mathematical models reproducing traffic flows are relevant to provide extra equations (constraints), some authors have been concerned about the quality of models that reproduce the traffic behaviour. Wang et al. [98] provide a new linear integer programming model for the placement of sensors to maximize the reduction in the uncertainties in route flow estimates. The model assumes that a general underlying traffic loading model, the route choice set from each OD pair, is known and prior route flows and their reliabilities are given.

Similarly, Chen et al. [99] develop strategies in the screen-line-based traffic location problem model for selecting additional traffic counts for improving OD trip table estimation using a SUE principle.

Multiobjective Measures. The observability, estimation, and prediction problems are essentially multiobjective. Thus, several authors dedicate work to the problem of how to deal with multiobjective traffic problems. This is the case of Brenninger-Göthe et al. [100] who present some multiobjective programming formulations for estimating OD matrices and show how multiobjective theory can be used in the interpretation of the problem or Fei and Mahmassani [69] who present a multiobjective model, which considers link information gains and OD demand coverage to locate a minimal number of passive point sensors in a roadway network subject to budget limits.

Bayesian Methods. Among statistically based methods to analyze traffic flows, Bayesian methods appear very useful because they allow us to include the expert knowledge by means of prior distributions. In this context, Maher [101] proposes a method based on Bayesian statistical inference that allows complete flexibility in the degree of belief placed on the prior estimate of the OD trip matrix and also in different parts of the prior estimate; Castillo et al. [102, 103] propose an extended gamma-shifted model to estimate traffic flows. The reconstruction of the sample flows can be done exactly or approximately, depending on the intensity of the plate scanning sampling procedure. To this end, a generalized least squares technique is used together with the conservation and flow definition laws. A Bayesian approach using special conjugate families is proposed that allows us to estimate different traffic flows, such as route, OD pair, scanned link, or counted link flows.

To end this section, we consider the problems of error recovery and some contributions related to some special solving algorithms.

Error Recovery. The incompatibility problems derived from data replications and the distortion produced by measurement errors led some authors to consider the error recovery problem. For example, Lo et al. [104] review the use of statistical modeling in the estimation of OD matrices from traffic counts and discuss statistical models that consider explicitly the presence of measurement and sampling errors in the observed link flows treating the link choice proportions as random variables; Castillo et al. [50, 105] and Sánchez-Cambronero et al. [106] provide models that permit error identification and recovery.

Branch-and-Bound Methods. Some particular cases, such as the cordon-screen lines, have been treated by Yang et al. [107] who dealt with the optimal selection of cordon-screen lines for traffic census study in road networks that can be stated as how to select the optimal locations of a given number of counting stations to separate as many OD pairs as possible and how to determine the minimum number of counting stations and their locations required for separating all OD pairs. The authors provide a solution scheme that combines a shortest route based column generation procedure and a branch-and-bound technique.

Genetic Algorithms. Since the resulting optimization problems are in some cases too complex, some authors suggest the use of genetic algorithms, such as Chootinan et al. [108], who consider the traffic counting location problem for the purpose of OD trip matrix estimation and determine the number and locations of counting stations that yield maximal coverage and minimal resources. A distance-based genetic algorithm is used to solve the proposed biobjective traffic counting location problem by explicitly generating the nondominated solutions.

Similarly, Cipriani et al. [109] deal with two different heuristic approaches for solving the problem of optimal location of traffic count sections. The first method applies deterministic rules on OD flows and OD pair coverage, but the second uses a genetic algorithm.

Finally, as a summary, we include Table 3 with a long list of contributions, together with the different data, variables, constraints, and objective functions used and the illustrative or real examples utilized. Table 2 provides the symbols and notation used in Table 3.

4.2. Dynamic Case. Most of the works in the existing literature deal with the static and ignore the dynamic case, which is a stochastic process estimation problem. Thus, some research

TABLE 2: Table 3 legend.

	Data
CP	Choice probability
L	Link data
OD	OD data
P	Route data
PA	Route knowledge
PI	Prior information
	Variable
CP	Choice probability
L	Link variable
OD	OD variable
P	Route variable
SL	Sensor location
	Constraint
BC	Bounded capacity
CE	Conservation equation
CP	Choice probability
FD	Flow definition
FM	Flow model constraint
NN	Nonnegativity
OC	Observation constraint
OT	Other constraints
	Metric (objective function)
BM	Bayesian model
CR	Correlation
ET	Entropy
FAI	Flow amount of information
FC	Flow covered
GLS	Generalized least squares
IN	Information
LH	Likelihood
MA	Mean absolute relative error (MARE)
MO	Multiobjective function
MP	Maximum possible relative error (MPRE)
MR	Matrix rank
MUV	Maximum observability variable
NS	Number of sensors
SD	Synthetic dispersion measure
TDS	Total demand scale
TPE	Trace of posterior estimates
UM	Unobserved variability
WMV	Wardrop minimum principle
	Examples
CN	Campania region network
CR	Ciudad Real network
CU	Cuenca network
EN	Edwards network
FB	Fishbone
FN	Fuorigrotta network
GN	Gateshead network
HC	Helena city
HS	Hanshin expressway network
IN	Irvine network
KA	Kowloon area
LN	Leicester network
MN	Maryland chart network
ND	Nguyen-Dupuis network
NP	30-node problem
SF	Sioux Falls network
T	Toy
TPE	Triangle park
RT	Rotterdam

TABLE 3: A summary of some of the main contributions to sensor locations and flow estimation in traffic networks in the static or steady case.

Paper	Approach type	Sensor type	Data	Variables	Constraints	Metric	Examples
Church and ReVelle [95]	OD based	Link count	L	L, OD	FD, OC	FC	NP
Branston [5]				Review			
Van Zuylen and Willumsen [65]	OD based	Link count	L, PI	OD	FD, OC	LH, ET	T
Bell [91]	OD based	Link count	L	L, OD	OC, FD	LH	—
Maher [101]	OD based	Link count	PI	OD	FD, OC	BM	—
Nihan and Davis [110]	OD based	Link count	L	OD, L	FD, OC	—	—
Spiess [61]	OD based	Link count	L	OD	FD, OC	LH	—
Brenninger-Göthe et al. [100]	OD based	Link count	L, CP	OD	CP	MO, ET	—
Nihan and Davis [62]	OD based	Link count	L	OD	FD, OC	LH, GLS	—
Yang et al. [51]	OD based	Link count	L	L, OD	FD, OC	MP	T
Bierlaire and Toint [88]	OD based	Link count	PI, L	L, OD	FD, OC	GLS	T
Lo et al. [104]	OD based	Link count	L, CP	CP, L, OD, CP	CP, FM	LH, BM	T
Vardi [63]	OD based	Link count	L	OD	FD, OC	LH	T
Abrahamsson [6]				Literature survey			
Yang and Zhou [13]	Route based, OD based	Link counts	CP, PA, L	SL, L, OD	CP, FD, OC	MP, FC	T, SF
Asakura et al. [89]	OD based	Plate scan, AVI	L	L, OD	FD, OC	GLS	HS
Lo and Chan [94]	OD based	Link count	L	CP	CP, FM	GLS	T
Chootinan et al. [108]	OD based	Link count	L	OD, SL	FD, OC	NS, GLS	T, SF
Gan et al. [54]	OD based	Link count	L	L, OD	FD, OC	GLS, NS, MP, TDS	T
Gentili and Mirchandani [4]	Route based	Active sensor	L	SL	FD, OC	NS	T
Cipriani et al. [109]	OD based	Link count	L	SL, CP	FD, OC, CP	MP	CN
Ehlert et al. [47]	OD based	Traffic counts	L, CP	L, OD	CP, OC	IN, MP, FC, NS	GN
Papola and Marzano [16]	OD based	Link count	L, OD	L, OD	FD, OC	LH, GLS	FN
Yang et al. [107]	OD based	Link count	L, OD	OD, L, SC	FD, OC	NS	T, SF
Chen et al. [99]	OD based	Link count	L, OD	L, OD, SL	OC, FD	NS	HC
Castillo et al. [56]	OD based	Link count	OD, L, CP	SL, L, OD	FD, CP, OC	NS, MR	ND, T
Castillo et al. [83]	OD based	Link count	L, OD, CP	SL, L, OD	FD, CP, OC	NS, MR	ND, T
Castillo et al. [50]	Route based	Link count	PI, L, OD, CP	SL, L, OD, P	FM, OC, CP, FD	NS, GLS	T, ND
Castillo et al. [21]	OD based, Bayesian	Link count	L, CP	OD, L	CE, NN, CP, FM	WMV, CR, BM	ND, CR
Viti et al. [52]	OD based	Link count	L, CP	SL, L	FD, OC, CP	MP	SF
Hu et al. [74]	Route based	Link count	L, PA	SL, L	FD, OC	MR, NS	PAR, FB, SF, OT
Marzano et al. [90]	OD based	Link count	L	L, OD	FD, OC	LH	—
Castillo et al. [105]	Route based	Plate scan	PA, L, OD	SL	OT	NS	ND, CU
Castillo et al. [84]	Route based	Link count, Plate scan	L, OD	SL, L, OD, P	FD, OC	MR, NS	T, PAR, ND
Larsson et al. [97]	OD based	Link count	L	OD, SL	FD	FC, GLS, NS	SF, EN
Mínguez et al. [22]	OD and route based	plate scan	PI, L, P, OD, PA	SL, L, OD, P	FD, OC	GLS, MA, MP	CU, ND
Zhou and List [66]	OD based	Link count	CP, PI	SL, L	CP	TPE	T, IN, TPE
Castillo et al. [77]	Route based	Plate scan	L, PA, OD	L, SL, OD, P	FD, OC	NS, MR	PAR, ND, CU
Castillo et al. [86]	Route based	Plate scan	CP, L, OD, PA	SL, L, OD, P	FD, OC, OT	NS, MR	T, CU
Fei and Mahmassani [69]	OD based	Link count	L	OD, SL	FD, OC	MO, FC, NS	MN

TABLE 3: Continued.

Paper	Approach type	Sensor type	Data	Variables	Constraints	Metric	Examples
Sánchez-Cambronero et al. [106]	Route based	Plate scan	PI, L, PA, OD	L, OD, P	FD, OC	GLS, LH	ND, T
Castillo et al. [60]	Route based	Plate scan	L, PA, OD	SL, L, OD, P	FD, OC	FAI, NS, MR	T, ND, CU
Gentili and Mirchandani [14]	Route based	Link count	PI, L, CP, OD	SL, L, OD, P	FD, OC, CP	NS, FC, MP	T
Ng [57]	Node based	Link count	L	L, SL	CE, OC	NS, MR	PAR, FB
Parry and Hazelton [64]	OD based	Link count	L, CP	OD, L	FM, CP	LH	T, LN
Simonelli et al. [67]	OD based	Link counts	L	SL, L, OD	FD, OC	SD	T
Wang et al. [98]	OD based	Link count	L, PI	SL, L, P	FD, OC	GLS, NS	—
Castillo et al. [72]	Node and route based	Link count	L, P, OD, PA	SL, L, OD, P	CE, NN, OC, FD	MR, NS	PAR
Castillo et al. [102]	Route based, Bayesian	Plate scan	PI, L, OD, PA	L, OD, P	OC, FD	MA, GLS, BM, LH	T, ND
Castillo et al. [12]	OD based	Plate scan, Link count	L, CP, OD, PA	SL, L, OD, P	FD, OC, CP	MP, NS, GLS	T, ND, CU
He [59]	Node based	Link count	L	L, SL	CE, OC	NS, MR	PAR, FB, SF
Rinaldi et al. [85]	OD based	Link count	L	SL, L, OD	FD, OC	MUV	—
Castillo et al. [73]	Route based	Link count	L, PA, OD	SL, L, OD, P	CE, NN, FD, OC	MR, NS	PAR, ND, OT
Viti et al. [71]	Route based	Link count	L	SL, L, OD	FD, OC	MUV	RT, PAR

is needed in this direction. One of the few examples of flow estimation in the dynamic case corresponds to the case of time-dependent OD matrix estimation.

Time-Dependent OD Matrix Models. Time-dependent proposals for traffic models were suggested by Ashok and Ben-Akiva [111] who provide a framework for real-time estimation/prediction of time-dependent OD matrices and apply the model to the Massachusetts Turnpike and Ashok and Ben-Akiva [112] who present models for the estimation and prediction of time-dependent OD matrices. The key contribution of the proposed approach is the explicit modeling and estimation of a stochastic assignment matrix between time-dependent OD flows and link volumes. The assignment matrix is assumed to depend on travel times and route choice fractions in the network; Peterson [113] deals with the problem of estimation of the OD matrices based on observed link counts and considers the time-independent and time-dependent cases.

As indicated by Barcelò et al. [114], time-dependent OD matrices are a key input to dynamic traffic models. Microscopic and mesoscopic traffic simulators are relevant examples of such models, which are used in the design of traffic systems. Typical approaches to the time-dependent OD estimation are based either on mathematical programming approaches or on Kalman filter models. However, new data types, such as AVI, plate scanning techniques, and detection of mobile devices, make available traffic data of higher quality allowing for new modeling hypothesis leading to more computationally efficient algorithms. In this context, Barcelò et al. [114] extend the previous research on Kalman filtering approaches for OD estimation to more complex topologies of urban networks, where alternative path choices between origins and destinations are available.

5. The Flow Prediction Problems

Models used to solve flow prediction problems incorporate simulation of the traffic behavior; that is, they permit analyzing how changes in the behaviour of the vehicles and infrastructure modify the flows; thus, the models are an important element to be considered.

5.1. Static Models. Some examples of static models for flow predictions are given next.

Boyce and Janson [115] analyze the effect of different link congestion functions on equilibrium assignment; Bell [116] proposes two logit assignment methods that work without route knowledge as alternatives to Dial's algorithm; some models considering multiple class users with overtaking were analyzed in Castillo et al. [117, 118]. Multiclass users with different concerns about travel time are discussed in Castillo et al. [119].

5.2. Dynamic Models. There is a wide collection of dynamic models used for flow prediction. We include below some of them.

Prior Information Based Models. One way to predict flows and flow evolution with time in the dynamic case is storing this information in previous periods and trying to identify the trends from this information. At a given instant, we can know the traffic characteristics at other locations in previous instants, and, looking for similarities or using inference, we can predict future traffic flows. For example, Castillo et al. [29] develop a stochastic demand dynamic traffic model to predict traffic variables (link travel times, flows, densities, etc.) and their time evolution in real networks. The model assumes the variables to be generalized beta variables such that when they are marginally transformed to standard normal all of them become multivariate normal. This gives sufficient degrees of freedom to reproduce (approximate) the considered variables at a discrete set of time-location pairs. The parameters of the model are learned based on previous observations. The model provides point or confidence intervals or the density of the variable being predicted. A closed formula for the conditional future variable values (link travel times or flows), given the available past variable information, is provided. Since only local information is relevant to short-term link flow predictions, the model is applicable to very large networks.

Mathematical Programming Based Traffic Assignment Models. Merchant and Nemhauser [31] consider a dynamic traffic assignment model as a nonlinear and nonconvex mathematical program. Necessary optimality conditions that require marginal costs for all the same OD routes that are being used to be equal are shown to be a generalization of the optimality conditions of the static traffic assignment problem. They also show that the behavior of the dynamic model under static demand conditions is a generalized version of the static model.

Friesz et al. [120] consider two formulations of the dynamic traffic assignment problem, system and user optimizations on a single node network using optimal control theory, and offer the first dynamic generalization of Beckmann's equivalent optimization problem for static user optimized traffic assignment in the form of an optimal control problem.

Jayakrishnan et al. [121] present a dynamic traffic assignment model with traffic flow relationships based on a bilevel optimization framework. Using the modified Greenshields speed-density relationship, they derive a link-cost function that is monotonically nondecreasing and convex with respect to density.

Szeto and Lo [122] compare dynamic traffic assignment models with point and physical queues and discuss their implications together with their solution existence and uniqueness.

Carey and Subrahmanian [123] model link flows considering the flow rate when the vehicle enters and exits the link and knock-on effects from traffic ahead of the link and suggest dividing each link into a travel link followed by a queue link.

Cell Transmission Type Models. The cell transmission model (CTM) is a simple representation of traffic flow that is consistent with the kinematic wave theory under all traffic conditions to predict traffic's evolution over time and space, including transient phenomena such as the building, propagation, and dissipation of queues.

One of the first and better known models is Daganzo's cell transmission model (CTM). Daganzo [36, 37, 124] presents the cell transmission model and introduces a numerical procedure for networks, assuming that a time-varying origin-destination OD matrix is given and that the proportion of turns at every junction is known or the best routes to each destination from every junction are known at all times. The difference equations are shown to be the discrete analog of the differential equations arising from a special case of the hydrodynamic model of traffic flow reproducing shockwaves in a simple way and showing that the model can mimic the stop-and-go traffic within moving queues. Daganzo [125] shows that if the kinematic wave model of freeway traffic flow in its general form is approximated by a particular type of finite difference equation, the finite difference results converge to the kinematic wave solution despite the existence of shocks in the latter. This result is shown not to hold for other commonly used finite difference schemes. An error analysis is also performed.

Lo [39] develops a dynamic traffic assignment formulation, referred to as DUO, based on a dynamic extension of Wardrop's principle. Traffic is modeled as a cell transmission model (CTM), which provides a convergent approximation to the Lighthill and Whitham and Richards (LWR) model and covers the full range of the fundamental diagram and transforms CTM in its entirety to a set of mixed-integer constraints.

Alecsandru [126] presents a series of enhancements to the original form of CTM such as topological enhancements and modifications to the flow advancing equation to allow variable cell lengths and nondiscrete movements of vehicles between cells and lane-changing behavior.

Zhong and Sumalee [127] develop a stochastic cell transmission model for the evolution of traffic flows on freeways taking into account various uncertainties. The model is formulated as a discrete time bilinear stochastic system.

Sumalee et al. [128] propose the stochastic cell transmission model, to model traffic flow density on freeway segments with stochastic demand and supply considering five operational modes corresponding to different congestion levels of the freeway segment, which are formulated as discrete time bilinear stochastic systems. The model captures the mean and standard deviation of density of the traffic flow and the propagation of it over time and space.

Carey et al. [129] consider a road link that consists of two adjacent homogeneous lanes and uses a modified cell transmission model to allow moving between lanes. The authors investigate how lane changing and congestion are affected by varying some behavioural parameters.

Multiclass Models. Szeto et al. [130] propose a cell-based multiclass dynamic traffic assignment problem that is formulated as a fixed point problem and considers the random evolution of traffic states, where travelers select routes based on perceived effective travel time, that is, the sum of mean perceived travel time and safety margin. The problem includes a Monte-Carlo-based stochastic cell transmission model to capture the effect of physical queues and the random evolution of traffic states.

Tuerprasert and Aswakul [131] present a generalization of the cell transmission model to a more generalized multiclass cell transmission model taking into account the several vehicle classes to introduce head-of-cell and end-of-cell vehicles into the model. Cascading, merging, and diverging scenarios of cells allow the model to be applied to general road networks.

Variational Models. These models are represented mathematically by variational equations. Marcotte [132] deals with the variational inequality problem formulated as a nonconvex, nondifferentiable optimization problem, shows that any stationary point is optimal, and proposes an algorithm that decreases the nondifferentiable objective monotonically.

Zhu and Marcotte [133] deal with the existence of solutions to a dynamic network equilibrium problem modeled as an infinite dimensional variational inequality and introduce a novel concept that strengthens the familiar concept of FIFO.

Wang and Jiang [134] discuss the problem of traffic equilibrium in which the cost of route is not the simple sum of the cost of links and establish a variational inequality model to characterize the solutions of traffic equilibrium problems.

Lo and Szeto [135] present a cell-based dynamic traffic assignment formulation that follows the ideal dynamic user optimal (DUO) principle through a variational inequality approach, which encapsulates a network version of the cell transmission model (CTM) and satisfies the FIFO conditions through the CTM. The results showed that the formulation was capable of capturing shockwaves and queue formation and dissipation.

Bliemer and Bovy [136] consider multiple user classes in a macroscopic dynamic traffic assignment model. Considering different asymmetries, such as interuser-class interaction and interspatial and intertemporal asymmetries, the model is specified as a (quasi) variational inequality problem.

Network Loading Models. Network loading models allow simulating the propagation of traffic profiles through link and networks. To this end, several models have appeared in the existing literature.

Astarita [32] proposes a dynamic network loading model, which contrary to exit functions is based on travel times. Two requirements for a satisfactory model are identified: allowance for overtaking on a link and flow propagation consistent with speed. The FIFO rule, not satisfied in this case, and its implications are discussed.

Wu et al. [28] formulate the continuous dynamic network loading problem as a system of functional equations.

Xu et al. [137] present a formulation of the continuous dynamic network loading problem where travel delays may be nonlinear functions of arc traffic volumes and propose a finite-step algorithm to obtain solutions.

Astarita et al. [138] investigate three methods for dynamic network loading: link-based travel time functions, the cell transmission model, and a link-based model derived from a simplified car-following relationship.

Rubio-Ardanaz et al. [139] present two improved event-based simulations numerical algorithms for the continuous dynamic network loading problem.

Dynamic conservation equations (16) and flow definition constraints (20), by means of $\theta_a^r(t)$ function, permit relating the flows at different locations and times to establish a flow balance under a FIFO assumption. These relations allow modelling the flow evolution through the network. For example, Castillo et al. [45] present a FIFO consistent model for the continuous dynamic network loading problem. The nonlinear link travel time functions are calculated at a finite set of times used to interpolate a monotone cubic spline for all times. The model consists of five units: route origin flow wave definition, route wave propagation, congestion analysis, flow propagation, and inference engine. Next, the individual route profiles are propagated throughout routes using a conservation equation that stretches or enlarges the wave lengths and heights, depending on the degree of congestion at different links. Then, the individual route profiles are combined together to generate the link and node profiles. An iterative method is used until convergence.

Genetic Algorithms. Since the resulting optimization problems are in some cases too complex, some authors suggest the use of genetic algorithms, such as Lo [39] and Lo and Szeto [140], who use these algorithms to solve dynamic traffic assignment problem.

Neural Network Models. Celikoglu [141] proposes a neural network approximator that supports the dynamic network loading model to reproduce link flow dynamics on a sample network.

Other Models. G. C. K. Wong and S. C. Wong [142] present an extension of the Lighthill, Whitham, and Richards model with heterogeneous drivers where faster vehicles can overtake slower ones under the uncongested condition and the congested condition and slower vehicles could slow down the faster ones.

Blumberg and Bar-Gera [143] aim to achieve consistent integration between the behavior of traffic at different roadway elements and the propagation of flows along routes focusing in particular on the order of arrival to nodes. The authors propose the concept of anticipated arrival order (AAO) using a traffic flow behavior that follows the kinematic wave model with a triangular flow-density relationship to describe the number of vehicles expected to choose a certain downstream link at a node as a function of the total number of vehicles at this node. An iterative model permits solving the problem.

Long et al. [144] develop new formulations for dynamic traffic assignment of the step function (SF) and linear interpolation (LI) link travel time models in which the profiles of cumulative flows are piecewise linearized and both models are used to approximate cumulative flows over time. One of the formulations ensures causality, strong FIFO, and travel time continuity.

Kalman Filter Models. Kalman filter models and extensions were proposed by Okutani and Stephanedes [145] who propose two models employing Kalman filtering theory for predicting short-term traffic volume; Barcelò et al. [114] extend the previous research on Kalman filtering approaches for freeway OD estimation to more complex topologies of urban networks where alternative route choices between origins and destinations are available.

6. Conclusions

In this paper a state-of-the-art review of observability together with estimation and prediction methods has been performed. The three problems have been analyzed considering them as optimization problems or their associated systems of equations. The roles of data, constraints, and objective functions or metrics have been discussed to embed the existing literature under an optimization paradigm.

The main conclusions of this paper are the following:

(1) The basic equations or constraints arise from the topology of the network, the different flow definitions, and nonnegativity or capacity constraints, which can be stated in a natural way because they must always hold.

(2) Data arise as the second fundamental information to solve the three stated problems. They can be direct observations obtained from passive or active counters or other means or prior information on previous traffic flows.

(3) Since the two previous data items are not sufficient and lead to infinitely many solutions, a metric must be used to select among all feasible solutions one that is considered as the best under the metric criterion. Selecting an adequate metric is crucial to have a good solution to the traffic problem.

(4) When the objective functions or metrics are not sufficient to provide unique solutions, a hierarchical objective function considering several criteria (metrics) can be used (e.g., using weights).

(5) While the observability and the flow estimation problems have been widely analyzed for the static case in the existing literature, these problems for the dynamic case still require a lot of attention. In particular, observability methods and estimation of $\tau_a(t)$ function that provides the link travel time functions together with traffic volumes at some locations are interesting topics for future research.

(6) Special care must be taken with redundancy because it can lead to the nonexistence of feasible solutions. In this case, least squares equations or similar equations should replace the direct observation equations to allow for possible errors.

(7) Use of dimensionless equations is much better than other alternatives. This has been illustrated with expression (8), devoted to generalized least squares, where, instead of absolute flow values, relative flow values were used.

(8) In the static observability problem, algebraic tools, such as null spaces, are the basic tools to be used and the path information is crucial to obtain an important reduction in the number of links required for partial of complete link flow observability. Node based approaches provide interesting and easily obtainable links but they need not be sharp.

(9) The consideration of some model parameters, such as link choice proportions, as known values is not specially recommended, because they are large numbers and due to the difficulties in guessing realistic values.

(10) The use of cumulative flows in dynamic cases arises as a promising alternative that avoids the use of partial derivatives and permits replacing differential equations by functional equations. This alternative deserves further research.

(11) As indicated in this paper, passive counters provide a very limited and an insufficient amount of information. Thus, active counters must be used if relevant information about path information and good predictions are required.

(12) The optimal location of counters leads to important savings in cost and allows optimizing the resulting information. Several works dealing with this problem have been discussed.

(13) As indicated in Section 2.2.1, some constraints suggested for the dynamic case lead to contradiction and must be avoided.

(14) The FIFO rule plays an important role in building an important group of dynamic models but can be replaced by considering other types of models, such as multiclass models with overtaking possibilities.

(15) Finally, as indicated in this paper, two different classes of models, point and physical queue models, must be differentiated depending on the aim of the analysis.

Notations

$\mathcal{A}$: Set of counted links
$\mathcal{C}$: Subset of counted links
cp_a: Capacity of link a per unit time
$c_a(v_a)$: Link a travel time function
$C_a(v_a)$: Integral of the link a travel time function
$d_a(t)$: Travel time of a vehicle that enters link a at time t
f_r: Flow on route r
$F_r(t)$: Cumulated flow that has entered route r at time t
$g_a(x)$: Exit function giving the rate at which traffic exits link a as a function of link traffic volume x
h: Number of finite (bounded) holes
$\mathcal{I}_n$: Set of links starting at node n
$I_n(t)$: Cumulated flow that has entered node n at time t
k_{jam}: Jam density
k: Flow density
k_0: Number of added links to convert the maximal planar network into a nonplanar hole-generated network
n_c: Number of cameras or sensors
$n_c^{\max}$: Available number of cameras
$n_{\max}$: Maximum number of sensors per route
$\mathcal{N}$: Set of nodes
$\mathcal{L}$: Set of links
$\mathcal{O}$: Subset of observed flows
$\mathcal{O}_n$: Set of links ending at node n
$O_n(t)$: Cumulated flow that has left node n at time t
$p_{rw}(t)$: Proportion of cumulated users choosing route r in OD pair w at time t
Q: Inflow capacity
$\mathcal{R}$: Set of routes
$\mathcal{R}_{\text{obs}}$: Subset of routes that we want to observe
r: Route
$\mathcal{R}(w)$: Set of routes in OD pair w
$\psi_r(t)$: Travel time of route r at time t
τ_w: Flow that has selected a route in OD pair w
$\mathcal{T}$: Set of selected times
$\mathcal{T}^*$: Set of times at which data are collected
$T_w(t)$: Cumulated flow that has entered a route in OD pair w at time t
$u_a(t)$: Rate of flow entering link a at time t
$\mathcal{U}$: Subset of unobserved flows
$U_a(t)$: Cumulated flow at the entrance of link a at time t
v_a: Flow on link a
$v_a^{\max}$: Practical flow capacity of link a
$\hat{v}_a$: Counted (observed) flow on link a
V: Free-flow speed
$V_a^i(t)$: Cumulated flow that has entered link a at time t
$\widehat{V}_a^i(t)$: Counted cumulated flow that has entered link a at time t
$V_a^o(t)$: Cumulated flow that has left link a at time t
$w_a(t)$: Rate of flow exiting link a at time t
$\widehat{w}_s$: Flow associated with the plate scanned link combination s
W: Speed of the backward shock wave or backward disturbances propagation speed
$\mathcal{W}$: Set of OD pairs
$W_a(t)$: Cumulated flow at the exit of link a at time t
$\widehat{W}_s(t)$: Total plate scanned observations in subset s at time t
$x_a(t)$: Traffic volume on link a at time t
y_r: Binary variable that equals one if route r is observed and zero otherwise
z_a: Binary variable that equals one if link a contains a sensor and zero otherwise
α: Positive coefficient
δ_r^a: Element of the link-route incidence matrix, which equals one if link a is on route r and equals zero otherwise
ε: Small dimensionless number
η_r^s: Element of route-scanned combination incidence matrix for route r, which equals one if route r contains the subset $s \subseteq \mathcal{S}$ of scanned links and no more scanned links and zero otherwise

$\theta_a^r(t)$: Entry time to link a on route r of a vehicle that enters route r at time t

ρ_1, ρ_2: Nonnegative numbers that measure the relative weight we give to the prior information of route and OD flows with respect to link flows

θ_w: Parameter that scales the perceived travel time in OD pair w

κ: Positive coefficient

κ_s^a: Being one if link a appears in the link combination s and zero otherwise

ϖ^f: Covariance matrix associated with route flows

ϖ^ℓ: Covariance matrix associated with link flows

ϖ^τ: Covariance matrix associated with OD flows

ρ_{aw}: Proportions of OD flow τ_w using link a

ξ_{rw}: The element of the route-OD incidence matrix, which equals 1 if route r belongs to OD pair w and zero otherwise

τ_a: Travel time on link a

$\tau_a(t)$: Link a travel time of a vehicle that enters the link at time t

τ_{0a}: Free-flow travel time on link a.

Conflict of Interests

The authors declare that there is no conflict of interests regarding the publication of this paper.

References

[1] V. P. Gil Jiménez and M. J. Fernández-Getino García, "Simple design of wireless sensor networks for traffic jams avoidance," *Journal of Sensors*, vol. 2015, Article ID 380794, 7 pages, 2015.

[2] C. Liu, H. Zheng, D. Yu, and X. Xu, "A novel method of adaptive traffic image enhancement for complex environments," *Journal of Sensors*, vol. 2015, Article ID 516326, 9 pages, 2015.

[3] W. Kong, L. Zhou, Y. Wang, J. Zhang, J. Liu, and S. Gao, "A system of driving fatigue detection based on machine vision and its application on smart device," *Journal of Sensors*, vol. 2015, Article ID 548602, 11 pages, 2015.

[4] M. Gentili and P. B. Mirchandani, "Locating active sensors on traffic networks," *Annals of Operations Research*, vol. 136, no. 1, pp. 229–257, 2005.

[5] D. Branston, "Link capacity functions: a review," *Transportation Research Part B*, vol. 10, no. 4, pp. 223–236, 1976.

[6] T. Abrahamsson, "Estimation of origin-destination matrices using traffic counts literature survey," Interim Report IR98-021, International institute for Applied System Analysis, Computer Science Press, Laxenburg City, Laxenburg, 1998.

[7] R. Montoya-Zamora, J. A. Romero, and R. Guzmán-Cruz, "Comparison between heuristics and methodologies to solve the NSLP," *International Journal of Engineering Science and Technology*, vol. 3, no. 7, pp. 5929–5939, 2011.

[8] R. M. Li, S. M. Tang, and F. Y. Wang, "A review of dynamic user-optimal variational inequality traffic assignment," *Journal of Transportation Systems Engineering and Information Technology*, vol. 6, no. 2, pp. 83–90, 2006.

[9] S. Peeta and A. K. Ziliaskopoulos, "Foundations of dynamic traffic assignment: the past, the present and the future," *Networks and Spatial Economics*, vol. 1, no. 3-4, pp. 233–265, 2001.

[10] D. Boyce, D.-H. Lee, and B. Ran, "Analytical models of the dynamic traffic assignment problem," *Networks and Spatial Economics*, vol. 1, no. 3-4, pp. 377–390, 2001.

[11] W. Y. Szeto and H. K. Lo, "Dynamic traffic assignment: review and future," *Journal of Transportation Systems Engineering and Information Technology*, vol. 5, pp. 85–100, 2005.

[12] E. Castillo, M. Nogal, A. Rivas, and S. Sánchez-Cambronero, "Observability of traffic networks: optimal location of counting and scanning devices," *Transportmetrica B: Transport Dynamics*, vol. 1, no. 1, pp. 68–102, 2013.

[13] H. Yang and J. Zhou, "Optimal traffic counting locations for origin-destination matrix estimation," *Transportation Research—Part B: Methodological*, vol. 33, no. 2, pp. 109–126, 1998.

[14] M. Gentili and P. B. Mirchandani, "Locating sensors on traffic networks: models, challenges and research opportunities," *Transportation Research Part C: Emerging Technologies*, vol. 24, pp. 227–255, 2012.

[15] E. Cascetta, *Transportation Systems Engineering: Theory and Methods*, Kluwer Academic Publishers, Boston, Mass, USA, 2001.

[16] A. Papola and V. Marzano, "How can we trust in the OD matrix correction procedure using traffic counts?" in *Proceedings of the European Transport Conference (ETC '06)*, Strasbourg, France, 2006.

[17] E. Cascetta, "Estimation of trip matrices from traffic counts and survey data: a generalized least squares estimator," *Transportation Research Part B*, vol. 18, no. 4-5, pp. 289–299, 1984.

[18] E. Cascetta and S. Nguyen, "A unified framework for estimating or updating origin/destination matrices from traffic counts," *Transportation Research Part B*, vol. 22, no. 6, pp. 437–455, 1988.

[19] J. Doblas and F. G. Benítez, "An approach to estimating and updating origin-destination matrices based upon traffic counts preserving the prior structure of a survey matrix," *Transportation Research Part B: Methodological*, vol. 39, no. 7, pp. 565–591, 2005.

[20] S. C. Wong, C. O. Tong, K. I. Wong et al., "Estimation of multiclass origin-destination matrices from traffic counts," *Journal of Urban Planning and Development*, vol. 131, no. 1, pp. 19–29, 2005.

[21] E. Castillo, J. M. Menéndez, and S. Sánchez-Cambronero, "Traffic estimation and optimal counting location without path enumeration using Bayesian networks," *Computer-Aided Civil and Infrastructure Engineering*, vol. 23, no. 3, pp. 189–207, 2008.

[22] R. Mínguez, S. Sánchez-Cambronero, E. Castillo, and P. Jiménez, "Optimal traffic plate scanning location for OD trip matrix and route estimation in road networks," *Transportation Research Part B: Methodological*, vol. 44, no. 2, pp. 282–298, 2010.

[23] X. P. Li and Y. F. Ouyang, "Reliable sensor deployment for network traffic surveillance," *Transportation Research—Part B: Methodological*, vol. 45, no. 1, pp. 218–231, 2011.

[24] T. L. Friesz, D. Bernstein, N. J. Mehta, R. L. Tobin, and S. Ganjalizadeh, "Day-to-day dynamic network disequilibria and idealized traveler information systems," *Operations Research*, vol. 42, no. 6, pp. 1120–1136, 1994.

[25] G. E. Cantarella and E. Cascetta, "Dynamic processes and equilibrium in transportation networks: towards a unifying theory," *Transportation Science*, vol. 29, no. 4, pp. 305–329, 1995.

[26] H. K. Lo and W. Y. Szeto, "Modeling advanced traveler information services: static versus dynamic paradigms," *Transportation*

Research Part B: Methodological, vol. 38, no. 6, pp. 495–515, 2004.

[27] E. Cascetta, D. Inaudi, and G. Marquis, "Dynamic estimators of origin-destination matrices using traffic counts," *Transportation Science*, vol. 27, no. 4, pp. 363–373, 1993.

[28] J. H. Wu, Y. Chen, and M. Florian, "The continuous dynamic network loading problem: a mathematical formulation and solution method," *Transportation Research—Part B: Methodological*, vol. 32, no. 3, pp. 173–187, 1998.

[29] E. Castillo, M. Nogal, J. M. Menéndez, S. Sánchez-Cambronero, and P. Jiménez, "Stochastic demand dynamic traffic models using generalized beta-gaussian bayesian networks," *IEEE Transactions on Intelligent Transportation Systems*, vol. 13, no. 2, pp. 565–581, 2012.

[30] M. Bierlaire, "The total demand scale: a new measure of quality for static and dynamic origin-destination trip tables," *Transportation Research Part B*, vol. 36, no. 9, pp. 837–850, 2002.

[31] D. K. Merchant and G. L. Nemhauser, "Optimality conditions for a dynamic traffic assignment model," *Transportation Science*, vol. 12, no. 3, pp. 200–207, 1978.

[32] V. Astarita, "A continuous time link model for dynamic network loading based on travel time function," in *Transportation and Traffic Theory: Proceedings of the 13th International Symposium on Transportation and Traffic Theory (ISTTT '96)*, J. B. Lesort, Ed., pp. 79–101, ISTTT, Lyon, France, 1996.

[33] T. L. Friesz, D. Bernstein, T. E. Smith, R. L. Tobin, and B. W. Wie, "A variational inequality formulation of the dynamic network user equilibrium problem," *Operations Research*, vol. 41, no. 1, pp. 179–191, 1993.

[34] J. E. Fernández and J. De Cea, "Flow propagation description in dynamic network assignment models," in *Proceedings of the 2nd Triennal Symposium on Transportation Analysis (TRISTAN '94)*, pp. 517–531, Capri, Italy, June 1994.

[35] J. H. Wu, Y. Chen, and M. Florian, "The continuous dynamic network loading problem: a mathematical formulation and solution method," in *Proceedings of the 3rd Euro Working Group Meeting on Urban Traffic and Transportation*, Barcelona, Spain, 1995.

[36] C. F. Daganzo, "The cell transmission model: a dynamic representation of highway traffic consistent with the hydrodynamic theory," *Transportation Research Part B*, vol. 28, no. 4, pp. 269–287, 1994.

[37] C. F. Daganzo, "The cell transmission model, part II: network traffic," *Transportation Research Part B: Methodological*, vol. 29, no. 2, pp. 79–93, 1995.

[38] W. Y. Szeto and H. K. Lo, "A cell-based simultaneous route and departure time choice model with elastic demand," *Transportation Research Part B: Methodological*, vol. 38, no. 7, pp. 593–612, 2004.

[39] H. K. Lo, "A dynamic traffic assignment formulation that encapsulates the cell transmission model," in *Transportation and Traffic Theory*, A. Ceder, Ed., pp. 327–350, 1999.

[40] J. M. Rubio-Ardanaz, J. H. Wu, and M. A. Florian, "A numerical analytical model for the continuous dynamic network equilibrium problem with limited capacity and spill back," in *Proceedings of the IEEE Intelligent Transportation Systems Conference*, pp. 263–267, Oakland, Calif, USA, August 2001.

[41] M. Kuwahara and T. Akamatsu, "Dynamic user optimal assignment with physical queues for a many-to-many OD pattern," *Transportation Research Part B: Methodological*, vol. 35, no. 5, pp. 461–479, 2001.

[42] H. K. Lo and W. Y. Szeto, "Road pricing modeling for hypercongestion," *Transportation Research Part A: Policy and Practice*, vol. 39, no. 7-9, pp. 705–722, 2005.

[43] M. Nogal, *Mathematical methods for traffic prediction [Ph.D. thesis]*, University of Cantabria, Santander, Spain, 2011.

[44] A. Calviño, *Some statistical and mathematical tools for traffic modeling [Ph.D. thesis]*, University of Cantabria, Santander, Spain, 2013.

[45] E. Castillo, J. M. Menéndez, M. Nogal, P. Jiménez, and S. Sánchez-Cambronero, "A FIFO rule consistent model for the continuous dynamic network loading problem," *IEEE Transactions on Intelligent Transportation Systems*, vol. 13, no. 1, pp. 264–283, 2012.

[46] G. Strang, *Introduction to Applied Mathematics*, Wellesley-Cambridge Press, Wellesley, Mass, USA, 1986.

[47] A. Ehlert, M. G. H. Bell, and S. Grosso, "The optimisation of traffic count locations in road networks," *Transportation Research Part B: Methodological*, vol. 40, no. 6, pp. 460–479, 2006.

[48] H. D. Sherali, N. Arora, and A. G. Hobeika, "Parameter optimization methods for estimating dynamic origin-destination trip-tables," *Transportation Research—Part B: Methodological*, vol. 31, no. 2, pp. 141–157, 1997.

[49] S. M. Eisenman, X. Fei, X. Zhou, and H. Mahmassani, "Number and location of sensors for realtime network traffic estimation and prediction: a sensitivity analysis," TRB Paper 06-2408, Transportation Research Board CD-ROM Paper Preprints, 2006.

[50] E. Castillo, J. M. Menéndez, and P. Jiménez, "Trip matrix and path flow reconstruction and estimation based on plate scanning and link observations," *Transportation Research Part B: Methodological*, vol. 42, no. 5, pp. 455–481, 2008.

[51] H. Yang, Y. Iida, and T. Sasaki, "An analysis of the reliability of an origin-destination trip matrix estimated from traffic counts," *Transportation Research—Part B*, vol. 25, no. 5, pp. 351–363, 1991.

[52] F. Viti, C. Tampere, W. Verbeke, and B. Immers, "Sensor locations for reliable travel time prediction and dynamic management of traffic networks," *Transportation Research Record*, vol. 2049, pp. 103–110, 2008.

[53] A. Chen, P. Chootinan, and W. W. Recker, "Examining the quality of synthetic origin–destination trip table estimated by path flow estimator," *Journal of Transportation Engineering*, vol. 131, no. 7, pp. 506–513, 2005.

[54] L. Gan, H. Yang, and S. C. Wong, "Traffic counting location and error bound in origin-destination matrix estimation problems," *Journal of Transportation Engineering*, vol. 131, no. 7, pp. 524–534, 2005.

[55] M. Gentili, *New models and algorithms for the location of sensors on traffic networks [Ph.D. thesis]*, Dipartimento di Statistica, Probabilità e Statistiche Applicate, Università di Roma "La Sapienza", Roma, Italy, 2002.

[56] E. Castillo, P. Jiménez, J. M. Menéndez, and A. J. Conejo, "The observability problem in traffic models: algebraic and topological methods," *IEEE Transactions on Intelligent Transportation Systems*, vol. 9, no. 2, pp. 275–287, 2008.

[57] M. W. Ng, "Synergistic sensor location for link flow inference without path enumeration: a node-based approach," *Transportation Research—Part B: Methodological*, vol. 46, no. 6, pp. 781–788, 2012.

[58] M. W. Ng, "Partial link flow observability in the presence of initial sensors: solution without path enumeration," *Transportation Research—Part E: Logistics and Transportation Review*, vol. 51, pp. 62–66, 2013.

[59] S.-X. He, "A graphical approach to identify sensor locations for link flow inference," *Transportation Research Part B: Methodological*, vol. 51, pp. 65–76, 2013.

[60] E. Castillo, A. Rivas, P. Jiménez, and J. M. Menéndez, "Observability in traffic networks. Plate scanning added by counting information," *Transportation*, vol. 39, no. 6, pp. 1301–1333, 2012.

[61] H. Spiess, "A maximum likelihood model for estimating origin-destination matrices," *Transportation Research—Part B: Methodological*, vol. 21, no. 5, pp. 395–412, 1987.

[62] N. L. Nihan and G. A. Davis, "Application of prediction-error minimization and maximum likelihood to estimate intersection O-D matrices from traffic counts," *Transportation Science*, vol. 23, no. 2, pp. 77–90, 1989.

[63] Y. Vardi, "Network tomography: estimating source-destination traffic intensities from link data," *Journal of the American Statistical Association*, vol. 91, no. 433, pp. 365–377, 1996.

[64] K. Parry and M. L. Hazelton, "Estimation of origin-destination matrices from link counts and sporadic routing data," *Transportation Research—Part B: Methodological*, vol. 46, no. 1, pp. 175–188, 2012.

[65] H. J. Van Zuylen and L. G. Willumsen, "The most likely trip matrix estimated from traffic counts," *Transportation Research Part B: Methodological*, vol. 14, no. 3, pp. 281–293, 1980.

[66] X. Zhou and G. F. List, "An information-theoretic sensor location model for traffic origin-destination demand estimation applications," *Transportation Science*, vol. 44, no. 2, pp. 254–273, 2010.

[67] F. Simonelli, V. Marzano, A. Papola, and I. Vitiello, "A network sensor location procedure accounting for o-d matrix estimate variability," *Transportation Research—Part B: Methodological*, vol. 46, no. 10, pp. 1624–1638, 2012.

[68] P. K. N. Yim and W. H. K. Lam, "Evaluation of count location selection methods for estimation of O-D matrices," *Journal of Transportation Engineering*, vol. 124, no. 4, pp. 376–382, 1998.

[69] X. Fei and H. S. Mahmassani, "Structural analysis of near-optimal sensor locations for a stochastic large-scale network," *Transportation Research Part C: Emerging Technologies*, vol. 19, no. 3, pp. 440–453, 2011.

[70] F. Viti and F. Corman, "A novel approach to the sensor location problem for measuring the observed network flow variability," in *Proceedings of the 5th International Symposium of Traffic Network Reliability*, Hong Kong, December 2012.

[71] F. Viti, M. Rinaldi, F. Corman, and C. M. J. Tampèreb, "Assessing partial observability in network sensor location problems," *Transportation Research—Part B: Methodological*, vol. 70, pp. 65–89, 2014.

[72] E. Castillo, A. Calviño, J. M. Menéndez, M. P. Jiménez, and A. Rivas, "Deriving the upper bound of the number of sensors required to know all link flows in a traffic network," *IEEE Transactions on Intelligent Transportation Systems*, vol. 14, no. 2, pp. 761–771, 2013.

[73] E. Castillo, A. Calviño, H. K. Lo, J. M. Menéndez, and Z. Grande, "Non-planar hole-generated networks and link flow observability based on link counters," *Transportation Research Part B: Methodological*, vol. 68, pp. 239–261, 2014.

[74] S.-R. Hu, S. Peeta, and C.-H. Chu, "Identification of vehicle sensor locations for link-based network traffic applications," *Transportation Research Part B: Methodological*, vol. 43, no. 8-9, pp. 873–894, 2009.

[75] E. Castillo, A. Cobo, F. Jubete, and R. Pruneda, *Orthogonal Sets and Polar Methods in Linear Algebra: Applications to Matrix Calculations, Systems of Equations, Inequalities, and Linear Programming*, John Wiley & Sons, New York, NY, USA, 1999.

[76] E. Castillo, A. Cobo, F. Jubete, R. Pruneda, and C. Castillo, "An orthogonally based pivoting transformation of matrices and some applications," *SIAM Journal on Matrix Analysis and Applications*, vol. 22, no. 3, pp. 666–681, 2000.

[77] E. Castillo, I. Gallego, J. M. Menéndez, and P. Jiménez, "Link flow estimation in traffic networks on the basis of link flow observations," *Journal of Intelligent Transportation Systems: Technology, Planning, and Operations*, vol. 15, no. 4, pp. 205–222, 2011.

[78] H. Mori and S. Tsuzuki, "A fast method for topological observability analysis using a minimum spanning tree technique," *IEEE Transactions on Power Systems*, vol. 6, no. 2, pp. 491–500, 1991.

[79] E. Castillo, F. Jubete, R. E. Pruneda, and C. Solares, "Obtaining simultaneous solutions of linear subsystems of inequalities and duals," *Linear Algebra and Its Applications*, vol. 346, no. 1–3, pp. 131–154, 2002.

[80] E. Castillo, A. Cobo, F. Jubete, R. E. Pruneda, and C. Castillo, "An orthogonally based pivoting transformation of matrices and some applications," *SIAM Journal on Matrix Analysis and Applications*, vol. 22, no. 3, pp. 666–681, 2000.

[81] M. J. Hodgson, "A flow-capturing location-allocation model," *Geographical Analysis*, vol. 22, no. 3, pp. 270–279, 1990.

[82] D. R. Morrison and S. E. Martonosi, "Characteristics of optimal solutions to the sensor location problem," *Annals of Operations Research*, 2014.

[83] E. Castillo, A. J. Conejo, J. M. Menéndez, and P. Jiménez, "The observability problem in traffic network models," *Computer-Aided Civil and Infrastructure Engineering*, vol. 23, no. 3, pp. 208–222, 2008.

[84] E. Castillo, I. Gallego, S. Sánchez-Cambronero, and A. Rivas, "Matrix tools for general observability analysis in traffic networks," *IEEE Transactions on Intelligent Transportation Systems*, vol. 11, no. 4, pp. 799–813, 2010.

[85] M. Rinaldi, F. Corman, and F. Viti, "A null space metric for the analysis of partial network observability in sensor location problems," in *Proceedings of the Annual Meeting of the Transportation Board*, Washington, DC, USA, January 2013.

[86] E. Castillo, P. Jiménez, J. M. Menéndez, A. Rivas, and I. Gallego, "A ternary-arithmetic topological based algebraic method for networks traffic observability," *Applied Mathematical Modelling*, vol. 35, no. 11, pp. 5338–5354, 2011.

[87] W. H. K. Lam and H. P. Lo, "Accuracy of od estimates from traffic counting stations," *Traffic Engineering and Control*, vol. 7, no. 1, pp. 105–114, 1990.

[88] M. Bierlaire and P. L. Toint, "Meuse: an origin-destination matrix estimator that exploits structure," *Transportation Research Part B*, vol. 29, no. 1, pp. 47–60, 1995.

[89] Y. Asakura, E. Hato, and M. Kashiwadani, "Origin-destination matrices estimation model using automatic vehicle identification data and its application to the Han-Shin expressway network," *Transportation*, vol. 27, no. 4, pp. 419–438, 2000.

[90] V. Marzano, A. Papola, and F. Simonelli, "Limits and perspectives of effective od matrix correction using traffic counts," *Transportation Research Part C: Emerging Technologies*, vol. 17, no. 2, pp. 120–132, 2009.

[91] M. G. Bell, "The estimation of an origin-destination matrix from traffic counts," *Transportation Science*, vol. 17, no. 2, pp. 198–217, 1983.

[92] D. P. Watling and M. J. Maher, "A statistical procedure for estimating a mean origin-destination matrix from a partial registration plate survey," *Transportation Research—Part B*, vol. 26, no. 3, pp. 171–193, 1992.

[93] D. P. Watling, "Maximum likelihood estimation of an origin-destination matrix from a partial registration plate survey," *Transportation Research—Part B*, vol. 28, no. 4, pp. 289–314, 1994.

[94] H.-P. Lo and C.-P. Chan, "Simultaneous estimation of an origin-destination matrix and link choice proportions using traffic counts," *Transportation Research Part A: Policy and Practice*, vol. 37, no. 9, pp. 771–788, 2003.

[95] R. Church and C. ReVelle, "The maximal covering location problem," *Papers of the Regional Science Association*, vol. 32, no. 1, pp. 101–118, 1974.

[96] X. Fei, H. S. Mahmassani, and S. M. Eisenman, "Sensor coverage and location for real-time traffic prediction in large-scale networks," *Transportation Research Record: Journal of the Transportation Research Board*, vol. 2039, pp. 1–15, 2007.

[97] T. Larsson, J. T. Lundgren, and A. Peterson, "Allocation of link flow detectors for origin-destination matrix estimation—a comparative study," *Computer-Aided Civil and Infrastructure Engineering*, vol. 25, no. 2, pp. 116–131, 2010.

[98] N. Wang, M. Gentili, and P. Mirchandani, "Model to locate sensors for estimation of static origin-destination volumes given prior flow information," *Transportation Research Record: Journal of the Transportation Research Board*, vol. 2283, pp. 67–73, 2012.

[99] A. Chen, S. Pravinvongvuth, P. Chootinan, M. Lee, and W. Recker, "Strategies for selecting additional traffic counts for improving O-D trip table estimation," *Transportmetrica*, vol. 3, no. 3, pp. 191–211, 2007.

[100] M. Brenninger-Göthe, K. O. Jörnsten, and J. T. Lundgren, "Estimation of origin-destination matrices from traffic counts using multiobjective programming formulations," *Transportation Research Part B*, vol. 23, no. 4, pp. 257–269, 1989.

[101] M. J. Maher, "Inferences on trip matrices from observations on link volumes: a Bayesian statistical approach," *Transportation Research: Part B*, vol. 17, no. 6, pp. 435–447, 1983.

[102] E. Castillo, P. Jiménez, J. M. Menéndez, and M. Nogal, "A Bayesian method for estimating traffic flows based on plate scanning," *Transportation*, vol. 40, no. 1, pp. 173–201, 2013.

[103] E. Castillo, J. M. Menéndez, S. Sánchez-Cambronero, A. Calviño, and J. M. Sarabia, "A hierarchical optimization problem: estimating traffic flow using gamma random variables in a Bayesian context," *Computers & Operations Research*, vol. 41, no. 1, pp. 240–251, 2014.

[104] H. P. Lo, N. Zhang, and W. H. K. Lam, "Estimation of an origin-destination matrix with random link choice proportions: a statistical approach," *Transportation Research—Part B: Methodological*, vol. 30, no. 4, pp. 309–324, 1996.

[105] E. Castillo, I. Gallego, J. M. Menéndez, and A. Rivas, "Optimal use of plate-scanning resources for route flow estimation in traffic networks," *IEEE Transactions on Intelligent Transportation Systems*, vol. 11, no. 2, pp. 380–391, 2010.

[106] S. Sánchez-Cambronero, E. Castillo, J. M. Menéndez, and P. Jiménez, "Dealing with error recovery in traffic flow prediction using Bayesian networks based on license plate scanning data," *Journal of Transportation Engineering*, vol. 137, no. 9, pp. 615–629, 2011.

[107] H. Yang, C. Yang, and L. Gan, "Models and algorithms for the screen line-based traffic-counting location problems," *Computers and Operations Research*, vol. 33, no. 3, pp. 836–858, 2006.

[108] P. Chootinan, A. Chen, and H. Yang, "A bi-objective traffic counting location problem for origin-destination trip table estimation," *Transportmetrica*, vol. 1, no. 1, pp. 65–80, 2005.

[109] E. Cipriani, G. Fusco, S. Gori, and M. Petrelli, "Heuristic methods for the optimal location of road traffic monitoring," in *Proceedings of the IEEE Intelligent Transportation Systems Conference (ITSC '06)*, pp. 1072–1077, Toronto, Canada, 2006.

[110] N. L. Nihan and G. A. Davis, "Recursive estimation of origin-destination matrices from input/output counts," *Transportation Research Part B: Methodological*, vol. 21, no. 2, pp. 149–163, 1987.

[111] K. Ashok and M. E. Ben-Akiva, "Dynamic od matrix estimation and prediction for real-time traffic management systems," in *International Symposium on the Theory of Traffic Flow and Transportation (12th: 1993: Berkeley, Calif.). Transportation and Traffic Theory*, C. E. Daganzo, Ed., pp. 465–484, Elsevier, Amsterdam, The Netherlands, 1993.

[112] K. Ashok and M. E. Ben-Akiva, "Estimation and prediction of time-dependent origin-destination flows with a stochastic mapping to path flows and link flows," *Transportation Science*, vol. 36, no. 2, pp. 184–198, 2002.

[113] A. Peterson, *The Origin-Destination Matrix Estimation Problem: Analysis and Computations*, Institutionen för Teknik och Naturvetenskap, 2007.

[114] J. Barcelò, L. Montero, L. Marqués, and C. Carmona, "Ict based estimation of time dependent od matrices," in *Proceedings of the 90th TRB Annual Meeting*, Washington, DC, USA, January 2011.

[115] D. Boyce and B. Janson, "The effect on equilibrium trim assignment of different link congestion functions," *Transportation Research Part B*, vol. 15, pp. 223–232, 1981.

[116] M. G. H. Bell, "Alternatives to Dial's logit assignment algorithm," *Transportation Research Part B*, vol. 29, no. 4, pp. 287–295, 1995.

[117] E. Castillo, A. Calviño, S. Sánchez-Cambronero, and H. K. Lo, "A multiclass user equilibrium model considering overtaking across classes," *IEEE Transactions on Intelligent Transportation Systems*, vol. 14, no. 2, pp. 928–942, 2013.

[118] E. Castillo, M. Nogal, A. Calviño, A. Rivas, and H. K. Lo, "A model for continuous dynamic network loading problem with different overtaking class users," *Journal of Intelligent Transportation Systems: Technology, Planning, and Operations*, vol. 17, no. 4, pp. 328–350, 2013.

[119] E. Castillo, A. Calviño, S. Sánchez-Cambronero, M. Nogal, and A. Rivas, "A percentile system optimization approach with and without path enumeration," *Computers & Operations Research*, vol. 40, no. 11, pp. 2711–2723, 2013.

[120] T. L. Friesz, J. Luque, R. L. Tobin, and B.-W. Wie, "Dynamic network traffic assignment considered as a continuous time optimal control problem," *Operations Research*, vol. 37, no. 6, pp. 893–901, 1989.

[121] R. Jayakrishnan, W. K. Tsai, and A. Chen, "A dynamic traffic assignment model with traffic-flow relationships," *Transportation Research Part C: Emerging Technologies*, vol. 3, no. 1, pp. 51–72, 1995.

[122] W. Y. Szeto and H. K. Lo, "Dynamic traffic assignment: properties and extensions," *Transportmetrica*, vol. 2, no. 1, pp. 31–52, 2006.

[123] M. Carey and E. Subrahmanian, "An approach to modelling time-varying flows on congested networks," *Transportation Research Part B: Methodological*, vol. 34, no. 3, pp. 157–183, 2000.

[124] C. F. Daganzo, "The cell-transmission model. Part I: a simple dynamic representation of highway traffic," Research Report UCBITS-PRR-93-7, Institute of Transportation Studies, University of California, Berkeley, Calif, USA, 1993.

[125] C. F. Daganzo, "A finite difference approximation of the kinematic wave model of traffic flow," *Transportation Research Part B: Methodological*, vol. 29, no. 4, pp. 261–276, 1995.

[126] C. D. Alecsandru, *A stochastic mesoscopic cell-transmission model for operational analysis of large-scale transportation networks [Ph.D. thesis]*, Department of Civil and Environmental Engineering, Louisiana State University, 2006.

[127] R. Zhong and A. Sumalee, "Stochastic cell transmission model: traffic state estimation under uncertainties," in *Traffic and Transportation Studies: Proceedings of the Sixth International Conference on Traffic and Transportation Studies*, B. Mao, Z. Tian, H. Huang, and Z. E. Gao, Eds., pp. 462–478, ASCE, Reston, Va, USA, 2008.

[128] A. Sumalee, R. X. Zhong, T. L. Pan, and W. Y. Szeto, "Stochastic cell transmission model (SCTM): a stochastic dynamic traffic model for traffic state surveillance and assignment," *Transportation Research—Part B: Methodological*, vol. 45, no. 3, pp. 507–533, 2011.

[129] M. Carey, C. Balijepalli, and D. Watling, "Extending the cell transmission model to multiple lanes and lane-changing," *Networks and Spatial Economics*, 2013.

[130] W. Y. Szeto, Y. Jiang, and A. Sumalee, "A cell-based model for multi-class doubly stochastic dynamic traffic assignment," *Computer-Aided Civil and Infrastructure Engineering*, vol. 26, no. 8, pp. 595–611, 2011.

[131] K. Tuerprasert and C. Aswakul, "Multiclass cell transmission model for heterogeneous mobility in general topology of road network," *Journal of Intelligent Transportation Systems: Technology, Planning, and Operations*, vol. 14, no. 2, pp. 68–82, 2010.

[132] P. Marcotte, "A new algorithm for solving variational inequalities with application to the traffic assignment problem," *Mathematical Programming*, vol. 33, no. 3, pp. 339–351, 1985.

[133] D. Zhu and P. Marcotte, "On the existence of solutions to the dynamic user equilibrium problem," *Transportation Science*, vol. 34, no. 4, pp. 402–414, 2000.

[134] J. Wang and Q. Jiang, "A variational inequality model for traffic assignment," in *Proceedings of the International Conference on Internet Technology and Applications (iTAP '11)*, Wuhan, China, August 2011.

[135] H. K. Lo and W. Y. Szeto, "A cell-based variational inequality formulation of the dynamic user optimal assignment problem," *Transportation Research Part B: Methodological*, vol. 36, no. 5, pp. 421–443, 2002.

[136] M. C. J. Bliemer and P. H. L. Bovy, "Quasi-variational inequality formulation of the multiclass dynamic traffic assignment problem," *Transportation Research Part B*, vol. 37, no. 6, pp. 501–519, 2003.

[137] Y. W. Xu, J. H. Wu, M. Florian, P. Marcotte, and D. L. Zhu, "Advances in the continuous dynamic network loading problem," *Transportation Science*, vol. 33, no. 4, pp. 341–353, 1999.

[138] V. Astarita, K. Er-Rafia, M. Florian, M. Mahut, and S. Velan, "Comparison of three methods for dynamic network loading," *Transportation Research Record*, vol. 1771, pp. 179–190, 2001.

[139] J. M. Rubio-Ardanaz, J. H. Wu, and M. Florian, "Two improved numerical algorithms for the continuous dynamic network loading problem," *Transportation Research—Part B: Methodological*, vol. 37, no. 2, pp. 171–190, 2003.

[140] H. K. Lo and W. Y. Szeto, "A cell-based dynamic traffic assignment model: formulation and properties," *Mathematical and Computer Modelling*, vol. 35, no. 7-8, pp. 849–865, 2002.

[141] H. B. Celikoglu, "A dynamic network loading model for traffic dynamics modeling," *IEEE Transactions on Intelligent Transportation Systems*, vol. 8, no. 4, pp. 575–583, 2007.

[142] G. C. K. Wong and S. C. Wong, "A multi-class traffic flow model: an extension of LWR model with heterogeneous drivers," *Transportation Research Part A: Policy and Practice*, vol. 36, no. 9, pp. 827–841, 2002.

[143] M. Blumberg and H. Bar-Gera, "Consistent node arrival order in dynamic network loading models," *Transportation Research Part B: Methodological*, vol. 43, no. 3, pp. 285–300, 2009.

[144] J. Long, Z. Gao, and W. Y. Szeto, "Discretised link travel time models based on cumulative flows: formulations and properties," *Transportation Research Part B: Methodological*, vol. 45, no. 1, pp. 232–254, 2011.

[145] I. Okutani and Y. J. Stephanedes, "Dynamic prediction of traffic volume through Kalman filtering theory," *Transportation Research—Part B: Methodological*, vol. 18, no. 1, pp. 1–11, 1984.

Permissions

All chapters in this book were first published in JS, by Hindawi Publishing Corporation; hereby published with permission under the Creative Commons Attribution License or equivalent. Every chapter published in this book has been scrutinized by our experts. Their significance has been extensively debated. The topics covered herein carry significant findings which will fuel the growth of the discipline. They may even be implemented as practical applications or may be referred to as a beginning point for another development.

The contributors of this book come from diverse backgrounds, making this book a truly international effort. This book will bring forth new frontiers with its revolutionizing research information and detailed analysis of the nascent developments around the world.

We would like to thank all the contributing authors for lending their expertise to make the book truly unique. They have played a crucial role in the development of this book. Without their invaluable contributions this book wouldn't have been possible. They have made vital efforts to compile up to date information on the varied aspects of this subject to make this book a valuable addition to the collection of many professionals and students.

This book was conceptualized with the vision of imparting up-to-date information and advanced data in this field. To ensure the same, a matchless editorial board was set up. Every individual on the board went through rigorous rounds of assessment to prove their worth. After which they invested a large part of their time researching and compiling the most relevant data for our readers.

The editorial board has been involved in producing this book since its inception. They have spent rigorous hours researching and exploring the diverse topics which have resulted in the successful publishing of this book. They have passed on their knowledge of decades through this book. To expedite this challenging task, the publisher supported the team at every step. A small team of assistant editors was also appointed to further simplify the editing procedure and attain best results for the readers.

Apart from the editorial board, the designing team has also invested a significant amount of their time in understanding the subject and creating the most relevant covers. They scrutinized every image to scout for the most suitable representation of the subject and create an appropriate cover for the book.

The publishing team has been an ardent support to the editorial, designing and production team. Their endless efforts to recruit the best for this project, has resulted in the accomplishment of this book. They are a veteran in the field of academics and their pool of knowledge is as vast as their experience in printing. Their expertise and guidance has proved useful at every step. Their uncompromising quality standards have made this book an exceptional effort. Their encouragement from time to time has been an inspiration for everyone.

The publisher and the editorial board hope that this book will prove to be a valuable piece of knowledge for researchers, students, practitioners and scholars across the globe.

List of Contributors

Quan-guo Lu, Jian-wu Yan and Xiao-cui Yuan
Jiangxi Province Key Laboratory of Precision Drive & Control, Nanchang Institute of Technology, Nanchang, China

Bo-wen Wang
State Key Laboratory of Reliability and Intelligence of Electrical Equipment, Hebei University of Technology, Tianjin, China

Ran Zhao
Jiangxi Province Key Laboratory of Precision Drive & Control, Nanchang Institute of Technology, Nanchang, China
State Key Laboratory of Reliability and Intelligence of Electrical Equipment, Hebei University of Technology, Tianjin, China

D. Cheneler
Department of Engineering, Lancaster University, Lancaster, UK

M. C. L. Ward
School of Engineering and the Built Environment, Birmingham City University, Birmingham, UK

José A. Hidalgo-López, Raquel Fernández-Ramos, Jorge Romero-Sánchez, José F. Martín-Canales and Francisco J. Ríos-Gómez
Departamento de Electrónica, Andalucía Tech, Universidad de Málaga, Campus de Teatinos, 29071 Málaga, Spain

Marek Badura
Department of Air Conditioning, Heating, Gas Engineering and Air Protection, Faculty of Environmental Engineering, Wrocław University of Science and Technology, 50-373 Wrocław, Poland

Piotr Batog
INSYSPOM, 54-427 Wrocław, Poland

Anetta Drzeniecka-Osiadacz and Piotr Modzel
Department of Climatology and Atmosphere Protection, Institute of Geography and Regional Development, Faculty of Earth Science and Environmental Management, University of Wrocław, 51-621 Wrocław, Poland

Antonella D'Alessandro, Andrea Meoni and Filippo Ubertini
Department of Civil and Environmental Engineering, Via G. Duranti 93, 06125 Perugia, Italy

Felipe P. Vista IV and Kil To Chong
Electronic Engineering Department, Chonbuk National University, Jeonju 54896, Republic of Korea

Jian Sun
School of Mechanical Engineering & Automation, Beihang University, Beijing 100191, China
The F.R.I. of Ministry of Public Security, Beijing 100048, China

Yongling Fu
School of Mechanical Engineering & Automation, Beihang University, Beijing 100191, China

Shengguang Li and Lin Tan
The F.R.I. of Ministry of Public Security, Beijing 100048, China

Jie He and Cheng Xu
School of Computer and Communication Engineering, University of Science and Technology Beijing, Beijing 100083, China

Gyeoungrok Lee and Kyunghyun Yoon
School of Computer Science and Engineering, Chung-Ang University, Seoul, Republic of Korea

Dongwann Kang
Faculty of Science and Technology, Bournemouth University, Poole, UK

Zhuohua Liu and Xiangyang Gong
State Key Laboratory of Networking and Switching Technology, Beijing University of Posts and Telecommunications, Beijing 100876, China

Wei Liu
State Key Laboratory of Networking and Switching Technology, Beijing University of Posts and Telecommunications, Beijing 100876, China
Beijing University of Posts and Telecommunications, Beijing 100876, China
State Key Laboratory of Information Security, Institute of Information Engineering, Chinese Academy of Sciences, Beijing 100093, China

Jin Wu
School of Automation, University of Electronic Science and Technology of China, Chengdu, China

Zeyi Chao, Yuke Yin and Bin Han
School of Electronic Information, Wuhan University, Wuhan, Hubei 430072, China

Fangling Pu
School of Electronic Information, Wuhan University, Wuhan, Hubei 430072, China
Collaborative Innovation Center of Geospatial Technology, Wuhan University, Whan, Hubei 430079, China

Xiaoling Chen
State Key Laboratory for Information Engineering in Surveying, Mapping and Remote Sensing (LIESMARS), Wuhan University, Wuhan, Hubei 430079, China

Luis Armando Carvajal Ahumada
Facultad de Ingeniería y Ciencias Básicas, Universidad Central, Bogotá 110311, Colombia
Servicio Nacional de Aprendizaje (SENA), TecnoParque nodo Bogotá, Bogotá, Colombia

Andrés Felipe Sandoval Cruz, Mario Alejandro García Fox and Oscar Leonardo Herrera Sandoval
Facultad de Ingeniería y Ciencias Básicas, Universidad Central, Bogotá 110311, Colombia

Barun Basnet and Junho Bang
Department of IT Applied System Engineering, Chonbuk National University, Jeollabuk-do, Republic of Korea

Ayse Gokcen Kavaz and Burak Barutcu
Energy Institute, Istanbul Technical University, 34469 Istanbul, Turkey

Nhan Duc Nguyen, Duong Trong Bui and Gu-Min Jeong
Department of Electrical Engineering, Kookmin University, Seoul, Republic of Korea

Phuc Huu Truong
Korea Institute of Industrial Technology, Ansan, Republic of Korea

Dandan Pang and Ming Wang
School of Information and Electrical Engineering, Shandong Jianzhu University, Jinan 250101, China
Shandong Provincial Key Laboratory of Intelligent Buildings Technology, Jinan 250101, China

Qingmei Sui and Yaozhang Sai
School of Control Science and Engineering, Shandong University, Jinan 250061, China

Rongyao Sun and Yikang Wang
School of Information and Electrical Engineering, Shandong Jianzhu University, Jinan 250101, China

Rasha M. Al-Eidan, Hend Al-Khalifa and Abdul Malik Al-Salman
College of Computer and Information Sciences, King Saud University, Riyadh, Saudi Arabia

Index

www.ingramcontent.com/pod-product-compliance
Lightning Source LLC
LaVergne TN
LVHW080717210926
848485LV00003B/481